**WITHDRAWN**
UTSA LIBRARIES

# Spatiotemporal Patterns in Ecology and Epidemiology

Theory, Models, and Simulation

# CHAPMAN & HALL/CRC
Mathematical and Computational Biology Series

**Aims and scope:**
This series aims to capture new developments and summarize what is known over the whole spectrum of mathematical and computational biology and medicine. It seeks to encourage the integration of mathematical, statistical and computational methods into biology by publishing a broad range of textbooks, reference works and handbooks. The titles included in the series are meant to appeal to students, researchers and professionals in the mathematical, statistical and computational sciences, fundamental biology and bioengineering, as well as interdisciplinary researchers involved in the field. The inclusion of concrete examples and applications, and programming techniques and examples, is highly encouraged.

**Series Editors**
Alison M. Etheridge
*Department of Statistics*
*University of Oxford*

Louis J. Gross
*Department of Ecology and Evolutionary Biology*
*University of Tennessee*

Suzanne Lenhart
*Department of Mathematics*
*University of Tennessee*

Philip K. Maini
*Mathematical Institute*
*University of Oxford*

Shoba Ranganathan
*Research Institute of Biotechnology*
*Macquarie University*

Hershel M. Safer
*Weizmann Institute of Science*
*Bioinformatics & Bio Computing*

Eberhard O. Voit
*The Wallace H. Couter Department of Biomedical Engineering*
*Georgia Tech and Emory University*

Proposals for the series should be submitted to one of the series editors above or directly to:
**CRC Press, Taylor & Francis Group**
Albert House, 4th floor
1-4 Singer Street
London EC2A 4BQ
UK

# Published Titles

**Bioinformatics: A Practical Approach**
Shui Qing Ye

**Cancer Modelling and Simulation**
Luigi Preziosi

**Computational Biology: A Statistical Mechanics Perspective**
Ralf Blossey

**Computational Neuroscience: A Comprehensive Approach**
Jianfeng Feng

**Data Analysis Tools for DNA Microarrays**
Sorin Draghici

**Differential Equations and Mathematical Biology**
D.S. Jones and B.D. Sleeman

**Exactly Solvable Models of Biological Invasion**
Sergei V. Petrovskii and Bai-Lian Li

**Introduction to Bioinformatics**
Anna Tramontano

**An Introduction to Systems Biology: Design Principles of Biological Circuits**
Uri Alon

**Knowledge Discovery in Proteomics**
Igor Jurisica and Dennis Wigle

**Modeling and Simulation of Capsules and Biological Cells**
C. Pozrikidis

**Niche Modeling: Predictions from Statistical Distributions**
David Stockwell

**Normal Mode Analysis: Theory and Applications to Biological and Chemical Systems**
Qiang Cui and Ivet Bahar

**Pattern Discovery in Bioinformatics: Theory & Algorithms**
Laxmi Parida

**Spatiotemporal Patterns in Ecology and Epidemiology: Theory, Models, and Simulation**
Horst Malchow, Sergei V. Petrovskii, and Ezio Venturino

**Stochastic Modelling for Systems Biology**
Darren J. Wilkinson

**The Ten Most Wanted Solutions in Protein Bioinformatics**
Anna Tramontano

Chapman & Hall/CRC Mathematical and Computational Biology Series

# Spatiotemporal Patterns in Ecology and Epidemiology

## Theory, Models, and Simulation

Horst Malchow, Sergei V. Petrovskii, and Ezio Venturino

Chapman & Hall/CRC
Taylor & Francis Group
Boca Raton  London  New York

Chapman & Hall/CRC is an imprint of the
Taylor & Francis Group, an **informa** business

Chapman & Hall/CRC
Taylor & Francis Group
6000 Broken Sound Parkway NW, Suite 300
Boca Raton, FL 33487-2742

© 2008 by Taylor & Francis Group, LLC
Chapman & Hall/CRC is an imprint of Taylor & Francis Group, an Informa business

No claim to original U.S. Government works
Printed in the United States of America on acid-free paper
10 9 8 7 6 5 4 3 2 1

International Standard Book Number-13: 978-1-58488-674-7 (Hardcover)

This book contains information obtained from authentic and highly regarded sources. Reprinted material is quoted with permission, and sources are indicated. A wide variety of references are listed. Reasonable efforts have been made to publish reliable data and information, but the author and the publisher cannot assume responsibility for the validity of all materials or for the consequences of their use.

Except as permitted under U.S. Copyright Law, no part of this book may be reprinted, reproduced, transmitted, or utilized in any form by any electronic, mechanical, or other means, now known or hereafter invented, including photocopying, microfilming, and recording, or in any information storage or retrieval system, without written permission from the publishers.

For permission to photocopy or use material electronically from this work, please access www.copyright.com (http://www.copyright.com/) or contact the Copyright Clearance Center, Inc. (CCC) 222 Rosewood Drive, Danvers, MA 01923, 978-750-8400. CCC is a not-for-profit organization that provides licenses and registration for a variety of users. For organizations that have been granted a photocopy license by the CCC, a separate system of payment has been arranged.

**Trademark Notice:** Product or corporate names may be trademarks or registered trademarks, and are used only for identification and explanation without intent to infringe.

---
**Library of Congress Cataloging-in-Publication Data**
---

Malchow, Horst, 1953-
  Spatiotemporal patterns in ecology and epidemiology : theory, models, and simulation / Horst Malchow, Sergei V Petrovskii and Ezio Venturino.
      p. cm. -- (Chapman & Hall/CRC mathematical and computational biology ; 17)
  Includes bibliographical references and index.
  ISBN 978-1-58488-674-7 (alk. paper)
  1. Ecology--Mathematical models. 2. Epidemiology--Mathematical models. I. Petrovskii, Sergei V. II. Venturino, Ezio. III. Title. III. Series.

QH541.15.M3M25 2008
577.01'5118--dc22                                            2007040410

---

**Visit the Taylor & Francis Web site at**
http://www.taylorandfrancis.com

**and the CRC Press Web site at**
http://www.crcpress.com

*To our loved ones*

# *Preface*

Dynamics has always been a core issue of all natural sciences. What are the driving forces and the "mechanisms" that result in motion of the system parts and/or lead to its evolution as a whole, whatever that system may be (particular cases range from a single rigid body to the human society)? What are the properties and scenarios of this motion and evolution? What can be the meaning or implication of this dynamics if considered in a wider context, e.g., through interaction with other systems or other sciences? These have been challenging and exciting problems for philosophers and scientists throughout at least 30 centuries.

The focus of interest has evolved, too. At the dawn of contemporary science, stationary processes were regarded as the essence of dynamics. Non-stationary motion was often either attributed to a specific cause (such as periodic forcing) or considered as a mere transient that was bound to die out after a relatively short relaxation time. The corresponding system's geometry was assumed to be smooth and regular. Those phenomena that did not fit into this philosophy, turbulent flow being the renowned example, were regarded as exotic and rare and, possibly, not self-sustained.

Although this mainstream thought of science has been challenged from time to time from as early as ancient Greece, it was not until the late twentieth century that it was widely realized that the dynamics of even very simple systems can be – and often is – completely different and much more complicated. Relaxation to steady states and periodic motion (with the limit cycle as its mathematical paradigm), which used to be the main elements of dynamics, were displaced by the concept of deterministic chaos. Smooth surfaces and simple curves gave way to objects with fractal properties. Moreover, chaos and fractals were eventually found (if not empirically, then at least theoretically) nearly everywhere, from lasers to animal behavior.

Even more importantly, it was realized that complex temporal dynamics and, especially, spatial structures do not just exist per se but can arise as a result of a system's self-organization. In an open system, i.e., a system with an inflow of mass and/or energy, the dynamics can become intrinsically unstable, resulting either in the formation of periodic spatial patterns or in a complicated turbulence-like spatiotemporal behaviour.

A similar evolution of concepts and ideas occurred in ecology and population biology. Ecologists have long been aware that population distribution in a natural environment is normally distinctly heterogeneous; however, that was usually regarded as a separate phenomenon that was not directly related to the main properties of population dynamics in time such as its persistence, reproductive success, etc. Correspondingly, earlier studies tended to focus mainly on the dynamics of "nonspatial" systems (i.e., systems where the spatial distribution of all factors and agents was regarded to be homogeneous under any circumstances). The results of the last two decades, however, proved that the impact of spatial dimensions can be crucial. The dynamics of a spatially extended system can be qualitatively different from the dynamics of its nonspatial counterpart due to self-organized, "spontaneous" pattern formation.

It should be mentioned that recent progress in theoretical ecology would unlikely have become possible without extensive use of mathematical modeling. There are several reasons why a complete and thorough study of ecosystem dynamics is hardly possible if based only on field data collection. Field observations are often very expensive and field experiments can sometimes be dangerous for the environment. Moreover, a regular experimental study implies replicated experiments; however, this is hardly possible in ecology because of the virtual impossibility of reproducing the same initial and environmental conditions. Note that mathematical models have been used in ecology from as early as the nineteenth century (cf. the work by Malthus) but they became a really powerful research tool after the development of numerical simulation approaches and modern computers.

Importantly, although the spatial dimension of ecosystems dynamics is nowadays widely recognized, the specific mechanisms behind species patterning in space are still poorly understood and the corresponding theoretical framework is underdeveloped. In particular, existing textbooks and research monographs on theoretical/mathematical ecology, when addressing its spatial aspect, practically never go beyond the classical Turing scenario of pattern formation. This book is designed to fill this gap, in particular, by giving an account of the significant progress made recently (and published in periodic scientific literature) in understanding these issues through mathematical modeling and numerical simulations using some basic, conceptual models of population dynamics.

A special remark should be made regarding terminology. Apparently, the term "pattern" has originally appeared in application to spatial processes where some kind of heterogeneity is observed. In the context of this book, however, we use this term somewhat more broadly, embracing also the purely temporal dynamics of spatially homogeneous systems; cf. "patterns of temporal behaviour."

Another tendency in scientific periodics over the last decade has been convergence between ecology and epidemiology. Although it seems to be common knowledge that a disease can change population dynamics essentially, mathematical approaches to these issues remained distinctly different until recently. Remarkably, both population dynamics and disease dynamics exhibit many similar properties, especially with regards to pattern formation in space and time. This book provides a first attempt at a unified approach to population dynamics and epidemiology by means of considering a few "ecoepidemiological" models where both the basic interspecies interactions of population dynamics and the impact of an infectious disease are considered explicitly.

The book is organized as follows. Part I starts with a general overview of relevant phenomena in ecology and epidemiology, giving also a few examples of pattern formation in natural systems, and then proceeds to a brief synopsis of existing modeling approaches.

Part II deals with nonspatial models of population dynamics and epidemiology. We have already mentioned that the dynamics of spatial and corresponding nonspatial systems can be essentially different and the results of nonspatial analysis may, in some cases, be misleading. Nevertheless, it is also clear that the properties of nonspatial dynamics provide a certain "skeleton" important for a thorough understanding of spatiotemporal dynamics. Correspondingly, Part II gives a wide panorama of existing nonspatial approaches, starting from basic ideas and elementary models and eventually bringing the reader to the state-of-the-art in this area.

In Part III, we introduce space by means of including "diffusion" of the individuals and consider the main scenarios of spatial and spatiotemporal pattern formation in deterministic models of population dynamics.

Finally, in Part IV, we address the issue of interaction between deterministic and stochastic processes in ecosystem/epidemics dynamics and consider how noise and stochasticity may affect pattern formation.

When writing this book, we were primarily thinking about experienced researchers in theoretical and mathematical ecology and/or in relevant areas of applied mathematics as the "target audience," and that affected its structure and style. In particular, from the beginning of Part I we use some advanced terminology from applied dynamical systems, mathematical modelling, and differential equations, which requires the reader to have at least some basic education in these topics (even in spite of the fact that much of that terminology is actually explained later in the text). However, we do hope that many researchers from neighboring fields and also postgraduate students will find this book useful as well; in order to encourage them to read it, we give enough calculation details. For the same purpose, the presentation of some introductory items is, at times, made on a rather elementary level.

A considerable part of the results included into this book was obtained in numerical simulations. Therefore, although numerical results by no means can be regarded as an adequate substitute to rigorous analysis, we think that it will be only fair if the reader is given an opportunity to reproduce the main results and (which is probably even more important) to make a deeper look into the system's dynamics in his/her own numerical experiments. For that purpose, a CD is attached to this book that contains many of the computer programs that we have used in our work. The programs are written in MATLAB; it must be mentioned here that MATLAB® and Simulink® are trademarks of The MathWorks, Inc., and are used with permission. The MathWorks does not warrant the accuracy of the programs on the CD. This

CD's use or discussion of MATLAB® and Simulink® software or related products does not constitute endorsement or sponsorship by The MathWorks of a particular pedagogical approach or particular use of MATLAB® and Simulink® software.

In conclusion, it is our pleasure to express our gratitude to numerous people who helped this book to appear through many fruitful discussions and helpful comments, both during manuscript preparation and during the equally important time preceding this work. We are particularly grateful to Ulrike Feudel, Nanako Shigesada, Michel Langlais, Michael Tretyakov, Lutz Schimansky-Geier, Vitaly Volpert, Andrey Morozov, Frank Hilker, Jean-Christophe Poggiale, Hiromi Seno, Bai-Lian (Larry) Li, Herbert Hethcote, Alexander Medvinsky, Joydev Chattopadhyay, Olivier Lejeune, Michael Sieber, Guido Badino, Francesca Bona, and Marco Isaia. S.P. is very thankful to his colleagues in the Department of Mathematics of the University of Leicester for their continuing encouragement and support. E.V. thanks the Max–Planck–Institut für Mathematik in Bonn, where, during an informal visit, parts of this book were written. E.V. is also very much indebted to the persons who long ago introduced him to the fascinating field of mathematical modeling, especially to Brian Conolly, Edward Beltrami, and James Frauenthal. Last but not least, we all are thankful to Sunil Nair, publisher of mathematics and statistics, CRC Press/Chapman & Hall, for inviting us to write this book and for his patience and encouragement.

H. Malchow, S. Petrovskii, E. Venturino<br>
Osnabrück–Leicester/Birmingham–Torino

# Contents

## I   Introduction — 1

**1   Ecological patterns in time and space** — 3
   1.1   Local structures — 3
   1.2   Spatial and spatiotemporal structures — 6

**2   An overview of modeling approaches** — 11

## II   Models of temporal dynamics — 19

**3   Classical one population models** — 21
   3.1   Isolated populations models — 21
       3.1.1   Scaling — 25
   3.2   Migration models — 27
       3.2.1   Harvesting — 30
   3.3   Glance at discrete models — 44
   3.4   Peek into chaos — 46

**4   Interacting populations** — 49
   4.1   Two-species prey–predator population model — 50
   4.2   Classical Lotka–Volterra model — 57
       4.2.1   More on prey–predator models — 58
       4.2.2   Scaling — 59
   4.3   Other types of population communities — 60
       4.3.1   Competing populations — 60
       4.3.2   Symbiotic populations — 62
       4.3.3   Leslie–Gower model — 64
       4.3.4   Classical Holling–Tanner model — 65
       4.3.5   Other growth models — 66
       4.3.6   Models with prey switching — 66
   4.4   Global stability — 68
       4.4.1   General quadratic prey–predator system — 71
       4.4.2   Mathematical tools for analyzing limit cycles — 72
       4.4.3   Routh–Hurwitz conditions — 74
       4.4.4   Criterion for Hopf bifurcation — 75
       4.4.5   Instructive example — 76
       4.4.6   Poincaré map — 77

|  | 4.5 | Food web | 80 |
|---|---|---|---|
|  | 4.6 | More about chaos | 87 |
|  | 4.7 | Age-dependent populations | 91 |
|  |  | 4.7.1 Prey–predator, age-dependent populations | 95 |
|  |  | 4.7.2 More about age-dependent populations | 96 |
|  |  | 4.7.3 Simulations and brief discussion | 108 |

## 5 Case study: biological pest control in vineyards — 111
- 5.1 First model — 112
  - 5.1.1 Modeling the human activity — 114
- 5.2 More sophisticated model — 116
  - 5.2.1 Models comparison — 122
- 5.3 Modeling the ballooning effect — 124
  - 5.3.1 Spraying effects and human intervention — 134
  - 5.3.2 Ecological discussion — 134

## 6 Epidemic models — 137
- 6.1 Basic epidemic models — 137
  - 6.1.1 Simplest models — 139
  - 6.1.2 Standard incidence — 141
- 6.2 Other classical epidemic models — 145
- 6.3 Age- and stage-dependent epidemic system — 147
- 6.4 Case study: Aujeszky disease — 151
- 6.5 Analysis of a disease with two states — 156

## 7 Ecoepidemic systems — 165
- 7.1 Prey–diseased-predator interactions — 165
  - 7.1.1 Some biological considerations — 176
- 7.2 Predator–diseased-prey interactions — 178
- 7.3 Diseased competing species models — 184
  - 7.3.1 Simulation discussion — 189
- 7.4 Ecoepidemics models of symbiotic communities — 191
  - 7.4.1 Disease effects on the symbiotic system — 194
  - 7.4.2 Disease control by use of a symbiotic species — 195

## III Spatiotemporal dynamics and pattern formation: deterministic approach — 197

## 8 Spatial aspect: diffusion as a paradigm — 199

## 9 Instabilities and dissipative structures — 205
- 9.1 Turing patterns — 206
  - 9.1.1 Turing patterns in a multispecies system — 218
- 9.2 Differential flow instability — 223
- 9.3 Ecological example: semiarid vegetation patterns — 231

|  |  |  |
|---|---|---|
| | 9.3.1 Pattern formation due to nonlocal interactions | 236 |
| 9.4 | Concluding remarks | 245 |

## 10 Patterns in the wake of invasion  247
10.1 Invasion in a prey–predator system .............. 248
10.2 Dynamical stabilization of an unstable equilibrium ...... 260
    10.2.1 A bifurcation approach ................. 261
    10.2.2 Comparison of wave speeds .............. 266
10.3 Patterns in a competing species community .......... 269
10.4 Concluding remarks ....................... 277

## 11 Biological turbulence  281
11.1 Self-organized patchiness and the wave of chaos ....... 283
    11.1.1 Stability diagram and the hierarchy of regimes .... 290
    11.1.2 Patchiness in a two-dimensional case .......... 294
11.2 Spatial structure and spatial correlations ........... 296
    11.2.1 Intrinsic lengths and scaling .............. 300
11.3 Ecological implications ..................... 308
    11.3.1 Plankton patchiness on a biological scale ........ 308
    11.3.2 Self-organized patchiness, desynchronization, and the paradox of enrichment .................. 312
11.4 Concluding remarks ....................... 321

## 12 Patchy invasion  325
12.1 Allee effect, biological control, and one-dimensional patterns of species invasion ........................... 326
    12.1.1 Patterns of species spread ................ 331
12.2 Invasion and control in the two-dimensional case ....... 338
    12.2.1 Properties of the patchy invasion ............ 344
12.3 Biological control through infectious diseases ......... 350
    12.3.1 Patchy spread in SIR model ............... 353
12.4 Concluding remarks ....................... 358

# IV Spatiotemporal patterns and noise  363

## 13 Generic model of stochastic population dynamics  365

## 14 Noise-induced pattern transitions  369
14.1 Transitions in a patchy environment .............. 369
    14.1.1 No noise ......................... 370
    14.1.2 Noise-induced pattern transition ............ 370
14.2 Transitions in a uniform environment ............. 372
    14.2.1 Standing waves driven by noise ............. 373

## 15 Epidemic spread in a stochastic environment     375
    15.1 Model . . . . . . . . . . . . . . . . . . . . . . . . . . . 376
    15.2 Strange periodic attractors in the lytic regime . . . . . . . . 379
    15.3 Local dynamics in the lysogenic regime . . . . . . . . . . . 382
    15.4 Deterministic and stochastic spatial dynamics . . . . . . . . 383
    15.5 Local dynamics with deterministic switch from lysogeny to lysis   385
    15.6 Spatiotemporal dynamics with switches from lysogeny to lysis   390
        15.6.1 Deterministic switching from lysogeny to lysis . . . . . 391
        15.6.2 Stochastic switching . . . . . . . . . . . . . . . . . . 393

## 16 Noise-induced pattern formation     397

## References     403

## Index     443

# Part I

# Introduction

# Chapter 1

# Ecological patterns in time and space

## 1.1 Local structures

Ecological systems are open systems, highly nonlinear and, hence, one has to cope with all challenges of nonlinear, nonequilibrium dynamics. The simplest growth and interaction laws are already nonlinear and, together with the variable environment, drive ecosystems away from a static equilibrium, which is meant as thermodynamic equilibrium with maximum entropy. The dynamic or flux equilibria far from thermodynamic equilibrium are called steady states.

### Steady-state multiplicity

The most simple but important nonlinear effect is the emergence of *steady-state multiplicity*. Already the logistic growth of a single population has two of them, the unstable extinction and the stable carrying capacity. Bistability appears if this population is prey for a Holling type III predator like in the one-dimensional spruce budworm system (Ludwig et al., 1978; Wissel, 1985). Populations with strong Allee effect also have two stable steady states, contrary to logistic growth the extinction state and again the carrying capacity (Courchamp et al., 1999, 2000). Prey–predator interactions like in the Rosenzweig-MacArthur model (1963), with logistically growing prey and Holling type II predator may generate two alternative stable states in the presence of a top predator as planktivorous fish in Scheffer's plankton model (1991a). In a certain range of control parameters a stable constant and an oscillating state may also coexist. If the predator is of Holling type III, the system may become bistable through intra-specific competition in the predator population and even tristable by the mentioned top predator. However, the multiple stability masks a just as interesting property of this model type: it is *excitability*, i.e., supercritical perturbations of the only stable steady state may lead to an outbreak in the prey population with long relaxation time. This effect is used for modeling recurrent phytoplankton blooms (Truscott and Brindley, 1994) and outbreaks of infectious diseases (Malchow et al., 2005). However, back to steady-state multiplicity: Multiple states are known in rivers and lake ecosystems (Scheffer, 1998; Dent et al., 2002), but also in terrestrial such as semi-arid grazing systems (Rietkerk and van de Koppel, 2002) or in climate (Higgins et al., 2002). In a deterministic model in a uniform, changeless environment, the initial condition would determine which stable constant or oscillating state will be approached once and forever. But ecosystems are also exposed to noisy variability of conditions like climate and weather. If such fluctuations become supercritical, the systems may jump between alternative stable states. It is not only noise but also continuously changing environmental conditions, such as the current global warming crisis, that can lead to such jumps. The latter is related to the observations and theory of *regime shifts* in ecosystems, e.g. from clear to eutrophicated water or from wet to dry land (Wissel, 1981; Rietkerk, 1998; Scheffer et al., 2001; Scheffer and Carpenter, 2003; Foley et al., 2003; Rietkerk et al., 2004; Greene and Pershing, 2007). The return, if possible at all, is very slow, usually on a hysteresis loop.

## Regular population oscillations

Since Elton (1924; 1942), Lotka (1925), Volterra (1926a), Gause and Vitt (1934), and others, *oscillations* in populations bother experimental and theoretical ecologists. There are ongoing discussions about the underlying mechanisms. One side underlines the role of population interactions as predation or competition, the other side the control through environmental variability, and there are examples for both. Corresponding prominent cases are the cycles in a vertebrate prey–predator community of the collared lemming in the high-Arctic tundra in Greenland that is preyed by four predators (Gilg et al., 2003; Hudson and Bjørnstad, 2003; Gilg et al., 2006), the predation behind pine beetle oscillations in the southern United States (Turchin et al., 1999; Turchin, 2003) as well as the oscillations in different fish that are induced by climate fluctuations (Stenseth et al., 2002). However, the truth will be somewhere in between, as in the Canada lynx oscillations that are induced by stochastic climatic forcing and density-dependent processes (Stenseth et al., 1999), and we do not want to participate in this discussion. The importance of oscillations for biodiversity maintenance and noninvadibility has been stressed by Vandermeer (2006) with an earlier application to plankton (Huisman and Weissing, 1999).

## Irregular population oscillations

There is even more dispute about the existence and role of *deterministic chaotic oscillations* in ecology (Berryman and Millstein, 1989; Pool, 1989; Scheffer, 1991b; Ascioti et al., 1993; Hastings et al., 1993; Ellner and Turchin, 1995; Cushing et al., 2001; Rai and Schaffer, 2001; Cushing et al., 2003). Though there might be no convincing proof of chaos in wildlife, except for some signs in boreal rodents (Hanski et al., 1993) or in the epidemics of a few childhood diseases (Olsen et al., 1988; Olsen and Schaffer, 1990; Engbert and Drepper, 1994); there are examples in laboratory experiments (Dennis et al., 1997; Becks et al., 2005). We believe that there is chaos in popu-

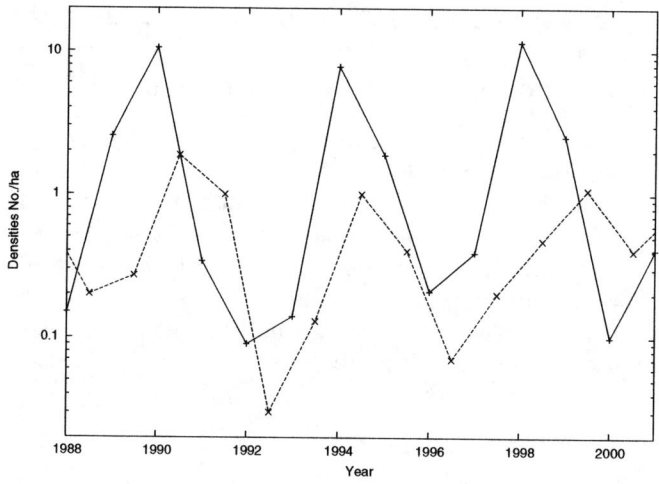

**FIGURE 1.1**: Prey–predator oscillations in lemmings (solid line) and stoats (dashed line). Data courtesy of Olivier Gilg, Helsinki.

lation dynamics and that it is masked by environmental and demographic noise. If one accepts the existence of intrinsic population oscillations, then the superposition of extrinsic forcings may naturally lead to quasiperiodic and aperiodic dynamics (Evans and Parslow, 1985; Truscott, 1995; Popova et al., 1997; Ryabchenko et al., 1997). However, we do not overemphasize the role of chaos in ecological systems; it is just another form of variability.

## Noise

Environmental variability is not purely deterministic, but also noisy. Therefore, the description by ordinary differential equations is always an approximation. One has to consider stochastic differential equations to account for the noise (Gardiner, 1985; Anishenko et al., 2003). There are noise-induced regime shifts between alternative stable states in ecosystems that are possible (Scheffer et al., 2001; Scheffer and Carpenter, 2003; Collie et al., 2004; Rietkerk et al., 2004; Steele, 2004; Freund et al., 2006) as well as counter-intuitive phenomena such as quasideterministic oscillations (Hempel et al., 1999; Neiman et al., 1999; Malchow and Schimansky-Geier, 2006), noise-enhanced stability, noise-delayed extinction, stochastic resonance (Freund et al., 2002), or noise-induced spatial pattern formation (García-Ojalvo and Sancho, 1999; Lindner et al., 2004; Spagnolo et al., 2004; Sieber et al., 2007).

## 1.2 Spatial and spatiotemporal structures

Ecology happens in time and space, and, therefore, its modeling also requires time and space. Not only growth and interactions but also spatiotemporal processes like random or directed and joint or relative motion of species as well as the variability of the environment must be considered. The interplay of growth, interactions, and transport causes the whole variety of spatiotemporal population structures that includes regular and irregular oscillations, propagating fronts, target patterns and spiral waves, pulses, and stationary as well as fuzzy dynamic spatial patterns.

### Diffusive fronts and spatial critical sizes

The simplest known spatiotemporal structures are diffusive invasion fronts of growing populations. For exponential growth, Luther (1906) estimated the front speed, which is numerically the same as the minimum speed of a logistically growing population (Fisher, 1937; Kolmogorov et al., 1937). The textbook example of the invasion of an exponentially growing population is the spread of muskrats in Europe (Skellam, 1951; Okubo, 1980; Okubo and Levin, 2001).

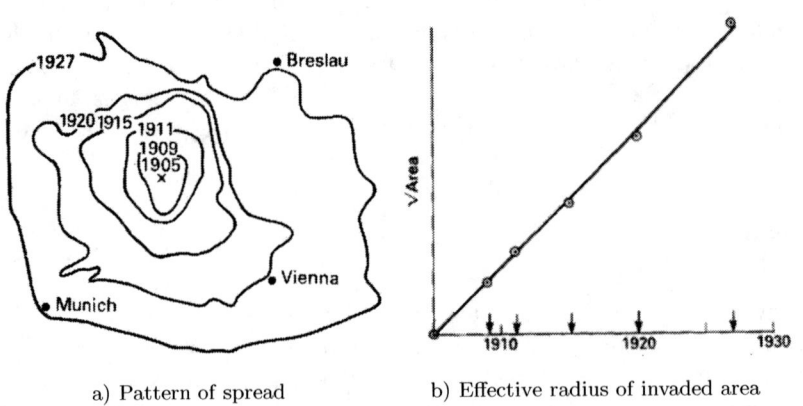

a) Pattern of spread    b) Effective radius of invaded area

**FIGURE 1.2**: Spread of muskrats over Europe. With permission from Skellam (1951).

For populations with multiple steady states, e.g., bistability in a population with a strong Allee effect, the spatial competition and spread of these states are known. In one spatial dimension, a critical radius of the spatial extension of a population can be defined (Schlögl, 1972; Nitzan et al., 1974; Ebeling and Schimansky-Geier, 1980; Malchow and Schimansky-Geier, 1985). Population patches greater than the critical size will survive, while the others will become extinct. However, bistability and the emergence of a critical spatial size do not necessarily require an Allee effect; logistically growing preys with a parametrized predator of type II or III functional response can also exhibit two stable steady states and the related hysteresis loops, cf. Ludwig et al. (1978); Wissel (1989).

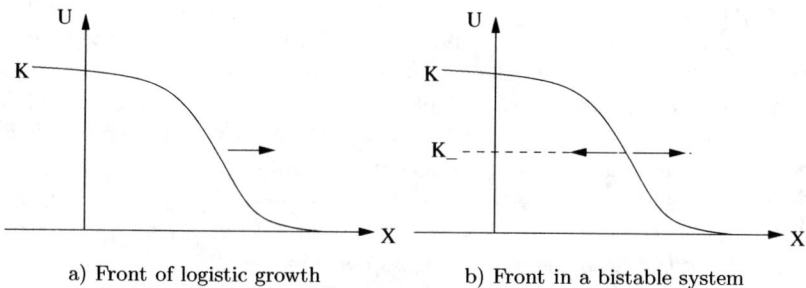

a) Front of logistic growth      b) Front in a bistable system

**FIGURE 1.3**: Diffusive fronts in systems with logistic growth and bistable dynamics. For logistic growth, the front moves to the right-hand side, and, finally, the space $X$ is filled with population $U$ at its capacity $K$. In bistable systems, the direction of the front depends on the initial condition: If the initial patch size exceeds a certain critical value, the picture will be the same as for logistic growth. However, if the initial patch is not large enough, the population will go extinct, though the local model would have predicted its survival.

## Diffusion-driven instabilities

Alan Turing (1952) was among the first to emphasize the role of nonequilibrium diffusion–reaction processes and patterns in biomorphogenesis. Since then, dissipative nonequilibrium mechanisms of spontaneous spatial and spatiotemporal pattern formation in a uniform environment have been of uninterrupted interest in experimental and theoretical biology and ecology. The interaction of at least two species with considerably different diffusion coefficients can give rise to spatial structure. A spatially uniform population distribution that is stable against spatially uniform perturbations (or in the local model without diffusion) can be driven to diffusive instability against spatially heterogeneous perturbations, e.g., a population wave or local outbreak, for sufficient differences of diffusivities. First, Segel and Jackson (1972) applied Turing's idea to a problem in population dynamics: the dissipative instability in the prey–predator interaction of algae and herbivorous copepods with higher herbivore motility. Levin and Segel (1976) suggested this scenario of spatial pattern formation as a possible origin of planktonic patchiness. Rietkerk et al. (2002) propose this mechanism as possible for the formation of tiger bushes; see Section 9.3.

### Differential-flow-induced instabilities

The Turing mechanism depends on the strong requirement of a sufficient difference of the diffusion coefficients. The latter neither exists for chemical reactions in aqueous solutions nor for micro-organisms in meso- and large-scale aquatic systems where the turbulent diffusion is relevant. Differential-flow-induced instabilities of a spatially uniform distribution can appear if flowing reactants or moving species like prey and predator possess different velocities, regardless of which one is faster. This mechanism of generating patchy patterns is more general. Thus, one can expect a wider range of applications of the differential-flow mechanism in population dynamics (Malchow, 2000b; Rovinsky et al., 1997; Malchow, 2000a). Conditions for the emergence of three-dimensional spatial and spatiotemporal patterns after differential-flow-induced instabilities (Rovinsky and Menzinger, 1992) of spatially uniform populations were derived (Malchow, 1998, 1995, 1996) and illustrated by patterns in Scheffer's model (1991a).

Complex spatial group patterns may also be generated by different animal communication mechanisms (Eftimie et al., 2007).

Also, vegetation can form a number of spatial patterns, especially in arid and semi-arid ecosystems. Usually, there is a combination of diffusive and advective mechanisms that yield gaps, labyrinths, stripes (tiger bush), or spots (leopard bush) (Lefever and Lejeune, 1997; Klausmeier, 1999; Lefever and Lejeune, 2000; Rietkerk et al., 2002, 2004).

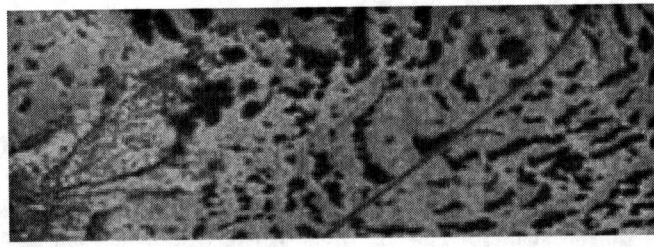

**FIGURE 1.4**: Tiger bush in Niger. Courtesy of Charlie Walthall, James R. Irons, and Philip W. Dabney, NASA Goddard Space Flight Center; see also Brown de Colstoun et al. (1996).

## Target patterns and spiral waves

Target patterns and spiral waves were first known from oscillating chemical reactions, cf. Field and Burger (1985), but have only later been observed as biologically controlled structures in natural populations. Spirals have been found to be important in models of parasitoid-host systems (Boerlijst et al., 1993). For other motile microorganisms, traveling waves like targets or spirals have been found in the cellular slime mold *Dictyostelium discoideum* (Gerisch, 1968; Keller and Segel, 1970; Gerisch, 1971; Segel and Stoeckly, 1972; Segel, 1977; Newel, 1983; Alt and Hoffmann, 1990; Siegert and Weijer, 1991; Steinbock et al., 1991; Vasiev et al., 1994; Ivanitskii et al., 1994; Höfer et al., 1995; Polezhaev et al., 2005). These amoebae are chemotactic species, i.e., they move actively up the gradient of a chemical attractant and aggregate. Chemotaxis is a kind of density-dependent cross-diffusion (Keller and Segel, 1971a,b). Agladze et al. (1993) have shown that colliding taxis waves may generate stationary spatial patterns, and they suggest this as an alternative to the classical Turing mechanism. Bacteria like *Escherichia coli* or *Bacillus subtilis* show a number of complex colony growth patterns (Shapiro and Hsu, 1989; Shapiro and Trubatch, 1991), some of them similar to diffusion-limited aggregation patterns (Witten and Sander, 1981; Matsushita and Fujikawa, 1990). Their emergence also requires cooperativity and active motion of the species, which has been modeled as density-dependent diffusion and predation (Kawasaki et al., 1995, 1997).

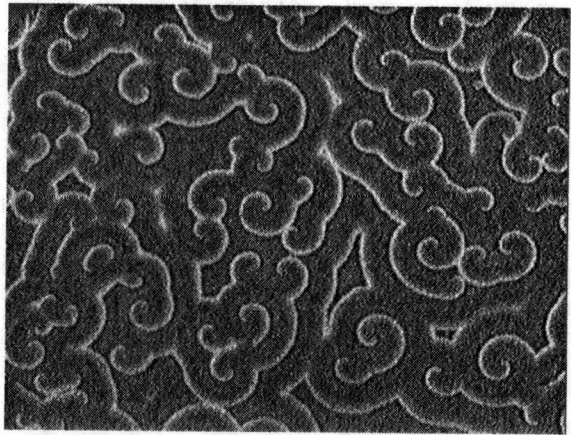

**FIGURE 1.5:** Spirals in an amoeba population (*Dictyostelium discoideum*). The base line of the photo is about 28.9 mm. Courtesy of Christiane Hilgardt and Stefan C. Müller, University of Magdeburg.

## New routes to spatiotemporal chaos

Space also provides new routes to chaotic dynamics. The emergence of diffusion-induced spatiotemporal chaos along a linear nutrient gradient has been found by Pascual (1993) in a Rosenzweig-MacArthur phytoplankton-zooplankton model. Chaotic oscillations behind propagating diffusive fronts are found in a prey–predator model (Sherratt et al., 1995, 1997). Furthermore, it has been shown that the appearance of chaotic spatiotemporal oscillations in a prey–predator system is a somewhat more general phenomenon and need not be attributed to front propagation or to an inhomogeneity of environmental parameters (Petrovskii and Malchow, 1999; Petrovskii and Malchow, 2001b; Petrovskii et al., 2003; Petrovskii and Malchow, 2001a; Petrovskii et al., 2005).

## Patterns of biological invasion and epidemic spread

A problem of increasing concern is the spread of non-native species and epidemic diseases (Drake and Mooney, 1989; Pimentel, 2002; Sax et al., 2005; Allen and Lee, 2006). Biological invasions are regarded as one of the most severe ecological problems, being responsible for the extinction of indigenous species, sustainable disturbance of ecosystems, and economic damage. Therefore, there is an increasing need to control and manage invasions. This requires an understanding of the mechanisms underlying the invasion process. Recently, many factors have been identified that affect the speed and the pattern of the spatial spread of an introduced species or pathogen, such as spatial heterogeneity, resource availability, stochasticity, environmental borders, predation, competition, infection, etc.; cf. recent reviews by Fagan et al. (2002), Hastings et al. (2005), Hilker et al. (2005), Holt et al. (2005), and Petrovskii et al. (2005). Mathematics, mathematical modeling, and computer science play an increasing if not central role in exploring, understanding and predicting these complex processes. This is especially true in studies of biological invasions where laboratory experiments cannot embrace appropriate spatial scales, and manipulative field experiments are, if at all ethically acceptable, very expensive, potentially dangerous, and lack the necessary time series.

The transmission dynamics of infectious diseases is one of the oldest topics of mathematical biology. As early as 1760, Daniel Bernoulli provided the first known mathematical result of epidemiology, that is, the defense of the practice of inoculation against smallpox (Brauer and Castillo-Chavez, 2001). The amount of works in this area has exploded in the last decades. Different aspects are dealt in the literature, from human health assessment to environmental assessment.

# Chapter 2

# An overview of modeling approaches

The great variety of different environmental settings and ecological interactions existing in mother nature requires, in order to make their theoretical study more effective, an equally broad range of modeling approaches. Quite typically, an adequate model choice depends not only on the ecosystem or community type, but also on the goals of the study, in particular, on the spatial and/or temporal scales where the given phenomena are developing. Discreteness of populations is a fundamental property (cf. Durrett and Levin, 1994), yet on a spatial scale much larger than the size of a typical individual description of the population dynamics by a continuous quantity such as the *population density* (the number of individuals of a given species per unit area or unit volume) was proved to be effective, e.g., see Murray (1989) and Shigesada and Kawasaki (1997). On a larger scale, however, environmental heterogeneity and habitat fragmentation become important, which may make space-discrete models more appropriate. A similar duality arises in the temporal dynamics. Moreover, due to the possible overlapping of spatial and/or temporal scales associated with different processes, sometimes a hybrid approach might be required that describes some of the processes continuously and some of them discretely; one example is given by a fish school feeding on plankton (Medvinsky et al., 2002).

The very first step to be done in choosing the model is a decision about the "state variables," i.e., the quantities that give sufficient (for the purposes of a given study) information about the state of the system. There is an apparent fundamental difference between the modeling approaches that take into account each individual separately, cf. "individual-based modeling," and the approaches that describe the system state in a collective way, e.g., by means of introducing the population density. Obviously, an individual-based approach[1] gives more information about the population system than an approach based on the population density; however, its disadvantage is that this information can rarely be obtained other than through extensive numerical simulations. On the contrary, the density-based models often allow rigorous mathematical analysis and analytical treatment. Another important distinction is that predictions of individual-based approaches are usually restricted to the spa-

---

[1] For the basics and state-of-the-art in that field, an interested reader is advised to check the books by DeAngelis and Gross (1992) and Grimm and Railsback (2005).

tial scales compatible with either size or motion of a single individual while density-based approaches are valid on a larger scale.

Having chosen a density-based approach, the next step is to decide whether details of system's spatial structure may be important. In case they are not, then one arrives at a *nonspatial model* where the population density or densities are functions of time but not of space. Specific mathematical settings for writing down the model depend on whether the dynamics of a given population is more adequately described as time-discrete or time-continuous. The former approach works better for the populations with nonoverlapping generations, the latter is valid when the generations are tangled. Throughout this book, we will mostly focus on the second case; a relevant mathematical technique is then given by ordinary differential equations:

$$\frac{dU_i(t)}{dT} = f_i(U_1, U_2, \ldots, U_n), \quad i = 1, \ldots, n, \qquad (2.1)$$

where $U_i$ is the population density of the $i$th species at time $T$, $n$ is the number of species in the community, and functions $f_i$ take into account effects of birth and mortality; in most biologically meaningful situations, the functions $f_i$ are nonlinear with respect to at least some of their arguments.

Now, a very subtle issue is the decision about how many equations should the system (2.1) contain and (a closely related question) what are the properties of functions $f_i$, which define the species responses and the types of interspecific interactions. Obviously, the population community even in a very simple ecosystem consists of dozens (more typically, hundreds or even thousands) of different species. Therefore, an idea to write a separate equation for each species is totally unrealistic. One way around this difficulty is to consider some "functional groups" instead of particular species. Most straightforwardly, these groups would correspond to different trophic levels. A classical example is given by phytoplankton and zooplankton; although in any natural aquatic ecosystem each of these two groups consists of many different species (sometimes interacting with each other in a very complicated way), ecologists use these rough "binary" description quite successfully, in both empirical and theoretical studies. Correspondingly, application of the model (2.1) to the plankton system dynamics would result in a two-species prey–predator system.

An alternative approach to minimize the number of equations in (2.1) is to focus on the dynamics of particular species. The reasons behind the choice of those "key species" depends on the ecosystem properties but also on the purpose of the study; for instance, in the case of biological invasion, one of them should obviously be the alien pest. The system (2.1) then may be reduced to either just a single equation or to a few-species system described by two or three equations, e.g., accounting for the given species and its immediate consumers or competitors. The impact of other species can be taken into account in an indirect way by means of either adjusting parameter values (e.g., introducing additional mortality rates in order to account for other predators) or by including additional terms into the equations, cf. "closure terms" (van den Bosch et al., 1988; Steele and Henderson, 1992a; Edwards and Yool, 2000).

Note that, while the predictive power of the few-species "conceptual" models is usually not very high, they are very important in a wider theoretical aspect because they make it possible to study the implications of basic interspecific interactions thoroughly.

It should be also mentioned that, apart from the population dynamics, the generic system (2.1) has been effectively used for modeling the dynamics of infectious diseases, up to a somewhat different meaning of the state variables (such as density of susceptibles instead of density of prey, etc.) and to the choice and meaning of the functions $f_i$, e.g., see Busenberg and Cooke (1981), Capasso (1993), and Dieckmann et al. (2002).

Once the model is specified, the next step is to reveal its main properties such as existence and stability of the steady states, existence of periodic solutions, solution boundedness, invariant manifold(s), etc. Careful analytical study of these issues often requires application of some rather advanced mathematical techniques; cf. Part II. The goal of this analysis is twofold. First, the model must be biologically reasonable and thus should exclude some obviously artificial situations (such as, for instance, population growth from a vanishingly small value of the population density). Second – and this is, in fact, the principal idea of a mathematical modeling approach – a change in the model properties with respect to a change in a certain *controlling parameter* is usually assumed to reflect the changes in the dynamics of the given natural (eco)system and thus has immediate biological implications.

The system (2.1) creates an appropriate modeling framework in the case of a "well-mixed" community in a homogeneous environment, i.e., when the community may be in all circumstances regarded as spatially homogeneous. Obviously, this is not always the case, and this affects the model choice. For instance, the spatial structure of a given population or community can be predefined by the environmental heterogeneity. In the case of small environmental gradients, a relevant mathematical model can still be space-continuous; however, in an extreme case of large environmental gradients or a fragmented habitat, a space-discrete approach will sometimes be more insightful. A mathematical model would then consist of a few systems such as (2.1) where different systems describe the dynamics of different subpopulations,[2] being coupled together due to migration between the habitats (Jansen and Lloyd, 2000; Jansen, 2001; Petrovskii and Li, 2001).

A separate branch of space-discrete models is made by metapopulation models (Gilpin and Hanski, 1991; Hanski, 1999), where the state of the "metacom-

---

[2] In the time-discrete case, a model would consist of coupled difference equations or maps; cf. Comins et al. (1992); Allen et al. (1993).

munity" may be described by variables other than population density, e.g., giving a proportion of all sites where the given species is present.

Another source of spatial heterogeneity is the formation of self-organized patterns as a result of inter- and intra-specific interactions, and this is going to be the main focus of this book.

Obviously, for the spatial aspect of population dynamics to be nontrivial, there must exist a mechanism of population redistribution in space. The most common one is due to individual motion. The motion can be either active (i.e., due to self-motion) or passive (e.g., when an individual of an air-borne species is carried about by wind). Also, motion can take place with or without a preferred direction. The simplest (but yet biologically meaningful) case is given by random isotropic motion, i.e., diffusion. In Chapter 8, we will talk in more details about diffusion in the population dynamics; for the moment we simply assume that the spatial aspects can be taken into account by adding the diffusion terms to Equations (2.1):

$$\frac{\partial U_i(\mathbf{R}, T)}{\partial T} = D_i \nabla^2 U_i(\mathbf{R}, T) + f_i(U_1, U_2, \ldots, U_n) \qquad (2.2)$$

($i = 1, \ldots, n$), where $\mathbf{R} = (X, Y, Z)$ is the position in space, $D_i$ is the diffusion coefficient of the $i$th species, and $\nabla^2$ is the Laplace operator:

$$\nabla^2 = \frac{\partial^2}{\partial X^2} + \frac{\partial^2}{\partial Y^2} + \frac{\partial^2}{\partial Z^2}.$$

The system (2.2) is a system of nonlinear partial differential equations and, as such, is a difficult mathematical object to study. Although some regular analytical approaches are available [e.g., see Petrovskii and Li (2006) and also Chapter 9 of this book], more often its properties are studied through computer simulations by means of solving Equations (2.2) numerically (cf. Thomas, 1995).

Correspondingly, the next important step is scaling. In analytical approaches it may be easier to work with the original systems (2.1) or (2.2) because it sometimes makes the interpretation of the results more straightforward. In numerical simulations, however, we actually work with numbers rather than with dimensional quantities. Therefore, it is more convenient to first transform a given model to a dimensionless form. Remarkably, it is always possible; the corresponding procedure is usually called either dimensions analysis or scaling. Indeed, the functions $f_i$ depend not only on the population densities but also on a number of parameters, such as the birth/death rate(s), population carrying capacity(-ies), etc. These parameters provide an intrinsic scale for each of the variables. Some typical examples showing how to do it in practice are given below; a very general description of the procedure[3] along

---

[3]It should be mentioned here that scaling by itself is a powerful method of analysis, and it often allows one to arrive at some important conclusions without even specifying the model; see Barenblatt (1996).

with its application to a wide range of problems can be found in Barenblatt (1996).

A considerable part of this book, especially in Parts III and IV, will be concerned with the dynamics of a prey–predator system continuous in space and time. According to the general theoretical framework introduced above, it is described by the following equations:

$$\frac{\partial U(\mathbf{R},T)}{\partial T} = D_1 \nabla^2 U(\mathbf{R},T) + P(U) - E(U,V) , \qquad (2.3)$$

$$\frac{\partial V(\mathbf{R},T)}{\partial T} = D_2 \nabla^2 V(\mathbf{R},T) + \kappa E(U,V) - \mu(V) , \qquad (2.4)$$

where $U$ and $V$ are the population densities of prey and predator, respectively. For biological reasons, the corresponding functions $f_1$ and $f_2$ are now split to separate terms such as prey population growth $P$, predation $E$, and predator mortality $\mu(V)$, and the coefficient $\kappa$ is called the predation efficiency or conversion rate.

Prey–predator systems have been at the focus of mathematical biology for several decades, starting from the works by Lotka and Volterra, yet there is still a lot of controversy regarding the optimal choice of the predator functional response to the prey density, e.g., see Arditi and Ginzburg (1989) and van Leeuwen et al. (2007). Presently, the forms that are used most often are the so-called Holling types II and III, that is,

$$\text{(a)} \quad E(U,V) = A \frac{UV}{U+H} \quad \text{and} \quad \text{(b)} \quad E(U,V) = A \frac{U^2 V}{U^2 + H^2} , \qquad (2.5)$$

respectively (cf. Murray, 1989), where $A$ is the predation rate and the parameter $H$ has the meaning of the half-saturation prey density.

Aiming here to give an instructive example, we choose Equation (2.5a) for predation, logistic growth for prey, and linear mortality for predator, $\mu(V) = MV$. Also, we focus on the one-dimensional case, i.e., on the system with one spatial dimension; extension of the scaling procedure to two- and three-dimensional cases is obvious. Equations (2.3)–(2.4) are then reduced to

$$\frac{\partial U(X,T)}{\partial T} = D_1 \frac{\partial^2 U}{\partial X^2} + \alpha U \left(1 - \frac{U}{K}\right) - A \frac{UV}{U+H} , \qquad (2.6)$$

$$\frac{\partial V(X,T)}{\partial T} = D_2 \frac{\partial^2 V}{\partial X^2} + \kappa A \frac{UV}{U+H} - MV . \qquad (2.7)$$

It is readily seen that the carrying capacity $K$ makes a convenient scale for the prey population density, $u = U/K$, where the new variable $u$ is dimensionless. In a similar way, the prey maximum per capita growth rate $\alpha$ gives a scale for the time, $t = \alpha T$. Introducing other dimensionless variables as $x = X(\alpha/D_2)^{1/2}$ and $v = VA/(\alpha K)$, Equations (2.6)–(2.7) take the following forms:

$$\frac{\partial u(x,t)}{\partial t} = \epsilon \frac{\partial^2 u}{\partial x^2} + u(1-u) - \frac{uv}{u+h}, \qquad (2.8)$$

$$\frac{\partial v(x,t)}{\partial t} = \frac{\partial^2 v}{\partial x^2} + k\frac{uv}{u+h} - mv, \qquad (2.9)$$

where $k = \kappa A/\alpha$, $m = M/\alpha$, $h = H/K$, and $\epsilon = D_1/D_2$ are dimensionless parameters. Correspondingly, the properties of $u(x,t)$ and $v(x,t)$ now depend not on all the original parameters separately but only on their combinations $k$, $m$, $h$, and $\epsilon$.

Note that (2.8)–(2.9) contain fewer parameters than the original Equations (2.6)–(2.7), i.e., four instead of eight. A decrease in parameter number due to the transition to dimensionless variables is a typical result. It gives another advantage of scaling: The less parameters a given model contains, the more effective is its study by means of numerical simulations.

An important remark that should be made here is that in most cases the choice of dimensionless variables is not unique. Indeed, talking about the prey–predator system (2.6)–(2.7), it is readily seen that we can introduce dimensionless variables as $t = T(\kappa AK/H)$, $x = X(\kappa AK/(D_2 H))^{1/2}$, $u = U/K$, and $v = V/(\kappa K)$. From Equations (2.6)–(2.7), we then obtain:

$$\frac{\partial u(x,t)}{\partial t} = \epsilon \frac{\partial^2 u}{\partial x^2} + au(1-u) - \frac{uv}{1+\alpha u}, \qquad (2.10)$$

$$\frac{\partial v(x,t)}{\partial t} = \frac{\partial^2 v}{\partial x^2} + \frac{uv}{1+\alpha u} - mv, \qquad (2.11)$$

where $\alpha = K/H$, $a = \eta HK/(\kappa A)$, $m = MH/(\kappa AK)$, and $\epsilon = D_1/D_2$.

Unfortunately, there are no accepted standards regarding the scaling procedure, and which of these two schemes to choose[4] is to a large extent a matter of personal preference. Different authors use different schemes, and that often makes quantitative comparisons between published results rather difficult, even in the case when the principal implications are clear.

Prey–predator interactions are ecologically meaningful but surely they do not exhaust all possible types of interspecific interactions. Another important case is given by competition. In contrast to the prey–predator system, where predation is beneficial for one species and detrimental for the other, competition hampers the population growth of both involved species. Mathematically, it is usually described by a bilinear function, i.e., a general time- and space-continuous model of a community of $N$ competing species looks as

---

[4]There can be some variations within each of these two approaches, such as using $D_1$ instead of $D_2$ for scaling $X$, or $X$ can be scaled to the domain length, etc.

follows:

$$\frac{\partial U_i(\mathbf{R}, T)}{\partial T} = D_i \nabla^2 U_i(\mathbf{R}, T) + \left( \alpha_i - \sum r_{ij} U_j \right) U_i , \qquad (2.12)$$

$$i = 1, \ldots, N$$

(cf. May, 1973), where $\alpha_i$ is the inherent population growth rate of the $i$th species and $\mathcal{R} = (r_{ij})$ is the so-called "community matrix." In a general case, $r_{ij} \neq r_{ji}$ because different species possess different abilities for competition.

The properties of the competing species system appear to be significantly different from those of a prey–predator system. In particular, it is readily seen that, contrary to a prey–predator system where cycles are generic, a nonspatial system of two competing species cannot have periodic solutions but only equilibrium states.

In agreement with the general idea of scaling, the number of parameters in (2.12) can be reduced (by means of choosing relevant scaling factors for the population densities, space, and time) from $N^2 + 2N$ to $N^2 + N - 2$. One particular case will be considered in Section 10.3.

Note that, in the general systems like (2.1) and/or (2.2), the state variables $U_1, \ldots, U_N$ must not necessarily be regarded as the densities of species 1 to $N$, respectively. Instead, at least some of them may have the meanings of the densities of *subpopulations* of the same species, provided that given species is in some way structured. In this book, we will be especially interested in one particular case when a given species is affected by an infectious disease, and thus its population is split to susceptible, infected, and removed/recovered parts.

Remarkably, a mathematical description of the disease dynamics appears to be similar (at least, in the simplest cases) to that of the prey–predator system. Indeed, denoting the densities of the susceptibles and infected as $S$ and $I$, respectively, instead of the system (2.3)–(2.4) we obtain

$$\frac{\partial S(\mathbf{R}, T)}{\partial T} = D_S \nabla^2 S(\mathbf{R}, T) + P(S, I) - \Sigma(S, I) - M_S S , \qquad (2.13)$$

$$\frac{\partial I(\mathbf{R}, T)}{\partial T} = D_I \nabla^2 I(\mathbf{R}, T) + \Sigma(S, I) - M_I I , \qquad (2.14)$$

where $\Sigma$ describes the disease transmission and other terms have the same meanings as above. In a general case, the "kinetics" of the disease transmission can be very complicated (cf. Fromont et al., 1998); however, having assumed that it is described by the so-called mass-action law, $\Sigma = \sigma SI$, and that only susceptibles can produce offspring, the system (2.13)–(2.14) is then reduced to the classical Lotka–Volterra prey–predator model.

Finally, the systems (2.1) and/or (2.2) can be of a mixed origin, i.e., some of the state variables may correspond to population densities and others to the subpopulations of the infected species. In that case, we arrive at an *ecoepidemiological* system. We will provide a detailed consideration of the corresponding models in Chapter 7.

All the models mentioned above are deterministic in the sense that they are described by deterministic equations. As such, they largely neglect the impact of stochastisity and noise. To distinguish between the cases when stochastisity may or may not be important is a highly nontrivial problem. Indeed, although the population dynamics is intrinsically stochastic, in many cases it is very well described by the deterministic equations. In practice, the choice of the model (i.e., deterministic or stochastic) is often purely heuristic. Also, it of course depends on the goals of the study. In order to take into account possible effects of stochastisity and noise, one can either apply statistical modeling (cf. Czárán, 1998) or include stochastic terms/factors into the deterministic model. The latter approach will be used in Part IV of this book, where we make an insight into the issue and reveal, by means of considering several specific cases, how the system's dynamics can be modified by the impact of stochastisity.

# Part II

# Models of temporal dynamics

# Chapter 3

# Classical one population models

Population theory, with its long history, dates back to the Malthus model, formulated for economic reasons in the early nineteenth century, predicting the population problems due to an exponential growth not supported by unlimited resources. The basic equation was then corrected by Verhulst (1845; 1847) to compensate for the shortcomings of the earlier model. The logistic equation states instead the existence of a horizontal asymptote to which the population tends as time flows. This asymptote has a biological interpretation, namely the carrying capacity the environment possesses, to support the population described by the model.

Later on, in the twenties of the past century, the works by Lotka and Volterra extended the mathematical modeling to prey–predator situations (Lotka, 1925; Volterra, 1926a,b). Since then the field has developed, and the biomathematical literature is now very large.

In this chapter we present the basic models in population theory, at first considering an isolated population and in Section 3.2 models including migrations. The purpose of Section 3.1 is to introduce some basic notation and terminology, while in Section 3.2 the most elementary bifurcations will naturally arise from the analysis. We stress the fact that in some cases the effort is made to keep the considerations simple enough to help the less familiar reader to grasp the basic concepts. Also, occasionally very elementary examples are made in both this as well as in subsequent chapters. In Section 3.3, simple discrete versions of the basic one-population models are discussed, to pave the way for the topic of the following Section 3.4, a short introduction of chaos.

## 3.1 Isolated populations models

Consider a single population in an idealized situation, in which neither deaths nor migration occur. Since then the only process that can happen is reproduction, the population must be a function of time $T$, i.e., $U \equiv U(T)$. Taking the time unit to be months, for instance, suppose we count the individuals at different times and find the following results:

$$U(0) = 1000, \quad U(10) = 1200, \quad U(20) = 1440.$$

If we take the population differences first with respect to the origin and then with respect to the previous measurement, we find

$$U(10) - U(0) = 200, \quad U(20) - U(0) = 440, \quad U(40) - U(20) = 240.$$

One question we may ask is, how do we know the estimate size of the population at intermediate times, such as $U(1)$ and $U(2)$? An intuitive answer would then be $U(1) = 1020$, $U(2) = 1044$. The reason is that we tend to think linearly in terms of time. If 200 newborns occur in a time of 10 months, we expect 20 in just one. In other words, we are saying that the monthly increment in population is $\delta_U = 20$.

Suppose now we take a second population, $V(T)$, in the same environmental conditions, the situation from the previous case differing only in the initial conditions. We would then find

$$V(0) = 2000, \quad V(10) = 2400, \quad V(20) = 2880,$$

again as 1000 individuals contribute 200 newborns in 10 months. The reason is once again that we also think linearly in terms of the population size. However, in this case we have

$$V(10) - V(0) = 400, \quad V(20) - V(0) = 880, \quad V(40) - V(20) = 480.$$

Thus, comparing the examples, the monthly population increment $\delta_V = 40 \neq \delta_U$ must be a function of population size $V$, $\delta_V \equiv \delta_V(V)$.

Consider a third example, again in different environmental circumstances, but still not accounting for deaths and migrations, in which the population $F$ is counted at other times giving the following data:

$$F(T_0) = 100, \quad F(T_1) = 105, \quad F(T_2) = 110.$$

We may now ask which of the two populations $F$ or $U$ reproduces faster. Now, taking differences, we find

$$F(T_1) - F(T_0) = 5, \quad F(T_2) - F(T_1) = 5, \quad F(T_2) - F(T_0) = 10,$$

and these absolute numbers seem to be smaller than the previous ones, so that we may think that $F$ reproduces the slowest. But if time is now measured in weeks, so that $T_1 = 1$, $T_2 = 2$, and we make the proportions, clearly $F$ reproduces the fastest of all populations. Indeed, at the end of one month, keeping the same growth proportion, we find $F(1) = 120$, and at the end of 10 months then $F(10) = 300$, assuming here that newborns do not start to reproduce in the whole period taken into account. The absolute increment is the same as for $U$, but here the percentage increase is clearly different. Indeed, in 10 months the increase of $F$ was 200%, while the one of $U$ was only 20%. This indicates first of all that the population increment is also a function of time,

$$\delta_F \equiv \Delta F \equiv F(T + \Delta T) - F(T) \equiv \delta_F(T).$$

Secondly, we should take the reference value of the population into account, thus not speaking of absolute increases, but rather of percentage accruals, i.e., relative increases $\Delta F/F$.

The simplest form for the population population increment over time that we can write is then

$$\frac{1}{U}\frac{\Delta U}{\Delta T} \equiv \frac{1}{U}R(U) \equiv r_M(U) = r,$$

with $r_M = r$ denoting a constant. Dimensionally, $U$ is a number, and $R(U)$ is then a per capita rate, i.e., a per capita frequency of reproduction. The word "per capita" refers to the fact that we were naturally induced to divide by the whole population size $U$, to determine the individual reproduction rate. The function $r_M(U)$ is then the individual frequency of reproduction, in this case constant. Equivalently, the function $R(U)$ is thus a linear function of U, $R(U) = rU$. Finally, we can let the time interval tend to zero, to get a differential equation. We have thus obtained the Malthus model,

$$\frac{dU}{dT}(T) = rU(T), \tag{3.1}$$

in which case the (constant) function $r_M = r$ denotes the per capita instaneous reproduction rate. More generally, as seen before, it may be a function of time. The equation is dimensionally sound as on the left there is a rate measured in numbers by time, and on the right a frequency times a number. Its solution is easily obtained:

$$U(T) = U(0)\exp(rT), \tag{3.2}$$

with $U(0)$ denoting the initial value of the population. The problem with this model is that for $r > 0$, the solution goes to $\infty$, while for $r < 0$, it goes to zero.

To correct this behavior Verhulst proposed the following modification. The constant $r$ should be replaced by a function, namely, the simplest nonconstant function of the population, $r_L(U)$, which implicitly becomes a function of time. This means that it should be taken as a linear function of $U$, namely, $r_L \equiv r_L(U) = r(1 - \frac{U}{K})$. Notice indeed that the slope of this straight line must be negative, since otherwise $r_L$ would be bounded below by the earlier constant $r$, and therefore the solution of the corrected equation would also be bounded below by the exponential function of the Malthus equation; thus, for a positive slope, the modification will not solve the problem, because the solution would still diverge for large times. We thus have the logistic equation

$$\frac{dU}{dT}(T) = r\left(1 - \frac{U(T)}{K}\right)U(T), \tag{3.3}$$

in which now geometrically $r$ has the meaning of the intercept at the origin and therefore must be nonnegative; otherwise, as seen above, the population will vanish. The solution of (3.3) can be analytically evaluated, by separation of variables,

$$\frac{dU}{(1-\frac{U}{K})U(T)} = \frac{dU}{U(T)} + \frac{dU}{K(1-\frac{U}{K})}$$

$$= [d\ln(U) - d\ln(1-\frac{U}{K})] = d\ln(\frac{U}{1-\frac{U}{K}}) = rdT,$$

followed by integration, to give

$$\ln\frac{U}{1-\frac{U}{K}} = rT + C, \quad \frac{U}{1-\frac{U}{K}} = \exp(C)\exp(rT) \equiv C_0\exp(rT)$$

and finally

$$U(t) = \frac{C_0 K \exp(rT)}{K + C_0 \exp(rT)}. \tag{3.4}$$

Thus, for $T \to \infty$, the solution tends to a horizontal asymptote, $U \to K$. The value of this constant represents the population that the environment can support in the long run, and it is called the carrying capacity. It also represents the root of the linear function $r(U)$ in the $RU$ phase plane. Notice that $R(U) \geq 0$ for $0 \leq U \leq K$, and conversely for $U \geq K$. It thus follows that in the logistic equation, $\frac{dU}{dT} \geq 0$ for $0 \leq U \leq K$, and $\frac{dU}{dT} \leq 0$ for $U \geq K$, thus implying that $U$ must grow when it has a value below $K$, while it should decrease when it exceeds it. These qualitative results then confirm the analysis we performed above: In both cases the population ultimately tends to the value $K$. Notice finally that, as mentioned earlier, for $r < 0$ the solution obtained also shows that $U \to 0$ as $T \to \infty$, allowing us to then discard this case for the reproduction parameter.

A further second order correction consists in taking a parabola instead of a straight line for the reproduction function, namely,

$$r_A \equiv r_A(U) = r(U - B)(1 - \frac{U}{K}), \quad 0 < B < K,$$

where again we take $r > 0$ in view of the previous discussion, so that the differential equation now takes into account the impact of *the Allee effect*:

$$\frac{dU}{dT}(T) = r(U - B)(1 - \frac{U}{K})U(T). \tag{3.5}$$

In this model the reproduction function $r_A$ becomes negative for population values smaller than $B$. This models the well-known fact that for small populations that live in a large environment in which individuals cannot easily find each other, reproduction becomes difficult. The phase plane analysis shows that the equilibria now are the same as for the logistic model, the origin, and the carrying capacity $K$, but there is also the equilibrium at the value $B$.

The rate of change of the population is positive only for $B < U < K$, so that again $K$ is a stable equilibrium since $U$ tends to grow if it is $B < U < K$ and decreases to this value if it exceeds it. For the equilibrium $B$, the reverse is true, thus, as just seen, $U$ increases away from it if it is larger than $B$, but if $0 \leq U < B$, then the derivative is negative so that in such case $U \to 0+$, meaning that ultimately the population vanishes, but from positive values. Thus, in the above-described environment, the population will be wiped away when it drops below a threshold population value represented by $B$. A further difference with the logistic model is given by the fact that here $K$ is not a globally asymptotically equilibrium, while in the logistic case it is. Starting from any value of $U$, indeed, in the latter case $K$ is ultimately reached, while for the Allee model (3.5), this is not true for $0 \leq U \leq B$.

### 3.1.1 Scaling

Let us consider now the effect of scaling. Taking into account the Malthus model, with a net growth rate $r \in \mathbf{R}$ that may be positive or negative, let us rescale the variables by setting

$$U(T) = \beta u(t), \quad T = \alpha t. \tag{3.6}$$

The constants $\beta$ and $\alpha$ must be positive, because biologically we account only for nonnegative populations and since we look at the future evolution of the system. By suitably differentiating we find

$$\frac{dU(T)}{dT} = \frac{\beta}{\alpha} \frac{du(t)}{dt}, \tag{3.7}$$

and substituting into Malthus equation upon simplification we have

$$\frac{du(t)}{dt} = \alpha r u(t). \tag{3.8}$$

In view of $\alpha > 0$ and $r \in \mathbf{R}$, we cannot impose $r\alpha = 1$, but rather consider the two cases $r\alpha = \pm 1$, i.e., work separately with the two equations

$$\frac{du(t)}{dt} = u(t), \quad \frac{du(t)}{dt} = -u(t).$$

This is not practical, so in the case of unrestricted net growth rate for the Malthus model, adimensionalization does not bring too much of an advantage. In the case of the logistic model, instead we must have $r > 0$, otherwise the model would show the population to have a negative carrying capacity to keep the "crowding competition" term negative, i.e., to still ensure $-\frac{r}{K}U^2 < 0$. In any case, the growth of the population would always be negative, leading to its disappearance, and thus vanifying the benefit of the introduction of the correction term. In the logistic case with $r > 0$, scaling can be performed

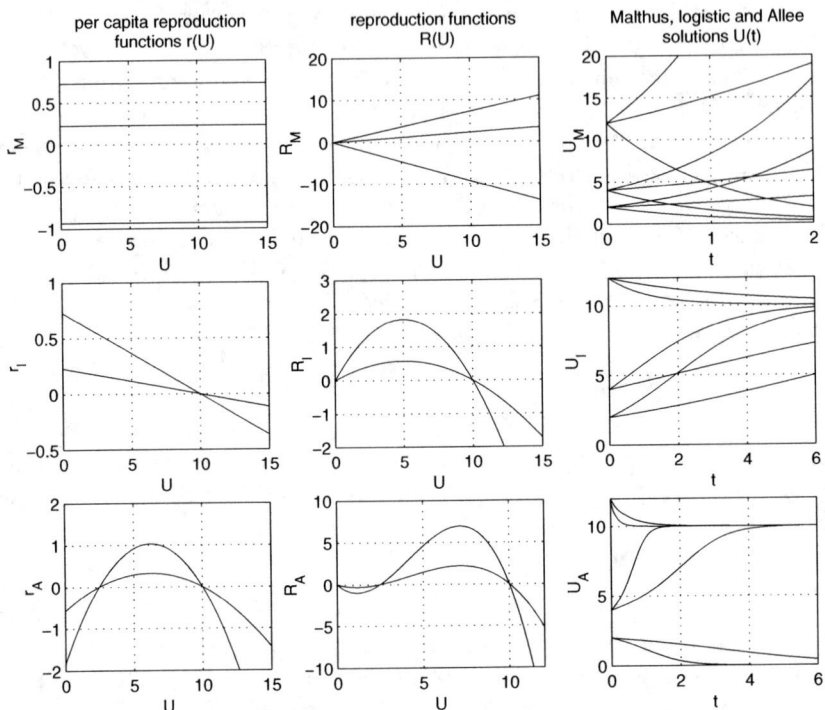

**FIGURE 3.1**: The rows refer to the three models considered: the first row to the Malthus model (3.1), the second one to the logistic one (3.3), and the third one to the Allee model (3.5). The first column contains the picture of the per capita reproduction function $r(U)$, the second one the reproduction function $R(U)$, and the last one the analytic solution of each model for three different initial conditions and for each value of the function $r(U)$ plotted in the corresponding row. In the logistic and Allee models, however, we do not show the case $r < 0$ as it is not meaningful. For the Malthus growth, a negative net growth function leads to the disappearance of the population. For the logistic model, all solutions tend to the horizontal asymptote; for the Allee model, this is the case only if the initial conditions are above the threshold value $B$, corresponding to the extra root, with respect to the logistic case, of the reproduction function in the second column.

with the same substitutions indicated above to get

$$\frac{du(t)}{dt} = \alpha r u(t)\left(1 - \frac{\beta u(t)}{K}\right) = u(1-u), \tag{3.9}$$

with the choice $\beta = K$ and $\alpha = r^{-1}$. For the Allee model (3.5), again assuming $r > 0$ and using the same substitutions, we find

$$\frac{du(t)}{dt} = \alpha r u(t)(\beta u(t) - B)\left(1 - \frac{\beta u(t)}{K}\right) = u(u-b)(1-u), \qquad (3.10)$$

choosing $\beta = K$, $\alpha = (Kr)^{-1}$ and setting $b = BK^{-1} < 1$.

## 3.2 Migration models

Consider now a population $U(T)$ that does not reproduce; rather, it is subject only to a constant rate migration $m$ and competition for resources. Its evolution equation then reads

$$\frac{dU}{dT} = m - U^2 \equiv f(U). \qquad (3.11)$$

From the logistic model we take the quadratic death rate term, while the constant term represents the external feed (or loss) of new individuals. Indeed, if we take $m < 0$, the first term represents an emigration term, and seeking equilibria, there is no way of zeroing out the right-hand side, as $U^2 = -m$ has no real roots. Thus the population $U$ will decrease at all times as in the $Uf$ phase plane, $f(U) = 0$ is a parabola always below the horizontal axis. For $m = 0$, i.e., no migration of any sort occurs, the same parabola will move upwards and become tangent to the $U$ axis at the origin. The origin will then become a semistable equilibrium, but the behavior of the model will essentially be unchanged. A more interesting situation arises for an immigration rate $m > 0$. In such a case there are two real roots, $U_\pm = \pm\sqrt{m}$. The parabola $f(U)$ is above the $U$ axis in between them, so that $f(U) \geq 0$ for $U_- \leq U \leq U_+$, and therefore $U$ grows in such an interval and decreases for $U \geq U_+$. This means that the equilibrium $U_+$ is stable, and conversely $U_- < 0$ is unstable, but since it is unfeasible, we disregard it. When $m$ increases, so does $U_+$. By drawing a picture of $U_+$ as a function of $U$ we obtain what is called a bifurcation diagram; see Figure 3.2.

We can also consider a more complex situation, a logistic model in which we allow the growth rate $r$ to be of either sign. In such a case, the model should not be written as (3.3) of Section 3.1 because the term containing the carrying capacity would become positive for $r < 0$, which does not correspond to the population pressure. We then write instead

$$\frac{dU}{dT} = aU - bU^2, \qquad (3.12)$$

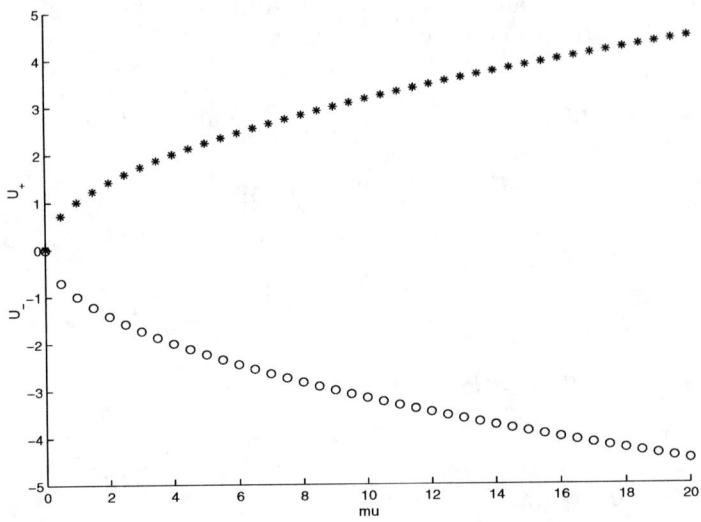

**FIGURE 3.2**: Bifurcation diagram for the logistic model (3.11), with immigration and no reproduction, $dU/dT = m - U^2 \equiv f(U)$. The stable feasible branch is represented by $*$; we also show the infeasible unstable branch $\circ$.

with $a \in \mathbf{R}$ and $b \geq 0$. Let us rescale the equations by introducing new positive parameters $\alpha$ and $\beta$, so that $U = \beta u$ and $T = \alpha t$ give

$$\frac{dU}{dT} = \beta \frac{du}{dt}\frac{dt}{dT} = \frac{\beta}{\alpha}\frac{du}{dt}$$

and the rescaled model becomes

$$\frac{du}{dt} = a\alpha u - b\alpha\beta u^2. \qquad (3.13)$$

Now we can choose the free parameters so as to simplify the equation, but the choice $a\alpha = 1$ is not possible, because in such a case the growth rate $a$ would be positive. If we want to account for $a \in \mathbf{R}$, we then need to take $b\alpha\beta = 1$, i.e., $\beta = \frac{1}{b\alpha}$, and therefore we must then set $a\alpha = \frac{a}{b\beta} \equiv \mu \in \mathbf{R}$. The rescaled model then reads

$$\frac{du}{dt} = \mu u - u^2 \equiv F(u). \qquad (3.14)$$

The equilibria are the origin and the point $u_m = \mu \in \mathbf{R}$. The parabola $F(u)$ is positive once again in between the roots, implying with the discussion as above that the right root is always stable and the left one is unstable. For $\mu < 0$, the right root is the origin, while for $\mu > 0$ it is the point $u_m > 0$. At $\mu = 0$, the origin is a double root, a semistable equilibrium in this case, as it seems stable if approached from the right and unstable from the left. If we draw the roots as function of the parameter $\mu$, we find the diagram of a transcritical bifurcation. Notice that for $\mu < 0$ the stable branch is given by the origin and the bisectrix is unstable, while for $\mu > 0$ the two branches exchange their stability properties. Clearly, however, for $\mu < 0$, the bisectrix would give an unfeasible equilibrium; see Figure 3.3.

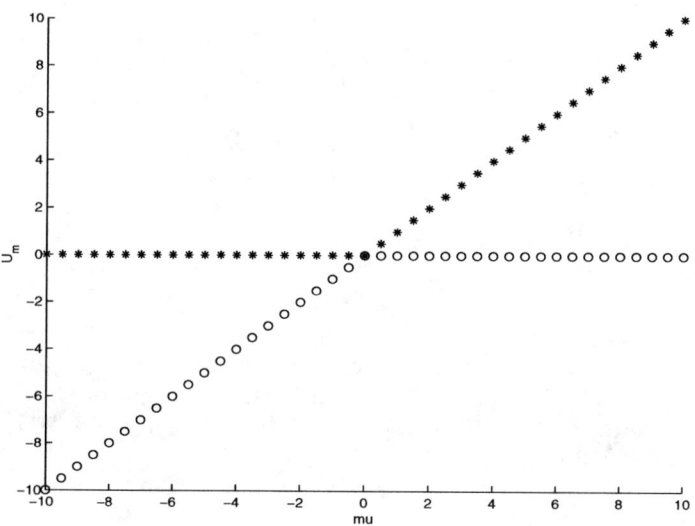

**FIGURE 3.3**: Logistic model (3.14) with reproduction and no immigration, $du/dt = \mu u - u^2 \equiv F(u)$. Bifurcation diagram as a function of reproduction rate $\mu$. The stable feasible branches are represented by $*$; the partly infeasible unstable branch is denoted by $\circ$.

We now consider a particular case of (3.5), namely,

$$\frac{dU}{dT} = \mu U - U^3, \tag{3.15}$$

with $\mu \in \mathbf{R}$, obtained from (3.5) again relaxing an assumption. We allow here a negative threshold $B$, so that $K + B = 0$, and rescale as indicated above, choosing $\alpha\beta^2 r = K$ and setting $\mu = -B\alpha r$. Note that, although (3.15) formally gives a particular case of the Allee model (3.5), the population growth described by (3.15) is no longer damped by the Allee effect. It is readily seen that for any $B \leq -K$, the right-hand side of (3.15) becomes a convex function and thus is qualitatively similar to the logistic growth. Correspondingly, Equation (3.15) is sometimes called a model with a "generalized logistic growth."

Figure 3.4 shows that we obtain a pitchfork bifurcation diagram.

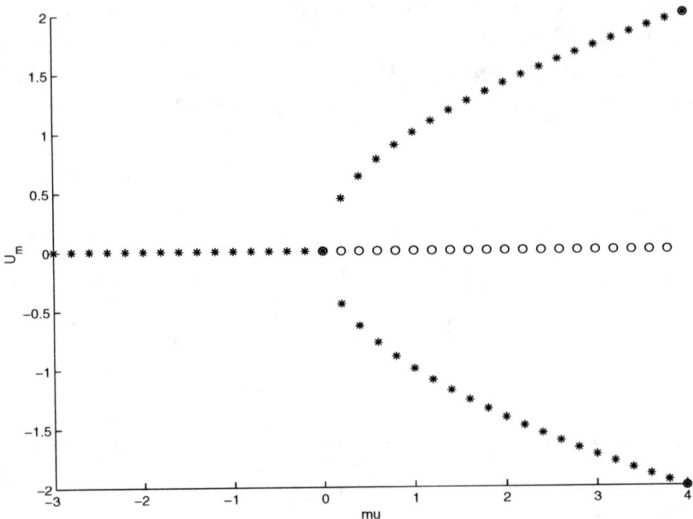

**FIGURE 3.4**: Bifurcation diagram for a generalized logistic growth (3.15), $dU/dT = \mu U - U^3$. The pitchfork bifurcation occurs at the origin. The stable branches are represented by $*$; there is one that is infeasible. The unstable branch is denoted by $\circ$.

### 3.2.1 Harvesting

We now consider a term $H(U)$ denoting harvesting of the population,

$$\frac{dU(T)}{dT} = R(U) - H(U).$$

The exploitation term can generally be chosen in one of the following ways:

$$H(U) = h, \quad H(U) = hU, \quad H(U) = h\frac{U}{A+U},$$

$$\text{or} \quad H(U) = h\frac{U^2}{A+U^2}, \tag{3.16}$$

where $A$ and $h$ are positive parameters. The first two expressions in (3.16) correspond to constant and linear exploitation, respectively, allowing for possibly unbounded harvests if the population is large enough. The last two assume an upper bound on the return, in a sense analogous to the limitations provided by the logistic equation. They differ essentially near the origin, the former starting with a slope at the origin, while the last one has zero derivative there. Compare Figures 3.11 and 3.13.

Combining now different types of harvestings of the population, let us look in detail at its consequences, with respect to the different growth mechanisms introduced before. For the Malthus model, with the net growth rate $\tilde{r} \in \mathbf{R}$ unrestricted in sign, we have

$$\frac{dU(T)}{dT} = \tilde{r}U(T) - h. \tag{3.17}$$

After adimensionalization again with $U(T) = \beta u(t)$, $T = \alpha t$, we find

$$\frac{du(t)}{dt} = \alpha \tilde{r} u(t) - \frac{\alpha}{\beta} h \tag{3.18}$$

and setting $\beta = \alpha h$, $r = \alpha \tilde{r}$, we finally end up with

$$\frac{du(t)}{dt} = r u(t) - 1, \quad r \in \mathbf{R}. \tag{3.19}$$

The right-hand side is thus linear in $u$, positive for $u > u_0 \equiv r^{-1}$, and negative conversely, so that $u$ grows if it has a larger value than $u_0$, and vice versa. This fact renders $u_0$, if feasible, i.e., for $r > 0$, an unstable equilibrium while the origin in a such case is a stable one, contrary to what happens in the classical model without harvesting. Thus exploitation changes the model's behavior. For $r < 0$, as for the original Malthus model, $O$ is always stable on biological grounds, as negative populations make no sense. Mathematically, it would not even be an equilibrium, as in a such case the only one would be $u_0 < 0$, which would be infeasible. The reader should compare the very simple bifurcation diagram as a function of $r$ reported in Figure 3.5 with the above considerations.

Consider now a linear harvesting term

$$\frac{dU(T)}{dT} = rU(T) - hU(T), \quad r \in \mathbf{R}. \tag{3.20}$$

Scaling as before, we find

$$\frac{du(t)}{dt} = \theta u(t), \quad \theta = \alpha(r - h) \in \mathbf{R}. \tag{3.21}$$

In this case, then, we see that the Holling type I exploitation leads to the same original Malthus model, with a different growth function. The corresponding bifurcation diagram as a function of $\theta$ is again elementary to construct. $O$ is unstable for $\theta > 0$, and conversely it is always stable; see Figure 3.6.

The Holling type II harvesting gives

$$\frac{dU(T)}{dT} = \tilde{r}U(T) - h\frac{U(T)}{A + U(T)}, \quad A, h, \tilde{r} \in \mathbf{R}. \tag{3.22}$$

Scaling in this case gives

$$\frac{du(t)}{dt} = \alpha u(t) \left[ \tilde{r} - \frac{h}{A + \beta u} \right],$$

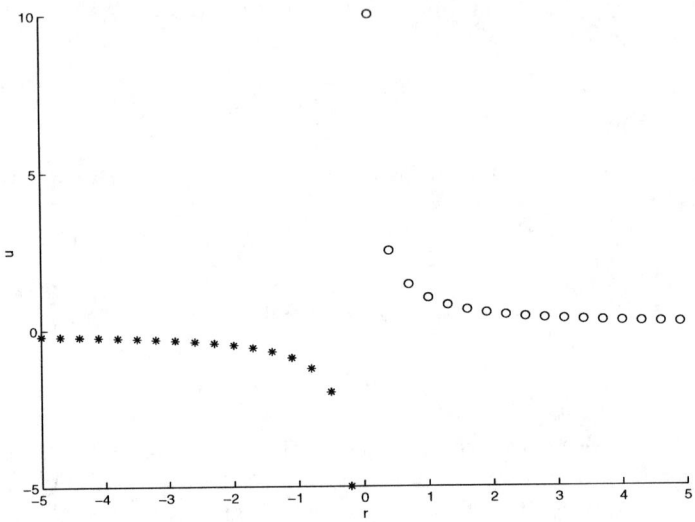

**FIGURE 3.5**: Bifurcation diagram for the Malthus model (3.19) with constant harvesting: stable branch ∗, unstable branch ○.

from which, setting $\beta = A$, $\alpha = \frac{A}{h}$, and $r = \alpha \tilde{r}$, we find

$$\frac{du(t)}{dt} = u\left[r - \frac{1}{1+u}\right]. \qquad (3.23)$$

The equilibria are once again $O$ and the point $u_* = \frac{1-r}{r}$, which is feasible for $0 < r < 1$. In such a case, however, it is unstable; for $r > 1$, the origin is unstable and for $r < 1$, it is stable. The bifurcation diagram is now composed by a branch of a hyperbola in addition to the line $u = 0$; see Figure 3.7.

We finally consider harvesting using the Holling type III form, on the simple Malthus model:

$$\frac{dU(T)}{dT} = \tilde{r}U(T) - h\frac{U^2(T)}{A + U^2(T)}, \quad A, h, \tilde{r} \in \mathbf{R}. \qquad (3.24)$$

The scaling procedure with $\beta = \sqrt{A}$, $\alpha = \beta h^{-1}$ and $r = \tilde{r}\alpha$ now leads to

$$\frac{du(t)}{dt} = ru - \frac{u^2}{1+u^2}. \qquad (3.25)$$

The equilibria in this case are $O$ once again and $u_\pm = \frac{1}{2r}(1 \pm \sqrt{1 - 4r^2})$, which exist and are feasible for $-\hat{r} < r < \hat{r}$, with $\hat{r} = \frac{1}{2}$. Analysis of the trajectories shows that $O$ for $r < 0$ is stable, for $0 < r < \hat{r}$ it becomes unstable, and there are also the two equilibria found above, of which $u_-$ is stable and $u_+$ is unstable, for $r > \hat{r}$ we find only $O$, which retains its instability. The

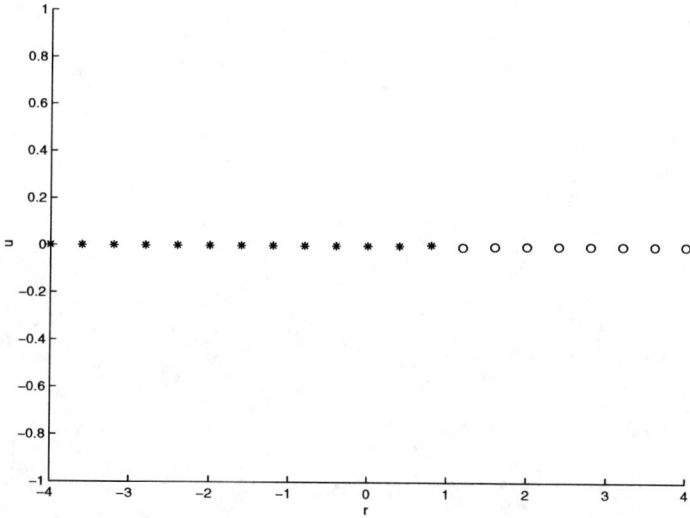

**FIGURE 3.6**: Bifurcation diagram for the Malthus model (3.21) with Holling type I harvesting: stable branch ∗, unstable branch ∘.

bifurcation diagram (see Figure 3.8) exhibits once again an unstable branch of a hyperbola in the interval $0 < r < \hat{r}$, which coalesces with the stable branch emanating from $O$ at the point $\hat{u} = 1$. The latter is easily calculated by imposing tangence between the straight line and the Holling type III term, yielding the equations

$$ru = \frac{u^2}{1+u^2}, \quad r = \frac{2u}{(1+u^2)^2}.$$

Substitution of one into the other easily leads to $1 + u^2 = 2$, from which the values of $\hat{u}$ and $\hat{r}$ are obtained, in agreement with what was formerly obtained.

We now will consider the logistic model subject to the various forms of exploitation. As mentioned before, here we will assume $r > 0$, which will give an alternative choice for the rescaling. In particular, we will construct bifurcation diagrams now as functions of the harvesting effort $h$. We have, under constant effort,

$$\frac{dU(T)}{dT} = rU(T)\left(1 - \frac{U(T)}{K}\right) - \tilde{h}, \quad \tilde{h} > 0. \tag{3.26}$$

Rescaling with the usual substitutions and setting $\beta = K$, $\alpha = r^{-1}$, and $h = \tilde{h}r^{-1}$ gives the adimensional model

$$\frac{du(t)}{dt} = u(1-u) - h. \tag{3.27}$$

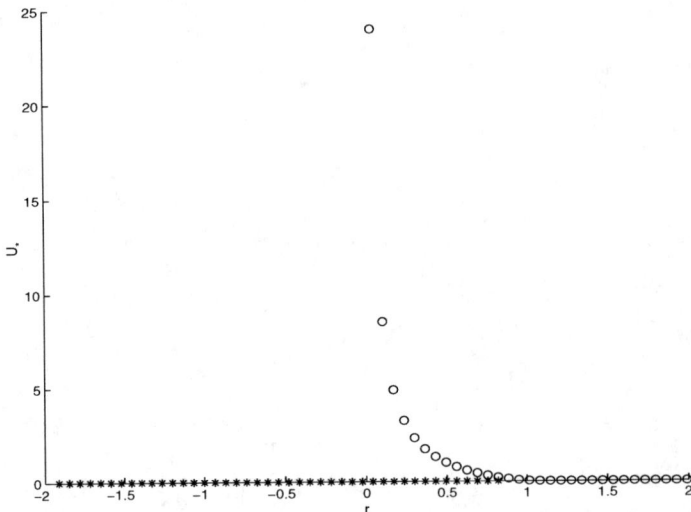

**FIGURE 3.7**: Bifurcation diagram for Malthus model with Holling type II harvesting (3.23): stable branch ∗, unstable branch ∘.

Clearly the right-hand side represents the logistic parabola through the origin and the normalized carrying capacity $K = 1$ pulled down by an amount $h$. Its roots are intersections of this parabola with the horizontal straight line at height $h$. They then move toward each other the higher $h$ becomes, retaining their stability properties, stable on the right and unstable on the left, until they coalesce and then disappear, at which point the origin becomes the only stable equilibrium of the system. The coalescence point concides with the vertex of the parabola, thus it occurs for $h = \frac{1}{4}$. The bifurcation diagram (Figure 3.9) thus shows a parabola with a stable and an unstable branch, while the origin for $h > 0$ is always stable. Biologically, this constant harvesting effort thus has on the equilibria the same effect of the Allee model without exploitation. Notice also that in fisheries models the level of the harvesting effort at the stable equilibrium gives the so-called maximum sustainable yield of the exploitation.

We now turn to Holling type I harvest namely,

$$\frac{dU(T)}{dT} = rU(T)\left(1 - \frac{U(T)}{K}\right) - \tilde{h}U(T), \quad \tilde{h} > 0. \tag{3.28}$$

Scaling gives

$$\frac{du(t)}{dt} = u(1 - h - u), \tag{3.29}$$

with the choices $\alpha r = 1$, $\beta = K$ and setting $h = \tilde{h}\alpha$. In this case, the stable equilibrim at $u = 1$ for $h = 0$ moves leftwards linearly with $h$, as it is given

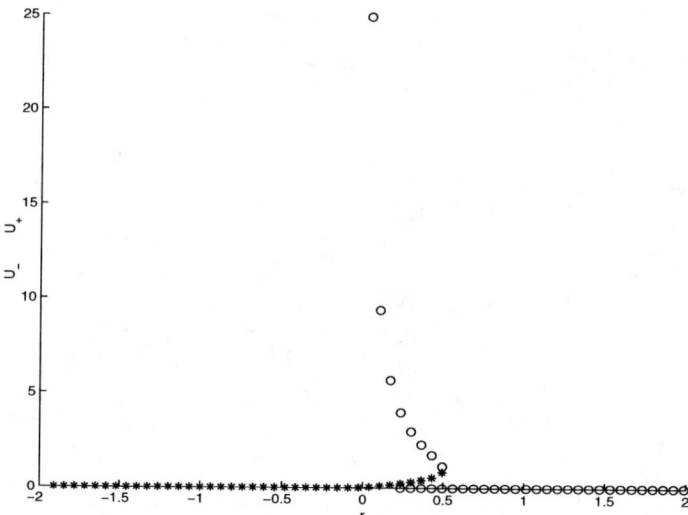

**FIGURE 3.8**: Bifurcation diagram for Malthus model with Holling type III harvesting (3.25): stable branch ∗, unstable branch ○.

by $u_s = 1 - h$, retaining its stability, until it hits the origin at $h = 1$. This corresponds to the logistic and harvesting curves to be tangent to each other at O. The origin that was formerly unstable at this point becomes stable and remains so for every $h \geq 1$. Figure 3.10 shows the corresponding bifurcation diagram.

The Holling type II exploitation gives

$$\frac{dU(T)}{dT} = rU(T)\left(1 - \frac{U(T)}{K}\right) - \tilde{h}\frac{U(T)}{A+U(T)}, \quad A, \tilde{h} > 0. \tag{3.30}$$

Upon rescaling we find for $\alpha = r^{-1}$, $\beta = A$, $\theta = \frac{A}{K}$,

$$\frac{du(t)}{dt} = u(1 - \theta u) - h\frac{u}{1+u} \equiv P(u) - \tilde{H}(u). \tag{3.31}$$

Notice that the slopes of the parabola $P$ and the hyperbola $\tilde{H}$ can be evaluated so that at the origin they are, respectively,

$$\frac{dP}{du} = 1 - 2\theta u, \quad \frac{d\tilde{H}}{du} = \frac{h}{1+u^2}, \quad \left.\frac{dP}{du}\right|_{u=0} = 1, \quad \left.\frac{d\tilde{H}}{du}\right|_{u=0} = h.$$

The equilibria are given by

$$u_\pm = \frac{1}{2\theta}\left[1 - \theta \pm \sqrt{(\theta+1)^2 - 4h\theta}\right].$$

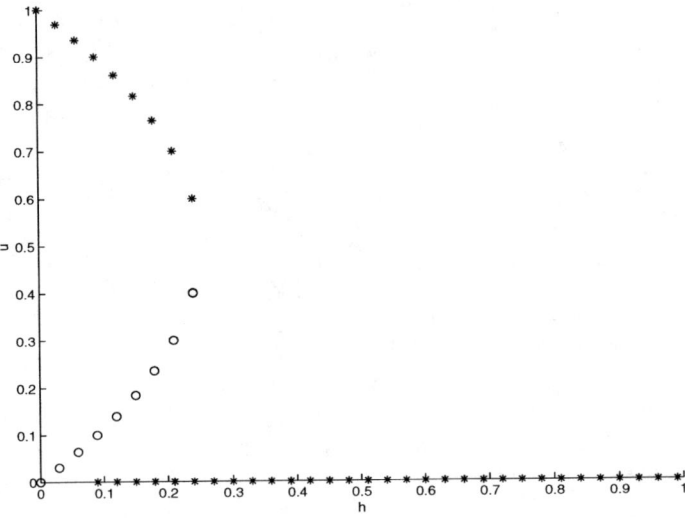

**FIGURE 3.9**: Bifurcation diagram for the logistic model (3.27) with constant harvesting: stable branch ∗, unstable branch ○.

They are real only for $h < h_s \equiv \frac{(\theta+1)^2}{4\theta}$. In such a case, the rightmost one, $u_+$, is stable, as the hyperbola given by the harvesting term lies above the logistic parabola for $u > u_+$, and conversely for $u < u_-$. The leftmost one, $u_-$, will instead be unstable. Feasibility is better described geometrically. We need to ensure that the curves $P(u)$ and $\tilde{H}(u)$ meet in the first quadrant. For low values of $h$, this is always ensured, $0 < h < 1$, but for $h > 1$, this happens only if the value of the horizontal asymptote of $\tilde{H}$ is below the vertex of the parabola $P$, of height $(4\theta)^{-1}$. Thus, for $1 < h < (4\theta)^{-1}$, there are the intersections $u_\pm$ described above, and for this to occur, naturally we need to require $\theta < \frac{1}{4}$. In the contrary case, no intersection between $P$ and $\tilde{H}$ is possible for $h > 1$, and the origin becomes the only equilibrium of the system. Figure 3.11 shows these situations. Notice also that the origin is always a stable equilibrium when $u_\pm$ are feasible, and is instead ustable for $1 > h > 0$. The bifurcation diagram is reported in Figure 3.12.

The logistic growth subject to Holling type III exploitation is ruled by the equation

$$\frac{dU(T)}{dT} = rU(T)\left(1 - \frac{U(T)}{K}\right) - \tilde{h}\frac{U^2(T)}{A + U^2(T)}, \quad A, \tilde{h} > 0. \qquad (3.32)$$

The rescaling is now done with $\alpha = r^{-1}$, $\beta = \sqrt{A}$, and setting $h = \frac{\tilde{h}}{r\sqrt{A}}$, $\phi = \frac{\sqrt{A}}{K}$, to get

$$\frac{du(t)}{dt} = u(1 - \phi u) - h\frac{u^2}{1 + u^2} \equiv u[\widehat{\ell}(u) - \widehat{H}(u)]. \qquad (3.33)$$

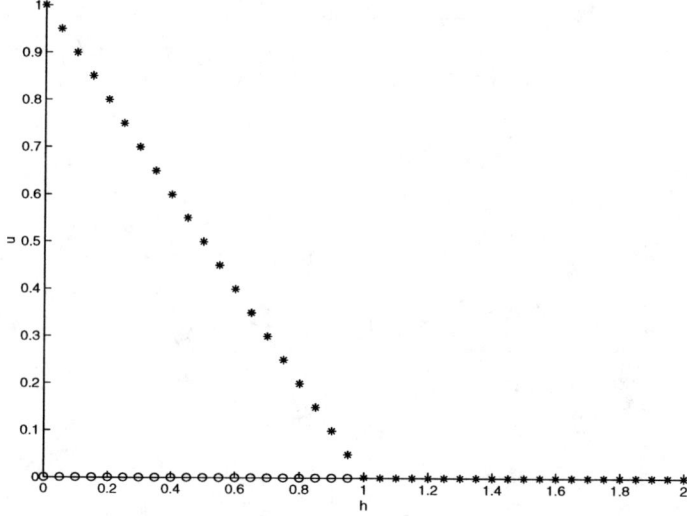

**FIGURE 3.10**: Bifurcation diagram for the logistic model (3.29) with linear harvesting: stable branch ∗, unstable branch ∘.

This shows that $O$ is again an equilibrium while the other ones come from solving $\widehat{\ell}(u) = \widehat{H}(u)$. The analysis of $\widehat{H}(u)$ shows that it has a maximum located at $(1, \frac{1}{2}h)$ and an inflection point at $Q \equiv (\sqrt{3}, \frac{\sqrt{3}}{4}h)$. We can then choose $\widehat{\ell}(u)$ through $Q$ and with a height at the origin located between the heights of the two former points, so that then $\frac{1}{2}h > 1 > \frac{\sqrt{3}}{4}h$. These give the interval $2 < h < \frac{4}{\sqrt{3}}$. The condition ensuring that $\widehat{\ell}(u)$ goes through $Q$ is easily found to be $\phi = \frac{1}{\sqrt{3}} - \frac{h}{4}$, and the constraint $\phi > 0$ is satisfied since it is equivalent to the former requirement $1 > \frac{\sqrt{3}}{4}h$. Thus, in these conditions there are three intersections, otherwise just one. The two different situations are shown in Figure 3.13. The origin is always an unstable equilibrium of the system. If there is only one other equilibrium, it is then stable (Figure 3.14). If there are three more, the intermediate of the two is unstable, while the other two are stable (see Figure 3.15).

Notice also that Figure 3.15 shows the phenomenon of hysteresis. Namely, starting from the population being exploited at the carrying capacity in absence of harvesting, $h = 0$, the equilibrium is the uppermost on the vertical axis. Increasing the exploitation, the stable upper branch of the bifurcation is followed, with the equilibrium decreasing steadily and slowly for increasing $h$. However, when $h \approx .85 \equiv h^c$, the upper critical level, the two stable and unstable branches coalesce, the upper equilibrium for larger harvesting efforts then disappears, and sudddenly the population moves toward the only stable equilibrium, which is now situated very near the horizontal axis. If we think

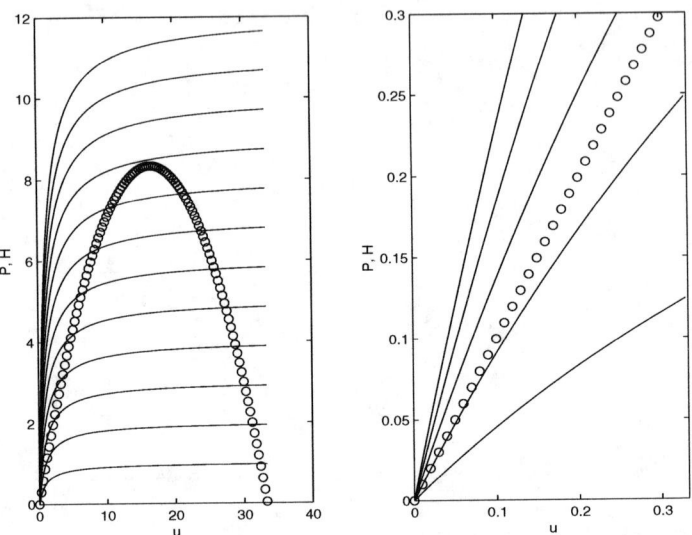

**FIGURE 3.11**: Diagram of the right hand-side for the logistic model (3.31) with Holling type II exploitation, on the left for $K = 33.3$, i.e., $\phi = 0.3$, and for integer harvesting rates $h \in \{1, \ldots, 10\}$. On the right is a blowup of the same picture around the origin for $h \in \{0.5, 1, 1.5, 2, 2.5\}$

now to go back to the previous situation only by diminishing the harvesting to values just below the critical one, say at $h = 0.8$, the reached stable equilibrium on the lower branch of the bifurcation diagram moves upwards, but only by a very small amount. This is, of course, provided that the whole ecosystem has not collapsed before due to some external disturbances that push the population $u$ to zero. Thus we have the counter-intuitive fact that it is not enough to backtrack a little to push the system to its previous situation. It is not until $h = 0.2 \equiv h_c$, the lower critical value, that $u$ recovers, as in such a case following toward the left the lower branch of the bifurcation diagram, the lower stable equilibrium vanishes and the system is pushed suddenly upwards to the stable branch near the carrying capacity. On increasing again the harvesting effort $h$, the cycle would then be repeated.

We finally turn to the analysis of the Allee model, coupled with harvesting:

$$\frac{dU(T)}{dT} = rU(T)(U(T) - B)\left(1 - \frac{U(T)}{K}\right) - \tilde{h}, \quad \tilde{h} > 0. \tag{3.34}$$

The adimensionalization procedure in this case, letting $\beta = K$, $\alpha = (Kr)^{-1}$, and $h = \tilde{h}K^{-2}r^{-1}$, $b = \frac{B}{K}$, gives the rescaled model

$$\frac{du(t)}{dt} = u(u - b)(1 - u) - h. \tag{3.35}$$

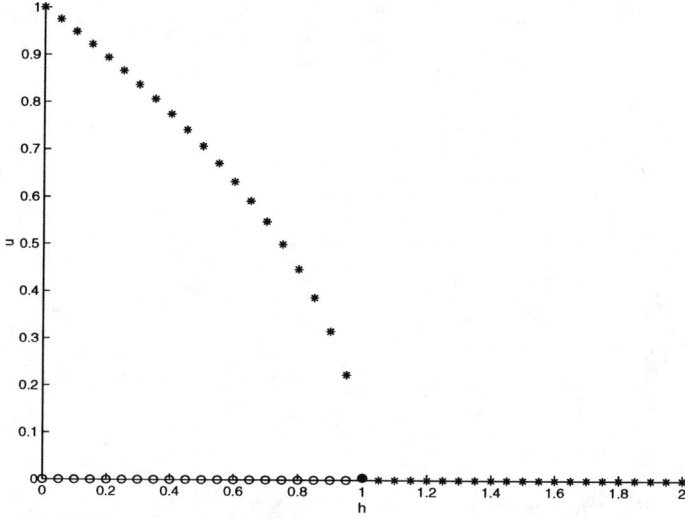

**FIGURE 3.12**: Bifurcation diagram for the logistic model (3.31) with Holling type II harvesting: stable branch ∗, unstable branch ∘.

The equilibria are found geometrically to move from the values $b$ and 1 toward each other, until they coalesce and disappear, as the cubic in the right-hand side is "pulled down" by an amount $h$. Mathematically, $O$ is not a root of the right-hand side, but biologically it remains a stable equilibrium of the system. The other two retain their type of stability, namely, the one emanating from $b$ is unstable and the one coming from 1 remains stable. Figure 3.16 shows the bifurcation diagram.

For linear exploitation, the unscaled model reads

$$\frac{dU(T)}{dT} = rU(T)(U(T) - B)\left(1 - \frac{U(T)}{K}\right) - \tilde{h}U(T), \quad \tilde{h} > 0. \tag{3.36}$$

Adimensionalization is carried out almost as above, with $\beta = K$, $\alpha = (Kr)^{-1}$, and $h = \tilde{h}(Kr)^{-1}$, $b = \frac{B}{K}$, to get

$$\frac{du(t)}{dt} = u[(u - b)(1 - u) - h]. \tag{3.37}$$

This shows that $O$ is always an equilibrium and stable since linearization (i.e., dropping the higher order terms, which vanish faster near 0) gives $\frac{du}{dt} \approx -(b+h)u$. Then there are two other, coming from "pulling down" the parabola with roots in $b$ and 1, so that these roots, as above, approach each other and retain their stability properties. The corresponding bifurcation diagram is given in Figure 3.17.

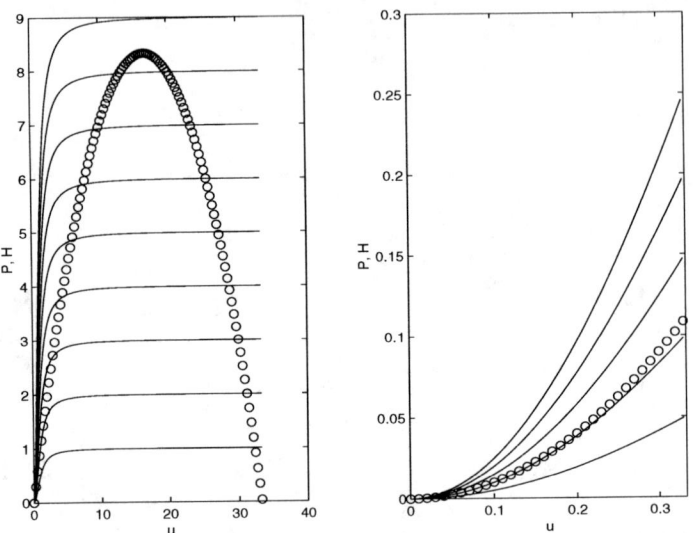

**FIGURE 3.13**: Diagram of right-hand side for Holling type III exploitation (3.33), on the left for $K = 33.3$, i.e., $\phi = 0.3$, and for integer harvesting rates $h \in \{1, \ldots, 10\}$. On the right is a blowup of the same picture around the origin for $h \in \{0.5, 1, 1.5, 2, 2.5\}$. The lowmost harvesting curve intersersects the reproduction curve at two feasible points, and the other ones at four such points.

For a Holling type II harvesting, the unscaled Allee model is

$$\frac{dU(T)}{dT} = rU(T)(U(T) - B)\left(1 - \frac{U(T)}{K}\right) - \tilde{h}\frac{U(T)}{A + U(T)}, \quad A, \tilde{h} > 0. \quad (3.38)$$

Rescaling is performed choosing $\beta = A$, $\alpha = (Ar)^{-1}$, and $h = \tilde{h}A^{-2}r^{-1}$, $b = \frac{B}{A}$, $\phi = \frac{A}{K}$, to find

$$\frac{du(t)}{dt} = u(u - b)(1 - \phi u) - h\frac{u}{1 + u}. \quad (3.39)$$

The equilibria in addition to the origin, stable, are two points moving toward each other for increasing $h$ from $b$ and $\phi^{-1}$, the former unstable and the latter stable. Then they coalesce and disappear, leaving only the origin as stable equilibrium. Figure 3.18 shows the bifurcation diagram.

The Allee model with Holling type III harvesting is

$$\frac{dU(T)}{dT} = rU(T)(U(T) - B)\left(1 - \frac{U(T)}{K}\right) - \tilde{h}\frac{U^2(T)}{A + U^2(T)} \quad (3.40)$$

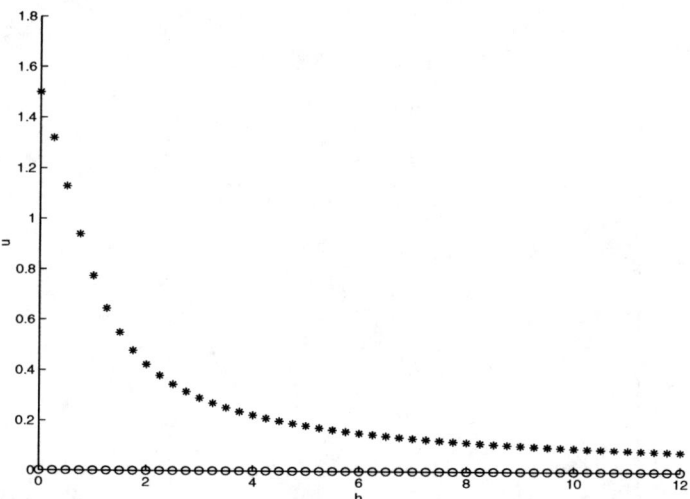

**FIGURE 3.14**: Bifurcation diagram for the logistic model (3.33) with Holling type III harvesting: case of only two feasible intersections between the reproduction and harvesting curves; stable branch ∗, unstable branch ∘; $K = 1.5$, i.e., $\phi = 0.75$.

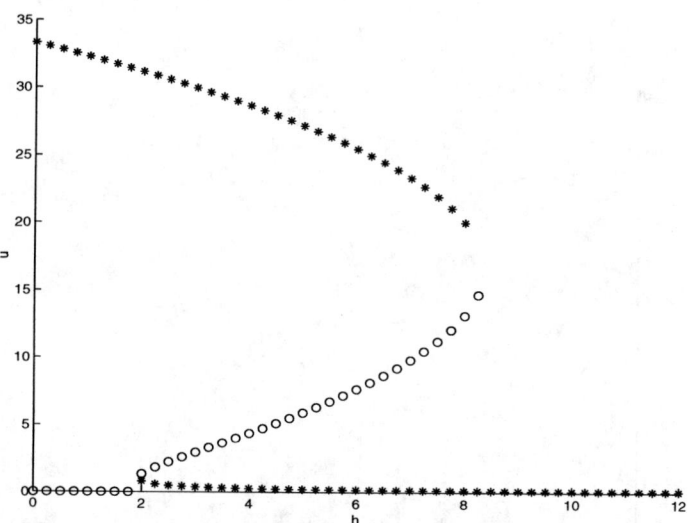

**FIGURE 3.15**: Bifurcation diagram for the logistic model (3.33) with Holling type III harvesting: case of four feasible intersections between reproduction and harvesting curves; stable branch ∗, unstable branch ∘; $K = 33.3$, i.e., $\phi = 0.3$.

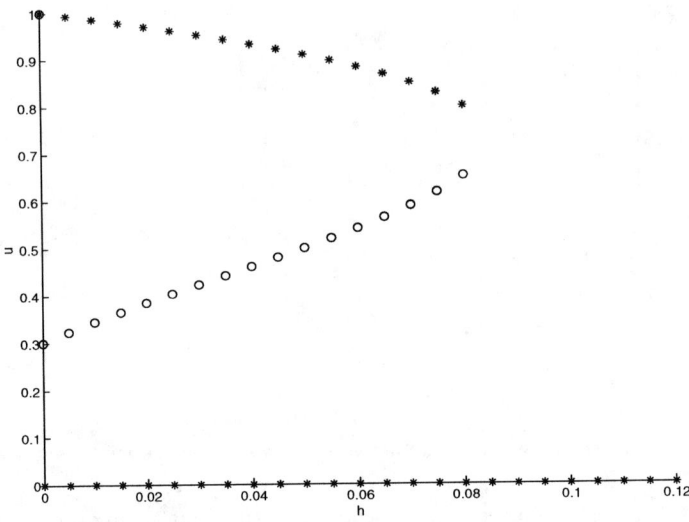

**FIGURE 3.16**: Bifurcation diagram for the Allee model (3.35) with constant harvesting: stable branch ∗, unstable branch ○.

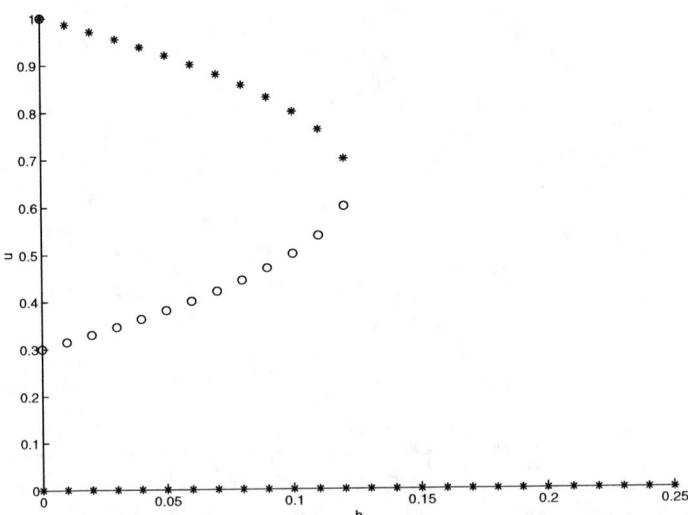

**FIGURE 3.17**: Bifurcation diagram for the Allee model (3.36) with linear harvesting: stable branch ∗, unstable branch ○.

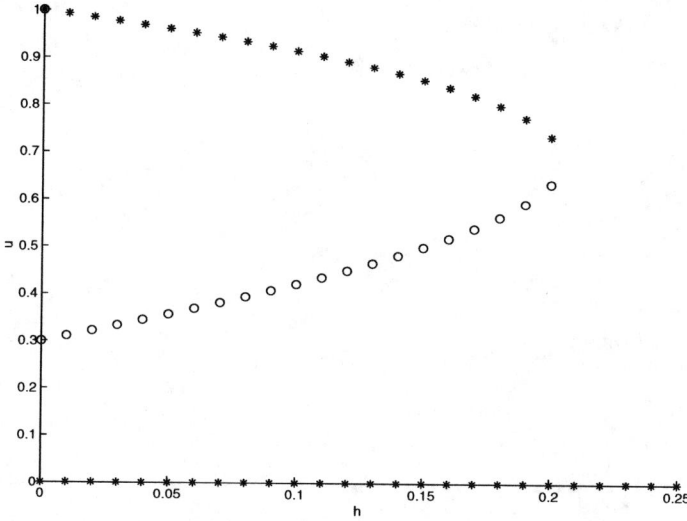

**FIGURE 3.18**: Bifurcation diagram for the Allee model (3.39) with Holling type II harvesting: stable branch *, unstable branch o.

(with $A, \tilde{h} > 0$), and rescaling via $\beta = \sqrt{A}$, $\alpha = (\sqrt{A}r)^{-1}$, and $h = \alpha\tilde{h}(\sqrt{A})^{-1}$, $b = \frac{B}{\sqrt{A}}$, $\phi = \frac{\sqrt{A}}{K}$, leads to

$$\frac{du(t)}{dt} = u(u-b)(1-\phi u) - h\frac{u^2}{1+u^2}. \qquad (3.41)$$

The equilibria are again the origin and the two points moving toward each other from $b$ and $\phi^{-1}$, retaining their stability, until they coalesce and vanish. The bifurcation diagram is shown in Figure 3.19.

On comparing Figures 3.16–3.19, we notice that they are only quantitatively different, but their qualitative behavior is exactly the same. This agrees with previous remarks on the intersections of the reproduction and harvesting functions.

There is a further remark we ought to make. On comparing Figures 3.16 to 3.19 with Figure 3.15, where we have seen the hysteresis phenomenon occur, again qualitatively we see that the behavior for the Allee model is similar, but for one important feature. Namely, this model entails the sure disappearance of the population once the critical harvesting effort $h_C$ is reached. In this case, indeed, the lower branch of the bifurcation diagram lies exactly on the horizontal axis so that the population vanishes and thus cannot recover even by drastically reducing the exploitation down to $h = 0$. The only difference, as remarked above for the four cases of harvesting in the Allee situation, is the increasing value of $h_C$, namely, $h_C = 0.08$ for constant effort, $h_C = 0.12$ for Holling type I effort, $h_C = 0.2$ for Holling type II effort, and $h_C = 0.28$ for Holling type III effort.

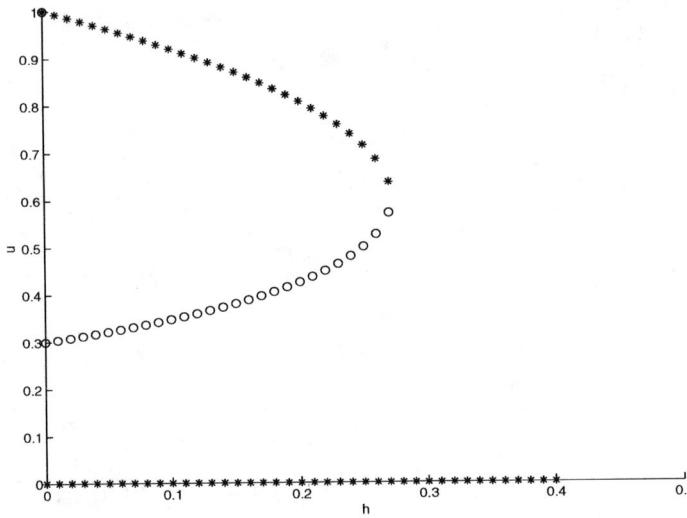

**FIGURE 3.19**: Bifurcation diagram for the Allee model with Holling type III harvesting (3.41): stable branch ∗, unstable branch ∘.

## 3.3 Glance at discrete models

We finally consider here models in which the population size is measured at fixed time intervals, in general corresponding to generation times. Let $T_n$ be the instants in time corresponding to the appearance of the new generation; it could be eggs disclosure for some insects, for instance, which may happen a few times during a good season. Therefore, the new generation of size $X_n \equiv X(T_n)$ may appear a number of times a year, depending on the species under consideration and the climactic factors. The quantity $X_{n+1} - X_n$ measures the change in the population size during one such typical generation, and in absence of migrations, it is the balance of births and deaths, namely, $X_{n+1} - X_n = bX_n - dX_n = rX_n$, where $b$ and $d$ denote the per capita birth and death rates per generation time. What we have described is a version of the discrete Malthus model,

$$X_{n+1} = aX_n, \quad a = 1 + r = 1 + b - d. \tag{3.42}$$

Attempting the solution of this difference equation by means of a discrete exponential function $X_n = \alpha^n$, we obtain the characteristic equation $\alpha = a$, from which the solution $X_n = X_0 \alpha^n$ follows. This model exhibits the same drawbacks as its continuous counterpart, namely, for $a < 1$ we find $X_n \to 0^+$, and for $a > 1$ we have $X_n \to +\infty$. But a new problem arises for $-1 < a < 0$, since now the solution oscillates taking negative values, which is clearly a nonbiological situation. A natural correction would require $X_{n+1} = \max\{aX_n, 0\}$.

An improvement would be to consider logistic-type models, of which the following are examples:

$$X_{n+1} = \frac{aX_n}{X_n + A}, \tag{3.43}$$

$$X_{n+1} = \frac{aX_n^2}{X_n^2 + A}, \tag{3.44}$$

$$X_{n+1} = X_n + aX_n\left(1 - \frac{X_n}{K}\right), \tag{3.45}$$

$$X_{n+1} = a\left(1 - \frac{X_n}{K}\right), \tag{3.46}$$

$$X_{n+1} = e^{a\left(1 - \frac{X_n}{K}\right)}, \tag{3.47}$$

$$X_{n+1} = aX_n\left(1 - \frac{X_n}{K}\right)^{-\beta}. \tag{3.48}$$

In general, these models are of the form $X_{n+1} = f(X_n)$, and if $X^*$ exists such that $X^* = f(X^*)$, so that it is a fixed point for the map $f$, then it is an equilibrium of the discrete system. We now consider a perturbation, $u_n = X_n - X^*$, so that for a suitable $\bar{X} \in CH\{X^*, X_n\}$, with the symbol $CH$ denoting the convex hull of the points $X^*$, $X_n$, i.e., the smallest interval that contains them, Taylor's formula gives

$$X_{n+1} - X^* = f(X_n) - f(X^*) \tag{3.49}$$

$$= \frac{df}{dX}(X^*)u_n + \frac{1}{2}\frac{d^2f}{dX^2}(\bar{X})u_n^2 \equiv \rho u_n + \frac{1}{2}\frac{d^2f}{dX^2}(\bar{X})u_n^2.$$

Since upon linearization we have $u_{n+1} = \rho u_n = \rho^2 u_{n-1} = \ldots = \rho^{n+1} u_0$, convergence occurs if

$$|\rho| < 1. \tag{3.50}$$

Finally, let us mention a very famous model, due to Fibonacci, that gives rise to a "second order" difference equation, namely, it involves the two previous generations to determine the next one, $x_{n+1} = x_n + x_{n-1}$, with initial conditions $x_0 = 0$, $x_1 = 1$. Its solution involves the golden ratio, which is related to the length of the side of the regular decagon, the polygon with 10 sides. Indeed, upon trying an exponential solution, we find the characteristic equation

$$\alpha^2 - \alpha - 1 = 0, \tag{3.51}$$

whose solutions are $\alpha_\pm = \frac{1}{2}[1 \pm \sqrt{5}]$. Now the solution of the Fibonacci model can be written as a linear combination of exponentials with these bases:

$$x_n = c_+ \alpha_+^n + c_- \alpha_-^n = c_+ \{\frac{1}{2}[1 + \sqrt{5}]\}^n + c_- \{\frac{1}{2}[1 - \sqrt{5}]\}^n$$

so that as $n \to \infty$, since $|\alpha_-| < 1$, we have

$$\frac{x_{n+1}}{x_n} \to \frac{1}{2}[1 + \sqrt{5}] = \alpha_+ \equiv \phi = 1.61803399.$$

In classical euclidean geometry, the golden ratio $\phi$ is determined by the point $\alpha$ that splits the (unit) segment in two parts, so as to satisfy the proportion $1 : \alpha = \alpha : (1-\alpha) = \phi$. On expanding, $\phi = \frac{\alpha}{1-\alpha} = \frac{1}{\alpha}$ and $\phi$ is the only positive solution of the resulting quadratic equation, i.e., (3.51), so that $\phi = \alpha_+$.

## 3.4 Peek into chaos

Let us specifically investigate Equation (3.45). We find

$$f(X) = (1+a)X - \frac{a}{K}X^2, \quad \frac{df}{dX} = 1 + a - \frac{2a}{K}X_n.$$

Calculating the fixed points of $f$, we find the equilibria of the equation, $X_* = 0$ and $X^* = K$. To study the stability, we are led to the characteristic equation and the consequent conditions $\rho = \frac{df}{dX}(X_*) = \frac{df}{dX}(0) = |1 + a| < 1$, which implies the stability condition $0 > a > -2$, so that for $a > 0$, the origin is unstable, and $\rho = \frac{df}{dX}(X^*) = \frac{df}{dX}(K) = |1 - a| < 1$, which gives instead $0 < a < 2$ for stability. Now, to see what happens for $a > 2$, we can consider the iterated map $X_{n+2} = f_2(X_n)$,

$$f_2(X) = f(f(X)) = (1+a)f(X) - \frac{a}{K}f(X)^2$$

$$= (1+a)^2 X - X^2(a+1)\frac{a(a+2)}{K} + \frac{2a^2}{K^2}(1+a)X^3 - \frac{a^3}{K^3}X^4.$$

To find the equilibria $f_2(X) = X$, we must then solve the following quartic equation:

$$aX\left(1 - \frac{X}{K}\right)\left[\left(\frac{a}{K}\right)^2 X^2 - a(a+2)\frac{X}{K} + a + 2\right] = 0,$$

which has the roots $X = 0$, $X = K$, and

$$X_\pm = \frac{K^2}{2a^2}\left\{\frac{a}{K}(a+2) \pm \left[\frac{a^2}{K^2}(a+2)^2 - \frac{4}{K^2}(a+2)a^2\right]^{\frac{1}{2}}\right\}$$

$$= \frac{K}{2a}\left[(a+2) \pm \sqrt{a^2 - 4}\right]. \tag{3.52}$$

The latter are real for $a \geq 2$. Moreover, observe that

$$f(X_+) = X_+ + aX_+\left(1 - \frac{X_+}{K}\right) = (1+a)X_+ - \frac{a}{K}X_+^2$$

$$= (1+a)\frac{a+2}{2a}K + \frac{1+a}{2a}K\sqrt{a^2-4}$$

$$- \frac{a}{K}\frac{K^2}{4a^2}\left[(a+2)^2 + a^2 - 4 + 2(a+2)\sqrt{a^2-4}\right]$$

$$= \sqrt{a^2-4}\left[\frac{1+a}{2a}K - \frac{K}{2a}(a+2)\right] + \frac{K}{2a}\left[(1+a)(2+a) - (a^2+2a)\right]$$

$$= \frac{K}{2a}\left[(2+a)(a+1-a) - \sqrt{a^2-4}\right] = \frac{K}{2a}\left[2+a - \sqrt{a^2-4}\right] = X_-.$$

Since we have a fixed point for the iterated map, it follows that

$$X_+ = f_2(X_+) = f(f(X_+)) = f(X_-), \quad \text{i.e.} \quad X_- f(X_+), \quad X_+ = f(X_-).$$

We have thus found a periodic solution for $f(X)$, of period 2, for $a > 2$. We investigate now the stability of this periodic solution. Recalling the convergence condition (3.50) for fixed point iterates of the map $f_2(X)$, let us differentiate $f_2(X) - X = 0$. Since $X_+$ solves the quadratic equation (3.52), we find

$$\frac{df_2}{dX}(X_+) - 1 = a\left(X_+ - \frac{X_+^2}{K}\right)\left[2\frac{a^2}{K^2}X_+ - \frac{a}{K}(a+2)\right]$$

$$= \frac{K}{2}\left(a+2+\sqrt{a^2-4}\right)\frac{1}{2a}\left(a-2-\sqrt{a^2-4}\right) \cdot$$

$$\cdot \frac{a}{K}\left[\left(a+2+\sqrt{a^2-4}\right) - a - 2\right]$$

$$= \frac{1}{4}\left[a^2 - \left(2+\sqrt{a^2-4}\right)^2\right]\sqrt{a^2-4}$$

$$= \frac{1}{4}\left[a^2 - 4 - 4\sqrt{a^2-4} - (a^2-4)\right]\sqrt{a^2-4} = 4 - a^2.$$

Hence $\frac{df_2}{dX}(X_+) = 5 - a^2$ and the condition for stability then becomes $|5-a^2| < 1$, so that the cycle of period 2 for the map $f$ is stable for the parameter range $2 < a < \sqrt{6}$.

The question now arises as to what happens for $a > \sqrt{6}$. Again a periodic solution of period 4 can be found that is stable for $\sqrt{6} < a < 2.544$. Iterating the question and the procedure, a periodic solution of period 8 arises hereafter, which is stable for $2.544 < a < 2.564$. On continuing this procedure, we obtain the so-called period doubling route to chaos. It can be shown that if $a_n$ is the parameter value for which a periodic solution of period $2^n$ appears, then

$$\lim_{n \to \infty} \frac{a_{n+1} - a_n}{a_{n+2} - a_{n+1}} = 4.6692016... ,$$

this limiting value being known as the Feigenbaum constant.

# Chapter 4

# Interacting populations

The Lotka–Volterra model takes its name from the authors of the first investigations of interacting populations. Volterra (1926a; 1926b; 1931) based his considerations on the unexpected results of the field observations of the biologist D'Ancona (1954) on the fishing catches in the Adriatic sea after the First World War. Other models were developed later to avoid the main drawback of the original Lotka–Volterra model, namely, the neutral stability of its interior equilibrium point. This makes the solutions sensitive to variations of the initial conditions of the system, and therefore biologically they are not very satisfactory. These modifications include at first quadratic models, which incorporate a logistic growth term for at least one of the two populations, and then also other nonlinear terms, such as the effect of satiation in feeding.

Competing situations and species living in symbiosis are other examples of interacting populations, extended then to more complicated food webs in which the species at a certain level in the chain is the predator of those in the levels below and a prey for the higher level ones.

To model the quadratic interactions between different populations, we proceed similarly to what is done for the logistic term in a single population growth. If one individual of the first population $U$ wanders in the environment, the meetings with individuals of the second population $V$ will occur at a certain per capita rate $\alpha$. Thus the total encounters of that specific individual per unit time are $\alpha V$. But this must be multiplied by every individual of species $V$, so that the total number of interactions in the time unit between the two populations is then $\alpha UV$. More generally, the interactions for the population $V$ are described by a per capita rate $f(U,V)V$. The quadratic case then corresponds to a Holling type I term. If we want to model the saturation effect experienced by predators when the prey abound, we should then take for $f$ a Holling type II or III function. The latter correspond to what in operations research is called the law of diminishing returns, for the same amount of effort the gain is larger if the effort is small, and tends to a plateau for increasing efforts.

In this chapter we consider basically the Lotka–Volterra model and its modifications. In addition to presenting the models, we also develop the main mathematical techniques needed for the analysis of these systems, by discussing very simple situations in detail. We start with an example in Section 4.1 that mathematically illustrates the local stability analysis and the tools needed for it. In Section 4.2 a more in-depth discussion of the Lotka–Volterra model is related and also used for a more in-depth discussion of the scaling procedure, also in view of the later parts of the book. Section 4.2.1 presents another modification of the prey–predator system and the other interesting biological situations of competing and symbiotic species. Quadratic and other classical nonlinear models are discussed. The main mathematical tool for studying the global stability of these models is the Lyapunov function, presented in Section 4.3.1. Then the question of the investigation of possible limit cycles is addressed, with Section 4.3.2 giving positive and negative criteria.

The last two sections are devoted to a step further, beyond the two-species models. In Section 4.3.3, a four-level food web is considered, from minerals through plants and herbivores up to the top predators, the carnivores. Section 4.4 describes a very important problem in ecology, namely, the preservation of cultures, in this case vineyards, by using ecological controls.

From now on all parameters appearing in the models are assumed to be nonnegative constants, unless otherwise specified.

## 4.1 Two-species prey–predator population model

We consider an idealized ecosystem in which only two populations are accounted for, and one may be hunted by the second one. More specifically, let the prey and the predator populations be denoted by $U(T)$ and $V(T)$, respectively. We assume that the environment provides enough resources for the former to ensure its Malthusian growth. The predators instead show logistic growth, where $r$ denotes their reproduction rate and $K$ represents the environment carrying capacity. These assumptions imply that the predators have sources of food other than the prey $U$. In the absence of interactions, the two populations would grow independently as follows:

$$\frac{dU}{dT} = \alpha U(T), \qquad \frac{dV}{dT} = rV(T)\left[1 - \frac{V(T)}{K}\right].$$

Assume now that the predation occurs at rate $b$, while the conversion rate into new predators is described by the parameter $c$. Combining the above equations, we thus obtain

$$\frac{dU}{dT} = U(\alpha - bV), \qquad \frac{dV}{dT} = rV\left(1 - \frac{V}{K}\right) + cUV. \tag{4.1}$$

To analyze this system, we first find the nullclines, i.e., the lines in the phase plane on which each population does not change. Their intersections also give the equilibria of the system, i.e., the points at which both populations do not change. The two nullclines are respectively defined by the equations $\frac{dU}{dT} = 0$,

$\frac{dV}{dT} = 0$, and these in turn give for the equilibria the system of algebraic equations

$$U(\alpha - bV) = 0, \qquad V\left[r\left(1 - \frac{V}{K}\right) + cU\right] = 0. \qquad (4.2)$$

Now, from the first one, we get either $U(T) \equiv U_0 = 0$ or $V(T) \equiv V_2 = \frac{\alpha}{b}$. In the former case, from the second one, we have either $V(T) \equiv V_0 = 0$ or $V(T) \equiv V_1 = K$. The second choice on the first equation gives, upon substitution into the second one, $U(T) \equiv U_2 = \frac{r}{c}(\frac{\alpha}{bK} - 1)$. Thus we have the three equilibria

$$E_0 \equiv (U_0, V_0) \equiv (0, 0), \quad E_1 \equiv (U_0, V_1) \equiv (0, V_1), \quad E_2 \equiv (U_2, V_2).$$

Clearly $E_0$ and $E_1$ are always feasible, while to have a meaningful solution for $E_2$ we need the populations to be nonnegative, and this entails the restriction $K < \frac{\alpha}{b}$, which is then the feasibility condition for $E_2$.

## Local Stability Analysis

We now study the long-term behavior of the model near the equilibria. To do so, we need to linearize it. The procedure consists in defining new variables in the neighborhood of each equilibrium point and then disregarding perturbations of higher order, so that the system becomes easily integrable. Of course the solutions we find are valid only as a first approximation near the equilibrium, but nevertheless they tell a lot about the model's behavior.

Consider $E_0$ at first. We investigate what happens to the system in a neighborhood of the origin. As $U, V$ in this case are small, namely $U, V < 1$, then $U^2, V^2, UV < U, V$ so that $U^2, V^2, UV << 1$. These second order perturbations are much smaller than the original ones and can be disregarded as mentioned earlier. The system then simplifies to

$$\frac{dU}{dT} = \alpha U(T), \qquad \frac{dV}{dT} = rV(T). \qquad (4.3)$$

This is easy to solve, giving $U(T) = U(0)\exp(\alpha T)$, $V(T) = V(0)\exp(rT)$. In fact, it is so easy because it is uncoupled. Since $\alpha, r > 0$, the solutions will grow, thus the trajectories will wander away from the origin, even though we chose $U(0) = U_0 \approx 0$, $V(0) = V_0 \approx 0$.

Look now at $E_1$. In this case, $U$ is still small, but $V$ lies near the value $V_1$. It is therefore better to introduce the perturbation $y = V - V_1$, for which $\frac{dy}{dT} \equiv \frac{dV}{dT}$. Again we have $y^2, yU < y$, so that $y^2, yU << 1$ and the latter can be disregarded. The first equation of the linearized system in a such case reads

$$\frac{dU}{dT} = U(T)[\alpha - b(y + V_1)] = U(T)[\alpha - bV_1] = U(T)[\alpha - bK],$$

while the second one is

$$\frac{dy}{dT} = r(y+V_1)\left[1 - \frac{(y+V_1)}{K}\right] + cU(T)(y+V_1)$$

$$= r(y+K)\left[1 - \frac{(y+K)}{K}\right] + cU(T)(y+K)$$

$$= -r(y+K)\frac{y}{K} + cKU(T) = -ry + cKU(T).$$

In this case the system is not so easy to solve. But substitution is still a viable method. Solve the first equation, which is still independent of $y$, to get $U(T) = U(0)\exp[(\alpha - bK)T]$. Substituting into the second equation, we find

$$\frac{dy}{dT} = -ry + cKU(0)\exp[(\alpha - bK)T]. \tag{4.4}$$

As no powers higher than one, nor product of unknown functions appear in it, this is a linear constant coefficient, first order nonhomogeneous ordinary differential equation. Notice that each expression has its own meaning, namely, the coefficients are indeed constants, not functions of time; the highest derivative is the first one; there is a term independent of the unknown $y$ making the equation nonhomogeneous, i.e., carrying it to the right-hand side; the latter is nonzero, i.e., nonhomogeneous; and, finally, no partial derivatives arise in the formulation. This classification may be useful for understandig how to solve it, if one wishes to look for the method in books. Instead, we will proceed in a constructive way. If we forget about the nonhomogeneous part, the solution is easily found to be

$$y(T) = k\exp(-rT), \tag{4.5}$$

$k$ denoting an arbitrary integration constant. This is just the solution of the homogeneous equation. To find the one for our problem, we use the method of *variation of parameters*, which amounts to letting $k$ become a function of time, $k \equiv k(T)$. Differentiation and substitution into the original nonhomogeneous equation give for the left-hand side

$$\frac{dy}{dT} = \frac{d[k(T)\exp(-rT)]}{dT} = \frac{dk}{dT}\exp(-rT) + k(T)\frac{d[\exp(-rT)]}{dT}$$

$$= \frac{dk}{dT}\exp(-rT) - rk(T)\exp(-rT),$$

while for the right-hand side we have

$$-ry(T) + cKU(T) = -rk(T)\exp(-rT) + cKU(0)\exp[(\alpha - bK)T].$$

Observe that two terms on the two sides of the equation are equal. Thus, equating and simplifying, we find

$$\frac{dk}{dT} = cK\exp(rT)U(0)\exp[(\alpha - bK)T] = cKU(0)\exp[(\alpha + r - bK)T].$$

Solving this last differential equation for $k$, we have
$$k(T) = C + \frac{cKU(0)}{\alpha + r - bK} \exp[(\alpha + r - bK)T],$$
where $C$ denotes an arbitrary constant of integration. Substituting back into (4.5), we have
$$y(T) = \left[ C + \frac{cKU(0)}{\alpha + r - bK} \exp[(\alpha + r - bK)T] \right] \exp(-rT)$$
$$= C \exp(-rT) + \frac{cKU(0)}{\alpha + r - bK} \exp[(\alpha - bK)T].$$
This function solves the original nonhomogeneous equation, a fact that can be verified by backsubstitution into (4.4).

We have thus shown that $y$ tends to zero exponentially if $\alpha - bK < 0$, while it tends to infinity exponentially if $K < \frac{\alpha}{b}$. Hence $E_1$ is stable in the former case, as $U$ tends always to zero, while it is unstable in the latter. The trajectories in the latter case diverge along the $y$-axis, i.e., the $V$-axis. Note that the condition for stability of $E_1$ is the very same condition for the *infeasibility* of $E_2$. Thus when the latter is infeasible, then trajectories approach $E_1$, while if $E_1$ is unstable, then the trajectories most likely will approach $E_2$, as the latter would now be feasible, and stable, as we will show shortly.

We now turn to the more complicated analysis for the case of $E_2$. Linearization entails the definition of two new variables, the perturbations in each component $x = U - U_2$ and $y = V - V_2$. We then obtain
$$\frac{dx}{dT} = (x + U_2)[\alpha - b(y + V_2)] = -by(x + U_2) = -bU_2 y,$$
$$\frac{dy}{dT} = \{r[1 - \frac{(y + V_2)}{K}] + c(x + U_2)\}(y + V_2).$$
The second one can be further processed as follows:
$$\frac{dy}{dT} = \left\{ r\left[1 - \frac{(y + \frac{\alpha}{b})}{K}\right] + c\left[x + \frac{r}{c}\left(\frac{\alpha}{bK} - 1\right)\right] \right\} \left(y + \frac{\alpha}{b}\right)$$
$$= -r\frac{y}{K}\left(y + \frac{\alpha}{b}\right) + r\left(y + \frac{\alpha}{b}\right) - \frac{\alpha r}{bK}\left(y + \frac{\alpha}{b}\right)$$
$$+ cx\left(y + \frac{\alpha}{b}\right) + r\left(\frac{\alpha}{bK} - 1\right)\left(y + \frac{\alpha}{b}\right)$$
$$= -r\frac{y}{K}\left(y + \frac{\alpha}{b}\right) - \frac{\alpha r}{bK}\left(y + \frac{\alpha}{b}\right) + cx\left(y + \frac{\alpha}{b}\right) + \frac{\alpha r}{bK}\left(y + \frac{\alpha}{b}\right)$$
$$= -r\frac{y}{K}\left(y + \frac{\alpha}{b}\right) + cx\left(y + \frac{\alpha}{b}\right).$$
Finally, we have to analyze the linearized system
$$\frac{dx}{dT} = -bU_2 y, \qquad \frac{dy}{dT} = -\frac{\alpha r}{bK}y + \frac{\alpha c}{b}x. \qquad (4.6)$$

Let us introduce the perturbation vector $\mathbf{u} = (x,y)^T$. We can then write the above system in the form $\frac{d\mathbf{u}}{dT} = M\mathbf{u}$, where the matrix $M$ will now be determined. To do this in a very elementary but constructive way, let us observe that for a generic matrix $A$ with elements

$$A \equiv \begin{bmatrix} A_{1,1} & A_{1,2} \\ A_{2,1} & A_{2,2} \end{bmatrix},$$

upon postmultiplication by $\mathbf{u} \equiv (u_1, u_2)^T$, the row by column product gives the vector

$$(A_{1,1}u_1 + A_{1,2}u_2, \ A_{2,1}u_1 + A_{2,2}u_2)^T.$$

If we denote by $A_{.,j}$ the $j$th column of the matrix $A$, so that $A \equiv [A_{.,1}, A_{.,2}]$, then the above equations can be rewritten as $A\mathbf{u} = A_{.,1}u_1 + A_{.,2}u_2$. It is then apparent that the $j$th column of the matrix $A$ acts on the $j$th component of $\mathbf{u}$. To construct the matrix $M$, we have to look at its action on the vector $\mathbf{u}$, action coming from the equations of the system. From the first one, the component $u_1 \equiv x$ is seen to be absent, while $u_2 \equiv y$ gets multiplied by $-bU_2$. The only number that makes $u_1$ disappear is zero. Similarly, in the second equation, $x$ is multiplied by $\frac{ac}{b}$. Hence the first column of $M$ must be given by $[0, \frac{ac}{b}]^T$. In the second equation, moreover, $y$ has the coefficient $-\frac{ar}{bK}$, so that the second column of $M$ must be given by $[-bU_2, -\frac{ar}{bK}]^T$. In summary,

$$A = \begin{bmatrix} 0 & -bU_2 \\ \frac{ac}{b} & -\frac{ar}{bK} \end{bmatrix}.$$

If we consider carefully the matrix just obtained, we find that it coincides with the Jacobian of the original system (4.1) evaluated at $E_2$. This is a general result. Any time we linearize around an equilibrium, we may skip all the steps undertaken above and just consider directly the Jacobian of the model. This will be done repeatedly in the following chapters.

To get an uncoupled system, let us introduce the change of basis matrix $W$, yet unknown, and a new vector of unknowns $\mathbf{z}$, such that $\mathbf{u} = W\mathbf{z}$, for which the system is uncoupled. We must then have

$$W\frac{d\mathbf{z}}{dT} = MW\mathbf{z}, \qquad \frac{d\mathbf{z}}{dT} = W^{-1}MW\mathbf{z} \equiv \Lambda\mathbf{z},$$

with $\Lambda = diag(\lambda_1, \lambda_2)$. Here the entries of this diagonal matrix as well as those of $W$ are not known. To find these unknowns, let us once again identify the matrix $W$ by its columns, $W \equiv [W_{.,1}, W_{.,2}]$. For an arbitrary vector $\mathbf{z}$, it then follows

$$MW\mathbf{z} = W\Lambda\mathbf{z},$$

that is,

$$M[W_{.,1} \ W_{.,2}] = [W_{.,1} \ W_{.,2}]\Lambda \equiv [W_{.,1} \ W_{.,2}][L_{.,1} \ L_{.,2}],$$

where $L_{.,j}$ denotes the $j$th column of $\Lambda$ and therefore is the $\lambda_j$ multiple of the $j$th vector in the standard basis of the euclidean space, in view of the action of a matrix over a vector described above,

$$L_{.,j} = (0,\ldots,0,\lambda_j,0,\ldots,0)^T = \lambda_j (0,\ldots,0,1,0,\ldots,0)^T = \lambda_j \mathbf{e}_j.$$

Indeed, once again, the action of the $j$th column of the matrix $W$ is on the $j$th component of $L_{.,j}$ of the vector $L_{.,j}$, since the latter is the $\lambda_j$ multiple of the $j$th vector of the standard euclidean basis $\mathbf{e}_j$. Hence the $j$th column gets multiplied by $\lambda_j$, i.e., $W\Lambda \equiv [\lambda_1 W_{.,1}\ \lambda_2 W_{.,2}]$, so that, equating columns, we find $MW_{.,1} = \lambda_1 W_{.,1}$ and $MW_{.,2} = \lambda_2 W_{.,2}$.

More in general, for $j = 1, 2$, we have to solve the problem

$$MW_{.,j} = \lambda_j W_{.,j}. \tag{4.7}$$

This is is the well-known linear algebra eigenvalue problem. Geometrically, a direction $\mathbf{z}$ is sought, which is invariant under the action of the matrix $\Lambda$, i.e., the resulting vector $\Lambda \mathbf{z}$ through the eigenvalue $\lambda$ can change length but not the direction. Note that a zero eigenvalue is allowed. Indeed, the resulting null vector is assumed to possess every direction in space. This follows from the fact that it must be orthogonal to any arbitrary vector $\mathbf{w}^*$ in space, since $\mathbf{0}^T \mathbf{w}^* = 0$. However, a null eigenvector makes no sense, as it trivially satisfies the eigenvalue equation $M\mathbf{0} = \lambda \mathbf{0}$ but gives no indication as to which direction is being left invariant by the matrix. To solve the eigenvalue problem, rewrite the equation as $(M - \lambda I)\mathbf{w} = \mathbf{0}$, having set $W_{.,j} = \mathbf{w}$. This is a homogeneous system, for which we seek a nonzero solution, in view of the above discussion on eigenvectors. We need further tools from linear algebra.

The main result that we are now going to quote is the summary of several equivalent statements, the proofs of which are easily found in elementary linear algebra texts.

### Fundamental Theorem of Linear Algebra

For the system $A\mathbf{x} = \mathbf{b}$, the following statements are all equivalent to each other:

- The matrix $A$ is nonsingular, i.e., it has an inverse $A^{-1}$.

- The nonhomogeneous system, i.e., the system with $\mathbf{b} \neq \mathbf{0}$, is uniquely solvable.

- $\det A \neq 0$.

- The homogeneous system, i.e., the system with $\mathbf{b} = \mathbf{0}$, which has clearly always the trivial solution $\mathbf{x} = \mathbf{0}$, also possesses infinitely many nontrivial solutions.

- The column rank of the matrix is maximal.

- The row rank of the matrix is maximal.
- The eigenvalues of the matrix are nonzero.
- The right-hand side **b** belongs to the subspace spanned by the columns of the matrix $A$.
- All the columns of the matrix are linearly independent.
- All the rows of the matrix are linearly independent.

Thus, since our system is homogeneous and since we seek the eigenvectors, i.e., its nontrivial solutions, we must impose that the matrix $M - \lambda I$ be singular. Using the statements above, we impose that its determinant must vanish, $\det(M - \lambda I) = 0$. This gives only an algebraic equation, namely, the characteristic equation $(M_{1,1} - \lambda)(M_{2,2} - \lambda) - M_{1,2}M_{2,1} = 0$, making the problem easier to tackle. Substituting the values of our situation, we find $\lambda^2 + \frac{\alpha r}{bK}\lambda + \alpha c U_2 = 0$. This algebraic equation has two roots,

$$\lambda_1 = -\frac{\alpha r}{2bK} + \Delta^{\frac{1}{2}}, \quad \lambda_2 = -\frac{\alpha r}{2bK} - \Delta^{\frac{1}{2}}, \quad \Delta \equiv \left(\frac{\alpha r}{2bK}\right)^2 - \alpha c U_2.$$

Observe that the second term in the discriminant $\alpha c U_2$ is always positive, so that we are really subtracting it from the square. Thus the square root of $\Delta$ is smaller than $\frac{\alpha r}{2bK}$, and this entails that $\lambda_1$ cannot be positive. Certainly $\lambda_2 < 0$, so that both roots, which are either real or complex conjugate, depending on the discriminant, always have a negative real part, $\Re(\lambda_i) < 0$. They are real if

$$\alpha r \left(\frac{1}{2bK}\right)^2 - \left(\frac{\alpha}{bK} - 1\right) > 0.$$

In both cases, however, the solutions will be either decaying exponentials, for real eigenvalues, or complex conjugate exponentials. In the former case, the solutions apart from arbitrary constants are $z_1 = \exp(-\lambda_1 T)$, $z_2 = \exp(-\lambda_2 T)$. In such a case, the equilibrium is stable, as the perturbations approach zero. Indeed, if this is the case for **z**, so it is for the nonsingularly transformed vector $\mathbf{u} = W\mathbf{z}$. In the case of complex conjugate roots, Euler's formula allows the computation of the result, $\exp(i\theta) = \cos(\theta) + i\sin(\theta)$. Thus, for $j = 1, 2$, if $\lambda_j = \mu_j + i\nu_j$, we have $\mathbf{z_j} = \exp(\lambda_j T) = \exp[(\mu_j + i\nu_j)T] \equiv \exp(\mu_j)[\cos(\nu_j T) + i\sin(\nu_j T)]$. The trajectories in a neighborhood of $E_2$ in this case approach the equilibrium following spirals, as indicated by the trigonometric terms. The trajectories are indeed spirals as they are "almost" periodic, in the sense that after a time of $2\pi$ from a given instant $T^*$ they are in the same position with respect to the $x$-axis, making with this axis the very same angle as they were at time $T^*$. Hence they have gone around $E_2$ once. But their distance from the equilibrium at time $T^*$ was $\exp(\mu_j T^*)$ and at time $T^* + 2\pi$ it is $\exp[\mu_j(T^* + 2\pi)]$, i.e., it has shrunk. Recall that $\mu_j < 0$ in view of the above analysis. This remark will be the heart of the method of Section 4.4.6.

## Summarizing discussion

Implications of any model predictions should be interpreted in terms of the science where it stems from. Let us therefore consider what we have learnt from the above analysis. Suppose that $U$ represents some kind of nuisance species that we would like to fight with biological means, i.e., using one of its natural predators $V$. The problem may then be that in the environment where $U$ is present, other species appetible to $V$ may also be present. In such a case, $V$ has a choice on which species to feed upon. The interior equilibrium we have found expresses the fact that if the carrying capacity $K$ for the predators $V$ is not high enough, they survive by feeding partly on the other populations present and partly on $U$, but the latter survives in the environment and settles to the value $U_2$. On the other hand, if $K$ is raised high enough, the equilibrium moves on the $V$-axis, i.e., the predator will bring the pest $U$ to extinction. Thus, in order to eliminate the latter, if we cannot act directly on the number of their natural predators, we could at least provide them with enough alternative food sources in order to raise their environment carrying capacity $K$ so that they are then able to wipe out the pest $U$.

## 4.2 Classical Lotka–Volterra model

In contrast to the model presented earlier, the classical Lotka–Volterra considers an aquatic environment, in which the sharks have as their food source the other fish, which all together constitute the prey. In their absence, the predators exhibit a negative growth rate, $-e$. This assumption is indeed the main difference with the model of Section 4.1. In the classical model, the equations read

$$\frac{dU}{dT} = U(T)[\alpha - bV(T)], \quad \frac{dV}{dT} = V(T)\left[-e + cU(T)\right]. \quad (4.8)$$

If one carries out the linearization procedure outlined before, about the equilibrium $(\frac{e}{c}, \frac{\alpha}{b})$ in the $UV$ phase space, we would find in this case purely imaginary eigenvalues $\lambda = \pm i\sigma$, $\sigma = \sqrt{\alpha e}$, since the Jacobian and characteristic equations are now

$$M \equiv \begin{bmatrix} 0 & -\frac{1}{c}eb \\ \frac{1}{b}c\alpha & 0 \end{bmatrix} \quad \text{and} \quad \lambda^2 + e\alpha = 0,$$

respectively.

The basis for the linearized solution is then given by $\exp(\pm\sigma T) = \cos(\sigma T) \pm i\sin(\sigma T)$. Near the equilibrium point the solution trajectories are thus closed curves, giving the so-called neutral stability. If two different initial conditions are considered, the trajectories starting from each one of them will return to the initial values after a certain time. In other words, perturbing the initial values, the trajectories do not approach the equilibrium, they rather remain confined to a suitable neighborhood of this point. From a practical point of view this is not satisfactory, as biologically an equilibrium should attract or repel trajectories and not be too sensitive to the initial conditions.

### 4.2.1  More on prey–predator models

Now we consider a more general model than the original Lotka–Volterra model, with the aim of overcoming the weakness it possesses, namely, the neutral stability of its equilibrium. In order to do that, we will assume that the environment for the prey is no longer unlimited, that a carrying capacity exists to limit their growth. Let the prey and predator populations be again expressed respectively by the functions $U(T)$ and $V(T)$. As for the classical Lotka–Volterra system, assume the prey are the sole food source for the predators, so that the mortality rate for the latter is exponential in absence of the former. This is then a different environment than the one discussed in Section 4.1. The system is therefore

$$\frac{dU}{dT} = U(a - bV - eU), \quad \frac{dV}{dT} = V(-c + dU). \tag{4.9}$$

The equilibria are the origin and then the points $E_1 = (\frac{a}{e}, 0)$ and $E_2 \equiv (U^*, V^*)$, with $U^* = \frac{c}{d}$, $V^* = \frac{1}{bd}(ad - ce)$. This nontrivial equilibrium point is feasible for $ad > ce$. The system's Jacobian is easily evaluated,

$$J = \begin{pmatrix} a - bV - 2eU & -bU \\ dV & -c + dU \end{pmatrix}. \tag{4.10}$$

It follows that, when feasible, $E_2$ is also asymptotically stable, as the eigenvalues of the Jacobian are

$$\lambda_\pm = -\frac{eU^*}{2} \pm \sqrt{\left(\frac{eU^*}{2}\right)^2 - bdU^*V^*} = -\frac{ce}{2d} \pm \frac{1}{2}\sqrt{\frac{c^2e^2}{d^2} + \frac{4c}{d}(ce - ad)}$$

and thus always have a negative real part. The point $E_2$ is then a focus or a node, depending on whether the argument of the root is negative or positive. Both conditions can arise, as by fixing all parameters, except for the prey net birth rate, we see that for $a = 0$ or very small, the eigenvalues are real, thus giving a node, while for $a \to +\infty$ they become complex conjugate, and $E_2$ would thus be a focus. Worthy of notice is also the fact that should $E_2$ become infeasible, the boundary equilibrium $E_1 = (\frac{a}{e}, 0)$ would then be the new and only equilibrium. Its eigenvalues are $-a$, $\frac{1}{e}(ad - ce)$. These entail an important consequence, indeed that the infeasibility condition of $E_2$ implies $E_1$ to be stable, and vice versa. Thus the $\omega$-limit set of this model is given only by the points $E_2$ and $E_1$. Notice indeed that the origin is a saddle since the trajectories escape from it along the $U$-axis. Thus the community can never become extinct, as the prey are prevented from vanishing. On the other hand, it is possible either that both species survive at levels prescribed by $E_2$, or that only the predators disappear, at the equilibrium $E_1$.

### 4.2.2 Scaling

We want to review what was already done in the case of the one-population models, to simplify the equations. In the later chapters we will appreciate this technique, since it is powerful enough to allow substantial reduction of the number of parameters appearing in a system of equations. Bearing in mind the system (4.8), let us introduce new dependent and independent variables by means of the equations $U(T) = Au(t)$, $V(T) = Bv(t)$, $T = Ct$, with $A$, $B$, $C$ denoting positive constants to be determined. Taking derivatives, the chain rule yields

$$\frac{dU}{dT} = A\frac{du}{dt}\frac{dt}{dT} = \frac{A}{C}\frac{du}{dt}, \qquad \frac{dV}{dT} = B\frac{dv}{dt}\frac{dt}{dT} = \frac{B}{C}\frac{dv}{dt}.$$

Thus back substitution into (4.8) gives

$$\frac{du}{dt} = Cu(t)[\alpha - bBv(t)], \qquad \frac{dv}{dt} = Cv(t)\left[-e + cAu(t)\right]. \qquad (4.11)$$

Now we want to choose the scaling constants so as to minimize the number of parameters in the rescaled system. We can thus choose to set $C\alpha = 1$ and $CbB = 1$ in the first equation in (4.11), thus giving $C = \frac{1}{\alpha}$, $B = \frac{\alpha}{b}$. If we now try to make use of the second equation in (4.11), setting $Ce = 1$, we see that it is not possible. Thus $\theta = \frac{e}{\alpha}$ is a rescaled parameter intrinsic to the model.

On the contrary, we can still use the left degree of freedom by setting $CcA = 1$, i.e., selecting $A = \frac{\alpha}{c}$. The rescaled model then reads

$$\frac{du}{dt} = u(t)[1 - v(t)], \qquad \frac{dv}{dt} = v(t)\left[-\theta + u(t)\right]. \qquad (4.12)$$

From the simulation point of view, let us illustrate the advantage of this procedure. Suppose we would like to investigate the model's behavior for a range of parameter values, saving the plots of the simulations. Fixing the original parameters to some values, say $b = 0.1$, $c = 0.1$, $e = 0.1$, we may let the last parameter vary in a certain range, say $\alpha = 0.1, 0.2, \ldots, 0.9, 1.0$, thus obtaining 10 figures. Then we need to change the value of another parameter, $b = 0.2$ say, and repeat the operation, obtaining another 10 figures. Repeating again the procedure for $b = 0.3, \ldots, 0.9, 1$, we thus have 100 figures, which we can arrange in a 10 by 10 table in a single page. Then we repeat the whole construction for the value $c = 0.2$, obtaining a second "page" with another 100 figures, and then a third one for $c = 0.3$, and so on up to $c = 1.0$. In this way a "chapter" of 10 pages and 1000 figures is obtained, all corresponding to $e = 0.1$. We then let $e = 0.2$ and again repeat the procedure, getting a second chapter, and so on, until for $e = 1.0$ we have completed a book of 10000 figures. Scaling instead allows us just to run the simulation for a range of values of the new parameter $\theta$, so that all the behaviors of the model contained in the former simulations can be traced back to some of the latter, for suitable parameter combinations. Indeed, looking at the definition of $\theta$, it is clear that the simulations for the pairs $\alpha = 0.1$, $e = 0.2$, $\alpha = 0.3$, $e = 0.6$ and $\alpha = 0.4$, $e = 0.8$ all correspond to the same figure in the last set obtained for $\theta = 2$. The advantage in terms of computational costs and easiness of reading the results is apparent.

## 4.3 Other types of population communities

Prey–predator interactions are among the most important in ecological communities, yet they of course are not the only ones possible. In this section we give an overview of some other types of interspecific interactions as well as some other and complementary modeling approaches.

### 4.3.1 Competing populations

Competing species can be either predators directly competing among themselves or species that survive in the same ecosystem exploiting the same resources, e.g., sharing a common pasture. It is known that among wild ungulates, direct competition for the same pastures arises between the red deer (*Cervus elaphus hippelaphus*) and the roe deer (*Capreolus capreolus*), the former suffering the most as the other wild ungulates are more generalist in the habitat selection as well as in feeding habits. The roe deer, in Mediterranean areas, also suffers from the presence of the fallow deer (*Dama dama*). An example for exploitation of the same trophic niches, in the Gran Paradiso National Park in the Northwestern Alps in Italy, is given by goats and the wild chamois (*Rupicapra rupicapra*) and ibex (*Capra ibex*). Another instance involves skuas and seagulls populations, which compete for food, since their trophic niches overlap. Skuas (*Stercorarius* spp.) predates on seagulls' newborns, while the seagulls predate on eggs of the other species. In the National Park of the Tuscan Islands, in the Tirrenian Sea, competion for nidification occurs among seagulls (*Larus audouinii* and *Larus cacchinnans*). Moreover, the latter predates upon newborns and adults of the former.

To model situations of this type, let the population sizes be denoted $P(T)$ and $Q(T)$. The classical competition model then reads

$$\frac{dP}{dT} = P[a - bP - cQ], \qquad \frac{dQ}{dT} = Q[d - eP - fQ]. \qquad (4.13)$$

Each equation states the logistic reproduction of the corresponding species, and its negative interaction with the other one is expressed by the negative sign of the mixed term. The equilibria analysis for (4.13) shows that the nullcline for $P$ intersects the coordinate axes at $\bar{P} \equiv P_1 = \frac{a}{b}$ and $\bar{Q} = \frac{a}{c}$, and the one for $Q$ intersects them at $\check{P} = \frac{d}{e}$ and $\check{Q} = \frac{d}{f}$.

Since for very small values of both populations the dominant terms in (4.13) are the positive constants, it is immediately found that the equilibrium located at the origin $\hat{E}_0 \equiv (P_0, Q_0) \equiv (0,0)$ is unstable, while the equilibrium $\hat{E}_1 \equiv (P_1, Q_1) \equiv (\bar{P}, 0) \equiv \left(\frac{a}{b}, 0\right)$ is conditionally stable, namely, if the condition $\check{P} < \bar{P}$ holds, i.e.,

$$bd < ae. \tag{4.14}$$

For $\hat{E}_2 \equiv (P_2, Q_2) \equiv (0, \check{Q}) \equiv \left(0, \frac{d}{f}\right)$ stability holds for $\bar{Q} < \check{Q}$, i.e.,

$$af < cd. \tag{4.15}$$

The nontrivial equilibrium $\hat{E}_3 \equiv (P_3, Q_3) \equiv \left(\frac{af-cd}{bf-ce}, \frac{bd-ae}{bf-ce}\right)$ exists if the two nullclines intersect each other, which happens if the condition $bf \neq ce$ is satisfied. But this intersection has to lie in the first quadrant, and the condition under which this nontrivial equilibrium is feasible then becomes the requirement that $bf - ce$, $bd - ae$, and $af - cd$ all possess the same sign, i.e., either one of the following alternative statements holds:

$$\frac{b}{e} > \frac{a}{d} > \frac{c}{f}, \text{ i.e. } \check{P} > \bar{P}, \ \bar{Q} > \check{Q} \text{ and } \text{sl}\left(\bar{P}\bar{Q}\right) < \text{sl}\left(\check{P}\check{Q}\right), \tag{4.16}$$

$$\frac{b}{e} < \frac{a}{d} < \frac{c}{f}, \text{ i.e. } \check{P} < \bar{P}, \ \bar{Q} < \check{Q} \text{ and } \text{sl}\left(\check{P}\check{Q}\right) < \text{sl}\left(\bar{P}\bar{Q}\right), \tag{4.17}$$

where "sl" stands for the slope.

The Jacobian of the system is

$$J = \begin{pmatrix} a - 2bP - cQ & -cP \\ -eQ & d - eP - 2fQ \end{pmatrix}. \tag{4.18}$$

The eigenvalues at $\hat{E}_3$ are then easily found,

$$\lambda_{1,2} = \frac{1}{2}\frac{\alpha \pm \sqrt{\Delta}}{bf - ce}, \quad \alpha = -b(af - cd) - f(bd - ae), \tag{4.19}$$

$$\Delta = \alpha^2 - 4(af - cd)(bd - ae)(bf - ce).$$

The feasibility conditions (4.16) imply that

$$\frac{\alpha}{bf - ce} \equiv -bP_3 - fQ_3 < 0,$$

from which it follows that at least one of the eigenvalues is necessarily negative. From (4.16) it follows that $\Delta < \alpha^2$, so that $\lambda_2 \equiv \frac{1}{bf-ce}\left(\alpha + \sqrt{\Delta}\right) < 0$. In such a case, the equilibrium $\hat{E}_3$ is stable and the whole ecosystem thrives, with both populations at nonzero level. Since only one eigenvalue can be made positive, upon destabilization of $\hat{E}_3$ we would get a saddle. For this to occur, we need the conditions (4.17). It then follows that the saddle implies the existence through it of a separatrix in the phase plane, dividing it into basins of attraction, one each for the remaining equilibria $\hat{E}_1$ and $\hat{E}_2$, which are then mutually exclusive. Thus only one of the two species survives, while the other one becomes extinct. As to which is which, the answer depends on the initial condition of the system. Trajectories emanating from a point in the basin of the equilibrium $\hat{E}_1$ where only the $P$'s thrive tend asymptotically to this point, thus wiping out the $Q$ population, and vice versa.

The Russian biologist Gause in his famous laboratory experiments has determined the evolution of two species of yeast, the $P$ population of *Saccharomyces cerevisiae* and the $Q$ population of *Schizosaccharomyces kefyr*. When living independently from each other (cf. Renshaw, 1991), he found the following growth laws:

$$P = \frac{13}{1 + e^{3.32816 - 0.21827t}}, \quad Q = \frac{5.8}{1 + e^{2.47550 - 006069t}}.$$

When allowed to interact, however, the two species became competitors and their evolution would follow the empirical findings

$$\frac{dP}{dT} = r_P P(1 - S_{PP}P - S_{PQ}Q), \quad \frac{dQ}{dT} = r_Q Q(1 - S_{QQ}Q - S_{QP}P),$$

where

$$r_P = 0.21827, \quad s_{PP} = 0.01679, \quad s_{PQ} = 3.15 \times S_{PP} = 0.05289$$
$$r_Q = 0.06069, \quad s_{QQ} = 0.01046, \quad s_{QP} = 0.439 \times S_{QQ} = 0.00459.$$

### 4.3.2 Symbiotic populations

Some scientists argue that symbiotic communities do not develop easily in ecological time, but rather represent evolutionary interactions. Therefore, population models to take this into account should contain different time scales. However, their remark may be easily questioned by considering one of the most elementary facts in nature, namely, the role that insects, in particular bees, have in the fecundation of flowers (e.g., see Boucher, 1985, p. 85). An entire chapter is devoted to this topic in Boucher (1985, p. 145), which contains a discussion of some aspects of evolution in a mutualistic environment.

Other well-known mutualistic interactions in nature are found in a wealth of different enviroments, such as in diatom mats in the ocean, between mangroves and root borers, spiders and parasitic wasps, invertebrates and their epibionts, and corals and fish. These as well as many other examples are illustrated for instance in the book by Boucher (1985). Benefit models for symbiotic communities have been considered in Keeler (1985) for interactions involving myrmecochory, i.e., the phenomenon for which ants find and bury seeds while the plants producing the latter receive benefit by their dispersal. But more common examples include the pollination of plants by several insects, the mycorrizhal fungi, the fungus-gardening ants, the mixed feeding flocks of birds, and the other classical cases of the anemone-damselfish, and ant-plant interactions at extrafloral nectaries. Moths (of gene Tegeticula) pollinating yuccas have also been considered. Of importance is the activity of birds removing and disseminating in the environment the seeds of *Casearia corymbosa* in Costa Rica (cf. Jantzen et al., 1980). Algae-herbivore mutualistic interactions are studied in Porter (1976), Porter (1977). These are symbiotic communities as some algae survive the digestive process of the herbivore. While the herbivore feeds, the benefit for the undigested algae lies in the possibility of absorption of nutrients released by the herbivores.

The mutualistic model can elementarily be written by means of quadratic ordinary differential equations, as was done for the other interacting species models. We account for the logistic behavior of each independent population but then combine it with additional interaction terms, which are beneficial to both populations, i.e., improve the growth rate of each species and therefore carry a positive sign. Thus let $P(T)$ and $Q(T)$ denote the two population sizes. The model is then

$$\frac{dP}{dT} = P(a - bP + cQ), \qquad \frac{dP}{dT} = Q(d + eP - fQ). \qquad (4.20)$$

Thus the equations express logistic growth for each individual population, and the positive sign for the mixed terms $PQ$ reflects the remarks introducing the system, i.e., denote the common benefit from interactions among the different species.

The Jacobian of (4.20) is

$$J = \begin{pmatrix} a - 2bP + cQ & cP \\ eQ & d + eP - 2fQ \end{pmatrix}, \qquad (4.21)$$

and it is easily seen to differ from (4.18) in the sign of the off-diagonal terms and part of the diagonal ones.

The equilibria of the system are four. We list them together with their eigenvalues, the signs of which allow their stability classification.

The origin, $E_0 \equiv (0,0)$, has eigenvalues $a$ and $d$; thus it is always unstable, a repeller along both axes for the model trajectories. The point $E_1 \equiv \left(\frac{a}{b}, 0\right)$ is unstable as its eigenvalues are $-a, \frac{1}{b}(ae + db)$. Also, $E_2 \equiv \left(0, \frac{d}{f}\right)$ is unstable, again in view of the positivity of the second eigenvalue, $-d, \frac{1}{f}(af + cd)$.

The interior equilibrium, $E_3 \equiv \left(\frac{af+cd}{bf-ce}, \frac{ae+bd}{bf-ce}\right)$, is feasible for $bf - ce > 0$, and it has the eigenvalues

$$\lambda_{3,\pm}^{(d)} \equiv \frac{1}{2}\left\{-(bP_3 + fQ_3) \pm \left[(bP_3 + fQ_3)^2 - 4(bf - ce)P_3Q_3\right]^{\frac{1}{2}}\right\}. \quad (4.22)$$

Hence, in view of the feasibility condition, the real parts of the eigenvalues are negative, thus always providing stability. As no other point is stable, the model dynamics must then be drawn to this equilibrium, which is then globally asymptotically stable. Furthermore, at the level provided by $E_3$, survival of both species is at a higher level than the one specified by the equilibrium for each single species, this being the heart of mutualism. Thus $E_3$ is more favorable than the equilibria provided by the carrying capacity of each species, and which each population would enjoy in the absence of its symbiotic counterpart. Finally, for $bf < ce$, the $\omega$-limit set of the system is the point at $\infty$, i.e., all trajectories would tend to larger and larger values for both $P$ and $Q$. We would thus have an "explosion" of the populations, when $E_3$ becomes infeasible. On the other hand, it is easy to show that the trajectories are bounded if the feasibility condition for the interior equilibrium is satisfied, namely, $bf > ce$. It is enough to sketch the two nullclines in the phase plane and afterwards analyze the sign of the flow to show that it would point "inwards" toward the origin on any curve whose points $(P, Q)$ cut a set in the first quadrant containing $O$ and $E_3$ and lie enough away from the origin.

### 4.3.3  Leslie–Gower model

As usual, we denote by $U(T)$ and $V(T)$ respectively the prey and predator sizes. The Leslie–Gower classical model assumes that when predation is not considered the prey population exhibits logistic growth, so that there are limited resources in the environment, leading to a carrying capacity $K$ and net birth rate $r$. The basic feature of the Leslie–Gower model is the assumption that the predator's carrying capacity is variable, and in particular proportional to the prey size. The proportionality constant $h$ indicates the prey amount needed to feed a predator in equilibrium conditions:

$$\frac{dU}{dT} = rU\left(1 - \frac{U}{K}\right) - cUV, \qquad \frac{dV}{dT} = aV\left(1 - h\frac{V}{U}\right). \quad (4.23)$$

In adimensional form, it becomes

$$\frac{du}{dt} = u(1 - u - v), \qquad \frac{dv}{dt} = v\left(\delta - \beta\frac{v}{u}\right), \qquad \delta = \frac{a}{r}, \qquad \beta = \frac{ah}{cK^2r}. \quad (4.24)$$

The full analysis of this system can be found in Hsu and Huang (1995). Note only that there are two equilibria, $E_* \equiv (1, 0)$ and $E^\dagger \equiv (u^\dagger \equiv \frac{\beta}{\beta+\delta}, v^\dagger \equiv \frac{\delta}{\beta+\delta})$. The Jacobian is

$$J = \begin{pmatrix} 1 - 2u - v & -u \\ \beta\frac{v^2}{u^2} & \delta - 2\beta\frac{v}{u} \end{pmatrix}. \quad (4.25)$$

At $E_*$, the eigenvalues are easily seen to be $-1$, $\delta$ so that this equilibrium is always unstable. At $E^\dagger$, the characteristic equation is $\lambda^2 + (\delta + u^\dagger)\lambda + \delta = 0$ and gives the eigenvalues

$$\lambda_\pm = \frac{1}{2}\left[-\delta - u^\dagger \pm \sqrt{(\delta + u^\dagger)^2 - \delta}\right]$$

so that $\lambda_\pm < 0$ always, thus $E^\dagger$ is a stable equilibrium. It is a focus if $\delta^2 + (2u^\dagger - 1)\delta + (u^\dagger)^2 < 0$, which then implies $1 - 2u^\dagger - \sqrt{1 - 4u^\dagger} < \delta < 1 - 2u^\dagger + \sqrt{1 - 4u^\dagger}$, and conversely it would be a node.

### 4.3.4 Classical Holling–Tanner model

The Holling–Tanner model for the prey–predator interaction assumes the same dynamics of the Leslie–Gower model for both species, except that the capturing term on prey is replaced by a saturation function $f(U, V)$, i.e., a function with a horizontal asymptote, May (1973), Renshaw (1991), and Tanner (1975). Models for it could be, for instance,

$$f(U,V) = \frac{m}{A+U}UV, \qquad f(U,V) = mV(1 - e^{\alpha U}).$$

In this case we consider the Holling type II functional response. The model is then

$$\frac{dU}{dT} = rU\left(1 - \frac{U}{K}\right) - \frac{m}{A+U}UV, \qquad \frac{dV}{dT} = V\gamma\left(1 - h\frac{V}{U}\right). \qquad (4.26)$$

A nondimensional form of system (4.26) can be obtained by setting

$$u \equiv \frac{1}{K}U, \quad v \equiv \frac{m}{rK}V, \quad t \equiv rT, \quad \delta \equiv \frac{\gamma}{r}, \quad \beta \equiv \frac{\gamma h}{m}, \quad a \equiv \frac{A}{K},$$

so that we get

$$\frac{du}{dt} = u(1-u) - \frac{1}{a+u}uv, \qquad \frac{dv}{dt} = v\left(\delta - \beta\frac{v}{u}\right). \qquad (4.27)$$

These can be studied geometrically as follows. The isoclines give respectively a parabola and a straight line through the origin

$$\frac{du}{dt} = 0: \quad v = (a+u)(1-v); \qquad \frac{dv}{dt} = 0: \quad v = \frac{\delta}{\beta}u. \qquad (4.28)$$

The parabola has the vertex at $(\hat{u}, \hat{v})$, $\hat{u} = \frac{1}{2}(1-a)$, $\hat{v} = \frac{1}{4}r(1+a)^2$, and it has positive height $a$ at the origin and a feasible root at $u_0 = 1$. On geometrical grounds, an intersection in the first quadrant always exists, so that a feasible equilibrium is always guaranteed. In fact, for (4.27), we also find the boundary equilibria $O$ and $\hat{E}_1(1, 0)$ in addition to the interior one $\hat{E}_2(u^*, v^*)$ with

$$u^* = \frac{1}{2}\left[D + \sqrt{D^2 + 4a}\right], \quad D = 1 - a - \frac{\delta}{\beta}, \quad v^* = u^*\frac{\delta}{\beta}. \qquad (4.29)$$

The unconditional existence of the latter is also clear from the algebraic representation, as the root contains a sum of positive quantities, that make it always larger than $D$ and render $u^*$ nonnegative. This model will be further analyzed for stability in Section 4.4.5.

### 4.3.5 Other growth models

Finally, we briefly mention a few other possible equations for the population growth. They are the Gompertz models:

$$\frac{dU}{dT} = \alpha U - \beta U \ln U, \qquad \frac{dU}{dT} = \begin{cases} \alpha U & , U < K \\ \alpha U - \beta U \ln \frac{U}{K} & , U \geq K \end{cases} \quad (4.30)$$

$$\frac{dU}{dT} = \alpha U^\gamma - \beta U^\gamma \ln U, \qquad \frac{dU}{dT} = \alpha U(\ln k - \ln U)^{1+p}. \quad (4.31)$$

The von Bertalanffy models are:

$$\frac{dU}{dT} = \alpha U^{\frac{2}{3}} - \beta U, \qquad \frac{dU}{dT} = \begin{cases} \alpha U^\gamma - \beta U & , \gamma < 1 \\ \alpha U - \beta U \ln U & , \gamma = 1 \\ \alpha U - \beta U^\gamma & , \gamma > 1 \end{cases} \quad (4.32)$$

Also, generalized or hyper-logistic are in use:

$$\frac{dU}{dT} = \alpha U^\gamma - \beta U^\delta, \qquad \frac{dU}{dT} = \alpha U^{1-p} - \frac{1}{k}(k - U)^{1+p}. \quad (4.33)$$

Analytical study of these models' properties can be found in literature; otherwise we leave it to a reader as an excellent exercise to practice the mathematical technique described earlier in this book.

### 4.3.6 Models with prey switching

"Higher dimensional" interacting population models have also been considered, where the "dimensionality" here refers to the number of species that are considered. This is also in preparation for more complex models to be analyzed at the end of this chapter, and later on in Chapter 7.

Models for prey exhibiting group defense have also been proposed, (see, for instance, Freedman and Wolkowicz (1986)):

$$\frac{dU}{dT} = Ug(U, K) - Vp(U), \qquad \frac{dV}{dT} = V(s - q(U)), \qquad s, K \in \mathbf{R}^+, \quad (4.34)$$

where $g$, $p$, $q$ are continuously differentiable functions. The first function represents the prey growth rate in absence of predators, $p$ is the predators' response function, while $q$ represents the predators' conversion function. The assumptions they obey are summarized as follows:

$$g(0, K) = g(K, K) = 0,$$

$$\frac{\partial g}{\partial U}(K, K) < 0, \quad \frac{\partial g}{\partial U}(U, K) \leq 0, \quad \frac{\partial g}{\partial K}(U, K) > 0,$$

$$p(0) = 0, \quad p(U) \geq 0,$$

$$\frac{dp}{dU} > 0 \text{ for } 0 \leq U < M, \quad \frac{dp}{dU} < 0 \text{ for } U > M,$$

$$q(0) = 0, \quad q(U) \geq 0, \quad q(M) > s,$$

$$\frac{dq}{dU} > 0 \text{ for } 0 \leq U < M, \quad \frac{dq}{dU} < 0 \text{ for } U > M,$$

where $M > 0$ denotes a suitable constant.

A model incorporating group defense in a community dwelling in a fragmented habitat was developed by Khan et al. (1998):

$$\frac{dU_1}{dT} = (\alpha_1 - \epsilon_1)U_1 + \epsilon_2 p_{21} U_2 - \frac{\beta_1 U_2^2 V}{U_1 + U_2}, \tag{4.35}$$

$$\frac{dU_2}{dT} = (\alpha_2 - \epsilon_2)U_2 + \epsilon_1 p_{12} U_1 - \frac{\beta_2 U_1^2 V}{U_1 + U_2},$$

$$\frac{dV}{dT} = V\left(-\nu + \frac{\delta_1 \beta_1 U_2^2}{U_1 + U_2} + \frac{\delta_2 \beta_2 U_1^2}{U_1 + U_2}\right),$$

where $U_1$ and $U_2$ denote the prey species living in the two separate habitats. The parameters are interpreted as follows. The predator response rates toward each species of prey are $\beta_i$, $i = 1, 2$; their respective conversion rates are $\delta_i$; $\epsilon_i$ denotes the inverse barrier strengths going out of habitat $i$; and $p_{i,j}$ is the probability of such a successful move. The specific growth rate of the prey when predators are absent is $\alpha_i$, and $\nu$ is the per capita death rate of predators.

Khan et al. (2002) also developed a model that incorporates a stage structure in addition to switching:

$$\frac{dU_1}{dT} = \beta U_2 - \mu_1 U_1 - \beta U_2(T - \tau)e^{-\mu_1 \tau} - \frac{b_1 U_1^2 V}{U_1 + U_2} + hU_1 U_2, \tag{4.36}$$

$$\frac{dU_2}{dT} = \beta U_2(T - \tau)e^{-\mu_1 \tau} - \mu_2 U_2^2 - \frac{b_1 U_2^2 V}{U_1 + U_2} + gU_1 U_2,$$

$$\frac{dV}{dT} = V\left(\frac{c_1 \beta_1 U_1^2}{U_1 + U_2} + \frac{c_2 \beta_2 U_2^2}{U_1 + U_2} - d\right)$$

(see also McNair, 1987), with $U_1$ and $U_2$ denoting the immature and adult populations, $\tau$ representing the time from birth to reach maturity, the term $\beta U_2(T - \tau)e^{-\mu_1 \tau}$ expressing the fraction of young individuals who reach the maturity stage, and $gU_1U_2$ and $hU_1U_2$ are interaction terms. The parameters have the following meanings: $\beta$ is the per capita birth rate of the prey species in stage 2, $\gamma$ is the maturation rate from stage 1 to stage 2, $\mu_i$ is the per capita death rate, $b_i$ is the capture rate of prey species of stage $i$, $c_i$ is the conversion factor for prey species of stage $i$, and $d$ is the per capita death rate of the predator.

Another switching similar model is presented in Tansky (1978):

$$\frac{dU_1}{dT} = U_1\left(E_1 - \frac{aV}{1+\left(\frac{U_2}{U_1}\right)^n}\right), \quad (4.37)$$

$$\frac{dU_2}{dT} = U_2\left(E_2 - \frac{bV}{1+\left(\frac{U_1}{U_2}\right)^n}\right),$$

$$\frac{dV}{dT} = V\left(-E_3 + \frac{aU_1}{1+\left(\frac{U_2}{U_1}\right)^n} + \frac{bU_2}{1+\left(\frac{U_1}{U_2}\right)^n}\right),$$

where the positive integer $n$ is a parameter.

A similar model was also considered by Khan et al. (2004):

$$\frac{dU_1}{dT} = g_1U_1\left(1 - \frac{U_1}{K_1}\right) + e_2p_{21}U_2 - \frac{\beta_1U_1U_2V}{U_1+U_2}, \quad (4.38)$$

$$\frac{dU_2}{dT} = g_2U_2\left(1 - \frac{U_2}{K_2}\right) + e_1p_{12}U_1 - \frac{\beta_2U_1U_2V}{U_1+U_2},$$

$$\frac{dV}{dT} = V\left(-\mu + \frac{(\delta_1\beta_1 + \delta_2\beta_2)U_1U_2}{U_1+U_2}\right),$$

where $K_i$ represents the carrying capacity of each prey; $g_i$ is the net effect of birth, death and migration rates; and the remaining parameters have interpretations as above, (4.36), namely, $e_i$ and $p_{ij}$ represent the strength of the barrier and the probability of moving to a different habitat, $\beta_i$ the predator responses to each prey, $\delta_i$ the conversion factors, and $\mu$ the per capita death rate of predators.

---

## 4.4 Global stability

Let us consider again the classical Lotka–Volterra system,

$$\frac{dU}{dT} = U(a - bV), \quad \frac{dV}{dT} = V(-m + cU). \quad (4.39)$$

In an attempt to find a closed form solution, we can calculate the slope of the trajectories in the $UV$ phase plane. Using the chain rule and exploiting (4.39), we get

$$\frac{dV}{dU} = \frac{dV}{dT} \times \left[\frac{dU}{dT}\right]^{-1} = \frac{V(-m+cU)}{U(a-bV)},$$

so that variables can be separated to yield

$$\int (a-bV)\frac{dV}{V} = \int (-m+cU)\frac{dU}{U}; \quad a\ln V - bV = -m\ln U + cU - h.$$

Therefore we can introduce the function

$$z = L(U,V) \equiv -a\ln V + bV - m\ln U + cU - h, \qquad (4.40)$$

which represents a surface defined on the phase space. The orbits of the model (4.39) are the loci of points $(U,V)$ in the phase space for which $L(U,V) = h$ for the constant $h \in \mathbf{R}$, i.e., the level curves $z = h$ of the surface $z = L(U,V)$; the larger $h$ is, the higher these levels are. In particular, given the initial condition $(U_0, V_0)$, the solution of the initial value problem is the level curve of $L = h_0$, where $h_0$ corresponds to

$$h_0 = -a\ln V_0 + bV_0 - m\ln U_0 + cU_0. \qquad (4.41)$$

To further study $L$, it is immediately seen that $\nabla L = 0$ at the equilibrium point of the system (4.39), $E_\infty \equiv (U_\infty, V_\infty)$ with $U_\infty = \frac{m}{c}$, $V_\infty = \frac{a}{b}$, and from this we define the value of the constant there as

$$h_\infty = -a\ln V_\infty + bV_\infty - m\ln U_\infty + cU_\infty \qquad (4.42)$$
$$= a\left(1 - \ln\frac{a}{b}\right) + m\left(1 - \ln\frac{m}{c}\right).$$

The shape of the surface can be investigated by intersecting it with vertical planes. On $V = \tilde{V}$, we easily find that $L \to +\infty$ both when $U \to +\infty$ as well as $U \to 0^+$. Similarly, on $U = \tilde{U}$, we have $L \to +\infty$ for $V \to 0^+$ and $V \to +\infty$. It appears that the equilibrium $E_\infty$ is then a minimum. To further support this statement, we consider the intersection with a generic plane $\bar{V} = rU + s$. Then

$$\tilde{L}(U) \equiv L(U, \bar{V}) \equiv -m\ln U + cU - a\ln(rU + s) + b(rU + s)$$

and we must distinguish several cases. For $r, s > 0$, as $U \to +\infty$ we have

$$\tilde{L} = U\left[c - m\frac{\ln U}{U}\right] + (rU + s)\left[b - a\frac{\ln(rU + s)}{rU + s}\right] \to +\infty.$$

The behavior $L \to +\infty$ can also be discovered for $U \to +\infty$ in the other cases. The function indeed needs to be investigated also on the remaining parts of the boundary of *its domain*, which is $\mathbf{R}_+ \times \mathbf{R}_+ = \{(U,V) : U > 0, ; V > 0\}$. By taking the other suitable limits, namely, for $r, s > 0$ as $U \to 0^+$, while for $r > 0$, $s < 0$ as $V \equiv rU + s \to 0^+$, we find again $L \to \infty$. The case $r < 0$, $s > 0$ needs more care as the limits at infinity cannot be taken; rather, we must evaluate $L$ when both $U \to 0^+$ and $rU + s \to 0^+$, but again the conclusion is $L \to \infty$. The surface $L$ thus appears to be a bowl, going up vertically on the coordinate axes and also growing as the points in the domain drift toward infinity. $E_\infty$ then appears to be its absolute minimum. To mathematically substantiate this claim, we can compute the Hessian matrix,

$$H_L = \begin{vmatrix} mU^{-2} & 0 \\ 0 & aV^{-2} \end{vmatrix} > 0,$$

to discover that the curvature of the surface is always positive, i.e., the surface is everywhere convex. Therefore $E_\infty$ really represents an absolute minimum. By (4.40) and (4.42) we would then have $L(E_\infty) = 0$. This is the first property of the Lyapunov function. We have also shown that $L(U, V) \geq L(E_\infty) = 0$, which constitutes its second property. Below we will discover finally its third and fundamental property, namely, that the system trajectories enter into this bowl. In the present case, calculating the tangential derivative of $L$ we find that it is tangent to the solution trajectories,

$$\frac{dL}{dT} = \frac{\partial L}{\partial U}\frac{dU}{dT} + \frac{\partial L}{\partial V}\frac{dV}{dT} = -\frac{m}{U}U(a - bV) + cU(a - bV)$$
$$- \frac{a}{V}V(-m + cU) + bV(-m + cU) = 0.$$

The function $L$ also allows the explicit analytical determination of the orbits. To this end, using the initial condition (4.41) coupled with the definition (4.40) of $L$, we have

$$-m(\ln U - \ln U_0) + c(U - U_0) = a(\ln V - \ln V_0) - b(V - V_0)$$

from which

$$c(U - U_0) + b(V - V_0) = \ln\left(\frac{V}{V_0}\right)^a + \ln\left(\frac{U}{U_0}\right)^m$$

and finally the closed forms of the orbits

$$\left(\frac{U}{U_0}\right)^m \left(\frac{V}{V_0}\right)^a = e^{c(U - U_0) + b(V - V_0)}.$$

To investigate further the latter, let us study small perturbations around the equilibrium, by defining $u = U - U_\infty = U - \frac{m}{c} \in \mathbf{R}$, $v = V - V_\infty = V - \frac{a}{b} \in \mathbf{R}$. Substitution into $L(U, V) = h$ brings

$$- m\left[\ln\frac{m}{c} + \ln\left(1 + \frac{cu}{m}\right)\right] + m\left(1 + \frac{cu}{m}\right)$$
$$- a\left[\ln\frac{a}{b} + \ln\left(1 + \frac{bv}{a}\right)\right] + a\left(1 + \frac{bv}{a}\right) = h.$$

# Interacting populations

Now, using Taylor's formula and keeping only terms up to the second order, we are led to

$$-m\left[\ln\frac{m}{c}+\frac{cu}{m}-\frac{1}{2}\left(\frac{cu}{m}\right)^2\right]+m\left(1+\frac{cu}{m}\right)$$
$$-a\left[\ln\frac{a}{b}+\frac{bv}{a}-\frac{1}{2}\left(\frac{bv}{a}\right)^2\right]+a\left(1+\frac{bv}{a}\right)=h.$$

Linearization and further simplifications give

$$\frac{m}{2}\left(\frac{cu}{m}\right)^2+\frac{a}{2}\left(\frac{bv}{a}\right)^2=-m-a+h+m\ln\frac{m}{c}+a\ln\frac{a}{b}$$

and, finally, using (4.42) we arrive at

$$\frac{u^2}{A^2}+\frac{v^2}{B^2}=H,\quad A^2=\frac{2m}{c^2},\quad B^2=\frac{2a}{b^2},\quad H=h-h_\infty\geq 0.$$

The trajectories around the equilibrium are therefore close to ellipses.

## 4.4.1 General quadratic prey–predator system

We consider now the prey–predator interactions with logistic growth in the prey, and competition among the predators, which do not have other food sources:

$$\frac{dU}{dT}=U(a-bV-\frac{U}{K_1}),\quad \frac{dV}{dT}=V(-m+cU-\frac{V}{K_2}),\qquad (4.43)$$

with nonboundary equilibrium at $E^*=(U^*,V^*)$, where

$$U^*\equiv K_1\frac{a+mbK_2}{1+bcK_1K_2},\quad V^*\equiv K_2\frac{-m+acK_1}{1+K_1bcK_2},$$

feasible for $acK_1>m$. Let us consider once again the perturbations $u=U-U^*$, $v=V-V^*$ and proceed as follows:

$$\frac{du}{dv}=-\frac{u+U^*}{v+V^*}\frac{K_2}{K_1}\frac{u+bvK_1}{cK_2u-v},\quad \frac{du}{u+U^*}=\frac{cK_2u-v}{K_2}=-\frac{dv}{v+V^*}\frac{u+bvK_1}{K_1},$$

which is a separable equation,

$$cdu-cU^*\frac{du}{u+U^*}-\frac{v}{K_2}\frac{du}{u+U^*}=-\frac{u}{K_1}\frac{dv}{v+V^*}-bdv+bV^*\frac{dv}{v+V^*},$$

which upon integration gives

$$cu-cU^*\ln\frac{u+U^*}{U^*}-\frac{v}{K_2}\ln(u+U^*)=-\frac{u}{K_1}\ln(v+V^*)-bv+bV^*\ln\frac{v+V^*}{V^*}+C,$$

where the term $C$ represents the constant of integration. This expression suggests to define the following Lyapunov function:

$$L(u,v) \equiv cU^* \left[ \frac{u}{U^*} - \ln\left(1 + \frac{u}{U^*}\right)\right] + bV^*\left[\frac{v}{V^*} - \ln\left(1 + \frac{v}{V^*}\right)\right].$$

The property of the logarithm, $x \geq \ln(1+x)$ for any $x > 0$, implies that $L(u,v) \geq 0$, with $L(0,0) = 0$. This of course corresponds to the vanishing of $L$ at the equilibrium point in the original coordinates. Finally,

$$\nabla L(u,v) \equiv \left(c - cU^*\frac{U^*}{u+U^*}\frac{1}{U^*}, b - bV^*\frac{V^*}{v+V^*}\frac{1}{V^*}\right)^T$$

$$= \left(\frac{cu}{u+U^*}, \frac{bv}{v+V^*}\right)^T.$$

It thus follows that the directional derivative along the solution trajectories of (4.43) is

$$\frac{dL}{dT} = \nabla L \cdot \left(\frac{du}{dT}, \frac{dv}{dT}\right)^T = \frac{cu}{u+U^*}(u+U^*)\left(-bv - \frac{u}{K_1}\right)$$

$$+ \frac{bv}{v+V^*}(v+V^*)\left(+cu - \frac{v}{K_2}\right)$$

$$= -\frac{1}{K_1}cu^2 - \frac{1}{K_2}bv^2 < 0.$$

This result thus shows that the trajectories make an obtuse angle with the (outward) normal to the level surfaces $L(u,v) = h$, i.e., they enter into the bowl defined by the Lyapunov function independently of their initial values, i.e., the equilibrium $E^*$ is indeed a global attractor of the system (4.43).

### 4.4.2 Mathematical tools for analyzing limit cycles

Here we want to mention some results in planar dynamical systems that are useful in the investigation of solutions' sustained oscillations; see Hirsch and Smale (1974) and Strogatz (1994).

**Poincaré–Bendixson theorem**

A closed orbit $\Gamma$ of a planar dynamical system is a nonempty subset of the phase plane, $\Gamma \subset \mathbf{R}^2$, not containing equilibrium points.

The Poincaré–Bendixson theorem states that if $\Gamma \subset \mathbf{R}^2$ is a closed and bounded set, and $\frac{d\mathbf{x}}{dT} = \mathbf{f}(\mathbf{x}) \in C^1(\mathbf{R}^2)$ is a vector field defined on an open set containing $\Gamma$, if $\Gamma$ does not contain equilibrium points, and if there is a trajectory $L \subset \Gamma$ contained in the orbit, then $L$ is a closed orbit or tends to a closed orbit (i.e., a limit cycle). In other words, in such conditions $\Gamma$ contains at least a limit cycle.

## Interacting populations

The theorem can be exploited by determining a positively invariant set $\Omega$ for the dynamical system, i.e., such that the system trajectories can only enter into $\Omega$, and then if possible trying to destabilize the equilibria that are found inside $\Omega$. The application of the above theorem would then show that at least a stable limit cycle exists around the equilibrium.

### Dulac's criterion

The next criterion is negative, in the sense that if satisfied it shows that limit cycles do not exist; see Strogatz (1994). It is essentially a consequence of Green's theorem.

Let $A \subset \mathbf{R}^2$ be a simply connected region in the plane, i.e., it does not contain holes, and let $f$ and $g$ be continuously differentiable functions in $A$, i.e., mathematically $f, g \in C^1(A)$. Given the system

$$\frac{dx}{dT} = f(x,y), \quad \frac{dy}{dT} = g(x,y), \tag{4.44}$$

if a function $\beta(x,y) \in C^1(\mathbf{R}^2)$ exists for which the function

$$\frac{\partial}{\partial x}[\beta(x,y)f(x,y)] + \frac{\partial}{\partial y}[\beta(x,y)g(x,y)]$$

is of one sign in $A$, then in $A$ no closed orbit can exist.

In general this result is utilized by taking $A \equiv \mathbf{R}^2$ and $\beta(x,y) = (xy)^{-1}$. As an example, let us consider the quadratic competition model

$$\frac{dU}{dT} = \alpha U \left(1 - \frac{U}{K}\right) - aUV, \quad \frac{dV}{dT} = rV \left(1 - \frac{V}{H}\right) - bUV.$$

With the above choice for $\beta$ we find

$$\frac{\partial}{\partial U}\left\{\frac{1}{UV}\left[\alpha U\left(1 - \frac{U}{K}\right) - aUV\right]\right\} + \frac{\partial}{\partial V}\left\{\frac{1}{UV}\left[rV\left(1 - \frac{V}{H}\right) - bUV\right]\right\}$$
$$= -\frac{\alpha}{KV} - \frac{r}{HU} < 0.$$

It follows then that quadratic competition models do not possess limit cycles, and this fact clearly stems from the logistic terms in the equations, the only ones surviving after the above differentiation.

### Potential systems

A related negative result (Strogatz, 1994) states that for potential systems, limit cycles do not exist; in other words, if a function $Z(\mathbf{x})$ exists for which the field $\mathbf{f}(\mathbf{x})$ is the gradient of $Z$, $\mathbf{f}(\mathbf{x}) = \nabla Z(\mathbf{x})$, i.e., $Z$ is a potential function for the field, then no limit cycle for the dynamical system can exist.

As an example let us consider two logistically growing and not interacting populations,

$$\frac{dU}{dT} = \alpha U\left(1 - \frac{U}{K}\right), \quad \frac{dV}{dT} = rV\left(1 - \frac{V}{H}\right).$$

Considering

$$Z(U,V) = \frac{\alpha}{2}U^2 - \frac{\alpha}{3K}U^3 + \frac{r}{2}V^2 - \frac{r}{3H}V^3,$$

the above condition is immediately verified. Of course each population evolves naturally toward its own carrying capacity undisturbed by the other one, these equilibria being stable in view of the results on the single-population models, and in the two-dimensional phase plane the corresponding equilibrium must be stable. The corresponding Jacobian is a diagonal matrix that when evaluated at the equilibrium $(K, H)$, has negative entries, namely, $J_{(K,H)} = diag(-\alpha, -r)$.

### 4.4.3 Routh–Hurwitz conditions

Stability of an equilibrium in dynamical systems as seen earlier is related to the negative sign of its eigenvalues, if real, or of their real part if complex. A criterion to determine the signs of the roots of a polynomial equation is the Routh–Hurwitz criterion. Namely, for the following polynomial equation the related determinants are defined:

$$\sum_{i=0}^{n} a_i \lambda^i = 0, \quad D_1 = a_{n-1}, \quad D_2 = \begin{vmatrix} a_{n-1} & a_{n-3} \\ a_n & a_{n-2} \end{vmatrix}, \quad (4.45)$$

$$D_3 = \begin{vmatrix} a_{n-1} & a_{n-3} & a_{n-5} \\ a_n & a_{n-2} & a_{n-4} \\ 0 & a_{n-1} & a_{n-3} \end{vmatrix}, \quad \ldots, \quad D_n = \begin{vmatrix} a_{n-1} & a_{n-3} & a_{n-5} & \ldots & 0 & 0 \\ a_n & a_{n-2} & a_{n-4} & \ldots & 0 & 0 \\ 0 & a_{n-1} & a_{n-3} & \ldots & 0 & 0 \\ & & \ldots & & & \\ 0 & 0 & 0 & \ldots & 0 & a_0 \end{vmatrix},$$

with the clear understanding that $a_j \equiv 0$ for $j < 0$ or $j > n$. The criterion states that all the roots have negative real parts if and only if all the $D_j$'s are positive.

There is a variant of the above criterion, the Liénard–Chipart criterion, which essentially states that for stability if all coefficients are positive, $a_i > 0$ for $i = 0, \ldots, n$, only the determinants $D_{n-1}, D_{n-3}, D_{n-5}, \ldots$ need to be positive.

It is useful to see what the above statements amount to for polynomials up to degree 4. Observe that by a change of sign of the whole polynomial, the assumption $a_n > 0$ can always be made. Then, for a linear polynomial with root $\lambda_1$, trivially we need only

$$D_1 = a_0 > 0, \quad \lambda = -\frac{a_0}{a_1} < 0. \quad (4.46)$$

The most useful cases, also for the subsequent chapters, are the low degree polynomials. For a quadratic, the conditions reduce to

$$a_0 > 0, \quad a_1 > 0. \tag{4.47}$$

Geometrically, they imply that the height and the slope at the origin are both positive, so that the vertex lies to its left, and the possible roots occur only on the left-half plane. For a cubic they are

$$a_0 > 0, \quad a_2 > 0, \quad a_2 a_1 > a_0 a_3, \tag{4.48}$$

and for a quartic they are

$$a_0 > 0, \quad a_3 > 0, \quad a_2 a_3 > a_1 a_4, \quad a_1(a_2 a_3 - a_1 a_4) > a_0 a_3^2. \tag{4.49}$$

### 4.4.4 Criterion for Hopf bifurcation

In general, the Routh–Hurwitz conditions are needed to determine the presence of negative roots for an equation. When the latter is the characteristic equation of an equilibrium point, the criterion tells us the stability of the associated equilibrium. If the latter depends on one parameter $\mu$, it is possible that its stability changes with modifications of this parameter value. To determine this, without explicitly calculating the zeros of the characteristic equation, another criterion is available; see Liu (1994). Let this characteristic polynomial be

$$P_\mu(\lambda) = \sum_{i=0}^{n} p_i(\mu) \lambda^i$$

and define, with the same conventions $p_j(\mu) \equiv 0$ for $j < 0$ or $j > n$ used above,

$$L_n(\mu) = \begin{pmatrix} p_1(\mu) & p_0(\mu) & \cdots & 0 \\ p_3(\mu) & p_2(\mu) & \cdots & 0 \\ \cdots & \cdots & & \\ p_{2n-1}(\mu) & p_{2n-2}(\mu) & \cdots & p_n(\mu) \end{pmatrix}, \tag{4.50}$$

and denote by $D_j$ the principal minors of order $j$ of the above matrix. If there is a locus of equilibria $(x(\mu), \mu)$ emanating from $x(\mu_0) = x_0$ for the dynamical system $\frac{dx_\mu}{dT} = f(x_\mu)$, the sufficient conditions for a simple Hopf bifurcation are that the Jacobian of the system at the equilibrium, $J_x[f_{\mu_0}(x_0)]$, possesses two purely imaginary eigenvalues, $\lambda(\mu)$, while all the other roots have negative real parts. Moreover,

$$\frac{d\Re(\lambda(\mu_0))}{d\mu} \neq 0. \tag{4.51}$$

The criterion then states that the above sufficient conditions are equivalent to

$$p_0(\mu_0) > 0, \quad D_1(\mu_0) > 0, \quad \ldots, \quad D_{n-2}(\mu_0) > 0, \tag{4.52}$$

$$D_{n-1}(\mu_0) = 0 \quad \text{and} \quad \frac{dD_{n-1}(\mu_0)}{d\mu} \neq 0.$$

### 4.4.5 Instructive example

As an example of the above techniques, we consider again the Holling–Tanner model introduced in Section 4.3.4. The Jacobian in this case is

$$J = \begin{pmatrix} 1 - 2u - av\frac{1}{(a+u)^2} & -u\frac{1}{a+u} \\ \beta\frac{v^2}{u^2} & \delta - 2\beta\frac{v}{u} \end{pmatrix}. \tag{4.53}$$

The origin and $\hat{E}_1$ are easily seen to be always unstable, having at least a positive eigenvalue. Using (4.29), the Jacobian at the equilibrium $\hat{E}_2 = (u^*, v^*)$ can be rewritten as

$$J^* = \begin{pmatrix} \frac{u^*}{a+u^*}\left(\frac{\delta}{\beta} - \sqrt{D^2 + 4a}\right) & -u^*\frac{1}{a+u^*} \\ \beta\frac{\delta^2}{\beta} & -\delta \end{pmatrix}. \tag{4.54}$$

The Routh–Hurwitz criterion for the quadratic characteristic equation (4.47) in this case requires $-\text{Tr}(J^*) > 0$ and $\det(J^*) > 0$. Easily, we find

$$\det(J^*) = \frac{u^*\delta\sqrt{D^2 + 4a}}{a + u^*} > 0.$$

Moreover, for the trace of the above matrix, using again (4.29), we must have

$$-\text{Tr}(J^*) = \frac{u^*}{a + u^*}\left(\sqrt{D^2 + 4a} - \frac{\delta}{\beta}\right) + \delta > 0. \tag{4.55}$$

If the abscissa of the equilibrium lies to the right of the abscissa of the vertex of the parabola (4.28), namely, $u^* > \hat{u}$, i.e. $\sqrt{D^2 + 4a} - \frac{\delta}{\beta} > 2a$, then (4.55) holds and $E_2$ is then stable. For $u^* < \hat{u}$ instead the condition (4.55) must be checked for stability.

The equilibrium $\hat{E}_2$ can be further analyzed on geometrical grounds. Let us consider a rectangle $R$ with vertex at the origin and the opposite one $(u_*, v_*)$ with $u_* > u_0 = 1$, where we recall that $u_0$ is the positive root of the isocline $\frac{du}{dT} = 0$, which as we know is a parabola (4.28), and $v_* > \hat{v}$, the height of its vertex. It is easily seen that the flow of the dynamical system enters it on the two sides that do not lie on the coordinate axes. Therefore, $R$ is a positively invariant set. All its equilibria are unstable if (4.55) is violated. In this case, by applying the Poincaré–Bendixson theorem, a limit cycle must arise around $\hat{E}_2$.

Figure 4.1 reports some results for chosen parameter sets. In the first column $E_2$ is stable since $u^* > \hat{u}$. The remaining columns show some cases for $u^* < \hat{u}$ and in particular two limit cycles

To complete the discussion, we mention two more recent results for the Holling–Tanner model. It has been shown to possess two bifurcation points (Braza, 2003); when the latter nearly coincide, the stable branches of the periodic solution connect, but by changing the growth rate, the Hopf-bifurcation

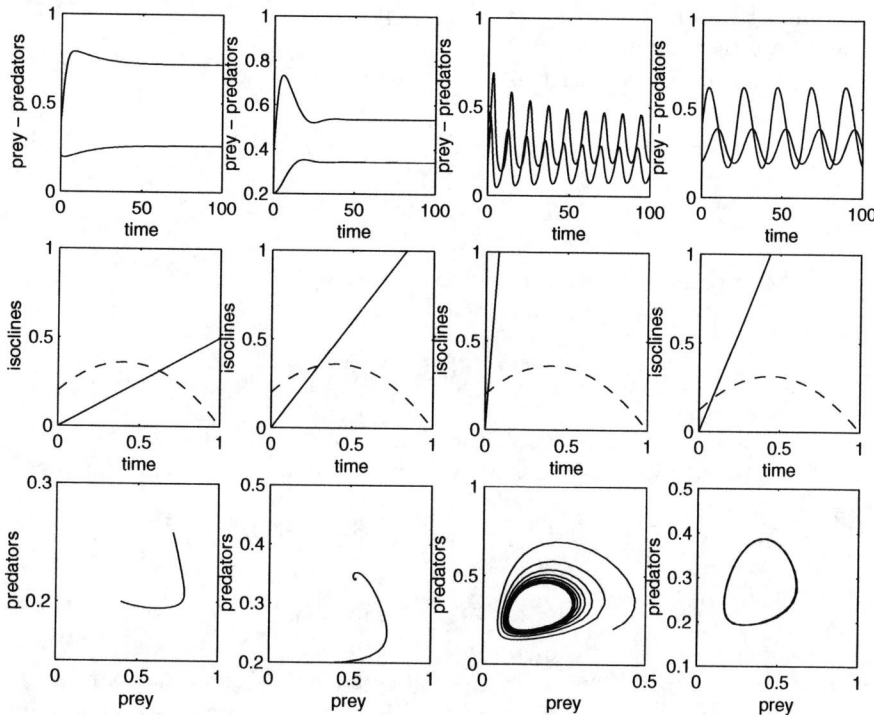

**FIGURE 4.1**: Interior equilibrium and limit cycles for the Holling–Tanner model. Top: populations vs time; center: isocline diagram; bottom: phase plane plots. Parameter values for first column are $a = 0.2$, $\beta = 0.1$, and $\delta = 0.05$. Parameter values for second column are $a = 0.2$, $\beta = 0.1$, and $\delta = 0.12$. Parameter values for third column are $a = 0.2$, $\beta = 0.1$, and $\delta = 1.3$. Parameter values for fourth column: $a = 0.12$, $\beta = 0.1$, and $\delta = 0.23$.

points separate, the limit cycle becomes unstable and this may possibly lead to population outbreaks.

Global stability can be shown by constructing a suitable Lyapunov function; for details see Hsu and Huang (1995). The analysis leads to the following conclusions. Both species can be shown to persist globally in the ecosystem. Moreover, if the prey carrying capacity is small, then both populations tend to constant values, and no periodic behavior is possible. For the predator net birth rate larger than the prey one, there is no limit cycle. A limit cycle exists if the maximal consumption is neither too small nor too large.

### 4.4.6 Poincaré map

We outline now another mathematical technique useful for analyzing the stability of limit cycles, but on a rather sophisticated and artificial biological example. Essentially we consider a prey–predator model that includes the Allee effect and migration.

With the usual notation let us consider the model

$$\frac{dU}{dT} = A + BU + CU^2 - U^3 + f(U, V), \qquad (4.56)$$
$$\frac{dV}{dT} = H + KV + LV^2 - V^3 + g(U, V),$$

where the interaction terms are assumed to be

$$f(U, V) = -V[1 + (U - U_*)(V - 2V_*)], \qquad (4.57)$$
$$g(U, V) = U[1 + (V - V_*)(U - 2U_*)]$$

(a generalized populations dependence on the predation rate and the conversion factor), and the coefficients are chosen in the following particular way:

$$A = V_* - U_* + U_*^3 + U_*V_*^2, \quad B = 1 - 3U_*^2 - V_*^2, \qquad (4.58)$$
$$C = 3U_*(>0), \quad H = U_*^2V_* + V_*^3 - U_* + V_*,$$
$$K = -1 - U_*^2 - 3V_*^2, \quad L = 3V_*(>0).$$

One of course can ask what the quantities $U_*$ and $V_*$ are, and here is the second artificial aspect of the model. They are supposed to represent an equilibrium value for the two populations. Substituting (4.58)–(4.57) into (4.56) and expanding we find

$$\frac{dU}{dT} = -(V - V_*) + (U - U_*) - (U - U_*)^3 - (U - U_*)(V - V_*)^2, \qquad (4.59)$$
$$\frac{dV}{dT} = U - U_* + V_* - V - (U - U_*)^2(V - V_*) - (V - V_*)^3.$$

We now introduce $x = U - U_*$ and $y = V - V_*$ as perturbations around the equilibrium value, so that (4.59) turns to

$$\frac{dx}{dT} = -y + x - x^3 - xy^2 = -y + x(1 - x^2 - y^2), \qquad (4.60)$$
$$\frac{dy}{dT} = x + y - x^2y - y^3 = x + y(1 - x^2 - y^2).$$

At this point, the introduction of polar coordinates becomes natural. On differentiating $r^2 = x^2 + y^2$ and $\theta = \arctan(y/x)$, we find

$$\frac{dr}{dT} = \frac{x}{r}\frac{dx}{dT} + \frac{y}{r}\frac{dy}{dT},$$
$$\frac{d\theta}{dT} = \frac{x}{x^2 + y^2}\frac{dy}{dT} - \frac{y}{x^2 + y^2}\frac{dx}{dT} = \frac{x}{r^2}\frac{dy}{dT} - \frac{y}{r^2}\frac{dx}{dT}.$$

On using (4.60), in our case we find

$$\frac{dr}{dT} = \frac{x}{r}[-y + x(1-r^2)] + \frac{y}{r}[x + y(1-r^2)] \tag{4.61}$$

$$= \frac{x^2}{r} - \frac{x^2 r^2}{r} + \frac{y^2}{r} - \frac{y^2 r^2}{r} = r(1-r^2),$$

$$\frac{d\theta}{dT} = \frac{x}{r^2}[x + y(1-r^2)] - \frac{y}{r^2}[-y + x(1-r^2)] = \frac{1}{r^2}(x^2 + y^2) = 1.$$

On integrating the second one, clarly $\theta = T + T_0$, thus the phase is tied to the flow of time, this already suggesting some kind of oscillatory behavior. Using partial fractions on the former,

$$\frac{dr}{r} + \frac{1}{2}\frac{dr}{1-r} - \frac{1}{2}\frac{dr}{1+r} = dT,$$

so that, upon integration,

$$\ln\frac{r}{\sqrt{1-r^2}} = T + C, \quad \frac{r}{\sqrt{1-r^2}} = \tilde{K}e^T, \tag{4.62}$$

$$\frac{r^2}{1-r^2} = Ke^{2T}, \quad r(T) = \frac{\tilde{K}e^T}{\sqrt{1+Ke^{2T}}},$$

with $\tilde{K}^2 = K$. Notice that for $T \to \infty$ starting from the initial condition inside the unit circle, we have

$$r(T_0) = r(0) = \frac{\tilde{K}}{\sqrt{1+K}} < 1, \quad r(T) \to \frac{\tilde{K}}{\sqrt{K}} = 1. \tag{4.63}$$

To better understand this behavior, let us consider a time interval $T_1 - T_0 = 2\pi$. The phase then returns to its original value $\theta(T_1) = \theta(T_0) + 2\pi \equiv \theta(T_0)$, but the point in the plane moves away from the origin, because its distance from the origin is now larger:

$$r(T_1) = r(2\pi) = \frac{\tilde{K}}{\sqrt{K + e^{-4\pi}}} > r(T_0) = r(0),$$

which is a true inequality, reducing itself to $1 + K > \exp(-4\pi) + K$. Now, the map $P : \mathbf{R} \to \mathbf{R}$ defined by $P(r_k) = r_{k+1}$ is the Poincaré map, which can be constructed on any arbitrary line with a given phase $\theta_*$. Another important remark that we make at this point concerns the unit circle. It is easily seen from the first equation of (4.61) that $r = 1$ is an equilibrium solution of the dynamical system. Therefore, because trajectories cannot be crossed by the existence and uniqueness theorem, the above reasoning with the Poincaré map shows that the unit circle can only be approached from the inside. To examine also the behavior from the outside, since (4.63) implies $r(0) < 1$, we need to rewrite the solution of (4.62) in another form for $r_0 > 1$:

$$r^2(T) = 1 - \frac{1}{1+Ke^{2T}}, \quad r_0^2 = 1 - \frac{1}{1+K}, \quad K = \frac{1}{\frac{1}{r_0^2}-1},$$

so that, finally,

$$r^2(T) = \frac{Ke^{2T}}{1+Ke^{2T}} = \frac{1}{1+e^{-2T}\left(\frac{1}{r_0^2}-1\right)}, \qquad (4.64)$$

$$r(T) = \left[1+e^{-2T}\left(\frac{1}{r_0^2}-1\right)\right]^{-\frac{1}{2}}.$$

Now it is easily seen that the Poincaré map satisfies $P(r_0) > 1$, and also $r_1 = P(r_0) < r_0$. Indeed, the two conditions respectively become

$$r_1 = \left[1+e^{-4\pi}\left(\frac{1}{r_0^2}-1\right)\right]^{-\frac{1}{2}} > 1^{-\frac{1}{2}} \equiv 1; \quad \left[1+e^{-4\pi}\left(\frac{1}{r_0^2}-1\right)\right]^{-1} < r_0^2,$$

$$1+e^{-4\pi}\left(\frac{1}{r_0^2}-1\right) > r_0^{-2}, \quad (1-e^{-4\pi})\left(1-\frac{1}{r_0^2}\right) > 0,$$

which are always verified. One final consideration. While the above computations have been carried out in almost every detail to double check the assertions, a simple consideration could have led to the same conclusions. Indeed, observe that the system rotates in view of the second equation of (4.61), while the right-hand side of the first one is a cubic polynomial $p_3(r)$ with a negative biologically infeasible root $r = -1$, the second one located in the origin and the positive one giving the unit circle as already remarked above, $r = 1$. On plotting it, we find that $p_3(r) > 0$ for $0 < r < 1$, so that the origin must be an unstable equilibrium, while $r = 1$ in polar coordinates is a stable point. But that means, in cartesian coordinates, that the unit circle is stable. Thus we have found a stable limit cycle.

As a final biological interpretation of this result, the equilibrium $(U_*, V_*)$ of the original system is unstable and around it there is a stable limit cycle $(U_* - U)^2 + (V_* - V)^2 = 1$. In these considerations note that the feasibility of both has not been analyzed, but just assumed.

## 4.5 Food web

Natural environments are not confined to contain only two populations, rather in general many species intermingle. Their interactions may be of all the kinds previously discussed, i.e., of competitive nature, or mutually beneficial. It turns out that in some circumstances a population that is a predator of another one is itself the prey of a higher trophic level predator. In this section we describe a very general model that attempts a more complex description of reality. It is meant as a brief introduction to food chains. We address a closed ecosystem, in which trophic levels from a top predator down to the mineral nutrients on which the plants feed are accounted for. The analysis should assess its possible sustainable equilibria and their stability. The mineral substrate and homogeneously mixed populations on three trophic levels are the populations in the environment, representing vegetation, herbivore, and carnivore animals. The homogeneous mixing mathematically implies that the predation terms are taken to be bilinear, i.e., proportional to both interacting populations. This also holds for the interactions among plants and minerals, as the growth of plants depends with direct proportionality on the abundance of the mineral substrate, while the latter is depleted in a way directly proportional to the plants population. It is important to also model the mineral level in a closed system, as it represents the bottom level to which organic matter is recycled. Questions of time scales are disregarded in this presentation.

Let $M$ represent the minerals, $P$ the plants, $H$ the herbivores, and $C$ the carnivores. Keeping on following the general convention of the book that all the parameters are nonnegative unless otherwise stated, the model then is given by

$$\frac{dM}{dT} = m_1 P + m_2 H + m_3 C - b_1 PM, \tag{4.65}$$

$$\frac{dP}{dT} = b_2 PM - m_1 P - c_1 PH,$$

$$\frac{dH}{dT} = c_2 PH - m_2 H - d_1 CH,$$

$$\frac{dC}{dT} = d_2 CH - m_3 C.$$

From the bottom, i.e., the first equation, decaying organic matter is converted into minerals and they in turn are depleted by the plants' uptake at rate $b_1$ for their own growth. The second equation describes this uptake from the plants' viewpoint, at rate $b_2 < b_1$. Plants are also grazed by herbivores at rate $c_1$ and have an intrinsic mortality $m_1$. Similar considerations hold for the herbivores, feeding at rate $c_2$ on plants, and for carnivores feeding on the former at rate $d_2$. Notice that we do not assume logistic growth, in part to keep the mathematics tractable, in view of the nine parameters the model already contains, as well as to investigate the system's behavior assuming it to be *closed*, i.e., when the possible carrying capacities are dictated by the availability of the nutrients at the lower trophic level. Indeed, as the only food source is given by the trophic level just below the one considered, as long as the latter supports the former, the assumption of "unlimited" growth is sound.

To reduce the parameters, let us introduce new dependent variables as follows: $M = \tilde{\alpha} x_1(t)$, $P = \beta x_2(t)$, $H = \gamma x_3(t)$, $C = \tilde{\delta} x_4(t)$, $T = \pi t$, and define the new coefficients as

$$p_1 = \frac{b_2}{b_1}, \quad p_2 = \frac{m_1 c_1}{m_2 b_2}, \quad p_3 = \frac{c_2}{b_1}, \quad p_4 = \frac{m_1 d_1}{m_3 b_2}, \quad (4.66)$$

$$p_5 = \frac{m_2}{m_1}, \quad p_6 = \frac{m_1 d_2}{m_2 b_2}, \quad p_7 = \frac{m_3}{m_1}.$$

The simplified model containing only seven parameters can be stated as follows:

$$\frac{dx_1}{dt} = -x_1 x_2 + p_1 x_2 + x_3 + x_4, \quad \frac{dx_2}{dt} = x_2(x_1 - p_2 x_3 - 1), \quad (4.67)$$

$$\frac{dx_3}{dt} = x_3(p_3 x_2 - p_4 x_4 - p_5), \quad \frac{dx_4}{dt} = x_4(p_6 x_3 - p_7),$$

with its Jacobian matrix

$$V = \begin{pmatrix} -x_2^{(i)} & p_1 - x_1^{(i)} & 1 & 1 \\ x_2^{(i)} & x_1^{(i)} - p_2 x_3^{(i)} - 1 & -p_2 x_2^{(i)} & 0 \\ 0 & p_3 x_3^{(i)} & p_3 x_2^{(i)} - p_4 x_4^{(i)} - p_5 & -p_4 x_3^{(i)} \\ 0 & 0 & p_6 x_4^{(i)} & p_6 x_3^{(i)} - p_7 \end{pmatrix}, \quad (4.68)$$

which we have evaluated at the equilibrium $E_i \equiv (x_1^{(i)}, x_2^{(i)}, x_3^{(i)}, x_4^{(i)})$.

The model (4.67) admits five such feasible equilibria. The first three of these are boundary equilibria, namely: the origin $E_0 = (0,0,0,0)$, the point $E_1 \equiv (x_1^{(1)}, x_2^{(1)}, 0, 0)$ with arbitrary values of $x_1^{(1)}$, $x_2^{(1)}$, an equilibrium which however is very sensitive to environmental fluctuations, since it exists only for $p_1 = 1$, and $E_2 = (x_1^{(2)}, 0, 0, 0)$ also with $x_1^{(2)}$ arbitrary. All of them are always feasible.

We then find another boundary equilibrium

$$E_3 \equiv \left( \frac{p_1 p_2 p_5 - p_3}{p_2 p_5 - p_3}, \frac{p_5}{p_3}, \frac{p_1 p_5 - p_5}{p_2 p_5 - p_3}, 0 \right), \quad (4.69)$$

which is feasible if all its components are nonnegative, i.e., for either of

$$p_1 < 1 < \frac{p_3}{p_2 p_5}, \quad p_1 > 1 > \frac{p_3}{p_2 p_5}, \quad (4.70)$$

conditions that also imply $x_1^{(3)} \geq 0$. Thus (4.70) are necessary and sufficient for the feasibility of $E_3$.

The inner equilibrium is the most important for the sustainability of the whole ecosystem. Let us define $\alpha \equiv (p_2 p_5 - p_3)p_7 - (p_1 - 1)p_5 p_6$ and $\delta \equiv (p_1 - 1)p_4 p_6 - p_2 p_4 p_7 + p_3 p_6$. Then the equilibrium is

$$E_4 \equiv \left( \frac{p_2 p_7 + p_6}{p_6}, \frac{p_5 p_6 - p_4 p_7}{\delta}, \frac{p_7}{p_6}, \frac{\alpha}{\delta} \right). \quad (4.71)$$

The feasibility conditions for $E_4$ require that $x_2^{(4)} \geq 0$, $x_4^{(4)} \geq 0$, as the remaining components are obviously nonnegative. These conditions reduce to either

$$\alpha \geq 0, \quad \delta \geq 0, \quad p_5 p_6 \geq p_4 p_7 \tag{4.72}$$

or

$$\alpha \leq 0, \quad \delta \leq 0, \quad p_5 p_6 \leq p_4 p_7. \tag{4.73}$$

## Model properties

To establish the behavior of the system near each equilibrium point, we need to determine the signs of the eigenvalues of the Jacobian (4.68).

The first equilibrium $E_0$ has one zero eigenvalue $\lambda_1 = 0$, and the other ones negative, $\lambda_2 = -1, \lambda_3 = -p_5$ and $\lambda_4 = -p_7$. The origin is thus a stable equilibrium, neutral along the $x_1$-axis. Also, for $E_2$ there is a vanishing eigenvalue, $\lambda_1 = 0$, the other ones being $\lambda_2 = x_1^{(2)} - 1, \lambda_3 = -p_5$, and $\lambda_4 = -p_7$. Thus, if $\check{x}_1^{(2)} \leq 1$, the equilibrium is stable, else it is unstable.

The first eigenvalue of $E_3$ is $p_6 x_3^{(3)} - p_7$. The signs of the other ones can be obtained from the Routh–Hurwitz conditions applied to the characteristic equation. They give $x_2^{(3)} x_3^{(3)} (p_2 p_3 x_2^{(3)} + 1) > 0$, which is obviously true, and $p_3 x_2^{(3)} x_3^{(3)} (p_2 x_2^{(3)} - 1) > 0$. Thus $E_3$ is stable if and only if the first eigenvalue is negative, i.e., $p_6 x_3^{(3)} < p_7$ and the further condition $p_2 p_5 > p_3$ is satisfied, which falsifies the first equation in (4.70). Hence feasibility must be given by the second condition (4.70), which yields $p_1 > 1$.

Let $N(\lambda) \equiv x_2^{(4)}[\lambda^2(p_1 - x_1^{(4)}) + \lambda p_3 x_3^{(4)} + p_3 p_6 x_3^{(4)} x_4^{(4)}(1 + p_1 - x_1^{(4)})]$ and $D(\lambda) \equiv x_2^{(4)} + \lambda$. The root-finding problem for the characteristic equation at $E_4$ can be restated as the intersection of the two curves

$$K(\lambda) = \lambda^3 + \lambda(p_4 p_6 x_3^{(4)} x_4^{(4)} + p_2 p_3 x_2^{(4)} x_3^{(4)}), \quad R(\lambda) \equiv \frac{N(\lambda)}{D(\lambda)}.$$

Now $K$ has only a real root at the origin and $K'(\lambda) > 0$ for every $\lambda$, so that $K(\lambda) \sim \lambda^3$ as $\lambda \to \infty$. Instead, $R(\lambda) \sim x_2^{(4)}(p_1 - x_1^{(4)})\lambda + p_3 x_2^{(4)} x_3^{(4)}$, and also as $\lambda \to \infty$. The function $R(\lambda)$ has a vertical asymptote located at the zero $-x_2^{(4)}$ of $D(\lambda)$, and two zeros at the roots of $N(\lambda)$:

$$\lambda_\pm \equiv \frac{-p_3 x_3^{(4)} \pm \sqrt{p_3^2 (x_3^{(4)})^2 - 4(p_1 - x_1^{(4)}) p_3 p_6 x_3^{(4)} x_4^{(4)}(1 + p_1 - x_1^{(4)})}}{2(p_1 - x_1^{(4)})}.$$

Now it is easy to verify that $\lambda_\pm < 0$ if $p_1 > x_1^{(4)}$ and $1 + p_1 - x_1^{(4)} > 0$. On the other hand, for $p_1 < x_1^{(4)}$ and $1 + p_1 - x_1^{(4)} > 0$, $\lambda_+ < 0 < \lambda_-$. Finally, for $p_1 < x_1^{(4)}$ and $1 + p_1 - x_1^{(4)} < 0$, $\lambda_\pm > 0$. Also, notice that $R(0) = 1 + p_1 - x_1^{(4)} \equiv p_1 - p_2 x_3^{(4)}$.

From this analysis, there is always an intersection between $K$ and $R$, $K(\lambda^*) = R(\lambda^*)$ occurring for $\lambda^* > 0$, unless $p_1 < x_1^{(4)}$ and $1 + p_1 - x_1^{(4)} < 0$. We have thus determined easily verifiable sufficient instability conditions for $E_4$. The necessary and sufficient conditions for stability are instead given by the following three Routh–Hurwitz conditions:

$$p_6(p_3 + p_1 p_4) < p_4(p_2 p_7 + 1), \quad x_2^{(4)}[p_2 p_7 + p_6(1 - p_1)] > p_3 p_7, \quad (4.74)$$

$$p_6 x_4^{(4)}(x_2^{(4)} - p_4 x_3^{(4)}) + (1 - p_2)[p_3 x_3^{(4)} + x_2^{(4)}(p_1 - x_1^{(4)})] > 0.$$

Observe further that letting

$$A = x_2^{(4)} > 0, \quad B = p_1 - x_1^{(4)} = p_1 - p_2 x_3^{(4)} - 1,$$

$$D = x_3^{(4)}, \quad F = \frac{1}{p_4}[p_6 p_3 x_2^{(4)} - p_5 p_6],$$

the characteristic equation can be written as

$$\varphi(\lambda) = \sum_{i=0}^{4} a_i \lambda^i, \quad a_4 = 1, \quad m_3 = x_2^{(4)}, \quad m_2 = -AB + p_2 p_3 AD + p_4 DF,$$

$$m_1 = -p_3 AD + p_2 p_3 A^2 D + p_4 ADF, \quad a_0 = -ADF(p_3 + p_4 B).$$

Since $\varphi$ is a quartic, $\varphi(\lambda) \to +\infty$ as $\lambda \to \pm\infty$. Its inflection points $\varphi''(\lambda_\pm) = 0$ are real if $\Delta \equiv 9m_3^2 - 24a_4 > 0$, in which case it is easily seen that they lie in the negative left half-plane, as their abscissae are

$$\lambda_\pm = \frac{1}{12}[-3m_3 \pm \sqrt{\Delta}] \leq 0. \quad (4.75)$$

In such a case,

$$a_0 = \varphi(0) > 0, \quad m_1 = \varphi'(0) > 0 \quad (4.76)$$

are sufficient conditions for stability, ensuring that no positive eigenvalue exists. For $\Delta < 0$ instead, no inflection point exists, but again satisfying both (4.76) ensures once more the stability of $E_4$.

## Results and discussion

The equilibrium $E_4$ can also be investigated numerically, to find features that are analytically harder to show. Indeed for the parameters, $p_1 = 1.84$ (Figure 4.2) and $p_1 = 1.89$ (Figure 4.3), $p_2 = 0.6$, $p_3 = 0.4$, $p_4 = 4.25$, $p_5 = 0.9$, $p_6 = 0.555$, $p_7 = 0.9$ exhibit limit cycles. We then numerically try to modify the parameters one at a time to investigate the effects of this change. Changes toward smaller values of $p_1$ are generally reflected in oscillations of smaller amplitude around a smaller reference value. The value $p_1 = 0.5$ brings a collapse of the limit cycle first toward the corresponding underlying equilibrium value $E_4$, then a further decrease of $p_1$ leads to the equilibrium

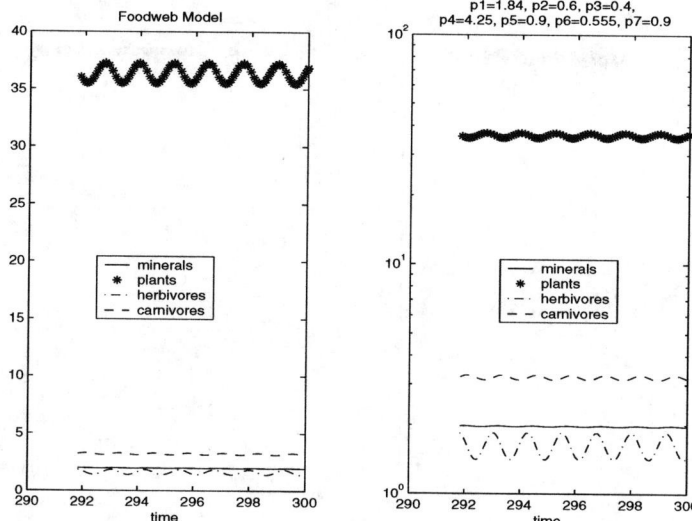

**FIGURE 4.2**: Persistent oscillations around the inner equilibrium $E_4$ shown in standard and semilogarithmic plots.

$E_3$ where only one population is wiped out, and then finally induces a further change also of this other equilibrium to the equilibrium $E_2$.

Changes toward higher values of $p_2$ damp out the oscillations so as to get back to $E_4$. Smaller values instead lead one of the species to grow toward very high values while still oscillating. Increasing $p_7$ instead gives us again the stable equilibrium $E_3$.

Decreasing $p_3$ from the reference value produces limit cycles, again around lower values of the equilibrium, and the smaller the value of $p_3$, the lower the equilibrium is. Increasing it instead to $p_3 = 1$, the oscillations disappear and at least two trophic levels prevail.

The parameters $p_4$ and $p_5$ finally affect much less the model behavior, although a bifurcation value at $p_5 = 8.33122$ exists, this time from the reference values $p_1 = 1.86$, $p_2 = 0.6$, $p_3 = 0.415$, $p_4 = 4.25$, $p_5 = 0.555$, $p_6 = 0.9$. Also from these latter values a bifurcation value $p_6 = 0.807$ is found driving the system back to $E_3$.

In summary, the results obtained are meaningful. Indeed, the neutral stability of the origin is easily explained since an empty closed system "by definition" remains empty as long as migrations into it are not allowed. At $E_2$ the ecosystem is lifeless and stable at an arbitrary level $x_1^{(2)}$. This is possible, however, only if the amount of the matter is below a threshold level. Thus the nutrients need to be at a minimum level, to avoid collapse of the plants and eventually of the ecosystem as a whole. An equilibrium environment is shown to exist, in which plants and herbivores can coexist in absence of carnivores, by the range of parameters leading to a stable $E_3$, namely, if $p_1 > 1$.

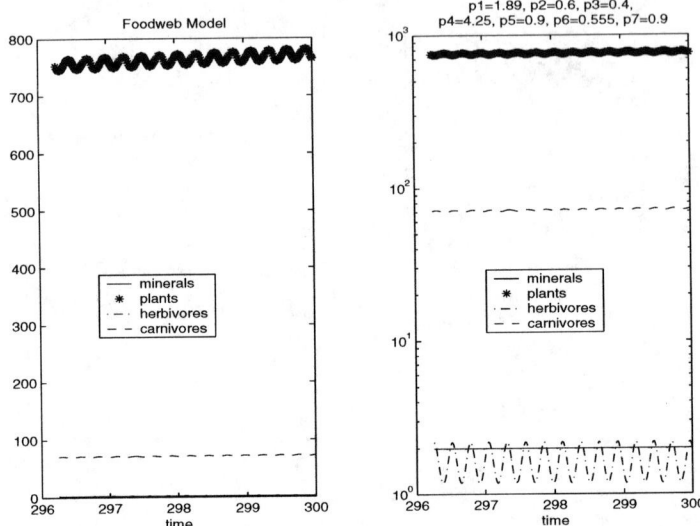

**FIGURE 4.3**: Change in the oscillations around the inner equilibrium $E_4$ due to a change in $p_1$, the cycles become wider, around a higher reference value; note that the vertical scale differs from the one in Figure 4.2.

The point where the whole ecosystem thrives is $E_4$. We have found conditions on its parameters ensuring feasibility, (4.72) or (4.73), necessary and sufficient stability conditions (4.74), and easier to verify sufficient conditions (4.76). However, the numerical simulations show that when $E_4$ is unstable, the model first exhibits sustained oscillations around this equilibrium and then trajectories may collapse into one of the former equilibria. When the dynamics is forced toward boundary equilibria, some of the populations disappear; in biological terms, we have loss of biodiversity. The ultimate failure of the system is thus represented by $E_1$, i.e., disappearance of life in some or all of its forms. This may very well be provoked by tampering with the model parameters by external factors, which may be due to "natural" causes, as wheather, sunlight, and so on, or man-made.

Although this study is far too elementary to draw any serious forecasts on the global environment, it nevertheless shows that sustained oscillations in the ecosystem are feasible and compatible with human activities, provided the latter do not induce too drastic changes in the parameters. A simple model like the one presented here thus shows that it is very important to assess our present stand, to understand whether the changes nowadays occurring on a worldwide scale are going to be permanent or are only part of a temporary and thus still reversible process.

## 4.6 More about chaos

In the previous sections, we have been dealing mostly with either a system's relaxation to a stable steady state or with periodic solutions due to a stable limit cycle. In a time-continuous population community with the number of species more than two, however, there can be a more complicated and more exotic type of dynamics which is known as (deterministic) chaos.

As an instructive example, following the work by Hastings and Powell (1991), let us consider a simple food web (or, rather, a food chain) that includes three trophic levels, i.e., a prey, its predator, and a top predator:

$$\frac{du}{dt} = u(1-u) - \frac{a_1 uv}{1 + b_1 u} , \qquad (4.77)$$

$$\frac{dv}{dt} = \frac{a_1 uv}{1 + b_1 u} - \frac{a_2 vw}{1 + b_2 v} - d_1 v ,$$

$$\frac{dw}{dt} = \frac{a_2 vw}{1 + b_2 v} - d_2 w$$

(in dimensionless variables), where $u$, $v$, and $w$ are the densities of prey, predator, and top predator, respectively, and $a_{1,2}$, $b_{1,2}$ and $d_{1,2}$ are parameters.

The system (4.77) is difficult to study analytically, but it is more or less straightforward to reveal its main properties by means of numerical simulations. A problem arising on this way is that it depends on as many as six parameters. Therefore, a comprehensive study would require thousands of simulation runs accomplished for different parameter sets. Alternatively, however, since our goal here is to give an example of chaotic dynamics rather than to fulfil a detailed study of the system properties, we can fix most of the parameters at a certain hypothetical value and study the system's behavior as a response to variations in the remaining "controlling parameter."

It is well-known (e.g., see Murdoch and Oaten, 1975, and also Section 10.1 of Chapter 10) that stability of a prey–predator system with Holling type II response of predator essentially depends on the half-saturation constant, and hence on parameter $b_1$. We can thus assume that $b_1$ is likely to control the properties of the tritrophic system (4.77) as well and choose it as the controlling parameter.

Numerical simulations show that for small values of $b_1$ the system normally possesses a stable steady state. With an increase in $b_1$, its stability is eventually broken through the Hopf bifurcation and a stable limit cycle appears; see the left-hand panel of Figure 4.4. With a further increase in $b_1$, the system undergoes a series of period-doubling bifurcations and the shape of the limit cycle becomes more and more complicated (cf. Figure 4.4b) until it turns into a strangely looking attractor that no longer bears any resemblance to a limit cycle; see Figure 4.5. Correspondingly, the oscillations of the population densities versus time become irregular, too.

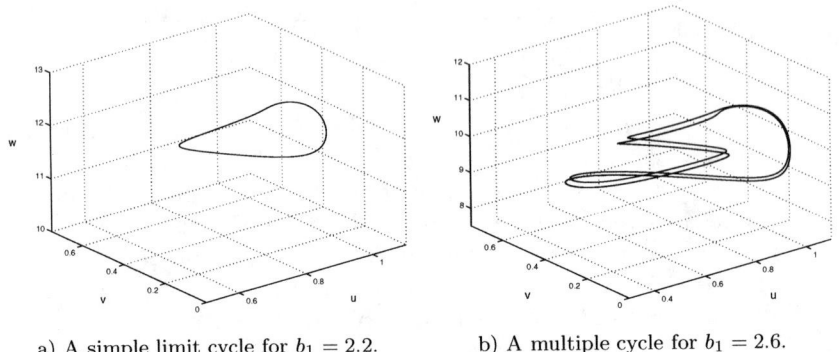

a) A simple limit cycle for $b_1 = 2.2$.    b) A multiple cycle for $b_1 = 2.6$.

**FIGURE 4.4**: Dynamics of system (4.77); see details in the text. Parameters are $a_1 = 5.0$, $a_2 = 0.1$, $b_2 = 2.0$, $d_1 = 0.4$, and $d_2 = 0.0086$.

A visually different shape of attractor, however, is surely not enough to conclude that the system dynamics has acquired any qualitatively new features. Indeed, theoretically speaking, the attractor shown in Figure 4.5 could still be a multiple cycle. What actually distinguishes chaos from a multiperiodic dynamics, however complicated the latter may be, is a different response to variations of the initial conditions. While a (multi)periodic limit cycle is stable with regards to such variations, a fundamental property of chaotic dynamics is a special type of sensitivity so that distance $d$ between any two initially closed trajectories grows with time exponentially. This property is usually quantified by the dominant Lyapunov exponent:

$$\lambda_D = \lim_{d(0) \to 0,\ t \to \infty} \frac{1}{t} \log \left[ \frac{d(t)}{d(0)} \right] \quad (4.78)$$

(cf. Nayfeh and Balachandran, 1995).

It is very difficult to prove chaotic behavior analytically, but it is relatively easy to demonstrate it numerically. Figure 4.6 shows the density of the top predator versus time calculated for two very close sets of initial conditions; $u_1(0) = u_2(0)$, $v_1(0) = v_2(0)$ and $w_1(0) - w_2(0) = 0.001$. It is readily seen that, while at an earlier stage of the system dynamics the trajectories remain close, the discrepancy between the solutions grows steadily so that for $t \simeq 2000$ it becomes on the same order as that of the solutions themselves. The corresponding value of $\lambda_D$ can then be roughly estimated as 0.01.

Runaways of system trajectories due to the system's sensitivity to the initial conditions make intermediate-term prediction of the system state impossible, and that may have profound implications. This is one of the reasons why the issue of chaos – in particular, in population dynamics and ecology – has been attracting a lot of attention over the last three decades; e.g., see May (1974),

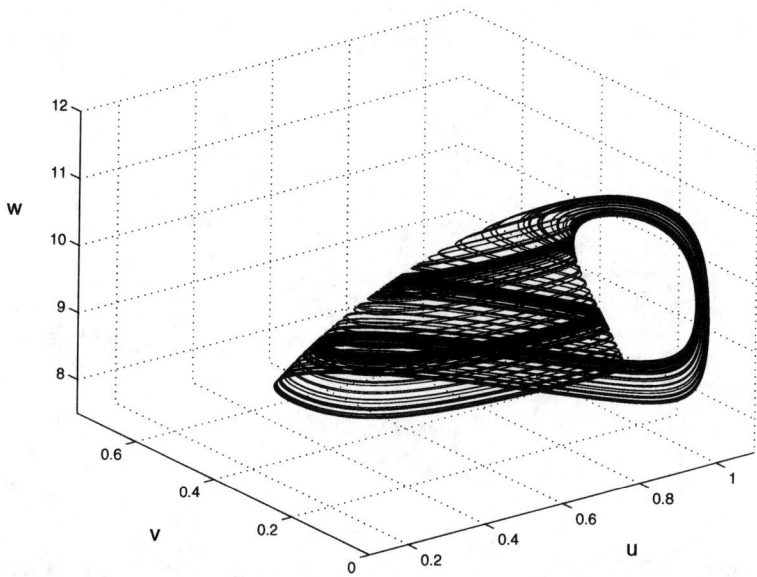

**FIGURE 4.5**: Chaotic dynamics of system (4.77) observed for $b_1 = 2.7$; other parameters are the same as in Figure 4.4. The shown trajectory is calculated for $5000 \le t \le 10000$; the trajectory for $0 < t < 5000$ was omitted in order to exclude the impact of transients.

Berryman and Millstein (1989), Scheffer (1991b), Hastings et al. (1993), and Cushing et al. (2003).

Another example of a population dynamics model exhibiting chaos is given by a Lotka–Volterra-type system with one predator and two preys (cf. Gilpin, 1979; Arneodo et al., 1980):

$$\frac{du_1}{dt} = (1 - r_{11}u_1 - r_{12}u_2 - r_{13}v)\,u_1\,, \qquad (4.79)$$

$$\frac{du_2}{dt} = (1 - r_{21}u_1 - r_{22}u_2 - r_{23}v)\,u_2\,,$$

$$\frac{dv}{dt} = (-1 + r_{31}u_1 + r_{32}u_2 - r_{33}v)\,v\,,$$

where $u_1$, $u_2$ and $v$ are the (dimensionless) densities of prey 1, prey 2, and predator, respectively, and all coefficients $r_{ij}$ are assumed to be nonnegative. Having chosen the predation rate $r_{31}$ as a controlling parameter and other parameters fixed at some hypothetical values, Schaffer and Kot (1986) showed that the system (4.79) exhibits onset of chaos through the period-doubling bifurcation scenario when $r_{31}$ increases to a certain critical value.

A further discussion of prerequisites and implications of chaos in "multi-

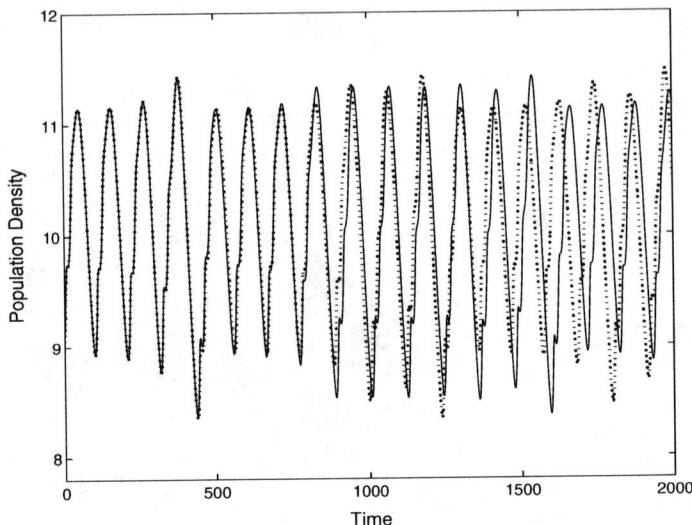

**FIGURE 4.6**: (See color insert.) Top predator density vs time: there is a growing discrepancy between two initially close trajectories ($w_1(0) - w_2(0) = 0.001$) in the chaotic regime of system (4.77). Parameters are the same as in Figure 4.5.

species systems"[1] can be found in Rinaldi and Feo (1999).

Note that, when looking for chaotic dynamics in a time-continuous model described by a system of ordinary differential equations, the condition that the number of interacting species must not be less than three is essential. Indeed, the solution uniqueness theorem, which is valid in a general case of $n$ species system (2.1) under some unrestrictive and biologically reasonable assumptions about the properties of the interaction terms $F_i$, obviously rules out a possibility of trajectories intersection. In the case of two-species system, where the phase space is reduced to a plane, it thus excludes existence of an attractor more complicated than a limit cycle. The situation is thus very much different from time-discrete systems where chaos can occur even in a single species model; see Section 3.4.

It should be mentioned here that, although there are many theoretical arguments showing that chaos should be a common phenomenon in population dynamics (and chaos has indeed been observed for at least two different laboratory populations; see Costantino et al., 1997; Becks et al., 2005), a conclusive evidence of chaos in ecological data is still lacking. This disagreement between theory and available data has been a subject of continuing intensive discussion and controversy; the two reasons that are usually mentioned as a

---

[1] In this context, systems with the number of species greater than two.

probable explanation why chaos is not immediately seen in ecological dynamics are the impact of noise and an insufficient length of ecological time series. The latter also raises a more fundamental concern that chaos in ecosystems dynamics cannot be seen at all (even if it may be an intrinsic property of a given system) because of the impact of transients (Hastings, 2001, 2004). As a whole, however, the question of whether mother nature is chaotic and if not, why not (cf. Berryman and Millstein, 1989; Hastings et al., 1993) yet remains largely open.

## 4.7  Age-dependent populations

This is probably the most theoretical section of the whole book. We start by illustrating a simple example in some of its fine details to describe the analytical technique used. The purpose of the later sections is to present the well-known contraction mapping principle on a nontrivial, rather sophisticated application. We have already encountered fixed points in the earlier chapter on a single discrete population. In this one the continuous case is discussed, allowing interactions among two populations. Excellent and thorough references on this topic for the interested reader are the books by Cushing (1998), Iannelli (1995), and Webb (1985).

Lotka (1956) and von Foerster (1959) in population theory and McKendrick (1926) in the context of epidemics introduced an age-dependent description of the population in which instead of the total population size as a function of time, they considered as the dependent variable a density function, which gives at every instant the number of individuals in the same age group, technically called a cohort, and looks at its evolution in time. The model is formulated as a hyperbolic partial differential equation of the first order.

Leslie (1945; 1948) considered a discrete system, in which each component of a vector describes the cohort in a certain age interval, while the evolution of the population is described by a matrix, the Leslie matrix, whose entries contain the mortality of each cohort.

We will consider only the continuous-time version, and describe now how to obtain the governing equations.

If $u(a, T)$ represents the density of the population, the total population is in this case provided by summing the density over all possible ages. Technically, two possibilities arise, namely, to assume that there is a maximum allowed age $A$, so that $u(a, T) \equiv 0$ for all $a \geq A$ at all times $T$, or simply to allow all possible ages up to $\infty$, in which case mathematically we need to assume the maternity to have compact support, so as to be able to meaningfully define the integral that follows. This second possibility will be used later in Section 6.3.

Thus

$$U(T) = \int_0^A u(a,T)\,da. \tag{4.80}$$

The above equation represents then the relationship of this model with the previous ones, which are independent of age.

If we now let a time $\delta T$ go by, we find that all individuals will be aged $a+\delta T$, apart from those who in the meantime passed away. This remark shows that the differentials of age and time are the same, $dT \equiv da$, or, integrating, $a = T - T_0$, where $T_0$ denotes the "birhdate" of the cohort. Thus, introducing the individual mortality rate $m(a,T)$, we find the balance equation:

$$\frac{u(a+\delta T, T+\delta T) - u(a,T)}{\delta T} \equiv \frac{u(a+\delta T, T+\delta T) - u(a+\delta T, T)}{\delta T}$$
$$+ \frac{u(a+\delta T, T) - u(a,T)}{\delta T} = m(a,T)u(a,T). \tag{4.81}$$

In general, the mortality function is independent of time, or better, it depends on time implicitly via the density $u(a,T)$ itself, or rather said, through the population $m \equiv m(a, U(T))$. As a first approach, however, for the next considerations, let us take it to be only age dependent, namely, $m \equiv m(a)$. Letting $\delta T \to 0$, we find the basic equation governing the age-dependent population models:

$$\frac{\partial u(a,T)}{\partial T} + \frac{\partial u(a,T)}{\partial a} + m(a)u(a,T) = 0. \tag{4.82}$$

The model is still incomplete, as an initial condition is needed, which in this case is provided by an initial density $u(a,0) = \phi(a)$, but above all because we need to describe the process by which new individuals are added to the population. To this end, let us introduce the maternity function $\beta(a,T) \equiv \beta(a, U(T))$, again implicitly a function of time via the population size. It represents the reproduction of the cohort in the population aged $a$ at time $T$. Now the newborns can be calculated by summing the contributions of each individual reproduction function over all possible ages. The newborns $B(T)$ at time $T$ are thus given by

$$B(T) = \int_0^A \beta(a, U(T))u(a,T)\,da. \tag{4.83}$$

The model thus formulated is an initial and boundary value problem, in the sense that the "boundary" is expressed by the fact that age can never be negative, and the new individuals coming into the population are the newborns of age 0; they are accounted for by the boundary term (4.83).

The characteristics of (4.82) are straight lines, as $dT = da$, which in parametric form are written $T = s + T_0$, $a = s$. Notice that the tangent unit vector to these lines is $\mathbf{t} = \frac{1}{\sqrt{2}}(1,1)^T$. Remembering the meanings of age and time differentials, observe that

$$ds^2 = dT^2 + da^2, \quad ds = \sqrt{2}da. \tag{4.84}$$

Consider $\hat{u}(s) \equiv u(s, s + T_0)$, i.e., the density along the characteristics. In general, the partial differential operator on the left of (4.82) represents the derivative of the restriction of the function $u$ to the caracterists,

$$Du(s,T) \equiv \lim_{\delta a \to 0} \frac{u(s+\delta a, T+\delta a) - u(s,T)}{\delta a} = \frac{d\tilde{u}(s)}{ds}. \tag{4.85}$$

Notice on the other hand that the partial differential operator on the left-hand side of (4.82) is related to the directional derivative of $u$ along the characteristics $D_t u$. Indeed, since

$$Du = \frac{\partial u}{\partial T} + \frac{\partial u}{\partial a} = \nabla u \cdot (1,1)^T = \sqrt{2}\nabla u \cdot \mathbf{t} \equiv \sqrt{2}D_t u. \tag{4.86}$$

Equations (4.86) and (4.85) are related, representing the same concept, the difference being the use of the arclength.

To better motivate the use of the latter, let us ignore dependence on time in the mortality function, which is then only dependent on age. Thus $u(a,T)$ will be subject to $m(a)$, and $u(a+\delta a, T+\delta T)$ will be subject to $m(a+\delta T) = m(a+\delta a)$. But letting $\delta T$ go by, from $u(a_0, t_0) = u(s_0)$ measured along one of the characteristics, subject to the mortality $m(a_0)$ we arrive at $u(a_0+\delta a, T_0+\delta T) = u(s_0+\delta s)$. Recalling (4.84), the density at the point $(a_0+\delta a, T_0+\delta T)$ would be subject to the mortality $m(a_0+\sqrt{2}\delta a)$ rather than to the correct one $m(a_0+\delta a)$. This consideration tells us that along the characteristics we need to use as a parameter the arclength, i.e., the normalized $s$ via the factor $\sqrt{2}$, namely, $\sigma = \frac{1}{\sqrt{2}}s$.

With this position, we can then describe the evolution of the density $u$ as follows:

$$[u(\frac{s+\delta s}{\sqrt{2}}) - u(\frac{s}{\sqrt{2}})]\frac{\sqrt{2}}{\delta s} = -m(\frac{s}{\sqrt{2}})u(\frac{s}{\sqrt{2}}).$$

Upon taking the limit as $\delta\sigma \to 0$,

$$\frac{du}{d\sigma}(\sigma) = m(\sigma)u(\sigma), \quad \frac{du}{u} = m(\sigma)d\sigma,$$

and integration setting $s = \sqrt{2}\alpha$ then finally gives

$$\ln u(\bar{\sigma}) - \ln u(0) = -\int_0^{\bar{\sigma}} m(\alpha)d\alpha, \quad u(\bar{\sigma}) = u(0)\exp\left(-\int_0^{\bar{s}} m(\alpha)d\alpha\right).$$

This is the process known as integration along the characteristics of the first order partial differential equation; for a geometric interpretation see Zachmanoglou and Thoe (1976).

We now need to distinguish two cases. When the characteristic meets the initial times, i.e., $T_0 > 0$, notice that when we start counting time the cohort is already aged a. In such a case, the value of $u(s)$ is provided by the value of the initial distribution, $\psi(a)$, so that $\tilde{u}(0) = u(a,0) = \psi(a)$. For $T_0 < 0$, instead, when $T_0$ denotes the birthtime of the cohort, $\tilde{u}(0) = u(0,T_0) = B(T_0)$, and the latter is determined from (4.83). Thus, for $a > T$ and $a < T$, we respectively have

$$u(a,T) = \psi(a-T)\exp\left[-\int_0^{a-T} m(\alpha)d\alpha\right], \qquad (4.87)$$

$$u(a,T) = B(T-a)\exp\left[-\int_0^{T-a} m(\alpha)d\alpha\right]. \qquad (4.88)$$

Upon substitution into the population equation, we respectively find for $\xi = a - T$ and $\eta = T - a$,

$$U(T) = \int_T^A u(a,T)da + \int_0^T u(a,T)da$$
$$= \int_0^{A-T} u(\xi,T)d\xi + \int_0^T u(a,T)da.$$

The problem is then finally completely described by the pair of integral equations (4.83) and (4.89):

$$U(T) = \int_0^{A-T} \psi(\xi)\exp\left(-\int_0^{\xi} m(\alpha)d\alpha\right)d\xi \qquad (4.89)$$
$$+ \int_0^T B(\eta)\exp\left(-\int_0^{\eta} m(\alpha)d\alpha\right)d\eta.$$

By instead multiplying by the maternity function (4.87) and using again (4.83), with the very same steps we find

$$B(T) = \int_0^{A-T} \beta(\xi, U(T))u(\xi,T)d\xi + \int_0^T \beta(a, U(T))u(a,T)da, \quad (4.90)$$

which leads to the single integral equation

$$B(T) = \int_0^{A-T} \psi(\xi)\beta(\xi, U(T))\exp\left(-\int_0^{\xi} m(\alpha)d\alpha\right)d\xi \qquad (4.91)$$
$$+ \int_0^T B(\eta)\beta(T+\eta, U(T))\exp\left(-\int_0^{\eta} m(\alpha)d\alpha\right)d\eta.$$

It is also possible to introduce survival probabilities and relate this equation to the renewal equation of probability theory; for details see Cushing (1998).

## 4.7.1 Prey–predator, age-dependent populations

Let the nonnegative densities for the predator and prey be denoted by $v(a,T)$ and $u(a,T)$, assumed to have compact support with respect to age at every time. Then the populations are easily found by integration:

$$U(T) = \int_0^\infty u(a,T)\,da, \qquad V(T) = \int_0^\infty v(a,T)\,da. \qquad (4.92)$$

The growth of each population would obey (4.82), in the absence of the other one, with $\mu$ and $\nu$ representing the mortality functions for each species. The mortality function should in general be higher for young individuals, then decreasing to a minimum at some intermediate age, and finally increasing steadily as the individual becomes older.

To describe mutual relations, an extra mortality term has to be added to the equation of the prey. The "reward" that the predators obtain is in general assumed to be new population members, i.e., it should be accounted for in the newborn term. With respect to the classical age-independent prey–predator models, we replace here the proportionality constants in the interaction terms by functions of age. This allows us to distinguish encounters of a particular age group within one population and the individuals of the other species, considered as a whole:

$$\frac{\partial u}{\partial a} + \frac{\partial u}{\partial t} + \mu(a,U)u + \alpha_0(a)uV = 0, \qquad (4.93)$$

$$\frac{\partial v}{\partial a} + \frac{\partial v}{\partial t} + \nu(a,V)v = 0.$$

Letting $\beta_u$ and $\beta_v$ represent the maternity functions for each species, the birth rates are then given by

$$B_u(T) = u(0,T) = \int_0^\infty \beta_u(a,U(T))u(a,T)\,da, \qquad (4.94)$$

$$B_v(T) = v(0,T) = \int_0^\infty [\beta_v(a,V(T)) + \gamma_0(a)U(T)]v(a,T)\,da.$$

Here $\alpha_0(a)$ expresses the hunting function on a prey aged $a$, while $\gamma_0(a)$ is the return predators aged $a$ get from hunting. Since young and old individuals have less defenses, $\alpha_0(a)$ could be modeled with a minimum at some maturity age $M$ and then tending asymptotically to a limiting value, while $\gamma_0(a)$, since it represents the hunting skills of predators, should have a maximum at some intermediate age.

Suitable initial conditions are as follows:

$$u(a,0) = g(a), \quad v(a,0) = \eta(a). \qquad (4.95)$$

For mathematical tractability, as functions of $U$ and $V$, all the mortalities and maternities are assumed to have continuous partial derivatives and together with them to be bounded functions. Instead of pursuing analytically this model, we will now instead formulate and investigate a more general one.

## 4.7.2 More about age-dependent populations

We extend the model formulated in the previous section by considering an environment in which several population are present and interact, accounting for species that either live in symbiosis or compete for common resources, or such that some species are hunted by others. The approach follows closely Gurtin and McCamy (1974), and the results are related to those obtained via nonlinear semigroups in Webb (1985). Here we also allow nonlinear interactions, by letting both the maternity and mortality functions for each species be influenced by all the other species. The necessary assumption will of course be that these modified mortalities retain positive signs so that they will never generate newborns of some positive age. The modifications increase or decrease each species density according to the type of interaction with other species. Similar remarks hold for the maternities as well, which are always positive, i.e., never generating a negative number of newborns.

Let $x_i(a,t)$, $i = 1, ..., N$ represent the densities of the $N$ populations, assumed to be of compact support with respect to age at any given time $t$. This ensures that the population sizes are well defined:

$$X_i(t) = \int_0^\infty x_i(a,t)\,da. \tag{4.96}$$

Letting $\mathbf{X} \equiv (X_1, ..., X_N)^T$, the birth rates can then be defined as

$$B_i(t) = x_i(0,t) = \int_0^\infty \beta_i(a, \mathbf{X}(t))\,x_i(a,t)\,da. \tag{4.97}$$

Here $\beta_i$ represents the maternity function for the $i$th species. The model then becomes

$$\frac{\partial x_i}{\partial t} + \frac{\partial x_i}{\partial a} + \mu_i(a, \mathbf{X}(t))\,x_i(a,t) = 0, \tag{4.98}$$

with initial conditions

$$x_i(a,0) = \phi_i(a). \tag{4.99}$$

The maternity and mortality functions are assumed to be continuous and nonnegative on $\mathbf{R}^+ \times (\mathbf{R}^+)^N$. Moreover, for every multiindex $\alpha$, $0 \leq |\alpha| \leq P$ for some $P$, the maternity and mortality are bounded above, as functions of $\mathbf{X} \in (\mathbf{R}^+)^N$, together with their mixed partial derivatives:

$$\frac{\partial^\alpha \beta_i}{\partial X^\alpha}, \frac{\partial^\alpha \mu_i}{\partial X^\alpha}, \quad i = 1, ..., N.$$

As functions of age, for the maternity and mortality functions, the considerations of the previous section hold. Thus the following quantities are well defined:

$$\mu^0 = \sup\left(\mu_i(a, \mathbf{X}) : (a, \mathbf{X}) \in \Omega; \ i = 1, ..., N\right); \quad (4.100)$$

$$\mu' = \sup\left(\frac{\partial \mu_i}{\partial X_j}(a, \mathbf{X}) : (a, \mathbf{X}) \in \Omega; \ i, j = 1, ..., N\right); \quad (4.101)$$

$$\beta^0 = \sup\left(\beta_i(a, \mathbf{X}) : (a, \mathbf{X}) \in \Omega; \ i = 1, ..., N\right); \quad (4.102)$$

$$\beta' = \sup\left(\frac{\partial \beta_i}{\partial X_j}(a, \mathbf{X}) : (a, \mathbf{X}) \in \Omega; \ i, j = 1, ..., N\right). \quad (4.103)$$

Moreover, assume

$$b^* = \sup\left\{\beta_i(a, \mathbf{X}) : a \geq 0, X_j \geq 0, \ i, j = 1, ..., N\right\} \quad (4.104)$$

$$m^* = \inf\left\{\mu_i(a, \mathbf{X}) : a \geq 0, X_j \geq 0, \ i, j = 1, ..., N\right\}. \quad (4.105)$$

## Integral reformulation – local existence

For $i = 1, ..., N$, let $x_i$ denote a solution of the problem up to some time $T > 0$. Let us then set

$$\tilde{x}_i(h) = x_i\left(a^0 + h, t^0 + h\right), \quad \tilde{X}_i = X_i\left(t^0 + h\right), \quad \tilde{\mu}_i(h) = \mu_i\left(a^0 + h, \tilde{\mathbf{X}}(h)\right).$$

As it was shown at the beginning of this section, since

$$\frac{d\tilde{x}_i(h)}{dh} = \frac{\partial x_i}{\partial a} + \frac{\partial x_i}{\partial t},$$

the model becomes the following system of ordinary differential equations:

$$\frac{dx_i(h)}{dh} + \tilde{\mu}_i(h)\tilde{\mathbf{x}}_i(h) = 0.$$

Integrating and rephrasing,

$$\tilde{x}_i(h) = \tilde{x}_i(0)\exp\left(-\int_0^h \tilde{\mu}_i(s)\,ds\right),$$

$$x_i\left(a^0 + h, t^0 + h\right) = x_i\left(a^0, t^0\right)\exp\left(-\int_0^h \tilde{\mu}_i(s)\,ds\right).$$

For $a \geq t$, choose $a^0 = a - t$, $t^0 = 0$, $h = t$, and for $a \leq t$, let $a^0 = 0$, $t^0 = t - a$, $h = a$, to find the densities in terms of the initial and boundary conditions

$$x_i(a, t) = \phi_i(a - t)\exp\left(-\int_0^t \mu_i(a - t + s, \mathbf{X}(s))\,ds\right) \quad (4.106)$$

for $a \geq t$, and

$$x_i(a, t) = B_i(t - a)\exp\left(-\int_0^a \mu_i(s, \mathbf{X}(t - a + s))\,ds\right) \quad (4.107)$$

for $a \leq t$.

Let

$$K_i(\alpha, t, \mathbf{X}) = \exp\left(-\int_{t-a}^{t} \mu_i(\tau + t - \alpha, \mathbf{X}(\tau))\, d\tau\right),$$

$$L_i(v, t, \mathbf{X}) = \exp\left(-\int_{0}^{t} \mu_i(v + \tau, \mathbf{X}(\tau))\, d\tau\right).$$

Integrating (4.106) and (4.107) gives the total population sizes and repeating the operation on the same formulas multiplied by the maternities provides the total birth rates. Mathematically, then,

$$X_i(t) = \int_0^t B_i(u) K_i(t-u, t, \mathbf{X})\, du \qquad (4.108)$$

$$+ \int_0^\infty \phi_i(v) L_i(v, t, \mathbf{X})\, dv,$$

$$B_i(t) = \int_0^t B_i(u) \beta_i(t-u, \mathbf{X}) K_i(t-u, t, \mathbf{X})\, du \qquad (4.109)$$

$$+ \int_0^\infty \phi_i(v) \beta_i(t+v, \mathbf{X}) L_i(v, t, \mathbf{X})\, dv.$$

Thus our model is reduced to the system of integral equations (4.108)–(4.109).

Given instead $X_i(t)$ and $B_i(t)$ continuous and positive, satisfying equations (4.108)–(4.109) on $[0, T]$, $x_i(a, t)$ can be defined on $\mathbf{R}^+ \times [0, T]$ following (4.96). Then $x_i(a, t)$ is nonnegative since $\phi$ and $\beta$ are; also, $x_i(0, t) = B_i(t)$ and $x_i(a, 0) = \phi_i(a)$. Moreover, $x_i(a, t) \in L_1(\mathbf{R}^+)$ since $\mu_i$, $B_i$ and $X_i$ are continuous and the initial condition belongs to $L_1(\mathbf{R}^+)$. Thus (4.96) and (4.97) hold, by using the definition of $x_i(a, t)$ and (4.108)–(4.109). Finally, from the definition of $x_i(a, t)$, the partial derivatives exist so that (4.98) is satisfied.

Thus the original model is equivalent to the system of Volterra integral equations (4.108)–(4.109).

## Application of the fixed point method to local existence

A nontrivial application of an important result of functional analysis, the contraction mapping principle, or fixed point method, provides the local existence result. Let us fix once again $T > 0$ and define

$$\mathcal{C}^+[0, T] = \{f \in \mathcal{C}[0, T] : f \geq 0\}.$$

Formally solving the linear uncoupled part of (4.108)–(4.109) we get the birth rates $B_i$:

$$B_i(t) = (\mathcal{B}_i(T, \mathbf{X})(t)).$$

Here the operator $\mathcal{B}(T, \mathbf{X})$ depends on $\mathbf{X}$. Define $\mathcal{X}(T, \mathbf{X})$ by substituting the $i$th component of $\mathcal{B}(T, \mathbf{X})$ into the right-hand side of the nonlinear part of the system,

$$(\mathcal{X}_i(T, \mathbf{X})(t)) = \int_0^t (\mathcal{B}_i(T, \mathbf{X})(u)) K_i(t-u, t, \mathbf{X}) du$$
$$+ \int_0^\infty \phi_i(v) L_i(v, t, \mathbf{X}) dv.$$

Now $\mathcal{B}(T, \mathbf{X})$ and $\mathcal{X}(T, \mathbf{X})$ map the Banach space $(\mathcal{C}^+[0, T])^N$ equipped with the supremum norm $\|\cdot\|_T$ onto itself. To get a solution of (4.108)–(4.109) we need to show that they have a fixed point, i.e., that they are contractions. Let $r > 0$ be given and define

$$\Phi = \left[ \int_0^\infty \phi_1(a) da, ..., \int_0^\infty \phi_N(a) da \right],$$
$$\Sigma_T = \left\{ f : f \in (\mathcal{C}^+[0,T])^N \to \mathbf{R}^N, \|f - \Phi\|_T < r \right\}.$$

We now prove that $\mathcal{X}(T, \mathbf{X})$ maps $\Sigma_T$ onto itself and is a contraction.

For the first step consider the strip in $\mathbf{R}^+ \times \mathbf{R}^+$ given by

$$\Omega = \{(a, \mathbf{X}) : a \geq 0, X_k \geq 0, \|\mathbf{X} - \Phi\| \leq r, \ k = 1, ..., N\}.$$

Let $\mathbf{X} \in \Omega$. Recalling (4.100) from (4.109), it follows,

$$(\mathcal{B}_i(T, \mathbf{X})(t)) \leq \beta^0 \int_0^t (\mathcal{B}_i(T, \mathbf{X})(u)) du + \beta^0 \Phi_i, \qquad (4.110)$$

and then Gronwall's inequality gives

$$(\mathcal{B}_i(T, \mathbf{X})(t)) \leq \beta^0 \Phi_i \exp(\beta^0 t).$$

Moreover, because

$$\sup |L_i(v, t, \mathbf{X}) - 1| \leq \mu^0 T \exp(\mu^0 T),$$

for $T$ small enough we have

$$|(\mathcal{X}_i(T, \mathbf{X})(t)) - \phi_i| \leq \Phi_i \beta^0 T \exp(\beta^0 T) + \Phi_i \mu^0 T \exp(\mu^0 T) \leq r,$$

and combining the components also $\|\mathcal{X}(T, \mathbf{X})(t) - \Phi\| \leq r$. Thus $\mathcal{X}_i(T, \mathbf{X})$ maps $\Sigma_T$ onto itself.

To show that $\mathcal{X}(T, \mathbf{X})$ is a contraction, take now $\mathbf{X}', \mathbf{X}'' \in \Sigma_T$ and estimate as follows:

$$|\mathcal{X}_1 - \mathcal{X}_2| \equiv |(\mathcal{X}_i(T, \mathbf{X}')(t)) - (\mathcal{X}_i(T, \mathbf{X}'')(t))| \tag{4.111}$$

$$\leq \int_0^t |K_i(t-u, t, \mathbf{X}') - K_i(t-u, t, \mathbf{X}'')| (\mathcal{B}_i(T, \mathbf{X}')(u)) du$$

$$+ \int_0^t |K_i(t-u, t, \mathbf{X}'')| |(\mathcal{B}_i(T, \mathbf{X}')(u)) - (\mathcal{B}_i(T, \mathbf{X}'')(u))| du$$

$$+ \int_0^\infty \phi_i(v) |L_i(v, t, \mathbf{X}') - L_i(v, t, \mathbf{X}'')| dv.$$

Taylor's formula gives

$$|L_i(v, t, \mathbf{X}') - L_i(v, t, \mathbf{X}'')| \leq N\mu'T \|\mathbf{X}' - \mathbf{X}''\|_T \exp(2\mu'T) \tag{4.112}$$

$$|K_i(t-u, t, \mathbf{X}') - K_i(t-u, t, \mathbf{X}'')| \leq N\mu'T \|\mathbf{X}' - \mathbf{X}''\|_T \exp(2\mu'T). \tag{4.113}$$

The definition of $\mathcal{B}(T, \mathbf{X})$, letting $m_i(t) = \mathcal{B}_i(T, \mathbf{X}')(t) - \mathcal{B}_i(T, \mathbf{X}'')(t)$, gives the estimate

$$|m_i(t)| = \left| \int_0^\infty m_i(u) \beta_i(t-u, \mathbf{X}') K_i(t-u, t, \mathbf{X}'') du \right| + |f_i(t)|$$

$$\leq \beta^0 \int_0^\infty |m_i(u)| du + |f_i(t)|,$$

and from Gronwall's inequality we conclude that

$$|m_i(t)| \leq |f_i(t)| + \beta^0 \int_0^t |f_i(u)| \exp[\beta^0(t-u)] du. \tag{4.114}$$

Now

$$|f_i(t)| \leq \int_0^\infty \phi_i(v) |\beta_i(\mathbf{X}') L_i(\mathbf{X}') - \beta_i(\mathbf{X}'') L_i(\mathbf{X}'')| dv \tag{4.115}$$

$$+ \int_0^t (\beta_i(T, \mathbf{X}'')(v)) |\beta_i(\mathbf{X}') K_i(\mathbf{X}') - \beta_i(\mathbf{X}'') K_i(\mathbf{X}'')| dv.$$

Using (4.113), the last term can be estimated to give the upper bound:

$$\Phi_i N \beta^0 T \exp(2\beta^0 T) [\beta' + \beta^0 \mu' T \exp(2\mu^0 T)] \|\mathbf{X}' - \mathbf{X}''\|_T.$$

Using instead (4.112), an upper bound for the first term in (4.115) is obtained:

$$\Phi_i N \mu' T \exp(2\mu^0 T) \left[ \beta^0 + \left( \beta'/2 \right) \|\mathbf{X}' - \mathbf{X}''\|_T \right] \|\mathbf{X}' - \mathbf{X}''\|_T.$$

From (4.114) we have

$$|m_i(t)| \leq T k(T) \|\mathbf{X}' - \mathbf{X}''\|_T, \tag{4.116}$$

and for $T$ small enough, the right-hand sides of (4.112), (4.113), and (4.116) are small and combining them with (4.111) we find

$$|(\mathcal{X}_i(T,\mathbf{X}')(t)) - (\mathcal{X}_i(T,\mathbf{X}'')(t))| \leq T\tilde{k}(T)\|\mathbf{X}' - \mathbf{X}''\|_T < \|\mathbf{X}' - \mathbf{X}''\|_T.$$

Thus, as claimed, $\mathcal{X}(T,\mathbf{X})$ is a contraction for $T$ small enough.

## Boundedness

Integrate Equation (4.98) with respect to age,

$$\frac{dX_i}{dt} = \int_0^\infty \beta_i(a,\mathbf{X})x_i(a,t)\,da - \int_0^\infty \mu_i(a,\mathbf{X})x_i(a,t)\,da. \tag{4.117}$$

From (4.104)–(4.105) we have $K_i(\alpha,t,\mathbf{X})L_i(v,t,\mathbf{X}) \leq \exp(-m^*t)$ and using then (4.106), (4.107), (4.108), (4.117), and Gronwall's inequality for $0 \leq t \leq T$, the population and density growths are bounded above by

$$X_i(t) \leq \Phi_i \exp[(b^* - m^*)t], \tag{4.118}$$

$$\beta_i(t) \leq b^*\Phi_i \exp[(b^* - m^*)t], \tag{4.119}$$

$$x_i(a,t) \leq b^*\Phi_i \exp[-m^*a + (b^* - m^*)t], \quad a < t, \tag{4.120}$$

$$x_i(a,t) \leq \|\Phi\|_t \exp(-m^*a), \quad a > t. \tag{4.121}$$

We now consider the Taylor's expansions of the maternities and mortalities $\beta_i$ and $\mu_i$ up to order $P$. These expansions hold in the sphere $S(R)$ of radius $R = \max(\|\mathbf{X}(0)\|, \|\mathbf{X}^0\|)$, whose radius $\mathbf{X}^0$ will be determined later. Let $B^0 = \sup\left(\left|\gamma_i^{(\alpha)}(a)\right| : a \geq 0, \mathbf{X} \in S(R)\right)$, $\gamma_i^{(\alpha)}(a) = \frac{\partial^\alpha \gamma_i}{\partial \mathbf{X}^\alpha}(a)$, and $\gamma_i^{(\alpha)}(a) = \frac{\partial^\alpha \gamma_i}{\partial \mathbf{X}^\alpha}(a,c\mathbf{X})$, where $0 < c < 1$, for $|\alpha| = P$, and $\gamma_i = \beta_i - \mu_i$. Taking the sum over the multiindex $\alpha = (k_1,...,k_N)$, the system then becomes

$$\frac{dX_i}{dt} = \sum r_{i,\alpha}(t)X_1^{k_1}...X_N^{k_N}, \quad r_{i,\alpha}(t) = \int_0^\infty \frac{\gamma_i^{(\alpha)}x_i(a,t)}{R'X_i(t)B^0}\,da.$$

The main result of this section shows that for (4.117), the unit hypercube $\mathbf{X}^0$ is a positive invariant set.

Let $J(i) = \{\alpha = (k_1,...,k_N) : r_{i,\alpha}(t) > 0\}$ and observe that taking the sum over the multiindex $\alpha$, for every $t$ we have

$$0 < |r_{i,\alpha}(t)| < \frac{1}{R'}, \quad \sum r_{i,\alpha}(t) < 1.$$

Define the vectors

$$(w_{i,\alpha}(\mathbf{X}))_j = \delta_{ij}\left(X_1^{k_1},...X_i^{k_i-2},...,X_N^{k_N}\right)(1-X_i)X_i, \quad j = 1,...,N,$$

starting from $|\alpha| = P$ and then iterating the procedure up to $|\alpha| = 0$. At every such step the coefficients also get modified; for $|\alpha'| \leq |\alpha|$ we have

$$r'_{i,\alpha}(t) = r_{i,\alpha'}(t) + r_{i,\alpha}(t).$$

The solution $\mathbf{X}(t)$ of $\frac{d\mathbf{X}}{dT} = sgn\left(r'_{i,\alpha}(t)\right) w_{i,\alpha}(\mathbf{X})$ satisfies $X_j(t) = X_j(0)$ for $j \neq 1$, and $X_i(t) \longrightarrow 0$ if $sgn\left(r'_{i,\alpha}(t)\right) < 0$ while $X_i(t) \longrightarrow 1 = X_i^0$ conversely. Then, in a neigborhood of $t = 0$, $X_i(t) < \max\left(X_i(0), X_i^0\right)$. We can now replace system (4.117) by

$$\frac{dX_i}{dt} = \sum w_{i,\alpha}(t) r'_{i,\alpha}(t). \tag{4.122}$$

We finally observe that $r'_{i,0}(t) < 0$ since this condition is equivalent to

$$\sum r_{i,\alpha}(t) \leq \sum r_{i,\alpha}(t),$$

where the first sum is for $\alpha \in J(i)$, while in the second one $\alpha \notin J(i)$, and this condition holds by assumption. Hence, as time flows, the solution of

$$\frac{dX_i}{dt} = sgn\left(r'_{i,0}(t)\right) w_{i,0}(\mathbf{X}) X_i$$

always tends to zero. We can rewrite our system with $v \in \mathbf{R}^N$ as

$$\frac{d\mathbf{X}}{dT} = u_1(t) X_1(\mathbf{X}) + \ldots + u_k(t) X_N(\mathbf{X}) \quad \mathbf{X}(0) = \mathbf{X}_0, \quad 0 \leq u_i(t) \leq \frac{1}{N}.$$

Let $\phi_t(\mathbf{X})$ be its actual solution. Let $\psi_{it}(x, y)$ be the solution trajectories in the population space of the individual systems,

$$\frac{d\mathbf{X}}{dT} = X_i, \quad \mathbf{X}(0) = \mathbf{X}_0 \quad i = 1, \ldots, N.$$

Rewrite the system as

$$\sum \left(\frac{d\mathbf{X}}{dT} - \mathbf{X}_t(\mathbf{X})\right) u_i(t) + \left(1 - \sum u_i(t)\right) \left(\frac{d\mathbf{X}}{dT} - 0\right) = 0, \quad \mathbf{X}(0) = \mathbf{X}_0.$$

Then, by the very definition of $\phi_t$ and $\psi_{it}$, at every instant

$$\phi_t(\mathbf{X}) = \sum u_i(t) \psi_{it}(\mathbf{X}) + \left(1 - \sum u_i(t)\right) \mathbf{X}_0,$$

thus showing that the trajectories are ultimately confined to the unit hypercube.

### Global existence and uniqueness

The result of the previous section allows us to show now that the problem has a unique solution for all time.

Indeed, by the local existence result, given an initial distribution $(\ldots, X_i(0), \ldots)$, the solution exists up to time $T$, where $T$ depends on the initial data. Boundedness implies then that $X_i(T) < \max(g_0, X^*)$. Now repeat the argument with $(\ldots, X_i(T), \ldots)$ as the new initial condition. Since local uniqueness holds, any two solutions agree on an interval $[0, T)$, and by their continuity necessarily $T = \infty$.

## Stability of equilibrium distributions

We have seen in the previous chapters how to analyze the stability of equilibria of models governed by ordinary differential equations. Here we study the same question for structured models formulated by means of hyperbolic partial differential equations.

The equilibrium distributions of the system are in this case once again the time independent solutions, which thus must be functions dependent only on age, i.e.,

$$x_i(a,t) = x_i(a), \quad X_i(t) = X_i.$$

The system (4.98) thus simplifies to give

$$\frac{dx_i}{da} + \mu_i(a, \mathbf{X}) x_i(a) = 0 \qquad (4.123)$$

$$X_i = \int_0^\infty x_i(a)\, da \qquad (4.124)$$

$$B_i = x_i(0) = \int_0^\infty \beta_i(a, \mathbf{X}) x_i(a)\, da. \qquad (4.125)$$

The formal solution of this system is

$$x_i(a) = x_i(0) \exp\left(-\int_0^a \mu_i(s, \mathbf{X})\, ds\right) \equiv x_i(0) M_i(a, \mathbf{X}), \qquad (4.126)$$

where $M_i(a, \mathbf{X})$ denotes the survival function up to age $a$ of the $i$th population. The total population sizes are obtained by integration over age of (4.126):

$$X_i = B_i \int_0^\infty M_i(a, \mathbf{X})\, da. \qquad (4.127)$$

The average number of offsprings of an individual is then

$$R_i(\mathbf{X}) = \int_0^\infty \beta_i(a, \mathbf{X}) M_i(a, \mathbf{X})\, da,$$

and together with (4.125) and (4.126) we have

$$R_i(\mathbf{X}) = 1, \quad i = 1, ..., N, \qquad (4.128)$$

and in turn (4.128) implies (4.125), so that the necessary and sufficient condition for existence of the equilibrium age distribution is represented by the fact that each individual in the population has exactly one offspring. At equilibrium, then,

$$x_i^*(a) = B_i^* M_i(a, \mathbf{X}), \quad B_i^* = \frac{1}{\int_0^\infty M_i(a, \mathbf{X}^*)\, da} X_i^*.$$

We investigate the stability of this equilibrium by considering now the following perturbation. Let

$$x_i(a,t) = x_i^*(a) + y_i(a,t), \quad X_i(t) = X_i^* + Y_i(t).$$

From now on, the sums are always meant for $j, k = 1, ..., N$. Letting

$$K_{ij} = \int_0^\infty V(i,j)(a) x_i^*(a) \, da,$$

$$Z_i(t) = Y_i(t) \sum_j \int_0^\infty V(i,j)(a) y_i(a,t) \, da$$

$$+ \sum_{j,k} \int_0^\infty W(i,j,k)(a, \mathbf{X}^* + \mathbf{Y}q) [x_i^*(a) + y_i(a,t)] \, da,$$

$$A_i(a,t) = \sum_{j,k} Q(i,j,k)(a, \mathbf{X}^* + z\mathbf{Y}) Y_j Y_k x_i^*$$

$$+ \sum_j E(i,j)(a) Y_j y_i(a,t),$$

$$Q(i,j,k)(a, \mathbf{p}) = \frac{\partial^2 \mu_i}{\partial X_j \partial X_k}(a, \mathbf{p}),$$

$$E(i,j)(a) = \frac{\partial \mu_i}{\partial X_j}(a, \mathbf{X}^*),$$

$$V(i,j)(a) = \frac{\partial \beta_i}{\partial X_j}(a, \mathbf{X}^*),$$

$$W(i,j,k)(a, \mathbf{p}) = \frac{\partial^2 \beta_i}{\partial X_j \partial X_k}(a, \mathbf{p}), \quad 0 < q, z < 1.$$

From Equations (4.98) and (4.99) it follows that

$$\frac{\partial y_i}{\partial a} + \frac{\partial y_i}{\partial t} + \mu_i(a, \mathbf{X}^*) y_i(a,t) + A_i(a,t)(x_i^* + y_i) = 0, \quad (4.129)$$

$$Y_i(t) = \int_0^\infty y_i(a,t) \, da, \quad (4.130)$$

$$y_i(0,t) = \int_0^\infty \beta_i(a, \mathbf{X}^*) y_i(a,t) \, da + \sum_j Y_j(t) K_{ij} + Z_i(t), \quad (4.131)$$

$$\phi_i(a,0) - x_i^*(a) = n_i(a).$$

Observe that $A_i(a,t)$ and $Z_i$ are quantities of second order in $y_i$, $Y_i$, and $W(i,j,k)(a,\mathbf{p})$ and $Q(i,j,k)(a,\mathbf{p})$ are $o\left(\|\mathbf{p}\|^{-1}\right)$ uniformly as $a \longrightarrow 0$. Set

$D_i(a) = \mu_i(a, \mathbf{X}^*)$. We consider the linearized system

$$0 = \frac{\partial y_i}{\partial t} + \frac{\partial y_i}{\partial a} + D_i(a) y_i(a,t) + \sum_j E_{ij}(a) x_i^*(a) Y_j(t), \quad (4.132)$$

$$Y_i(t) = \int_0^\infty y_i(a,t) \, da,$$

$$y_i(0,t) = \int_0^\infty \beta_i(a, \mathbf{X}^*) y_i(a,t) \, da + \sum_j Y_j(t) K_{ij},$$

and for $g_i$, $c$ both complexes look for solutions in the form

$$y_i(a,t) = g_i(a) e^{ct}, \quad i = 1, ..., N.$$

Now (4.132) gives

$$0 = \frac{dg_i}{da} + [c + D_i(a)] g_i(a) + \sum_j E_{ij}(a) x_i^*(a) G_j, \quad (4.133)$$

$$G_i = \int_0^\infty g_i(a) \, da,$$

$$g_i(0) = \int_0^\infty \beta_i(a, \mathbf{X}^*) g_i(a) \, da + \sum_j G_j K_{ij}.$$

Letting

$$M_i^0(a) = M_i(a, \mathbf{X}^*), \quad M_i^0(s,a) = \frac{M_i^0(a)}{M_i^0(s)},$$

by integration we get the homogeneous system in the $2N$ unknowns $G_i$ and $g_i(0)$:

$$0 = g_i(0) \left[ 1 - \int_0^\infty \beta_i(a, \mathbf{X}^*) \exp(-ca) M_i^0(a) \, da \right] - \sum_j G_j K_{ij} \quad (4.134)$$

$$+ \sum_j G_j \int_0^\infty \beta_i(a, \mathbf{X}^*) \int_0^a E_{ij}(w) x_i^*(w) \exp-[(a-w)c] M_i^0(w,a) \, dw \, da,$$

$$0 = -g_i(0) \int_0^\infty \exp(-ca) M_i^0(a) \, da + G_i$$

$$- \sum_j G_j \int_0^\infty \int_0^a E_{ij}(w) x_i^*(w) \exp-[(a-w)c] M_i^0(w,a) \, dw \, da.$$

Now the above system gives the criterion to determine whether the equilibrium distribution is stable, namely, if the determinant $\Delta(c)$ of the system (4.134) has no roots with positive real part, then for suitable $m^0$ and initial condition $\Phi$, the solution as $t \longrightarrow \infty$ satisfies

$$\|\mathbf{X}(t) - \mathbf{X}^*\| = O\left(\exp-(m^0 t)\right), \quad (4.135)$$

$$\|\mathbf{x}(a,t) - \mathbf{X}^*\| = O\left(\exp-(m^0 t)\right).$$

Letting $b_i(t) = y_i(0,t)$, the formal solution of Equations (4.129)–(4.131) gives for $a < t$,

$$y_i(a,t) = M_i^0(a) b_i(t-a) \qquad (4.136)$$
$$+ \int_{t-a}^{t} M_i^0(z+a-t,a) A_i(z+a-t,z) dz$$
$$- \sum_j \int_{t-a}^{t} M_i^0(z+a-t,a) E_{ij}(z+a-t) x_i^*(z+a-t) Y_j(t) dz,$$

and for $a > t$,

$$y_i(a,t) = M_i^0(a-t,a) n_i(a) \qquad (4.137)$$
$$+ \int_0^t M_i^0(a-t+s,a) A_i(a-t+s,s) ds$$
$$- \sum_j \int_0^t M_i^0(a-t+s,s) E_{ij}(a-t+s) x_i^*(a-t+s) Y_j(s) ds.$$

Setting $\mathbf{v}(t) = (Y_1(t), \ldots, Y_N(t), b_1(t), \ldots, b_N(t))^T$, an age integration then gives the system

$$\mathbf{H}\mathbf{v}(t) + \int_0^t \mathbf{K}(t-s) \mathbf{v}(s) ds = \mathbf{f}(t), \qquad (4.138)$$

where, for $i = 1, \ldots, N$,

$$f_i(t) = \int_0^\infty M_i^0(s, s+t) n_i(s+t) ds$$
$$+ \int_0^t \int_0^\infty M_i^0(s, s+t-z) A_i(s, s+t-z) ds dz,$$
$$f_{N+i}(t) = Z_i(t) + \int_0^\infty \beta_i(s+t, \mathbf{X}^*) M_i^0(s, s+t) n_i(s+t) ds$$
$$+ \int_0^t \int_0^\infty M_i^0(s, s+t-z) A_i(s, z) \beta_i(s+t-z, \mathbf{X}^*) ds dz,$$

while, for $i, j = 1, \ldots, N$,

$$H_{ij} = \delta_{ij},$$
$$H_{N+i,j} = -K_j + \delta_{i+N, N+j},$$
$$K_{ij}(t) = \int_0^\infty M_i^0(s, s+t) E_{ij}(s) x_i^*(s) ds,$$
$$K_{i,N+j}(t) = -\delta_{ij} M_i^0(t),$$
$$K_{N+i,j}(t) = \int_0^\infty \beta_i(s+t, \mathbf{X}^*) M_i^0(s, s+t) E_{ij}(s) x_i^*(s) ds,$$
$$K_{N+i,N+j}(t) = -\delta_{ij} \beta_i(t, \mathbf{X}^*) M_i^0(t).$$

Now $M_i^0(a)$ has at most an exponential growth; since $E_{ij}$ and the maternity function are bounded above, the Laplace transform of the system of integral equations exists for $Re(p) > -D^* \equiv -\inf\{D_i(a) : a \geq 0, \ i = 1, ..., N\}$. Also, the condition $\det\left(\mathbf{H} + \hat{\mathbf{K}}(p)\right) = 0$ is equivalent to system (4.134), and thus it has no solutions with $Re(p) > 0$. For $p \longrightarrow \infty$ and $Re(p) > -D^*$, $\det\left(\mathbf{H} + \hat{\mathbf{K}}(p)\right) \longrightarrow 1$. A positive constant $m$ exists then such that $\det\left(\mathbf{H} + \hat{\mathbf{K}}(p)\right)$ does not vanish for $Re(p) > -m$. Hence $\mathbf{H} + \hat{\mathbf{K}}$ is analytically invertible in $Re(p) > -m$ and in view of $\hat{\mathbf{K}}_{ij}(p) \longrightarrow 0$ as $p \longrightarrow \infty$, the inverse is of the form $\mathbf{H}^{-1} + \hat{\mathbf{J}}(p)$, with $\hat{\mathbf{J}}(p)$ analytic in $Re(p) > -m$ and $\hat{J}_{ij}(p) \longrightarrow 0$ as $p \longrightarrow \infty$ and $\hat{\mathbf{J}}(p) = \mathbf{J}^0/p + O(p^{-2})$. $\mathbf{J}^0$ is an invertible constant matrix so that with the bounds $|\mathbf{J}_{ij}(t)| \leq const \ \exp(-mt)$, we have

$$\mathbf{J}(t) = \frac{1}{2\pi}\int_{-\infty}^{\infty} \exp[t(iz-m)]\hat{\mathbf{J}}(iz-m)\,dz.$$

The solution of system (4.138) is

$$\mathbf{v}(t) = \mathbf{H}^{-1}\mathbf{f}(t) + \int_0^t \mathbf{J}(t-s)\mathbf{f}(s)\,ds.$$

We finally need an estimate of $\|\mathbf{v}\|_2$ and $\|\mathbf{y}(\cdot,t)\|_1$. Easily,

$$\|\mathbf{f}\|_2 \leq C_1 \left\{\|\mathbf{n}\|_1 \exp(-tD^*) + \int_0^t \|A_i(\cdot,z)\|_1 \exp[-D^*(t-z)]\,dz + \|\mathbf{Z}\|_2\right\}.$$

From the convolution properties, then,

$$\|\mathbf{v}\|_2 \leq C_2\left[\|\mathbf{n}\|_1 e^{-tD^*} + \|\mathbf{Z}\|_2 + \int_0^t e^{-m(t-s)}\left[\|A_i(\cdot,s)\|_1 + \|\mathbf{Z}\|_2\right]ds\right],$$

$$\int_0^\infty |y_i(a,t)|\,da \leq C_3\|\mathbf{n}\|_1 e^{-tD^*} + C_4 \int_0^t e^{-D^*(t-z)}\left[\|\mathbf{Z}\|_2 + \|A_i(\cdot,z)\|_1\right]dz.$$

For $1 > \epsilon > 0$, we can find a $\delta'(\epsilon) < \epsilon$ such that if $\|\mathbf{v}\|_2 < \delta'$, we have $|W(i,j,k)|, |Q(i,j,k)| \leq \epsilon\|\mathbf{v}\|_2$. For $\sigma(t) = \|y(\cdot,t)\|_1 + \|\mathbf{v}\|_2$, it follows that

$$\|A_i(\cdot,t)\|_1, \|\mathbf{Z}\|_2 \leq C_5\left[\|\mathbf{v}\|_2\|y(\cdot,t)\|_1 + \epsilon\|\mathbf{v}\|_2\right] = C_5\sigma(t),$$

$$\sigma(t) \leq C_6\left[\exp(-D^*t)\|\mathbf{n}\|_1 + 2\epsilon\int_0^t \exp[-D^*(t-s)]\sigma(s)\,ds\right],$$

and Gronwall's inequality for $0 \leq t < T$ implies

$$\sigma(t) \leq K\exp(-D^*t + 2K\epsilon t)\|\mathbf{n}\|_1 = K\|\mathbf{n}\|_1\exp(-m^0 t),$$

which in turn gives

$$\|\mathbf{Y}(0)\|_2 \leq \|\mathbf{v}(0)\|_2 \leq \delta, \quad \delta = \min(\delta'(\epsilon), \delta'(\epsilon)/K).$$

If the initial data satisfy $\|\mathbf{n}\|_1 < \delta$, iterating on intervals of length $T$ we get $\|\mathbf{v}(t)\|_2 < \delta$ for every $t > 0$. Then (4.135) follows from the above estimates using (4.137) and (4.136).

### 4.7.3 Simulations and brief discussion

The model presented is clearly very difficult to study analytically. Numerical experiments are then needed to assess its behavior. The program uses a finite difference approximation scheme to discretize directly the system (4.92)–(4.95), adapting to the present model the Wendroff implicit approximation scheme (cf. Smith, 1979, p. 155). Given a rectangle in the $at$ phase space, the computational cell to approximate the value of the function at its center is made by its four vertices. Assuming that we know the information at the current time level and the left upper vertex, i.e., the previous age point at the next time, then it becomes an explicit scheme for the upper left vertex, i.e., the next age point at the next time. The problem is that at the next time level, it needs the first "upper left" vertex, which corresponds to the newborns at the next time, and this information analytically comes from the integral of the maternity at the same time level of the newborns, i.e., at the next time. This remark makes the algorithm once more implicit. However, in our implementation we bypassed this point, by simply evaluating the integral (4.83) at the current time step. This corresponds to artificially adding a numerical delay the size of which is the time step.

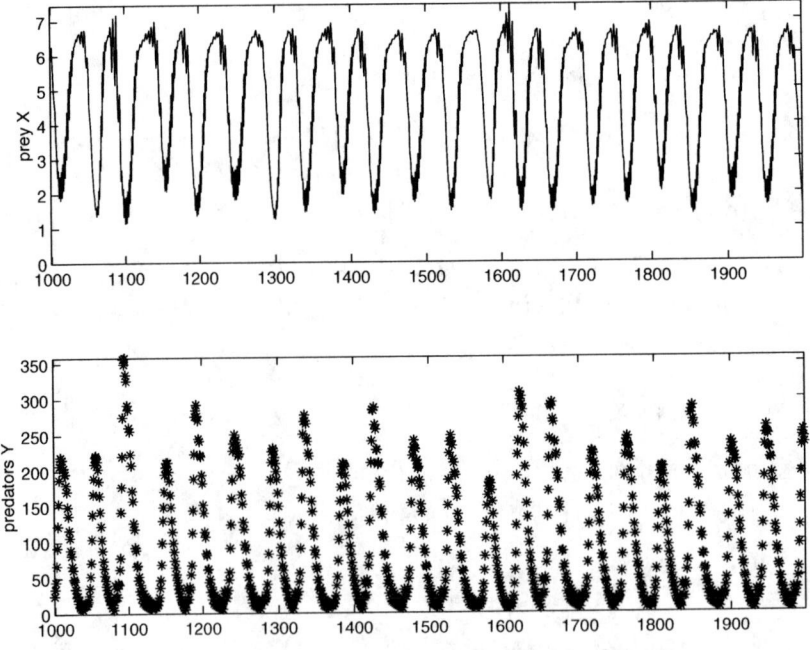

**FIGURE 4.7:** The two figures show the persistent oscillating behaviors of the solutions of the model: top, the prey; bottom, the predators.

Figure 4.7 shows results of numerical simulations obtained for parameters chosen in such a way that each species treated independently of the other one evolves toward a stable equilibrium. When the two interfere with each other by the predation mechanism, then the solution trajectory in the phase space shows a limit cycle; see Figure 4.7.

Thus the age-dependent prey–predator model exhibits radically different characteristics than the classical quadratic prey–predator model, as the latter, as shown in Section 4.4.1, always possesses asymptotically stable equilibria and limit cycles can never arise.

# Chapter 5

# Case study: biological pest control in vineyards

Vineyards are agroecosystems that strongly reflect man's historical relationship with the land. In Piedmont (NW Italy), as in other countries, this kind of agroecosystem is facing land abandonment and agricultural intensification. The latter process of transformation results in the progressive removal of natural landscape elements, in particular loss of woods in vineyard landscapes, thus contributing to loss of heterogeneity affecting biotic communities (Agger and Brandt, 1988). Wise (1993) stresses the importance of the spider assemblages in limiting insect pests in agroecosystems and remarks on the need for research on prey limitation in natural communities and agroecosystems. The landscape structure then needs to be considered both in spatial and temporal terms (Merriam, 1988). Spatial diversity is given by the number of spider habitat types available and landscape heterogeneity by how intermingled they are. Temporally, the landscape can also be described with respect to disturbance synchronization, e.g., by season and by how large a proportion of the fields are at the same time affected by management, crop rotation, land use, and husbandry (Thomas et al., 1990). Thus the landscape structure is not static. Spider communities are significantly affected by these factors, and different agroecosystem dynamics can show different spider assemblages (Isaia et al., 2006b).

The spiders' role in agroecosystems is not completely understood, but their impact is shown by several studies, e.g., see Wise (1993), Furuta (1977), and Nyffeler and Benz (1987). In more general terms, Nyffeler and Benz (1987), in reviewing several recent studies, support the conclusion that wandering spiders contribute significantly to the control of insect populations in soil/litter systems. The role of web weaving spiders as biological controllers is studied much more frequently by ecologists than wanderer spiders, since prey can be easily supplemented to web spinners, not moving their web sites frequently. Thus feeding behavior and interactions with conspecific and fecundity can be observed and measured directly and easily.

Here we consider first two mathematical models for understanding wanderer spiders as controllers of vineyard-infesting insects, expressly modeling the residual wood and green patches as the spiders' habitat in the otherwise homogeneuous landscape of the Langa Astigiana. These predators can move around among vineyards and woods, in which two different prey, i.e., two insect populations, live. Since vineyards predominate on woods, the insects affecting the cultures will be more abundant than the wood ones, so that in the more complex model spiders experience feeding satiation. Since they are able to move around and seek other food sources, they will then predate on the insect population living in the nearby green patches. In both models we then investigate the effect of human activities in this ecosystem, by also considering insecticide spraying and its effects.

We then consider a tranport phenomenon named ballooning for which young spiders are carried by the wind from vineyards into the surrounding woods, and vice versa, since this appears to be one of the key factors for the evolutionary success of spiders (Greenstone et al., 1987). It appears to be an example of an evolved behavior to avoid competition (Wise, 1993). Most ballooners are immature instars under 1.0 mg, but the most frequent long-range dispersers are the smaller adult Linyphiids. In view of the landscape not being a static environment, this immigration process becomes important as the spider population in a single field may need to recover several times every year. Experiments to understand population dynamics of Linyphiids are very labor intensive and expensive to carry out on a landscape scale (Wiens et al., 1993). From these considerations the need of modeling this phenomenon arises. The third mathematical model we propose takes into account this tranport phenomenon. A relevant feature consists in not taking space explicitly into account in the governing equations, in contrast to what is done elsewhere (Halley et al., 1996).

## 5.1 First model

Since we are considering wanderer spiders, we denote their population by $s(T)$ independently of the location in which they live. On the other hand, to model the differences in the landscape, the insects living in open fields and woods are denoted by $w(T)$, and by $v(T)$ we denote the population of parasites living in vineyards. Note that both are prey for these types of spiders, as they can move from one type of landscape to the other. The model is then given by a single predator population that feeds on two types of prey, not interacting with each other. It reads

$$\frac{ds}{dT} = s(-a + kbv + kcw), \quad (5.1)$$
$$\frac{dw}{dT} = rw\left(1 - \frac{w}{W}\right) - csw,$$
$$\frac{dv}{dT} = v(e - bs).$$

The first equation states as mentioned that $w$ and $v$ are prey for the spiders, but, moreover, since the latter die exponentially fast when the former lack, these two types of insects are then the only food source for spiders. The second equation models the wood-living insects, reproducing logistically. Notice that we assume a small carrying capacity $W$, in view of the reduced extension of their environment with respect to the vineyard. The third equation describes the Malthus growth of the parasites in the vineyard, since the latter is so large that we can take it as having unlimited resources. Predation is modeled by mass action law in each equation. Prey are turned into new spiders via the "efficiency" constant $0 < k < 1$.

We as usual investigate the feasible equilibria of (5.1) to find the origin $E_0 \equiv O$ and the points

$$E_1 \equiv (0, W, 0), \quad E_2 \equiv \left(\frac{e}{b}, 0, \frac{a}{bk}\right),$$
$$E_3 \equiv (s_3, w_3, 0) = \left(\frac{r}{c}\left(1 - \frac{a}{ckW}\right), \frac{a}{ck}, 0\right),$$
$$E_4 \equiv \left(\frac{e}{b}, W\left(1 - \frac{ce}{br}\right), \frac{1}{bk}\left[a - ckW\left(1 - \frac{ce}{br}\right)\right]\right).$$

The feasibility condition for $E_3$ is $a < ckW$, while those for $E_4$ are instead $br > ce$ and $abr > ckW(br - ce)$.

The local stability analysis shows that $E_0$ and $E_1$ are always unstable, their eigenvalues being respectively $r, -a, e$ and $-r, -a + ckW, e$. Those for $E_2$ are given by $b^{-1}(br - ce), \pm i\sqrt{ae}$, thus giving a stable or unstable center in the $s, v$ plane, depending on whether $br < ce$. The eigenvalues of $E_3$ are $\lambda_3^{(3)} = e - bs_3 < 0$ and the roots of the quadratic equation $\Psi(\lambda) \equiv ckW\lambda^2 + ra\lambda + ar(ckW - a) = 0$. Since $\Psi(0) > 0$ and $\Psi'(0) > 0$, we obtain two roots with negative real parts. Hence $E_3$ is stable if and only if $c^2kWe < br(ckW - a)$. Finally, the Routh–Hurwitz criterion applied to the characteristic equation of $E_4$,

$$b^2r\lambda^3 + br(br - ce)\lambda^2 + \lambda e\left[ab^2r + ckW(br - ce)(c - b)\right]$$
$$+ e(br - ce)[abr - ckW(br - ce)] = 0,$$

coincide with the two feasibility conditions of the same equilibrium $E_4$. Thus the latter is unconditionally stable when feasible.

Out of the five possible equilibria, the origin has two positive eigenvalues, so that the populations $w$ and $v$ increase. It is then always unstable. Similar results hold for $E_1$, in this case one of its three eigenvalue is positive, implying the growth of $v$.

The behavior of the system is thus determined by the remaining three equilibria. Notice that $E_4$ is always stable when feasible, $E_3$ is conditionally stable, and it is if and only if $E_4$ is infeasible. The point $E_2$ has one real and a pair of pure imaginary eigenvalues, so that it behaves like a center in the $s, v$ plane. If $br < ce$, the latter is stable, i.e., spiders and vineyard parasites tend to a Lotka–Volterra behavior in the $s, v$ plane around the projected point $(\frac{e}{b}, \frac{a}{bk})$.

Otherwise the system shows an oscillatory behavior moving away from the $s, v$ plane along the $w$ direction. These features can be tracked by the numerical simulations, which are not reported here. Notice also that there are strict interdependencies among the feasibility and stability of some of these nontrivial equilibria. More specifically, $E_2$ stable implies $E_4$ infeasible and $E_3$ unstable. Also, $E_4$ feasible implies the instability of $E_2$.

Summarizing, if $E_4$ is feasible, it is the only possible stable equilibrium. When it is infeasible because of $br < ce$ and $abr > ckW(br - ce)$, the equilibrium $E_2$ is a stable center. When it is infeasible because of $abr < ckW(br - ce)$ but $br > ce$, the point $E_3$ is feasible and stable. Note that the two infeasibility conditions for $E_4$, namely, $br < ce$ and $abr < ckW(br - ce)$, cannot occur at the same time. In conclusion, only one of the three equilibria at the time determines the final behavior of the system.

Numerical experiments further substantiate the theoretical findings, revealing the existence of oscillations in some cases. This indicates that the system seems to be persistent, although not at a steady state (see Figure 5.1), for the following parameter values: $a = 3.1$, $b = 2.87$, $c = .8$, $e = 3.5$, $r = 1$, $W = 5$, $k = 0.8$. However, for these values the equilibria are $E_2 \equiv (1.2195, 0, 1.3502)$ with real eigenvalue $\lambda_1^{[2]} = 0.0244$ and $E_4 \equiv (1.2195, 0.1220, 1.3162)$ with real eigenvalue $\lambda_1^{[4]} = -0.0242$. Thus, $E_2$ is unstable and repels the trajectories, which oscillate neutrally around it as we have theoretically seen they should do. The close stable equilibrium $E_4$ attracts them, but the eigenvalues are both small so that the dynamics is very slow, apparently giving rise to limit cycles, which, however, on a very much longer time scale should result in being damped toward $E_4$. In Figure 5.2 we report the findings on the spiders controlling the pests. This occurs for the parameter values $a = 3.1$, $b = 0.001$, $c = 0.2$, $e = 25$, $r = 1$, $W = 5$, $k = 0.8$, but the populations periodically rise to very high levels.

### 5.1.1 Modeling the human activity

The intervention in the ecosystem caused by man consists here in insecticide spraying, modeled by an impulse function, at particular instances in time. Let $\delta(T_i)$ denote Dirac's delta function and $T_i$, $i = 1, 2, \ldots$ be the spraying instants. Equations (5.1) are modified as follows:

$$\frac{ds}{dT} = s(-a + kbv + kcw) - hKq\delta(T_i) \qquad (5.2)$$
$$\frac{dw}{dT} = rw\left(1 - \frac{w}{W}\right) - csw - h(1-q)\delta(T_i)$$
$$\frac{dv}{dT} = v(e - bs) - hq\delta(T_i).$$

We allow for a part $1 - q$ of insecticide sprayed ending accidentally in the woods, while the fraction $q$ lands on the vineyards as aimed. Also, $h$ denotes the effectiveness against the parasites, the parameter $0 < K < 1$ models instead the smaller effect it supposedly should have on the spiders as the poison is meant to act on the pests and the former may be less affected.

The simulations are reported in Figures 5.3–5.6. Figure 5.3 is a reference picture, when no spraying is used. We see that the ecosystem thrives at least for the time shown. The effects of the spraying are then added. For a small fraction of insecticide landing in the woods, i.e., high $q = 0.9$, Figure 5.4, the wood insects survive, while the whole vineyard ecosystem is wiped away. In Figure 5.5, more adverse conditions are modeled where the whole ecosystem collapses, for a high fraction $q = 0.2$ of poison ending in the wood insects' habitat. The discontinuities are hardly seen in these cases as $K$ is small. Finally, Figure 5.6 shows that in some particularly nasty situations, even though the spraying may be aimed at controlling the vineyard pests, it is actually the spiders and the wood insects that are affected, as their populations vanish, while the pests remain uncontrolled and their growth explodes.

The results of our simulations show that spraying seems to destabilize the vineyards' pests explosions. These considerations are useful in the Langa Astigiana, where the small woods constitute a refuge for many spiders and parasites, other than the vineyards' own. Under natural conditions the ecosystem persists globally.

## 5.2 More sophisticated model

Wanderer spiders frequently confronted with a shortage of prey search for more productive microhabitats on the basis of prey abundance (Edgar, 1971; Kronk and Riechert, 1979; Morse and Fritz, 1982). Laboratory-field comparisons (Edgar, 1969; Hagstrum, 1970; Anderson, 1974) show that an increase in prey availability favors their growth rate and fecundity. If increases in these parameters tend to increase the average population, we can assume that wandering spiders are theoretically food limited. An increase of the population increases competition for territoriality among adults and thus favors their dispersal behavior. Researchers' tests show that spiders are not constantly food limited, but intrinsic limitations in their experimental design mean that food limitation cannot be ruled out for most of the species. More direct and indirect evidence points out the fact that spiders are frequently hungry, even to exhibit growth and reproduction rates below what is thought to be physiologically possible (Wise, 1993). There is evidence that spiders, for normal growth and reproduction, need to actively select prey so as to optimize the diet proportion of essential amino acids (Miyashita, 1968; Greenstone, 1979). The proportion of different types of prey in the diet and thus the need for

Case study: biological pest control in vineyards      115

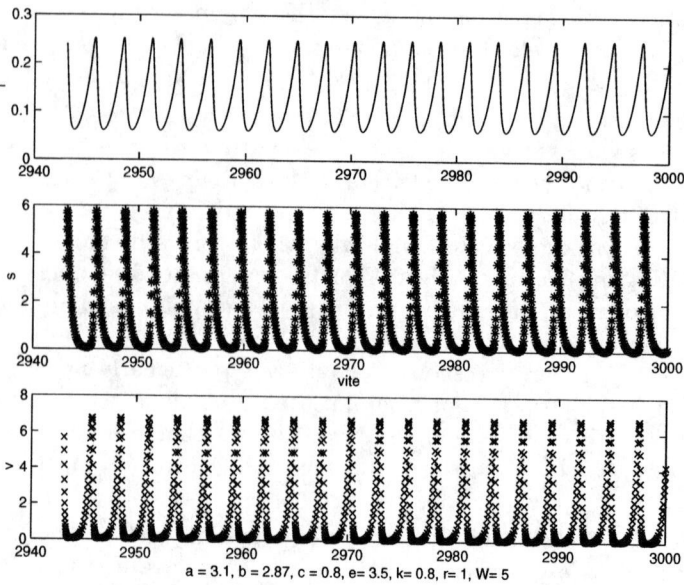

**FIGURE 5.1**: Wanderer spiders model: apparently persistent oscillations around the interior equilibrium.

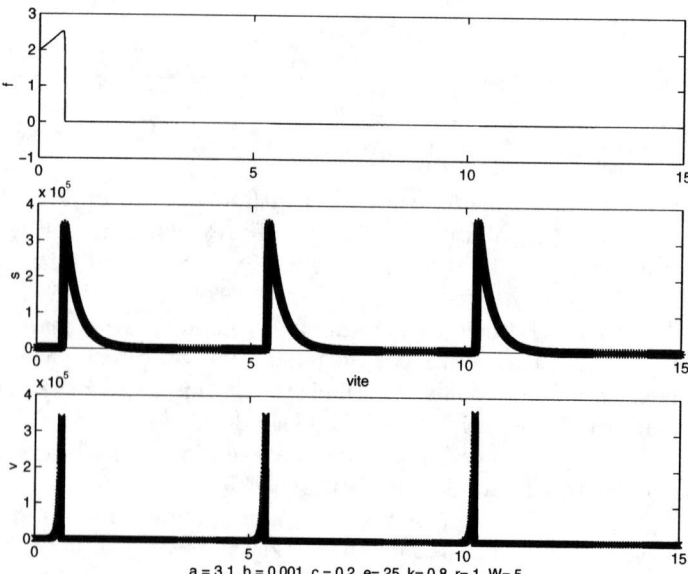

**FIGURE 5.2**: Wanderer spiders model: the spiders control the pests, although at very high levels.

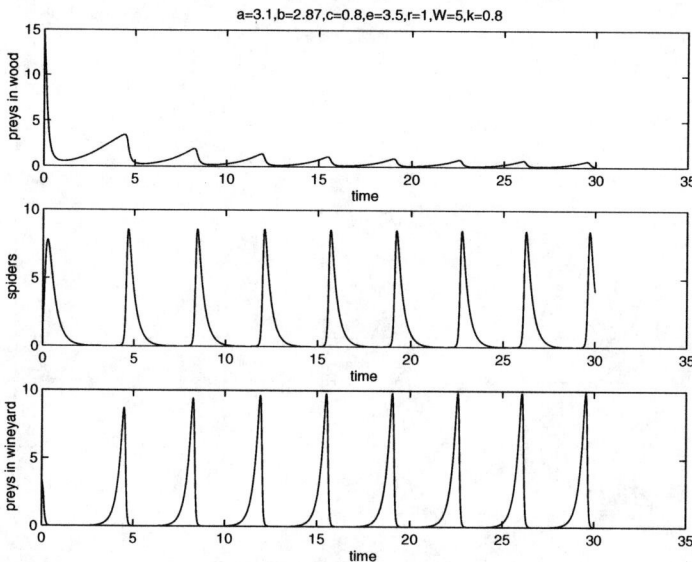

**FIGURE 5.3**: Wanderer spiders model-spraying effects: reference figure with no spray, persistent populations in ecosystem.

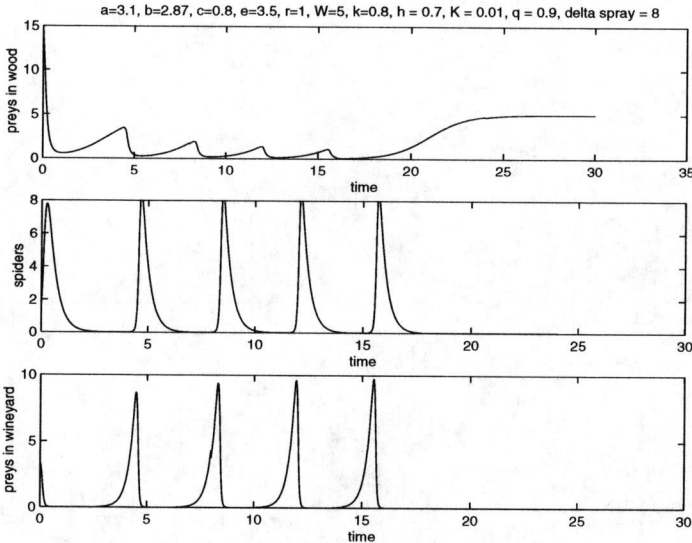

**FIGURE 5.4**: Wanderer spiders model-spraying effects: the wood insects survive, the vineyard ecosystem is wiped away.

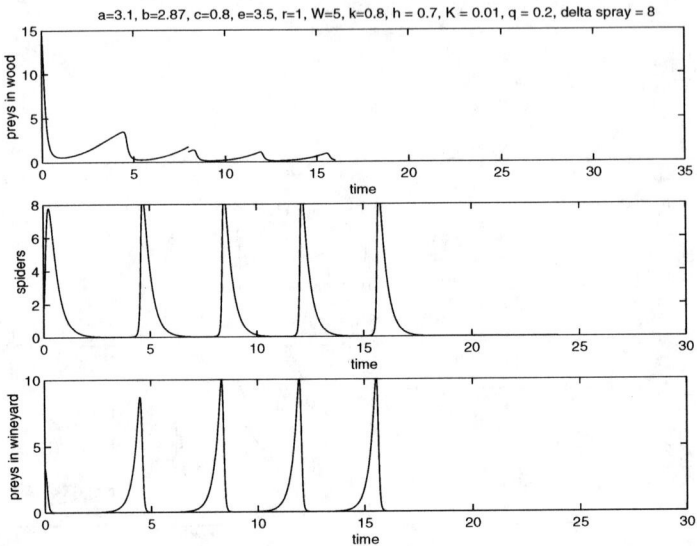

**FIGURE 5.5**: Wanderer spiders model-spraying effects: entire ecosystem collapse.

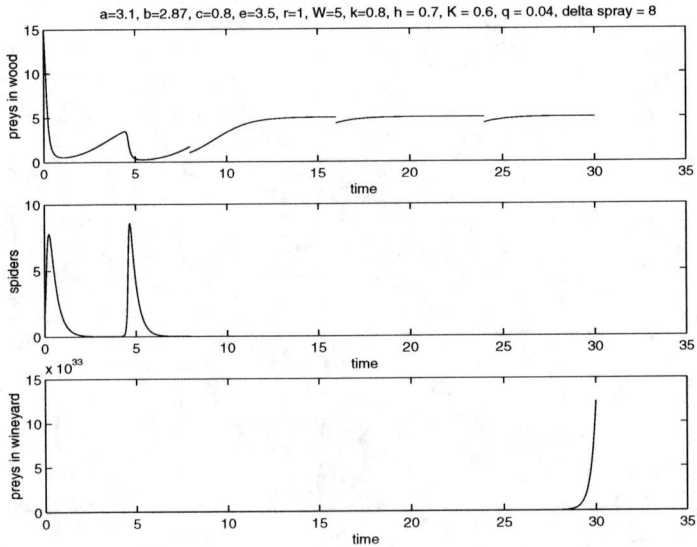

**FIGURE 5.6**: Wanderer spiders model-spraying effects: vineyard pests explosion, all spiders are wiped away.

# Case study: biological pest control in vineyards

searching for new microhabitats could then be due to the degree of satiation of spiders and to their rejection of certain types of prey.

We keep here the same notation previously used that led to (5.1). In the first equation we find again the dynamics of spiders, in which, as before, a negative Malthus growth describes their mortality, and the two nonnegative terms account for the growth due to predation on the two insect populations. But in contrast to (5.1), here we model the saturation effect the spiders experience when the vineyard insects are too numerous. In such a case, since the wanderers move around and search for other prey in the surroundings of the cultivated land, in the green niches around the vineyards inhabited by the $v$ insects, a suitable alternative prey source is found. Thus, basing our considerations on the field data analysis of Rypstra (1995) and Marc et al. (1999), we use Holling type II responses to model this feeding saturation effect. These terms of course reappear in the appropriate insect evolution equation, but for the rest the dynamics of all the insect populations considered is the same as in (5.1):

$$\frac{ds}{dT} = s\left(-a + \frac{kbv}{H+v} + kcw\right), \quad (5.3)$$

$$\frac{dw}{dT} = rw\left(1 - \frac{w}{W}\right) - csw,$$

$$\frac{dv}{dT} = v\left[e - \frac{bs}{H+v}\right].$$

## Analysis of the boundary equilibria

The first two feasible equilibria $E_0$ and $E_1$ of (5.1) are found again here, with the same inconditional instability due to the occurrence of the very same eigenvalues. The equilibrium point $E_2$ instead gets modified as follows: $\hat{E}_2 \equiv (\hat{s}_2, 0, \hat{v}_2) = \left(\frac{ekH}{bk-a}, 0, \frac{aH}{bk-a}\right)$. In this case, it exists only if $a < bk$, while for $E_2$ no such restriction arises. The last equilibrium point lying on the coordinate planes $\hat{E}_3$ is exactly $E_3$ of (5.1), with of course the same existence condition $a < ckW$.

Stability conditions of the last two equilibria differ however from those derived for (5.1). Notice indeed that for $\hat{E}_3$, although derived similarly to the former model, one of the eigenvalues is now $\frac{1}{H}(eH - bs_3)$ while the other two are the roots of the quadratic $Wkc\omega^2 + r(cWk - 2a)\omega + Wk^2cas_3 = 0$. Hence $E_3$ is stable if and only if $br(kcW - a) > c^2ekHW$ and $2a < ckW$. Notice that the presence of the half-saturation constant $H$ renders the satisfaction of the first condition more strict than for (5.1).

The eigenvalues of $\hat{E}_2$ are $r - \frac{1}{a}ce\hat{v}_2$ and those of the matrix

$$J = \frac{1}{a(H+\hat{v}_2)^2}\begin{pmatrix} abk\hat{v}_2(H+\hat{v}_2) - a^2(H+\hat{v}_2)^2 & bekH\hat{v}_2 \\ -ab\hat{v}_2(H+\hat{v}_2) & ae(H+\hat{v}_2)^2 - beH\hat{v}_2 \end{pmatrix}.$$

Necessary and sufficient conditions for its eigenvalues having negative real part are $\text{tr}(J) < 0$ and $\det(J) > 0$. Thus $\hat{E}_2$ is stable if and only if the following conditions are satisfied:

$$b\hat{v}_2(Hk + H + k\hat{v}_2) > a(\hat{v}_2 + H)^2, \quad bHe\hat{v}_2 + (a-e)a(H+\hat{v}_2)^2 > akb\hat{v}_2(H+\hat{v}_2).$$

## Nontrivial equilibrium $E_4$

Now, from the ecological standpoint, of particular interest is the system equilibrium preserving biodiversity, i.e., ensuring the coexistence of all species. The equilibrium point at which the ecosystem thrives is

$$\hat{E}_4 \equiv (\hat{w}_4, \hat{s}_4, \hat{v}_4) = \left( \frac{1}{b}e(H + \hat{v}_4), \frac{1}{br}[W(br - ce(H + \hat{v}_4))], \hat{v}_4 \right),$$

and $\hat{v}_4 > 0$ solves the following quadratic equation:

$$kc^2Wev^2 + (abr - kb^2r + 2kc^2WHe - kcWbr)v$$
$$+ aHbr - kcWHbr + kc^2WH^2e = 0.$$

Two positive roots arise if $bckrW < abr + 2c^2ekWH < bckrW + kb^2r$, while there is only one if $abr + kc^2WHe < kcWbr$. Thus the interior equilibrium is unique and feasible if and only if the latter condition holds and $br > ce(H+\hat{v}_4)$. This gives an upper bound on the location of this root.

For stability, let us analyze once again the characteristic equation at $\hat{E}_4$,

$$\lambda^3 + Q_1\lambda^2 + Q_2\lambda + Q_3 = 0. \tag{5.4}$$

The coefficients are

$$Q_1 = \frac{-bW\hat{s}_4\hat{v}_4 + rH^2\hat{w}_4 + 2rH\hat{w}_4\hat{v}_4 + r\hat{w}_4\hat{v}_4^2}{W(H+\hat{v}_4)^2},$$

$$Q_2 = \frac{s}{W(H+\hat{v}_4)^3}\left[ b^2kHW\hat{v}_4 - brH\hat{w}_4\hat{v}_4 - br\hat{w}_4(\hat{v}_4)^2 + kc^2W(H+\hat{v}_4)^3\hat{w}_4 \right],$$

$$Q_3 = \frac{bk\hat{w}_4\hat{s}_4\hat{v}_4}{W(H+\hat{v}_4)^3}\left[ brH - c^2WH\hat{s}_4 - c^2W\hat{s}_4\hat{v}_4 \right].$$

The Routh–Hurwitz criterion for $\hat{E}_4$ gives $Q_1 > 0$, $Q_3 > 0$ and $Q_1Q_2 - Q_3 > 0$. Here,

$$Q_1Q_2 - Q_3 = \frac{\hat{s}_4}{W^2(H+\hat{v}_4)^5} \cdot$$
$$\cdot \left[ r(kc^2W\hat{v}_4 + A)\hat{w}_4^2\hat{v}_4^4 + (b^2rW\hat{s}_4 + rB\hat{w}_4)\hat{w}_4\hat{v}_4^3 + D \right],$$

where $A = 5c^2kWH - br$, $B = 10Hc^2kW - 3br$, $C = rBH^2\hat{w}_4^2 + b^2HW\hat{s}_4(r\hat{w}_4 - bkW)$, and $D = \hat{v}_4^2 C + rAH^3\hat{w}_4^2\hat{v}_4 + rc^2kH^5W\hat{w}_4^2$.

We then find that $Q_1 > 0$ if and only if

$$rH^2\hat{w}_4 + 2rH\hat{w}_4\hat{v}_4 + r\hat{w}_4\hat{v}_4^2 > bW\hat{s}_4\hat{v}_4,$$

$Q_3 > 0$ if and only if

$$bHr > c^2HW\hat{s}_4 + c^2W\hat{s}_4\hat{v}_4,$$

and $Q_1Q_2 - Q_3 > 0$ if and only if

$$Q_1Q_2 - Q_3 \equiv r(c^2kW\hat{v}_4 + A)\hat{w}_4^2\hat{v}_4^4 + [b^2rW\hat{s}_4 + \hat{w}_4rB]\hat{w}_4\hat{v}_4^3 + D > 0. \quad (5.5)$$

Sufficient conditions for these conditions to hold are $10c^2kWH > 3br$ and $r\hat{w}_4 > bkW$. The latter should be satisfied together with the above feasibility conditions.

## Bifurcation analysis

We now investigate the stability behavior of the coexistence equilibrium as a function of the parameter $c$ using a relatively simple method. To ensure that the characteristic polynomial (5.4) has purely imaginary roots, it must have the form

$$(\lambda^2 + Q_2)(\lambda + Q_1) = 0, \quad (5.6)$$

which entails the equality $Q_1Q_2 = Q_3$ associated to (5.5). Also, a transversality condition must hold. Thus a value $c = c^*$ must exist such that

$$\text{(i)} \ g(c^*) = Q_1(c^*)Q_2(c^*) - Q_3(c^*) = 0, \quad \text{(ii)} \ \frac{d}{dc}\text{Re}(\lambda(c))|_{c=c^*} \neq 0. \quad (5.7)$$

The three roots of (5.6) are $\eta_1 = +i\sqrt{Q_2}$, $\eta_2 = -i\sqrt{Q_2}$, $\eta_3 = -Q_1$. In general, however, the roots are functions of $c$:

$$\eta_1(c) = \beta_1(c) + i\beta_2(c), \quad \eta_2(c) = \beta_1(c) - i\beta_2(c), \quad \eta_3(c) = -Q_1(c).$$

We now verify the transversality condition (ii) of (5.7) at $\eta_1$ and $\eta_2$. Substituting the general form of the root $\eta_j(c) = \beta_1(c) + i\beta_2(c)$ into (5.6) and taking the derivative, we find

$$K(c)\beta_1'(c) - L(C)\beta_2'(c) + A(c) = 0, \quad L(c)\beta_1'(c) + K(c)\beta_2'(c) + B(c) = 0,$$

with

$$K(c) = 3\beta_1^2(c) + 2Q_1(c)\beta_1(c) + Q_2(c) - 3\beta_2^2(c),$$
$$L(c) = 6\beta_1(c)\beta_2(c) + 2Q_1(c)\beta_2(c),$$
$$A(c) = \beta_1^2(c)Q_1'(c) + Q_2'(c)\beta_1(c) + Q_3'(c) - Q_1'(c)\beta_2^2(c),$$
$$B(c) = 2\beta_1(c)\beta_2(c)Q_1'(c) + Q_2'(c)\beta_2(c).$$

Since $L(c^*)B(c^*) + K(c^*)A(c^*) \neq 0$, we have

$$\left.\frac{d}{dc}(Re(\eta_j(c)))\right|_{c=c^*} = \left[\frac{LB+KA}{K^2+L^2}\right]_{c=c^*} \neq 0$$

and $\eta_3(c^*) = -Q_1(c^*) \neq 0$ so that the transversality condition holds. Thus a nondegenerate Hopf bifurcation occurs at $c = c^*$.

## Action of spraying and new simulations

Again human intervention in the ecosystem is modeled via insecticide spraying, which in the previous case mathematically corresponded to the addition of Dirac's delta functions at suitable instances in time to the system (5.1). It is modified here to take into account the fact that poisonous effects of the insecticide last in time. We assume then an exponential decay of these effects $\psi(T_i) = e^{-\alpha(t-T_i)}$, where $T_i \leq t$, $i = 1, 2, ..., N$ denote the spraying instances for (5.3). We then get

$$\frac{ds}{dT} = s(-a + \frac{kbv}{H+v} + kcw) - hKq\Sigma_{i=1}^{N}\psi(T_i), \quad (5.8)$$

$$\frac{dw}{dT} = rw(1 - \frac{w}{W}) - csw - h(1-q)\Sigma_{i=1}^{N}\psi(T_i),$$

$$\frac{dv}{dT} = v(e - \frac{bs}{H+v}) - hq\Sigma_{i=1}^{N}\psi(T_i).$$

The insecticide efficacy is modeled by the parameter $\alpha$, with the parameters $q$, $K$, $h$ retaining their meaning as for (5.2).

The behavior near the equilibria determined by the theoretical analysis can also be checked by means of suitable situations. In the absence of spraying, we show in Figure 5.7 the limit cycles obtained with the Hopf bifurcation described earlier. As far as the spraying is involved, general considerations on the collapsing of the ecosystem or some of its parts entirely similar to the ones of the previous model without saturation (5.1) also hold in this case, despite the different kind of modeling for the poison's effects used in this case. But for this more realistic situation (5.8) with insecticide effects lasting in time, the insects in the cultivated areas as well as the whole spider population become extinct even for very low parameter values.

### 5.2.1 Models comparison

A major difference between the two proposed models (5.1) and (5.3) is found in the cycles that can arise in the model introduced here, while they do arise in the simpler model. This is in agreement with recently collected vineyard field data, showing persistent oscillations for the spider population (Isaia et al., 2006a).

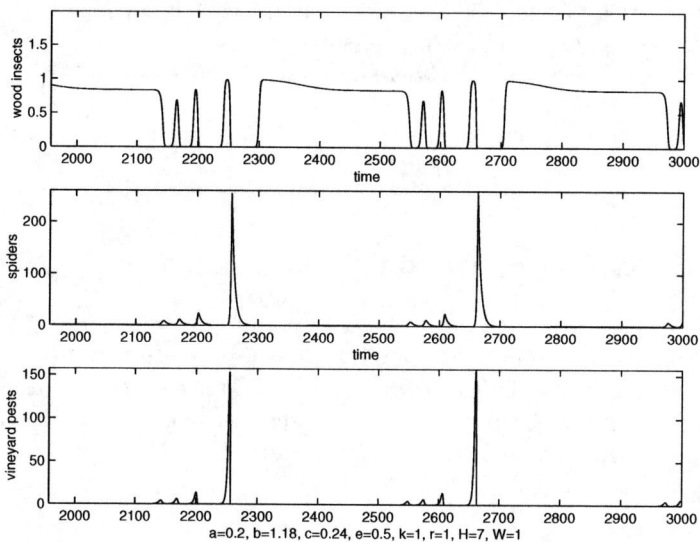

**FIGURE 5.7:** Limit cycles obtained by the Hopf bifurcation for the Holling type II saturation model (5.3).

From the numerical simulations, to maintain stability of the coexistence equilibrium point, the predation rates must be in an interval with lower and upper bounds respectively depending on the prey in the woods and the prey in the vineyards, i.e., the gap between the parameters $b$ and $c$ should not be too large. Should this difference be so large that the system oscillates, the stable equilibrium can be reobtained by an increase in the specific growth rate of the prey in the woods, i.e., the parameter $r$.

The method of spider augmentation led to 60% reduction of pesticide use, Maloney et al. (2003) and the effectiveness of biological control through spiders in rice fields and fruit orchards has been documented (Riechert and Bishop, 1990; Marc and Canard, 1997; Nyffeler and Benz, 1987). On the other hand, some insecticides are highly toxic for spiders, for instance, dimethoate gives 100% mortality for the lycosid *Trochosa ruricola* (Birnie et al., 1998) even when applied at lower concentrations than recommended, and methyl paranthion and pyrethroid cypermethring are poisons for the genus *Erigone* (Linyphiidae) (Brown et al., 1994; Huusela-Veistola, 1998).

These field observations should be coupled with our numerical experiments. In the simpler model (5.2) simulations showed that populations can recover after spraying if the atmospheric and environmental conditions are not too harsh. In the ecosystem (5.8), instead, even very low parameter values related to human intervention can permanently destabilize the ecosystem. Thus the results of these models agree with field data. Halley et al. (1996) discovered that if spraying is low and less frequently used, spiders can survive, but for larger quantities of the insecticide or if frequently used, chances are that the spider population will be wiped out. This is also documented by the mathematical models presented here.

## 5.3 Modeling the ballooning effect

The model we introduce here differs from the previous ones since in this context spiders are essentially stantial, so that we need to also distinguish the two spider populations living in the woods and in the vineyard, respectively, $s_w(t)$ and $s_v(t)$. We keep the same notation used earlier, but this time we also assume a vineyard insects carrying capacity $V$, with $V >> W$, to model the Langhe environment. The model then reads

$$\begin{aligned}
\frac{dw}{dT} &= bw(1 - \frac{w}{W}) - \ell w s_w, \\
\frac{ds_w}{dT} &= -cs_w + s_w[\tilde{\ell}w - \alpha w \tilde{\ell}\frac{V}{V+W}] + s_v \alpha v \tilde{k}\frac{W}{V+W}, \\
\frac{dv}{dT} &= av(1 - \frac{v}{V}) - kvs_v, \\
\frac{ds_v}{dT} &= -es_v + s_v[\tilde{k}v - \alpha v \tilde{k}\frac{W}{V+W}] + s_w \alpha w \tilde{\ell}\frac{V}{V+W}.
\end{aligned} \tag{5.9}$$

Here the first two equations express the fact that insects reproduce logistically and are subject to predation by the spiders living in the same environment. The two ecosystems are separate entities, as the web-making spiders tend to live in the same place, if there is no ballooning effect. Wind transport of young spiders is taken care of in the two equations modeling the growth of the spider populations. A fraction $\alpha$ of newborns is carried in the air and may land either in woods or surrounding vineyards with probability that is assumed proportional to the surface of the two patches, which in turn is related to their respective carrying capacities. The last two terms in the spiders' evolution equations describe this migration effect. The last one represents the immigration of newborns into the woods from the vineyards in the third equation and conversely in the fourth equation, while the third terms represent the emigration of wood-born spiders into the surrounding landscape in the third equation and the opposite way in the dynamics of the vineyard spiders. The predation efficiency of the latter is $\tilde{k}$. Notice that the transport effect of the wind is clearly the same for both species, while it is neglected for insects, as they are able to fly on their own and move toward the environment they prefer.

## Boundedness of the solutions

Now we are going to have a look at the properties of the model (5.9). The first step is to check solution boundedness.

The first two equations of (5.9) show that the two insect populations are bounded above:

$$w(t) \leq \max\{w(0), W\}, \quad \limsup_{t \to \infty} w(t) \leq W; \tag{5.10}$$
$$v(t) \leq \max\{v(0), V\}, \quad \limsup_{t \to \infty} v(t) \leq V.$$

Let us also define $\phi(T) = w + v + s_w + s_v$. Summing the equations (5.9) for any arbitrary $< \kappa \leq \min\{c, e, \ell - \tilde{\ell}, k - \tilde{k}\}$, we have

$$\frac{d\phi}{dT} + \kappa\phi \leq bw\left(1 - \frac{w}{W}\right) + av\left(1 - \frac{v}{V}\right)$$
$$- (c - \kappa)s_w - (e - \kappa)s_v - (\ell - \tilde{\ell} - \kappa)s_w - (k - \tilde{k} - \kappa)s_v$$
$$\leq bw\left(1 - \frac{w}{W}\right) + av\left(1 - \frac{v}{V}\right) \leq \frac{1}{4}[bW + aV] \equiv \bar{M}.$$

Then $\frac{d\phi}{dT} \leq -\kappa\phi + \bar{M}$, from which Gronwall's inequality shows that every population is bounded for all time:

$$\phi(t) \leq \exp(-\kappa t) + \frac{\bar{M}}{\kappa}[1 - \exp(-\kappa t)] \leq M. \tag{5.11}$$

## Long-term behavior

Now, we need to analyze the system's equilibria $E^{(i)} \equiv \left(w^{(i)}, v^{(i)}, s_w^{(i)}, s_v^{(i)}\right)$. The first three, the origin $E^{(0)}$ and the axial points with $w^{(1)} = W$, $w^{(2)} = V$, where the other components are zero, are easily ruled out, since they are unstable. Respectively, we find the eigenvalues

$$a, b, -c, -e;$$
$$a, -b, -e, -c + \tilde{\ell}\frac{W}{V + W}(V + W - \alpha V);$$
$$b, -a, -c, -e + \tilde{k}\frac{V}{V + W}(V + W - \alpha W).$$

The first interesting equilibrium is $E^{(3)} \equiv (W, V, 0, 0)$, whose eigenvalues are $-a, -b$, and the roots of the quadratic equation

$$\lambda^2 + \lambda\left[c - W\tilde{\ell} + e - V\tilde{k} + \frac{\alpha VW(\tilde{\ell} + \tilde{k})}{V + W}\right]$$
$$+ \left(c - W\tilde{\ell} + \frac{\alpha VW\tilde{\ell}}{V + W}\right)\left(e - V\tilde{k} + \frac{\alpha VW\tilde{k}}{V + W}\right) - \frac{V^2W^2\tilde{\ell}\tilde{k}\alpha^2}{(V + W)^2} = 0.$$

Sufficient conditions for which the Routh–Hurwitz criterion holds, so that $E^{(3)}$ is stable, are

$$c > W\tilde{\ell}, \quad e > V\tilde{k}. \tag{5.12}$$

The next two are still boundary equilibria, with only one zero component, and in them one of the two insect populations goes extinct. We find

$$E^{(4)} \equiv \left(0, \frac{e(V+W)}{\tilde{k}(V+W-\alpha W)}, \frac{a}{k}\frac{e\alpha W}{c(W+V-\alpha W)}\left[1 - \frac{e(V+W)}{\tilde{k}V(V+W-\alpha W)}\right], \right.$$
$$\left. \frac{a}{k}\left[1 - \frac{e(V+W)}{\tilde{k}V(V+W-\alpha W)}\right]\right) \tag{5.13}$$

$$E^{(5)} \equiv \left(\frac{c(V+W)}{\tilde{\ell}(V+W-\alpha V)}, 0, \frac{b}{\ell}\left(1 - \frac{c(V+W)}{\tilde{\ell}W(V+W-\alpha V)}\right),\right.$$
$$\left. \frac{bV\alpha}{\ell\,e}\left[\frac{c}{W+V-\alpha V}\right]\left[1 - \frac{c(V+W)}{\tilde{\ell}W(V+W-\alpha V)}\right]\right). \tag{5.14}$$

An easy verification based on the fact that $0 \le \alpha \le 1$ shows that the feasibility conditions reduce respectively to

$$\tilde{k}V(V+W-\alpha W) > e(V+W), \quad \tilde{\ell}W(V+W-\alpha V) > c(V+W). \tag{5.15}$$

For the stability analysis, let us calculate the Jacobian of the system $J^{(i)} \equiv J(E^{(i)})$:

$$\begin{pmatrix} b(1 - \frac{2w^{(i)}}{W}) - \ell s_w^{(i)} & 0 & -\ell w^{(i)} & 0 \\ 0 & a(1 - \frac{2v^{(i)}}{V}) - k s_v^{(i)} & 0 & -kv^{(i)} \\ (1-\alpha\frac{V}{V+W})\tilde{\ell}s_w^{(i)} & \alpha\frac{W}{V+W}\tilde{k}s_v^{(i)} & (1-\alpha\frac{V}{V+W})\tilde{\ell}w^{(i)} - c & \alpha\frac{W}{V+W}\tilde{k}v^{(i)} \\ \alpha\frac{V}{V+W}\tilde{\ell}s_w^{(i)} & \tilde{k}s_v^{(i)}(1-\alpha\frac{W}{V+W}) & \tilde{\ell}w^{(i)}\alpha\frac{V}{V+W} & \tilde{k}v^{(i)}(1-\alpha\frac{W}{V+W}) - e \end{pmatrix}$$

Thus, for equilibrium $E^{(4)}$, the eigenvalues are $-c$, $b - \ell s_w^{(4)}$, and the roots of the quadratic

$$\lambda^2 - \mathrm{tr}(\tilde{J}(E^{(4)}))\lambda + \det(\tilde{J}(E^{(4)})) = 0, \tag{5.16}$$

where

$$\tilde{J}(E^{(4)}) \equiv \begin{pmatrix} a(1 - 2\frac{v^{(4)}}{V}) - ks_v^{(4)} & -kv^{(4)} \\ \tilde{k}s_v^{(4)}(1-\alpha\frac{W}{V+W}) & \tilde{k}v^{(4)}(1-\alpha\frac{W}{V+W}) - e \end{pmatrix} \tag{5.17}$$

$$\equiv \begin{pmatrix} a(1 - 2\frac{v^{(4)}}{V}) - ks_v^{(4)} & -kv^{(4)} \\ \tilde{k}s_v^{(4)}(1-\alpha\frac{W}{V+W}) & 0 \end{pmatrix}.$$

The Routh–Hurwitz conditions for stability are $-\operatorname{tr}(J^{(4)}) > 0$ and $\det(J^{(4)}) > 0$, and in view of $0 \le \alpha \le 1$ we find

$$-\operatorname{tr}(J^{(4)}) \equiv k s_v^{(4)} - a + 2\frac{a}{V}v^{(4)} = ae\frac{V+W}{kV(V+W-\alpha W)} > 0,$$

$$\det(J^{(4)}) \equiv k\tilde{k}v^{(4)} s_v^{(4)} \frac{V+W-\alpha W}{V+W} > 0.$$

Stability of $E^{(4)}$ follows by just requiring

$$b < \ell s_w^{(4)}. \tag{5.18}$$

For $E^{(5)}$, the analysis is similar. We find the two eigenvalues, $-e$, $a - k s_v^{(5)}$, and then the eigenvalues of

$$\tilde{J}(E^{(5)}) \equiv \begin{pmatrix} b(1 - 2\frac{w^{(5)}}{W}) - \ell s_w^{(5)} & -\ell w^{(5)} \\ \tilde{\ell} s_w^{(5)}(1 - \alpha\frac{V}{V+W}) & \tilde{\ell} w^{(5)}(1 - \alpha\frac{V}{V+W}) - c \end{pmatrix} \tag{5.19}$$

$$\equiv \begin{pmatrix} -\frac{bc(V+W)}{\tilde{\ell}W(V+W-\alpha V)} & -\ell w^{(5)} \\ \tilde{\ell} s_w^{(5)}(1 - \alpha\frac{V}{V+W}) & 0 \end{pmatrix}$$

The Routh–Hurwitz criterion yields again

$$-\operatorname{tr}(\tilde{J}(E^{(5)})) = \frac{bc(V+W)}{\tilde{\ell}W(V+W-\alpha V)} > 0,$$

$$\det(\tilde{J}(E^{(5)})) = \ell\tilde{\ell} w^{(5)} s_w^{(5)} \frac{V+W-\alpha V}{V+W} > 0,$$

and stability of $E^{(5)}$ is ensured by

$$a < k s_v^{(5)}. \tag{5.20}$$

We analyze sufficient conditions for the existence of the coexistence equilibrium $E^{(6)} \equiv E^* = (w^*, v^*, s_w^*, s_v^*)$. Solving the first two equations of (5.9) in terms of $w$ and $v$, we find

$$w^* = \frac{W}{b}(b - \ell s_w^*), \quad v^* = \frac{V}{a}(a - k s_v^*). \tag{5.21}$$

For feasibility of $E^*$ the opposite conditions of (5.18) and (5.20), but both evaluated at $E^*$, must hold:

$$a > k s_v^*, \quad b > \ell s_w^*. \tag{5.22}$$

Let

$$A \equiv \frac{\ell}{b}\tilde{\ell}\frac{W}{V+W}(V+W-\alpha V) > 0,$$

$$\tilde{B} \equiv \frac{1}{2}[\tilde{\ell}\frac{W}{V+W}(V+W-\alpha V) - c] \in \mathbf{R},$$

$$C \equiv \frac{1}{2}\alpha\tilde{k}\frac{VW}{V+W} > 0, \quad D \equiv \frac{k}{a}\frac{VW}{V+W}\tilde{k}\alpha > 0,$$

$$E \equiv \frac{\ell}{b}\tilde{\ell}\frac{VW}{V+W}\alpha > 0, \quad F \equiv \frac{1}{2}\tilde{\ell}\alpha\frac{VW}{V+W} > 0,$$

$$\tilde{G} \equiv \frac{1}{2}[\tilde{k}\frac{V}{V+W}(V+W-\alpha W) - e] \in \mathbf{R},$$

$$H \equiv \frac{k}{a}\frac{V}{V+W}\tilde{k}(V+W-\alpha W) > 0.$$

Notice that in this context the feasibility of $E^{(4)}$ is equivalent to $\tilde{G} \geq 0$, and the one of $E^{(5)}$ corresponds to $\tilde{B} \geq 0$. Substitute now (5.21) into the last two equations of (5.9), to get two equations in $s_w$ and $s_v$,

$$-As_w^2 + 2\tilde{B}s_w + 2Cs_v - Ds_v^2 = 0, \tag{5.23}$$

$$-Es_w^2 + 2Fs_w + 2\tilde{G}s_v - Hs_v^2 = 0. \tag{5.24}$$

Both (5.23) and (5.24) are ellipses in the $(s_w, s_v) \equiv (x, y)$ plane, both crossing the origin. Notice that the origin here in $\mathbf{R}^2$ gives for the whole model the three equilibria we named $E^{(0)}$, $E^{(1)}$, and $E^{(2)}$. We now study the possible further intersections of the two ellipses.

The first ellipse with center $(\frac{\tilde{B}}{A}, \frac{C}{D})$ and vertices

$$\left(\frac{\tilde{B}}{A} \pm \frac{1}{AD}\sqrt{\tilde{B}^2D^2 + AC^2D}, \frac{C}{D}\right) \tag{5.25}$$

$$\left(\frac{\tilde{B}}{A}, \frac{C}{D} \pm \frac{1}{AD}\sqrt{A^2C^2 + A\tilde{B}^2D}\right)$$

intersects the positive vertical semiaxis at $P \equiv (x_P, y_P) = (0, 2\frac{C}{D})$ and the horizontal axis at $Q \equiv (x_Q, y_Q) = (2\frac{\tilde{B}}{A}, 0)$. Its derivative at the point $P$ is $y_1'(P) = \frac{\tilde{B}}{C}$, and at the origin it is $y_1'(O) = -\frac{\tilde{B}}{C}$. It is then easily seen to have the same sign as the abscissa $x_Q$ at $P$, while at the origin it always has its opposite sign.

The second one has center at $(\frac{F}{E}, \frac{\tilde{G}}{H})$, vertices

$$\left(\frac{F}{E}, \frac{\tilde{G}}{H} \pm \frac{1}{EH}\sqrt{\tilde{G}^2 E^2 + EF^2 H}\right) \tag{5.26}$$

$$\left(\frac{F}{E} \pm \frac{1}{EH}\sqrt{F^2 H^2 + E\tilde{G}^2 H}, \frac{\tilde{G}}{H}\right),$$

and a further intersection with the positive vertical semiaxis at $R \equiv (x_R, y_R) = (0, 2\frac{\tilde{G}}{H})$, with the horizontal axis at $S \equiv (x_S, y_S) = (2\frac{F}{E}, 0)$. The derivative at the point $R$ is $y_2'(R) = \frac{F}{\tilde{G}}$, and at the origin it is $y_2'(O) = -\frac{F}{\tilde{G}}$. It thus has the same sign as the height $y_R$ when evaluated at $R$ and opposite sign when evaluated at the origin. Thus both ellipses have axes parallel to the coordinate axes.

Four cases of sufficient conditions for the existence of $E^*$ are obtained from the above results:

1A) $\tilde{B} > 0$, $\tilde{G} > 0$; here existence and uniqueness of an intersection between $y_1$ and $y_2$ in the first quadrant, which gives a feasible $E^*$ if conditions (5.22) also hold, are guaranteed by either

$$CH > D\tilde{G} \text{ and } FA > E\tilde{B} \tag{5.27}$$

or, alternatively,

$$CH < D\tilde{G} \text{ and } FA < E\tilde{B}. \tag{5.28}$$

1B) $\tilde{B} > 0$, $\tilde{G} < 0$; we obtain only one condition,

$$FA > E\tilde{B}. \tag{5.29}$$

2A) For $\tilde{B} < 0$, $\tilde{G} > 0$, the condition now is

$$CH > D\tilde{G}. \tag{5.30}$$

2B) For $\tilde{B} < 0$, $\tilde{G} < 0$ comparing the derivatives of $y_1$ and $y_2$ at the origin, $y_1'(O) < y_2'(O)$, we impose an intersection in the first quadrant, in other words,

$$FC < \tilde{G}\tilde{B}. \tag{5.31}$$

Conditions (5.27) are always satisfied upon substituting the values of $A$, $\tilde{B}$, $C$, $D$, $E$, $F$, $\tilde{G}$. Indeed, they correspond to requiring $c > 0$ and $e > 0$ and geometrically to $x_S \geq x_Q$ and $y_P \geq y_R$, in which equality holds only for the cases $c = 0$ and $e = 0$, respectively. We find also that (5.29) and (5.30) hold always, together with (5.22). The latter imply that the intersection $E^*$ of the ellipses must lie within the square with vertices $OPUS$, where $U \equiv (x_S, y_P) \equiv (2\frac{F}{E}, 2\frac{C}{D}) \equiv (\frac{b}{\ell}, \frac{a}{k})$. But this square is inscribed into the first ellipse in the cases 1A and 1B corresponding to $x_Q > 0$, i.e., $\tilde{G} > 0$, so that in such cases $E^*$ is infeasible since one of the first two coordinates, $v, w$, is negative. In cases 2A and 2B the arc $OP$ of the first ellipse not containing $Q$ lies entirely in the square $OPUS$ and (5.30) and (5.31) guarantee an intersection. It follows that $E^*$ is always feasible. Figure 5.8 depicts such situations.

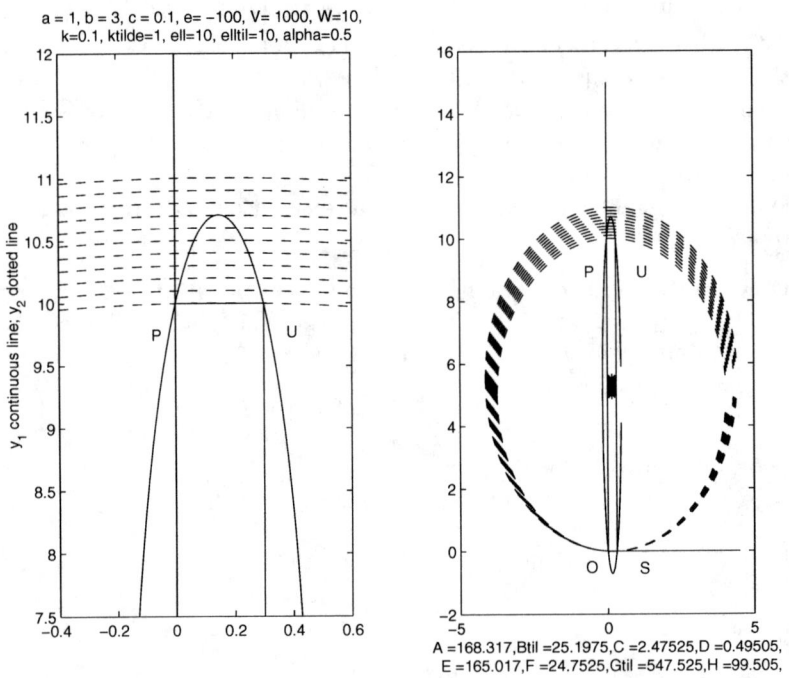

**FIGURE 5.8**: Illustration of the possible ellipses intersections; on the left, a blowup of the right picture.

Let us look now at some particular cases, starting from $e = 0$. Two new equilibria then arise, $\bar{E}^{(0)} \equiv (0, 0, 0, \bar{s}_v)$ for $\bar{s}_v \equiv y_P \equiv \frac{a}{b}$, or also $\bar{E}^{(1)} \equiv (W, 0, 0, \bar{s}_v)$. For $c = 0$ we find $\bar{E}^{(2)} \equiv (0, 0, \bar{s}_w, 0)$ and $\bar{E}^{(3)} \equiv (0, V, \bar{s}_w, 0)$, with $\bar{s}_w \equiv \frac{b}{\ell}$. Substituting into the Jacobian, we find the eigenvalues $a, b, -c, 0$ giving instability for $\bar{E}^{(0)}$ and $-b, a - k\bar{s}_v \equiv 0, -c + \tilde{\ell}W(1 - \alpha\frac{V}{V+W}), 0$ for $\bar{E}^{(1)}$. The stability condition for the latter follows if $\tilde{\ell}W(1 - \alpha\frac{V}{V+W}) < c$, i.e., $\tilde{B} < 0$, or also $x_Q < 0$. Case 2A makes $\bar{E}^{(1)}$ stable. The eigenvalues of $\bar{E}^{(2)}$ are $b - \ell\bar{s}_w \equiv 0, -a, 0, \tilde{k}\frac{V}{V+W}(V + W - \alpha W) - e$. Again, $\bar{E}^{(2)}$ is stable if $\tilde{G} < 0$, i.e., for $y_R < 0$, so that case 2A makes it unstable, while case 2B renders it stable. The equilibrium $\bar{E}^{(3)}$ is always unstable, as the eigenvalues are $0, a, 0, -e$.

Now as $e$ grows, the axis intersection between the ellipses moves away from the $s_v$-axis. It retains its stability and thus gives a stable manifold, the arc $OP$ of the ellipse not containing the point $Q$, where $E^*$ is located. It can be shown that for $e < 0$ this is the stable branch of a saddle-node bifurcation of an infeasible equilibrium. Similar remarks hold exchanging $e$ with $c$, $P$ with $S$, and $Q$ with $R$ and reversing the stability concepts.

Condition (5.31) is equivalent to the restriction $y_1'(O) = -\frac{\tilde{B}}{C} < -\frac{F}{\tilde{G}} = y_2'(O)$, i.e., to $FACH > E\tilde{B}\tilde{G}D > EFCD$, which yields $\frac{A}{E} > \frac{D}{H}$. Substituting the parameter values and simplifying, we obtain $(V+W-\alpha V)(V+W-\alpha W) > VW$. From this the true inequality $(1-\alpha)(V^2+W^2)+VW(1-\alpha)^2 > 0$ follows, so that (5.31) always holds. Case 2B always gives a feasible intersection.

## Hopf bifurcation

We also analyze in detail this case although the method is similar to the one used for the wanderer spiders models, since it applies here to a fourth degree equation, and the bifurcation condition (5.35) below differs from the one previously derived for a cubic equation. The characteristic polynomial in this case is the quartic in $\lambda$,

$$\sum_{i=0}^{4} b_i \lambda^i, \qquad (5.32)$$

and the following condition ensures that it has purely imaginary eigenvalues:

$$\sum_{i=0}^{4} b_i \lambda^i = (\lambda^2 + \eta^2)(\lambda^2 + \beta\lambda + \zeta) = 0. \qquad (5.33)$$

Expanding and equating coefficients of the same powers gives

$$b_3 = \beta, \quad b_2 = \zeta + \eta^2, \quad b_1 = \beta\eta^2, \quad b_0 = \eta^2\zeta, \qquad (5.34)$$

which in turn are equivalent to the following relationship for the coefficients of (5.32):

$$b_3 b_2 b_1 = b_3^2 b_0 + b_1^2. \qquad (5.35)$$

Let us denote the principal minors of order $k$ of the Jacobian by $M_k(J)$. Assuming that the following sums are taken over all possible principal minors of the given order, we then find $b_3 = -\text{tr}(J)$, $b_2 = \sum M_2(J)$, $b_1 = -\sum M_3(J)$, $b_0 = \det(J)$. As a bifurcation parameter we can take $e$, since it appears only in the element $J_{44}$ with a negative sign, or $c$ appearing only in $J_{33}$ also with negative sign. It turns out then that (5.35) is a cubic equation in $e$:

$$b_3 b_2 b_1 - b_3^2 b_0 - b_1^2 \equiv \sum_{i=0}^{3} c_i e^i. \qquad (5.36)$$

To guarantee the existence of a real positive root $e^\dagger$, the signs of the constant term and of the coefficient of $e^3$ must be opposite. The condition (5.35) can be rewritten as

$$\text{tr}(J)\sum M_2(J)\sum M_3(J) = (\text{tr}(J))^2 \det(J) + \left(\sum M_3(J)\right)^2. \quad (5.37)$$

Let $M_2[J_{(1:3,1:3)}]$ denote the principal minors of order two of the submatrix $J(1:3,1:3)$ of the Jacobian, formed by its first three rows and columns, and $\det[J_{(1:3,1:3)}]$ be the corresponding determinant. From (5.37) we extract the highest order terms containing $J_{44}$, obtaining

$$J_{44} \cdot J_{44}[J_{11} + J_{22} + J_{33}] \cdot J_{44}\{\sum M_2[J_{(1:3,1:3)}]\} = [J_{44}]^2 \cdot J_{44} \det[J_{(1:3,1:3)}].$$

The last term $(\sum M_3(J))^2$ gives only a contribution to the coefficient of $J_{44}^2$ and therefore has not been considered. The coefficient of $J_{44}^3$ is thus

$$c_3 \equiv [J_{11} + J_{22} + J_{33}][\sum_i \tilde{M}_2] - \det[J_{(1:3,1:3)}]. \quad (5.38)$$

The term not containing $e$ is

$$c_0 \equiv \sum_i M_2[J_{(1:3,1:3)}] \det[J_{(1:3,1:3)}] \quad (5.39)$$
$$- [J_{11} + J_{22} + J_{33}][\sum_i (-1)^i \hat{M}_3(J_{i4})],$$

$\hat{M}_3(J_{i4})$ being the minor of the element $J_{i4}$ in $J$. The existence condition of $e^\dagger > 0$ for which (5.33) holds is then

$$c_0 c_3 < 0. \quad (5.40)$$

To detect a Hopf bifurcation we need then only worry about the transversality condition. The roots of (5.33) are in general of the form

$$\lambda_1(e) = \xi_1(e) + i\xi_2(e), \quad \lambda_2(e) = \xi_1(e) - i\xi_2(e), \quad (5.41)$$

$$\lambda_\pm(e) = \frac{1}{2}\left[-b_3 \pm \sqrt{b_3^2 - \frac{4}{b_1}b_0 b_3}\right].$$

Differentiating the characteristic equation with respect to $e$ and denoting this by a prime, we get

$$\gamma_{11}\xi_1' - \gamma_{12}\xi_2' = \gamma_{10}, \quad \gamma_{12}\xi_1' + \gamma_{12}\xi_2' = \gamma_{20},$$

where

$$\gamma_{11} = 4\xi_1^3 - 12\xi_1\xi_2^2 + 3\beta(\xi_1^2 - \xi_2^2) + 2(\eta^2 + \zeta)\xi_1 + \eta^2\beta$$
$$\gamma_{12} = 12\xi_1^2\xi_2 - 4\xi_2^3 + 6\beta\xi_1\xi_2 + 2(\eta^2 + \zeta)\xi_2$$
$$\gamma_{10} = \beta'(3\xi_1\xi_2^2 - \xi_1^3) + [(\eta^2)' + \zeta'](\xi_2^2 - \xi_1^2)$$
$$\quad - \xi_1[(\eta^2)'\beta + (\eta^2)\beta'] - ((\eta^2)'\zeta + \eta^2\zeta')$$
$$\gamma_{20} = \beta'(\xi_2^3 - 3\xi_1^2\xi_2) - \xi_1\xi_2((\eta^2)' + \zeta') - \xi_2((\eta^2)'\beta + \eta^2\beta').$$

On solving for $\xi_1'$ and $\xi_2'$ in the above system, we find

$$\xi_1' = \frac{\gamma_{10}\gamma_{11} + \gamma_{20}\gamma_{12}}{\gamma_{11}^2 + \gamma_{22}^2}, \quad \xi_2' = \frac{\gamma_{20}\gamma_{11} - \gamma_{10}\gamma_{12}}{\gamma_{11}^2 + \gamma_{22}^2}.$$

Letting

$$m_2^\dagger = \sum M_2[J_{(1:3,1:3)}(e^\dagger)], \quad m_3^\dagger = \sum M_3[J(e^\dagger)]$$
$$t^\dagger = \operatorname{tr}[J(e^\dagger)] \qquad d^\dagger = \det[J(e^\dagger)],$$

the transversality condition $\xi_1'(e^\dagger) \neq 0$ in terms of the original entries in the Jacobian matrix is

$$(d^\dagger + m_2^\dagger)(m_2^\dagger t^\dagger - m_3^\dagger)(t^\dagger)^4 + 2(m_3^\dagger)^4[(t^\dagger)^2 + (m_3^\dagger)^3] \neq 0. \tag{5.42}$$

Numerical simulations show that the Hopf bifurcation does indeed arise, so that sustained oscillations are possible. Figure 5.9 supports our statement.

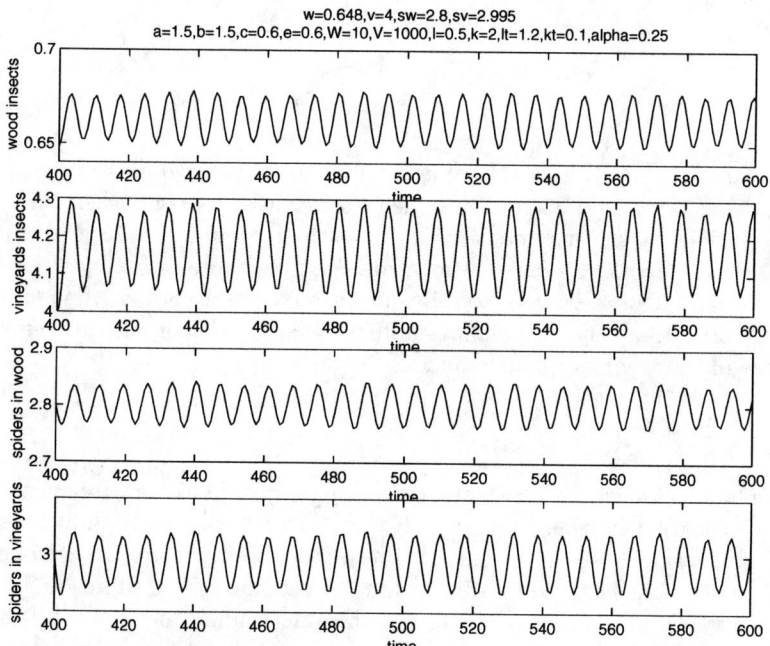

**FIGURE 5.9**: Sustained oscillations for the ballooning model (5.9); ecosystem survival.

### 5.3.1 Spraying effects and human intervention

Again we can investigate the results of insecticide spraying, similarly as done before. Observe that in general the vineyard is assumed to be sprayed from planes flying not too high above the ground and aiming at the vineyards, but due to the wind effect the insecticide may land on the woods as well. Also as already mentioned, the poison is meant to act on the insects and pests of the vineyards, but it may very well affect the spiders, too. Let $T_i$ denote the spraying instants, focusing on the "instant killing effect" of the poison; we thus obtain, as for the wanderer spiders models, the modified model

$$\frac{dw}{dT} = w\left[b\left(1 - \frac{w}{W}\right) - \ell s_w\right] - h\left(1 - q\right)\delta\left(T_i\right), \qquad (5.43)$$

$$\frac{dv}{dT} = v\left[a\left(1 - \frac{v}{V}\right) - k s_v\right] - hq\delta\left(T_i\right),$$

$$\frac{ds_w}{dT} = s_w\left[-c + \tilde{\ell}w\left(1 + \alpha\left(-\frac{V}{V+W}\right)\right)\right]$$
$$+ \frac{W}{V+W}\tilde{k}vs_v\alpha - K\left(1 - q\right)\delta\left(T_i\right),$$

$$\frac{ds_v}{dT} = s_v\left[-e + \tilde{k}v\left(1 + \alpha\left(-\frac{W}{V+W}\right)\right)\right]$$
$$+ \frac{V}{V+W}\tilde{\ell}ws_w\alpha - Kq\delta\left(T_i\right),$$

with parameters that have the same meaning as introduced in the corresponding subsection. Again, as the Dirac delta function suddenly pushes downward the solutions at the instants $T_i$, if such perturbation is large enough and if a basin of attraction of some other stable equilibrium exists, the trajectories tend toward this alternative equilibrium instead of the former. Thus the human intervention may very well destabilize the ecosystem, as already remarked in case of the wanderer spiders models.

### 5.3.2 Ecological discussion

We have discovered that the model presented contains three relevant equilibria that are conditionally stable, $E^{(3)}$, $E^{(4)}$, and $E^{(5)}$. In both at least one population vanishes. In the second, the wood insects disappear, causing loss of diversity. Thus this equilibrium should be avoided. The situation for $E^{(3)}$ is even worse, as it ia a spider-free equilibrium, leading to the disappearance of predators and survival only of possibly harmful insects. $E^{(5)}$ is desirable as it contains no vineyard-living prey. From the farmer's point of view, it is the equilibrium one should strive for. Finally, looking at the problem from the pespective of the ecologist, $E^* \equiv E^{(6)}$ shows the whole ecosystem to thrive, which is good for environmental biodiversity.

Stability of all these equilibria is always conditional, for $E^{(3)}$ being given by upper bounds on the woods and vineyard insects' carrying capacities, for $E^{(4)}$ and $E^{(5)}$ being ensured by keeping both spider populations high enough.

Around the ecological coexistence equilibrium $E^* \equiv E^{(6)}$ a Hopf bifurcation has been shown to arise, with sustained oscillations in time for all the model populations. Its consequences may be relevant for the ecosystem, since if the periodic solutions have too large amplitude, they may end up on one of the coordinate axes and therefore indicate the disappearance of at least one of the species and migration of the solution trajectories of the dynamical system toward one of the other equilibria, with possibly unpredictable consequences for the environment. This is to be noted, as some external unforeseeable circumstances, such as fast climatic variations, may very well result in a change of the spiders' death rate and lead to such phenomenon.

Similar remarks as for the previous models on the effects of human intervention with spraying hold here as well. In the case of a real application, a thorough study of the basins of attractions of the various stable equilibria is needed to find the impact of human intervention.

Field investigations demonstrate that phytophagous pests may have economically relevant outbreaks after treatments with pesticides that caused a reduction of the spiders feeding on them (see, e.g., Birnie et al. (1998), Putman and Herne (1966), Hukusima and Kondo (1962)), even to become extinct if insecticide is too high or the applications are too frequent (Mansour et al., 1980). The spider population reappears, however, if the interval between sprayings is sufficiently long.

Much richer and much more complex computer models aimed at the Danish farmland are those by Thomas et al. (2003), Thomas et al. (1990), Topping and Sunderland (1994), Topping (1999), and Thorbek and Topping (2005), where the heterogeneous landscape is simulated via a grid of cells for possibly different habitats. Spiders in various states of development are also considered (Topping et al., 2003). All these models, including the ones presented in this chapter, try to provide insights for the management of sustainable agricultural systems, using natural predators for pest control, possibly coupled with meteorological and farming data adapted to the various local conditions.

# Chapter 6

# Epidemic models

In this chapter we give a brief look at classical epidemics models. For the reader unfamiliar with these topics, this outline is meant as an introduction to some basic terminology, ideas, and equations that will be needed in Chapter 7 and in other parts of the book. Among the many review papers that give a more thorough picture of this field, we mention Hethcote (2000). But of course various chapters are devoted to it in several recent and classical books, for instance, Murray (1989), Anderson and May (1991), Brauer and Castillo-Chavez (2001), and Edelstein-Keshet (1988). The natural outcome of these studies is the determination of rational vaccination policies to prevent epidemics outbreaks and to control infectious diseases propagation (Dieckmann et al., 2002).

We illustrate first the basic classical epidemic models, introducing the terminology for the various subpopulations involved in the dynamical system. Then some more recent models are presented. In Section 6.3, a structured model is presented in which the biological age is accounted for together with the development stage reached by the diseased individuals. The final Section 6.4 presents a case study on an economically relevant disease affecting farmed animals.

## 6.1  Basic epidemic models

Mathematical epidemiology stems from the papers by McKendrick (1926) and von Foerster (1959), at about the same time of the original researches of Lotka and Volterra in population theory.

Basically, these epidemic models partition the population into several distinct classes, accounting for the individuals that may catch the disease, those that already are infected and can spread the infection, and those that are removed from the population. This last class can be formed by the recognized carriers, who are then quarantined so that they do not further spread the disease among the sound individuals, or by the people that have recovered from the disease, or even by those still affected by it but which are no more able to spread it. An important factor in infectious diseases, the one that mainly motivates the mathematical investigation of the situation, is the fact that very often the infectious people, those who can infect other individuals, are not recognized as being disease affected, since they do not yet have the symptoms. For instance, people affected by measles and chickenpox can infect other individuals only between the 10th and the 14th days of incubation, and the 11th and the 14th days respectively, and this occurs well before they are recognized as disease carriers. Different other possibilities also arise, such as accounting for an intermediate class of infected but not yet infectious individuals, or individuals that have become infected but for which the disease has not yet fully developed, and they are still in a stage preceding the one of being infectious. Such cases could be modeled for instance by introducing suitable delays in the differential equations governing the dynamics to simulate the incubation period of the disease. Later on in this chapter we will also consider a generalization in which age for the individuals and the "stage" of the disease are explicitly built into the model.

A second reason for the study of mathematical epidemiology is the development of meaningful vaccination policies, with the intent of eradicating the disease. This is possible because the so-called threshold theorems state the conditions under which the disease naturally vanishes.

The diseases we consider at first are the infectious ones, which are transmitted upon contact between the infectious and the susceptible individual. Other situations are possible, such as diseases transmitted by a vector, in general an insect or a worm, or viral diseases.

Now, we are going to list the basic assumptions leading to the classical epidemic model (e.g., see Frauenthal, 1980). They are discussed below and will be further analyzed and removed or changed in the following sections:

a) The disease is transmitted by contact between an infectious individual and a susceptible individual;

b) there is no latent period for the disease, hence a susceptible becomes instantaneously infected upon "successful" contact with a disease carrier;

c) all susceptible individuals are susceptible in the same way;

d) the infected individuals are all equally infectious; and

e) the population under consideration is fixed in size, so no demographics are accounted for in the classical models, this implying that no births or migration occur and all the deaths are taken into account.

A final word concerns the outcome of the disease. There are instances in which the disease is overcome and the individual remains immune for life, such as chickenpox, measles, mumps, and rubella. In other instances the individual recovers from the illness but may catch it again and again, as it happens for example for the common cold. In still more unfortunate cases, the disease, once contracted, cannot be overcome and is carried for life. AIDS represents such an example for humans, but in a later section we will analyze a specific example related to farming animals.

### 6.1.1 Simplest models

Consider at first what is called the SI epidemic model; as mentioned previously, it represents the case of the disease that, once contracted, cannot be overcome. Let $I(T)$ represent the sizes of the infectives at time $T$; see, e.g., Hethcote and Levin (1989, p. 204). Let $\tilde{\sigma}(T)$ denote the per infective capita rate of contacts leading to new infectives. Taking $\tilde{\sigma}(T) \equiv \sigma_0$, a simple balance consideration similar to the one that led to the Malthus model in population theory gives

$$I(T + \Delta T) = I(T) + \sigma_0 I(T) \Delta T,$$

and letting $\Delta T \to 0$, we obtain the simple model with an exponential solution:

$$\frac{dI}{dT} = \sigma_0 I(T), \quad I(T) = I(0) e^{\sigma_0 T}. \tag{6.1}$$

Clearly in this model the population is assumed to be infinite in size. In any case, the prediction is that every individual will eventually contract the disease, $I(T) \to \infty$ as $T \to \infty$. To construct a more meaningful model, let us now introduce the function $S(T)$ representing the sizes of the susceptibles. In this case, we then assume $S(T) + I(T) = N$, with $N$ constant, independent of time. Also, $\tilde{\sigma}$ now represents the rate of contacts leading to new infectives, per infective per susceptible. In other words, disease transmission is proportional to encounters among the individuals of the two subpopulations and the latter are proportional to the product of their sizes. Another way of viewing the situation is to say that $\tilde{\sigma}$, which was formerly constant, is becoming here a function of time, indirectly through $S(T)$, namely, $\tilde{\sigma}(T) \equiv \sigma S(T)$. Thus (6.1) gets modified as follows:

$$\frac{dS}{dT} = -\sigma I(T) S(T), \quad \frac{dI}{dT} = \sigma I(T) S(T). \tag{6.2}$$

On using the constraint on the fixed size total population, we can eliminate one variable, to obtain the logistic-like equation

$$\frac{dI}{dT} = \sigma I(T)[N - I(T)], \quad \frac{dI}{I} + \frac{dI}{N-I} = \sigma N dT$$

whose solution is then

$$I(T) = I(0) N \frac{e^{\sigma N T}}{N - I(0)(1 - e^{\sigma N T})} = \frac{I(0) N}{N e^{-\sigma N T} - I(0)(e^{-\sigma N T} - 1)} \to N$$

as $T \to \infty$. Thus the "final" number of infected individuals equals the total population size, i.e., again every one catches the disease. Notice also that as

$N \to \infty$ and $\sigma N \to \sigma_0$ we find $I(T) \to I(0)e^{\sigma_0 T}$, i.e., the model (6.2) in the limit as the population size becomes infinite recovers the solution of (6.1). It is also interesting to consider the epidemics curve, defined by how fast the disease propagates, and this is measured by the rate of change of infectives in the population

$$E(T) \equiv \frac{dI(T)}{dT} = \frac{\sigma(N - I(0))N^2 I(0) e^{\sigma NT}}{[N - I(0)(1 - e^{\sigma NT})]^2}.$$

The plots of the solutions and the epidemics curves are contained in Figure 6.1.

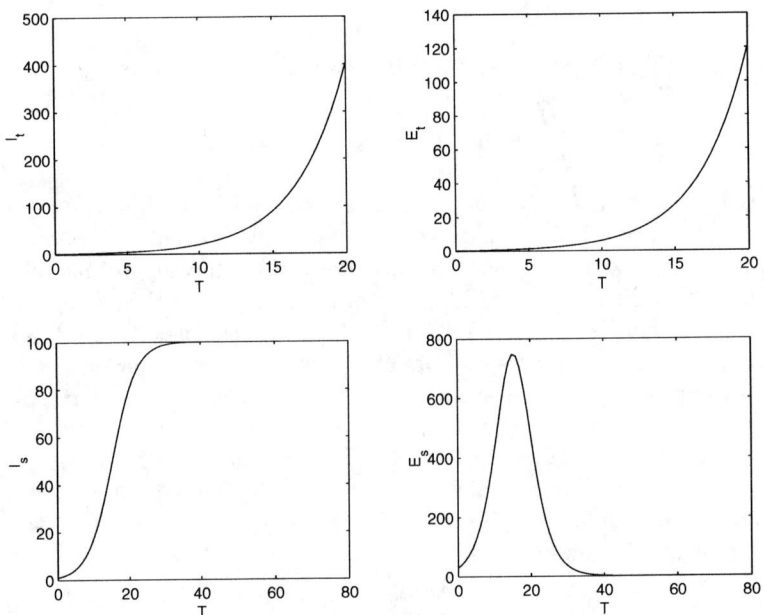

**FIGURE 6.1**: Top: trivial model; left: infectives curve $I(T)$; right: epidemics curve $E(T)$; bottom: SI model; left: infectives curve $I(T)$; right: epidemics curve $E(T)$.

The reason for which the whole population is ultimately affected is the fact that no individual can overcome the illness. Thus we finally allow the disease recovery, and introduce the class of removed individuals $R(T)$, which come from healing the infected at a per capita rate $\gamma$. The conservation condition in this case must then be written as $S(T) + I(T) + R(T) = N$. The complete model is then

$$\frac{dS}{dT} = -\sigma I(T)S(T), \quad \frac{dI}{dT} = \sigma I(T)S(T) - \gamma I(T), \quad \frac{dR}{dT} = \gamma I(T). \quad (6.3)$$

Notice that the infectives equation can be rewritten as

$$\frac{dI}{dT} = \sigma I(T)\left[S(T) - \frac{\gamma}{\sigma}\right], \qquad (6.4)$$

and therefore the epidemics outbreak, meaning the rapid increase of the infective numbers, is avoided and the disease disappears if $S(T) < \frac{\gamma}{\sigma}$, since then the derivative will be negative. $R_0 \equiv \frac{S(0)\sigma}{\gamma}$ is the basic reproduction number of the disease; if $R_0 < 1$ then the epidemics vanishes naturally. On solving formally (6.4) we find

$$I(T) = I(0)\exp\left[\sigma\int_0^T\left(S(t) - \frac{\gamma}{\sigma}\right)dt\right].$$

## 6.1.2 Standard incidence

In all the previous models the basic form of the incidence has been borrowed by the considerations made on chemical reactions (e.g., see Gray and Scott, 1990), namely, using the mass action law that takes the rate of encounters between the subpopulations to be constant $\sigma SI$. However, another form is in common use, named standard incidence. The basic idea is the fact that in a large population one individual cannot encounter all the others all the time; rather, it meets regularly with a small subset and with a larger number of people less frequently. The rate of encounters therefore should reflect this remark, and therefore be an inverse function of the population size. Once again, hidden here is the law of diminishing returns. This way of thinking involves then a nonlinearity, and the constant incidence rate now becomes $\sigma \equiv \sigma(N) = \sigma_0 \frac{1}{N}$. It thus apparently seems to make things mathematically worse. However, the contrary happens, as it allows us to reformulate the model not using the subpopulation *sizes* but using their *fractions*. Indeed, let us define

$$s = \frac{S}{N}, \quad i = \frac{I}{N}, \quad r = \frac{R}{N}, \quad s + i + r = 1. \qquad (6.5)$$

To determine their rates of change, observe for instance that the differentiation and carrying over the resulting fraction and nonlinear incidence into (6.2) gives

$$\frac{ds}{dT} = \frac{1}{N}\frac{dS}{dT} - \frac{S}{N^2}\frac{dN}{dT} \equiv \frac{1}{N}\frac{dS}{dT}, \quad \frac{ds}{dT} = \frac{1}{N}\sigma(N)IS = \sigma_0\frac{I}{N}\frac{S}{N} = \sigma_0 is.$$

As we will see, however, the nice property for which the rates of changes of the subpopulations $S$ and $I$ and the fractions $s$ and $i$ coincide does not hold in ecoepidemic situations, since in such cases the total population cannot be assumed to be fixed in size. From now on, we identify the constant $\sigma_0$ simply with $\sigma$. As an application, let us consider the classic SIR epidemic model in its formulation with sizes (Hethcote, 2000):

$$\frac{dS}{dT} = \mu N - \mu S - \sigma \frac{IS}{N}, \quad \frac{dI}{dT} = \sigma \frac{IS}{N} - \gamma I - \mu I, \quad \frac{dR}{dT} = \gamma I - \mu R. \quad (6.6)$$

Here the first term on the right in the first equation represents the newborns assumed to be born sound, i.e., belonging to the class of susceptibles. On introducing the fractions as indicated above, the model becomes

$$\frac{ds}{dT} = \mu - \mu s - \sigma is, \quad \frac{di}{dT} = \sigma is - (\gamma + \mu)i, \quad \frac{dr}{dT} = \gamma i - \mu r. \quad (6.7)$$

With respect to the population models, notice that in (6.7) the dynamics is confined to the unit simplex, $0 \leq i + s \leq 1$, i.e., the triangle in the $si$ phase space with vertices at the origin and at the unit points on the coordinate axes.

For more details on the next considerations, we refer the reader to the review by Hethcote (2000). A more general model is the MSEIR model, which represents a transition toward epidemics models including a demographic effect. The class of passively immune individuals $M(T)$ is introduced, these being the newborns of mothers that in some way have had the disease, and since they survived it, therefore acquired immunity and transmitted it to their offsprings. The newborns of the susceptible mothers are instead still susceptible, and therefore belong to class $S$. A possible transition from class $M$ to class $S$ is allowed at rate $\delta$. $E$ represents the class of exposed individuals, that are infected but not yet infectious, i.e., cannot spread the disease. If $d$ denotes the death rate for the population and $b$ its birth rate, the model reads

$$\frac{dM}{dT} = b(N - S) - (\delta + d)M, \quad (6.8)$$

$$\frac{dS}{dT} = bS + \delta M - \beta \frac{SI}{N} - dS,$$

$$\frac{dE}{dT} = \beta \frac{SI}{N} - (\epsilon + d)E,$$

$$\frac{dI}{dT} = \epsilon E - (\gamma + d)I,$$

$$\frac{dR}{dT} = \gamma I - dR,$$

$$\frac{dN}{dT} = (b - d)N,$$

since now the total population size is not constant, and therefore its dynamics must be accounted for. Using fractions with $s = 1 - m - e - i - r$ and $q = b - d$

and introducing the force of infection $\lambda = \beta i$, it becomes

$$\frac{dm}{dT} = (d+q)(e+i+r) - \delta m, \qquad (6.9)$$

$$\frac{de}{dT} = \lambda(1 - m - e - i - r) - (\epsilon + d + q)e,$$

$$\frac{di}{dT} = \epsilon e - (\gamma + d + q)i,$$

$$\frac{dr}{dT} = \gamma i - (d + q)r,$$

to be studied in the set $\{(m,e,i,r) : m \geq 0, e \geq 0, i \geq 0, r \geq 0, \ m+e+i+r \leq 1\}$. This model has only one endemic equilibrium, $E_* \equiv (m_*, e_*, i_*, r_*)^T$, with

$$m_* = \frac{d+q}{\delta+d+q}\left(1 - \frac{1}{R_0}\right), \qquad (6.10)$$

$$e_* = \frac{\delta(d+q)}{(\delta+d+q)(\epsilon+d+q)}\left(1 - \frac{1}{R_0}\right),$$

$$i_* = \frac{\epsilon\delta(d+q)}{(\gamma+d+q)(\delta+d+q)(\epsilon+d+q)}\left(1 - \frac{1}{R_0}\right),$$

$$r_* = \frac{\epsilon\delta\gamma}{(\gamma+d+q)(\delta+d+q)(\epsilon+d+q)}\left(1 - \frac{1}{R_0}\right),$$

where the basic reproduction number is related to the equilibrium number of susceptibles,

$$R_0 = \frac{\epsilon\beta}{(\gamma+d+q)(\epsilon+d+q)}, \quad s_* = \frac{1}{R_0}.$$

In this case the force of infection becomes positive if $R_0 > 1$,

$$\lambda = \delta(d+q)\frac{R_0 - 1}{\delta+d+q}.$$

The age structure previously introduced for structured populations in Chapter 4 can also be used in the context of epidemiology, as we will see in Section 6.3. Here we briefly mention an extension of the MSEIR model with age structure using the fractions, but introduced using the population *density* $U(a, T)$, via relationships of the form, e.g., $m(a,T)U(a,T) = M(a,T)$, to get

$$\frac{\partial m}{\partial a} + \frac{\partial m}{\partial T} = -\delta m, \qquad (6.11)$$

$$\frac{\partial s}{\partial a} + \frac{\partial s}{\partial T} = \delta m - \lambda(a,T)s,$$

$$\frac{\partial e}{\partial a} + \frac{\partial e}{\partial T} = \lambda(a,T)s - \epsilon e,$$

$$\frac{\partial i}{\partial a} + \frac{\partial i}{\partial T} = \epsilon e - \gamma i,$$

$$\frac{\partial r}{\partial a} + \frac{\partial r}{\partial T} = \gamma i,$$

$$\lambda(a,T) = b(a)\int_0^\infty \beta(w)i(w,T)\rho e^{-D(w)-qw}dw.$$

The boundary conditions at age 0 are homogeneous, except

$$m(0,T) = \int_0^\infty f(w)[1 - s(w,T)]e^{-D(w)-qw}dw,$$

$$s(0,T) = \int_0^\infty f(w)s(w,T)e^{-D(w)-qw}dw,$$

$$D(a) = \int_0^a d(w)dw.$$

Here again a steady state can be calculated, but we just mention the basic reproductive number in this case:

$$R_0 = \int_0^\infty \beta(w)\rho e^{-D(w)-qw-\gamma w}\int_0^w \epsilon e^{(\gamma-\epsilon)u}\int_0^u b(\chi)e^{\epsilon \chi}d\chi du dw.$$

A Lyapunov function can be determined, taking it in the following form, where the coefficients need to satisfy the following relationships:

$$L(T) = \int_0^\infty \alpha(a)e(a,T) + \kappa(a)i(a,T)da,$$

$$\frac{d\alpha}{da} = \epsilon[\alpha(a) - \kappa(a)], \qquad \frac{d\kappa}{da} = \gamma\kappa(a) - \rho\beta(a)e^{-D(a)-qa}.$$

Finally, observe that an average age $A$ of infection can be computed, at the endemic steady state, which for the SEIR and SIR models simplifies to

$$A = E[a] = \frac{\int_0^\infty a\lambda(a)e^{-\Lambda(a)-D(a)}da}{\int_0^\infty \lambda(a)e^{-\Lambda(a)-D(a)}da}, \qquad \Lambda(a) = \int_0^a \lambda(w)dw.$$

Many investigations have appeared on vaccination strategies, also in connection with models accounting for demographic changes. By way of example, we mention here only Greenhalgh (1992) and Pugliese (1990), and refer the reader to the bibliography of Hethcote (2000).

Let us finally mention pulse vaccination (d'Onofrio, 2002), in which at fixed time intervals $T_*$ a percentage of the population $s(nT_*^+) = (1-p)s(nT_*^-)$, $n \in \mathbf{N}_+$ is inoculated and acquires life-long immunity, and therefore migrates to the removed class. The model written for the subpopulation fractions reads

$$\frac{ds(T)}{dT} = m[1 - s(T)] - \beta(t)si, \quad (6.12)$$

$$\frac{di(T)}{dT} = ae(T) - (g+m)i(T),$$

$$\frac{de(T)}{dT} = \beta(t)s(T)i(T) - (a+m)e(T),$$

$$r(T) = [1 - s(T) - e(T) - i(T)].$$

Here $\rho$ represents the fraction of vaccinated susceptibles at times $T_n = nT_*$, $n \in \mathbf{N}_+$, $m$ is the mortality rate $a$ the inverse latent period of the disease, $g$ is the inverse infection period, and the contact rate $\beta$ is either constant or assumed to satisfy $\beta(T+1) = \beta(T)$, 1 being the time unit. In d'Onofrio (2002), stability of the limit cycle is investigated, and local and global asymptotic stability conditions of the eradication equilibrium are determined.

## 6.2 Other classical epidemic models

The model by Capasso and Serio (1978) was prompted by the cholera epidemics of 1973 in Bari, southern Italy. Here the incidence term in the SIR classical model is modified to take into account the response the susceptibles give when the epidemics starts to spread. Namely, they take measures not to be infected. This occurs when the number of infectives is high, and therefore it is replaced by a function of $I$. Two possible choices for the latter are possible. At first we can consider a saturation function, which reaches a saturation level with increasing $I$. But also, to model the psychological response of people, the incidence for low values of $I$ can be taken increasing to a maximum and then decreasing if $I$ increases further. A possible function that behaves this way could be provided for instance by the gamma function, $g(I) = I^n e^{-\alpha I}$. The model reads

$$\frac{dS}{dT} = -g(I)S, \quad \frac{dI}{dT} = g(I)S - \gamma I, \quad \frac{dR}{dT} = \gamma I. \quad (6.13)$$

In Capasso and Serio (1978) results on positivity, global existence, uniqueness, stability, and a threshold-like theorem are shown. The latter states that equilibria with $S \leq \rho^*$ are stable, with $\rho^* = \frac{\gamma}{g(0)}$. For global stability, a Lyapunov function is explicitly constructed, $L(S,I) = I + (S - \rho^*)H(S - \rho^*)$, where $H(x)$ denotes the Heaviside function. A further modification is then studied, in which an emigration term for susceptibles is introduced, thereby obtaining the model

$$\frac{dS}{dT} = -g(I)S - \lambda S, \quad \frac{dI}{dT} = g(I)S - \gamma I, \quad \frac{dR}{dT} = \gamma I + \lambda S. \qquad (6.14)$$

A model that includes a nonlinear incidence is considered in Liu et al. (1986), Liu et al. (1987), in which in simplified form we can write

$$\lambda(S,I) = \lambda I^p S^q. \qquad (6.15)$$

The model considered in Liu et al. (1986) is a SEIRS-type model, which reads for the fractions of the various subpopulations

$$\frac{dS}{dT} = -\lambda I^p S^q + \mu - \mu S + \delta R, \qquad (6.16)$$

$$\frac{dE}{dT} = \lambda I^p S^q - (\epsilon + \mu)E,$$

$$\frac{dI}{dT} = \epsilon E - (\gamma + \mu)I,$$

$$\frac{dR}{dT} = \gamma I - (\delta + \mu)R,$$

$$S + E + I + R = 1.$$

Another model is considered by Beretta and Kuang (1998), in which the infection is assumed to be caused by a viral agent, and the latter is explicitly built into the system equations. Let $P$ be the viral population and $N$ be the bacterial population that may be infected by the virus, and thus as usual is split into susceptibles $S$ and infected $I$. Only the susceptibles are assumed to reproduce according to a logistic equation, with carrying capacity $K$ and to which also the infected contribute in the population pressure term. There is a latent period in which the viruses replicate inside the infected bacteria, after which lysis occurs, i.e., the bacteria die and their virus contents, $b$ on average, are released into the environment, where the conditions are not ideal for their survival, so that the viral population experiences there a high death rate $\mu$. The equations then read

$$\frac{dS}{dT} = rS\left(1 - \frac{N}{K}\right) - \lambda SP, \qquad (6.17)$$

$$\frac{dI}{dT} = \lambda SP - \gamma I,$$

$$\frac{dP}{dT} = \gamma bI - \lambda SP - \mu P.$$

It is interesting to note here that upon summation of the right-hand sides, not all terms containing the incidence vanish. Upon rescaling, the system becomes

$$\frac{ds}{dt} = as(1-i-s) - sp, \quad \frac{di}{dt} = sp - \ell i, \quad \frac{dp}{dt} = -sp - mp + b\ell i. \quad (6.18)$$

The model is analyzed to find that it has only two equilibria, $E_0 \equiv (1,0,0)$, which is unstable, and the nontrivial equilibrium $E_f$ with all nonvanishing subpopulations. For the system the set $\Omega = \{s + i \leq 1, \ p \leq \frac{1}{m}b\ell\}$ is shown to be positively invariant. The equilibrium $E_f$ instead is shown to undergo a Hopf bifurcation.

## 6.3 Age- and stage-dependent epidemic system

In this section a step further for modeling the evolution of infectious diseases is presented, whose basic feature lies in an age description of the populations under scrutiny, following what was previously done for an age description of populations. Here we assume that the infection spreads among such structured population. However, the model is more general than that. In fact, since we have the possibility of describing the population by age cohorts, we also use this feature to account for the stage reached by the disease in the infected individuals. We therefore use the biological age for individuals, and introduce the "stage" of the disease, counting the time elapsed since contagion, hopefully up to the recovery. As it is the case for the age-structured populations, the individuals who just became infected are taken into account via boundary conditions. The same holds true for those who have just been recognized as disease carriers.

Before writing the model, we outline the changes needed in the basic assumptions of the classical model, previously stated in Section 6.1. Assumption a will still be maintained. Becoming infected depends on the age of the susceptible; in fact, children and elderly people contract diseases more easily than individuals of intermediate age because mature people have stronger natural defenses. Thus the age structure in the population removes assumption c. This will be suitably taken into account by the governing equations. To make the model closer to reality, we also remove assumption b. In the usual delay models the incubation period is generally taken as a fixed constant. We introduce instead the parameter $w$ to more adequately describe a variable time incubation period. Thus the elimination of assumption b allows us to model more realistically diseases such as measles and chickenpox. The carriers are not yet recognizable as infected (or infectious), but the ability of spreading the disease now is not the same for every individual, so that hypothesis d also no longer holds. Assumption e will be eliminated at a later stage, recalling that for the classical epidemics the inclusion of demographic dynamics has led to more general features (Gao and Hethcote, 1992; Mena-Lorca and Hethcote, 1992).

The population we model is described by a nonnegative density $n(a,t)$, a function with compact support for every $t$. As we know already from Section 4.7, the total population size is

$$N(t) = \int_0^\infty n(a,t)\,da. \tag{6.19}$$

Let $m(a,N)$ denote the biological mortality of the population; as in Section 4.7, it is assumed to be age dependent as well as a function of the total population size. As in the classical epidemic models, the population is partitioned into the three distinct classes of susceptibles, infected, and removed individuals. Again, these are nonnegative functions $s$, $i$, and $r$ of their arguments to be introduced below, with compact support at every fixed time. This partitioning is reflected in the following equation, in which the capitalized densities will be defined below:

$$R(a,t) + s(a,t) + I(a,t) = n(a,t). \tag{6.20}$$

The density $i(a,w,t)$ gives the number of individuals aged $a$ at time $t$ in the stage $w$ of the disease, i.e., who were infected $w$ units of time earlier. It evolves according to

$$\frac{\partial i}{\partial a} + \frac{\partial i}{\partial w} + \frac{\partial i}{\partial t} + (g(a,w) + m(a,N))\,i(a,w,t) = 0. \tag{6.21}$$

Here $g(a,w)$ represents the rate at which an infected aged $a$ and in stage $w$ of the disease is recognized as a disease carrier and thus migrates into the removed class. Thus the boundary condition for the latter takes into account the contributions over all the incubation periods:

$$r(a,0,t) = \int_0^\infty g(a,w)\,i(a,w,t)\,dw. \tag{6.22}$$

Notice that apparently the disease is not lethal. However, the case of an infected individual, not yet recognized as a carrier, who dies of the epidemics, will be accounted for as a death in the class of removed individuals in the stage 0 of the recovery.

The total populations in classes $r$ and $i$ are specified as usual by the integral over all possible recovery stages of their respective densities, which exist by the assumption of $n(a,t)$ being of compact support, since obviously $i(a,w,t), r(a,v,t) \leq n(a,t)$ for every $(a,t)$ and for every $w$, $v$, the latter denoting the recovery period,

$$I(a,t) = \int_0^\infty i(a,w,t)\,dw, \quad R(a,t) = \int_0^\infty r(a,v,t)\,dv. \tag{6.23}$$

Notice also that for $w \geq a$, or $v \geq a$, the above-mentioned densities are identically zero, since the stage of the disease, respectively, of the recovery period, cannot exceed the age of the individual.

The only possible outcomes for an individual in class $r$ are either the transition to class $s$ at rate $q$, or death due to the epidemics at rate $z$, or the disease-unrelated death. Thus, similarly to (6.21), we find

$$\frac{\partial r}{\partial a} + \frac{\partial r}{\partial v} + \frac{\partial r}{\partial t} + (q(a,v) + z(a,v) + m(a,N)) r(a,v,t) = 0. \tag{6.24}$$

An integration over all the recovery stages of the compactly supported density scaled via the function $z$ yields the total number of epidemics-related deaths for individuals aged $a$ at time $t$,

$$D(a,t) = \int_0^\infty z(a,v) r(a,v,t) \, dv. \tag{6.25}$$

For the evolution of the susceptibles, notice that, upon recovery, becoming again a susceptible, an individual aged $a$ still remains aged $a$ at the time of the transition. Thus the latter must be accounted for by a nonhomogeneous term, rather than by a boundary condition. A departure from the susceptible class occurs instead only if there is a biological death, or in case of contagion. The very particular case of death immediately following disease contraction is avoided by stating a loss in stage $r(a,0,t)$. Assumption a of the classical model is used here to write again a homogeneous mixing model. In our case, all the infectives of all ages are assumed to be equally able to spread the disease, if they are in the same stage. Hence we introduce the quantity

$$\widetilde{I}(w,t) = \int_0^\infty i(a,w,t) \, da \tag{6.26}$$

and find

$$\frac{\partial s}{\partial a} + \frac{\partial s}{\partial t} + s(a,t) \left( m(a,N) + \int_0^\infty p(a,w) \widetilde{I}(w,t) \, dw \right) \tag{6.27}$$
$$- \int_0^\infty q(a,v) r(a,v,t) \, dv = 0.$$

The function $p(a,w)$ is related to the stage $w$ of the infectives that cause the encounter, and the age $a$ of the susceptible in question. The "new" infectives come from the interaction term, i.e.,

$$i(a,0,t) = s(a,t) \int_0^\infty p(a,w) \widetilde{I}(w,t) \, dw. \tag{6.28}$$

To describe the reproduction of the population, thus removing assumption e of the classical model, we consider as in Section 4.7 the age-dependent maternity function $b(a,N)$. However, here we need to state what the effects of infected individuals are on offsprings. Infected parents can also affect the offsprings, but this might not at all be related to the epidemics in the sense that children of sick parents may not themselves carry the same disease, but may very well have other abnormalities. In the human case, this would possibly occur to children of a mother infected by rubella, for instance. Taking such situations into account is a hard task. We assume simply that only sound individuals reproduce and at birth newborns are sound. Thus

$$n(0,t) = s(0,t) = \int_0^\infty b(a,N) s(a,t)\, da. \tag{6.29}$$

Alternatively, every individual can reproduce, but again newborns are sound:

$$s(0,t) = \int_0^\infty b(a,N) n(a,t)\, da.$$

Suitable initial conditions must complement the above system, that is,

$$i(a,w,0) = I^0(a,w) \quad \text{for } a > w, \tag{6.30}$$
$$r(a,v,0) = R^0(a,v) \quad \text{for } a > v,$$
$$n(a,0) = n^0(a),$$
$$s(a,0) = S^0(a) = n^0(a) - \int_0^\infty R^0(a,v)\, dv - \int_0^\infty I^0(a,w)\, dw.$$

As is the case for the classical models, summing the governing equations (6.21), (6.24), (6.27) and using also (6.20), we obtain the description of the dynamics of the population. The equation governing the evolution of the density $n(a,t)$ is then

$$\frac{\partial n}{\partial a} + \frac{\partial n}{\partial t} + m(a,N) n(a,t) + D(a,t) = 0, \tag{6.31}$$

and this allows to deduce that the model is well posed. Notice that this equation differs from the classical Kermack–McKendrick or Von Foerster equation as it contains the migration term, i.e., the total loss due to disease-related deaths. To derive the global existence of the solutions, to apply the techniques of Gurtin and McCamy (1974) described in Section 4.7 requires us to handle this migration term. But an upper bound for $D(a,t)$ is easily obtained from (6.20):

$$I(a,t), \quad R(a,t), \quad s(a,t) \leq n(a,t) \tag{6.32}$$

for every $(a,t)$, $a \geq 0$ and $t \geq 0$. The natural hypotheses on the functions $m(a,N)$ and $z(a,v)$ require them to have a nonnegative infimum for $a \geq 0$, $v \geq 0$, and $N \geq 0$. Also, the maternity function $b(a,N)$ is assumed to be uniformly bounded above for all $a \geq 0$ and $N \geq 0$. Then the global existence result holds for the density $n$. The same result for the model is deduced from (6.32) and $I, R, s$ being nonnegative.

## 6.4 Case study: Aujeszky disease

Aujeszky disease (A.D.), caused by the Herpevirus 1 suis (ADV or SHV-1), although not lethal, affects several animal species, among which the hogs constitute the most relevant one for farmers. This disease is diffused in almost every European country with relevant economic consequences. In highly densely populated areas, the Italian law-required control measures are difficult to implement. In northwest Italy, the area of the towns of Villafalletto and Vottignasco in the Cuneo province within a surface of 38 km$^2$ hosts 90000 animals in 91 farms with density in some instances over 3500 individuals per square kilometer. In view of these facts, for modeling purposes we consider the area as a single giant epidemiologic unit. The seroprevalence dropped from the year 1997, in which the new vaccination law was issued, until year 2000 to reach the minimum value of 31.5%. Then it went up again to 54.7% in 2004 and afterwards remained stable. The mathematical model considered here studies the spread of the disease, with the aim of its eradication, if at all possible, or at least a reduction of the viral circulation.

The field measures give the following values for the measurable parameters. The size of the epidemiological unit is $N = 90000$. Data on the positivity of the farmed animals have been collected; the average natural mortality of sound animals is $\mu_s = 0.084$, the average disease-related mortality is $\mu_i = 0.087$, and the average birth rate is $\rho = 0.107$.

For the remaining parameters appearing in the mathematical model, namely the disease incidence $\beta$ and the (absence of) biosafety $\tau$, simulations have been performed. Among the latter, biosafety measures play an important role. These are represented by devices like continuous fencing of the farm, quarantine stalls, desinfestations, checked visitor access, filtering area between the stalls and other farming areas, and so on.

The disease, once contracted, affects the animal for its life. Hence the epidemics model we consider must be of SI type. The population is partitioned among the susceptible and infective classes as usual, but reproduces so that $S + I = N$, but here the population is not constant, rather $N \equiv N(T)$. The model is then described by the equations

$$\frac{dS}{dT} = \rho N - \mu_s S - \beta \frac{SI}{N} - \tau S, \quad \frac{dI}{dT} = -\mu_i I + \beta \frac{SI}{N} + \tau S. \qquad (6.33)$$

The first equation describing the susceptible evolution contains a reproduction term, coming from the whole population, as also diseased animals reproduce, but all newborns are considered to be sound. It also accounts for the intrinsic natural mortality $\mu_s$ of the susceptibles, the disease incidence term, and finally a term describing the absence of biosafety measures, for which the disease can be caught not by direct contact with a sick animal, but by vectors brought into the farm by external carriers. Notice that $\tau = 0$ corresponds to the highest safety measures. The second equation contains the disease-related mortality term, and the terms characterizing the new infected individuals: see, for instance, French et al. (1999).

Let us introduce normalized variables. As new variables we take the fractions of sound and infected individuals with respect to the total population,

$$s = \frac{S}{N}, \quad i = \frac{I}{N},$$

so that the population constraint becomes the line

$$\ell: \quad s + i = 1. \tag{6.34}$$

Since the latter is a function of time, the former will also be and their time derivatives have to take this into account. Thus, for instance,

$$\frac{ds}{dT} = \frac{1}{N}\frac{dS}{dT} - \frac{dN}{dT}\frac{S}{N^2} = \frac{1}{N}\frac{dS}{dT} - \frac{s}{N^2}\left(\frac{ds}{dT} + \frac{di}{dT}\right)$$

and similarly for $i$. Substituting from (6.33), we find

$$\frac{ds}{dt} = \mu_s s^2 + (\mu_i - \beta)si - (\mu_s + \tau + \rho)s + \rho, \tag{6.35}$$

$$\frac{di}{dt} = \mu_i i^2 + (\beta + \mu_s)si - (\rho + \mu_i)i + \tau s.$$

The equilibria of this system are then given by the equations

$$\mu_s s^2 + (\mu_i - \beta)si - (\mu_s + \tau + \rho)s + \rho = 0, \tag{6.36}$$

$$\mu_i i^2 + (\beta + \mu_s)si - (\rho + \mu_i)i + \tau s = 0.$$

These represent conic sections. In principle, they could be studied to determine the flow in the $si$ phase plane, but the advantage of the reduced model is that in this plane trajectories are confined to the straight line (6.34). We can just study then the flow on this "normalized population line" with normal $\mathbf{n} = (1,1)^T$. If it goes up, the angle $\theta$ that it makes with $\mathbf{n}$ is $0 < \theta < \pi$, while if it is downward, the angle will be $\pi < \theta < 2\pi$. To distinguish the two cases, notice that in the former we have $\sin\theta > 0$, while the converse situation holds in the latter. Hence, to calculate the angle $\theta$ that $\mathbf{n}$ makes with the flow, we can use the cross product

$$\frac{d\mathbf{x}}{dT} \equiv \left(\frac{ds}{dT}, \frac{di}{dT}\right)^T, \quad \frac{d\mathbf{x}}{dT} \times \mathbf{n} \equiv \left\|\frac{d\mathbf{x}}{dT}\right\| \|\mathbf{n}\| \sin\theta$$

so that

$$\operatorname{sgn}(\sin\theta) \equiv \operatorname{sgn}\left(\frac{d\mathbf{x}}{dT} \times \mathbf{n}\right),$$

and we are led to study the function

$$\Psi \equiv \mathrm{sgn}(\sin\theta) = \mathrm{sgn}\begin{vmatrix} \mathbf{i} & \mathbf{j} & \mathbf{k} \\ 1 & 1 & 0 \\ \frac{ds}{dT} & \frac{di}{dT} & 0 \end{vmatrix} \equiv \|\mathbf{k}\|\mathrm{sgn}\left(\frac{di}{dT} - \frac{ds}{dT}\right). \quad (6.37)$$

Upon substitution from (6.36) and (6.34) we obtain the equation $\Psi(s) = 0$, where explicitly

$$\Psi(s) \equiv \mu_i(1-s)^2 + s(1-s)(2\beta + \mu_s - \mu_i) \\ + s(\mu_s + 2\tau + \rho) - (1-s)(\rho + \mu_i) - \mu_s s^2 - \rho,$$

which is again a parabola. Recall that when such a function is positive, the projection onto the normalized population line of the flow shows that trajectories must move upwards toward increasing values of the infectives, i.e., the epidemic propagates. To assess this behavior, it is clearly easier to determine its zeros first, so that we need to solve the quadratic equation

$$s^2(\mu_i - \beta - \mu_s) + s(\beta - \mu_i + \mu_s + \tau + \rho) - \rho = 0,$$

whose solutions are

$$s_{1,2} = \frac{-(\mu_s + \tau + \rho + \beta - \mu_i) \pm \sqrt{\Delta_s}}{2(\mu_i - \beta - \mu_s)}. \quad (6.38)$$

These zeros will be real if the discriminant is nonnegative,

$$\Delta_s \equiv (\beta - \mu_i + \mu_s + \tau + \rho)^2 + 4\rho(\mu_i - \beta - \mu_s) \geq 0. \quad (6.39)$$

We need then to determine the region in the $\beta\tau$ parameter space in which this inequality is satisfied. Substituting the data for the known parameters and studying the equation corresponding to (6.39), we find

$$\phi(\beta, \tau) \equiv \beta^2 + \tau^2 - 0.22\beta + 0.208\tau + 2\beta\tau + 0.0121 = 0. \quad (6.40)$$

To understand its nature, let us consider the matrix of the quadratic form (6.40),

$$A \equiv (a_{j,\ell}) = \begin{pmatrix} 0.0121 & 0.104 & -0.11 \\ 0.104 & 1 & 1 \\ -0.11 & 1 & 1 \end{pmatrix}, \quad j, \ell = 0, 1, 2.$$

We then investigate the invariants of (6.40):

$$\Delta \equiv \det A = -0.045796, \quad \delta \equiv \begin{vmatrix} 1 & 1 \\ 1 & 1 \end{vmatrix} = 0.$$

Thus $\phi$ is a parabola. The slope of its axis is

$$\tan\alpha = -\frac{a_{1,1}}{a_{1,2}} = -1,$$

i.e., the angle it makes with the $s$-axis is $\frac{3}{4}\pi$. The vertex can be obtained by solving the system

$$a_{1,1}x_0 + a_{1,2}y_0 + (a_{1,1}a_{0,1} + a_{1,2}a_{0,2})\frac{1}{S} = 0,$$

$$\left[a_{0,1} + (a_{0,1}a_{2,2} - a_{1,2}a_{0,2})\frac{1}{S}\right]x_0$$
$$+ \left[a_{0,2} + \frac{1}{S}(a_{1,1}a_{0,2} - a_{0,1}a_{1,2})\right]y_0 + a_{0,0} = 0,$$

to give the point $(x_0, y_0) = (-0.031, 0.0253)$. The intersections with the $\beta$ coordinate axis are complex, while with the $\tau$-axis they reduce to a double root, namely, $\beta_1 \equiv \beta_2 = 0.11$. We also determine the intersections with the line $\beta = 1$ to find the equation

$$\tau^2 + 2.208\tau + 0.7921 = 0$$

with roots $\tau_1 = -1.757$, $\tau_2 = -0.451$. There are no real intersections with $\tau = 1$. From these considerations it follows that the discriminant is always positive in the unit square $\{0 \leq \beta \leq 1\} \times \{0 \leq \tau \leq 1\}$. Thus the discriminant $\Delta_s(\beta, \tau) \geq 0$ for all $0 \leq \beta \leq 1, 0 \leq \tau \leq 1$ and the roots (6.38) are

$$s_{\pm} \equiv s_{1,2} = \frac{\beta + \tau + 0.104 \pm \sqrt{(\beta + \tau + 0.104)^2 + 0.428(0.003 - \beta)}}{2(\beta - 0.003)},$$

always real and distinct. Now, for $\beta > 0.003$ we find $\Delta_s < \beta + \tau + 0.104$, so that $0 < s_-, s_+$, while for $\beta < 0.003$ we find $\Delta_s > \beta + \tau + 0.104$ and thus $s_+ < 0 < s_-$. The roots position with respect to the upper bound 1 can be directly checked, and the final inequalities we have are $0 < s_- < 1 < s_+$ for $\beta > 0.003$, and $s_+ < 0 < s_- < 1$ for $\beta < 0.003$.

Remembering that $\theta$ is the angle of the flow with the normal to the population line, then $\Psi(s) = \text{sgn}(\sin \theta) \geq 0$ means a downward flow on this line. Also, $\Psi(s) \geq 0$ holds always outside the roots interval. Thus, for the case $0 < s_- < 1 < s_+$, we have $\Psi(s) \geq 0$ for $s < s_-$ and conversely, so that trajectories approach $s_-$, while for $s_+ < 0 < s_- < 1$, we find $\Psi(s) \geq 0$ for $s_- < s$, again with trajectories approaching $s_-$. This entails that the only feasible equilibrium $s^-$ is always asymptotically stable. In the population space, this equilibrium is then

$$E^* \equiv (s_-, i_-) \equiv (s_-, 1 - s_-) \tag{6.41}$$

$$\equiv \left(\frac{-(\mu_s + \tau + \rho + \beta - \mu_i) - \sqrt{\Delta_s}}{2|\mu_i - \beta - \mu_s|}, 1 - \frac{-(\mu_s + \tau + \rho + \beta - \mu_i) - \sqrt{\Delta_s}}{2|\mu_i - \beta - \mu_s|}\right)$$

$$= \left(\frac{\beta + \tau + 0.104 - \sqrt{\Delta_s}}{2|\beta - 0.003|}, \frac{2|\beta - 0.003| - \beta - \tau - 0.104 + \sqrt{\Delta_s}}{2|\beta - 0.003|}\right),$$

Epidemic models

with $\Delta_s = (\beta - \mu_i + \mu_s + \tau + \rho)^2 - 4\rho|\mu_i - \beta - \mu_s| = (\beta + \tau + 0.104)^2 - .428|\beta - 0.003|$.

Now the goal would be to keep down the level of the infected population, or increase the one of the sound animals. If $\beta \to 0$, we find

$$i^*_{\beta=0} = \frac{0.006 - \tau - 0.104 + \sqrt{(\tau + 0.104)^2 - .428 \times 0.003}}{0.006}. \quad (6.42)$$

This condition is, however, essentially impossible to attain, as it imposes extremely strict measures within the farmed animals since we want to annihilate the intramural disease incidence. On the other hand, we could impose biosafety measures, to eliminate the imported disease carriers, in which case $\tau \to 0$, and we have

$$i^*_{\tau=0} = \frac{2|\beta - 0.003| - \beta - 0.104 + \sqrt{(\beta + 0.104)^2 - .428|\beta - 0.003|}}{2|\beta - 0.003|}. \quad (6.43)$$

In a more general case, we can try to minimize the surface

$$i^* \equiv i^*(\beta, \tau) \quad (6.44)$$

$$= \frac{2|\beta - 0.003| - \beta - \tau - 0.104 + \sqrt{(\beta + \tau + 0.104)^2 - .428|\beta - 0.003|}}{2|\beta - 0.003|}.$$

Note that the restriction of this surface to the coordinate planes is given by (6.42) and (6.43) above. By cutting this surface with planes parallel to the coordinate ones, we find similar behaviors, namely,

$$i^*_{\beta=h} = \frac{2|h - 0.003| - h - \tau - 0.104 + \sqrt{(h + \tau + 0.104)^2 - .428|h - 0.003|}}{2|h - 0.003|},$$

$$\quad (6.45)$$

$$i^*_{\tau=k} = \frac{2|\beta - 0.003| - \beta - k - 0.104 + \sqrt{(\beta + k + 0.104)^2 - .428|\beta - 0.003|}}{2|\beta - 0.003|},$$

$$\quad (6.46)$$

at "higher" levels. Note that the line $\beta = 0.003$ is a removable singularity, as after simplications the restriction of the surface to that line is described by the equation

$$i^*|_{\beta=0.003} = \frac{\tau}{\tau + \rho}.$$

We can better study the sections of $i^*$ with $\tau = k$ by taking the derivative with respect to $\beta$ to find

$$\left.\frac{di^*}{d\beta}\right|_{\tau=k}(\beta, k) = \frac{1}{2|\beta - 0.003|(\beta - 0.003)}[\beta + \tau + 0.104 - |\beta - 0.003|]. \quad (6.47)$$

Thus, for $\beta > 0.003$, the numerator of the partial derivative is $\tau + \rho > 0$ and for $\beta < 0.003$ it is $2\beta + \tau + 0.101 > 0$ in the unit square. Thus, the minimum of the surface is to be sought for $\beta = 0$. Also, the sections of $i^*$ with $\beta = h$ can be investigated by taking the derivatives with respect to $\tau$. We have

$$\left.\frac{di^*}{d\tau}\right|_{\beta=h}(h,\tau) = \frac{1}{2|h-0.003|}\left[-1 + \frac{h+\tau+.104}{\sqrt{(h+\tau+0.104)^2 - .428|h-0.003|}}\right]. \quad (6.48)$$

On verifying whether

$$\left.\frac{di^*}{d\tau}\right|_{\beta=h}(h,\tau) \geq 0,$$

we find $.428|h - 0.003| > 0$, which is clearly always true. Thus $i^*_{\beta=h}(h,\tau)$ is always an increasing function of $\tau$; cf. Figure 6.2. Thus the minimum will be achieved for $\tau = 0$. Combining this information with the former, we just need to check the value $i^*(0,0) = -0.0613$. The fact that at the origin the surface becomes negative induces us to also investigate the zero level curve $i^*(\beta, \tau) = 0$. Upon simplification we have

$$\beta + \tau = 0.003 + |\beta - 0.003|.$$

For $\beta > 0.003$, we find then $\tau = 0$, while for $\beta < 0.003$, the condition gives the straight line $\tau = 2(0.003 - \beta)$. Below this line, in the small triangle $\nabla$ containing the origin shown in Figure 6.3, the infected become negative, i.e., the infection level drops to zero. This agrees with the findings of Figure 6.2, in which the whole surface over the unit square is shown. In conclusion, the disease can be eradicated if the two parameters $\beta$ and $\tau$ are pushed down to values contained in $\nabla$. Eradication can be achieved by a combination of biosafety measures coupled with measures to contain the horizontal incidence $\beta$ of the disease. These conditions are, however, very difficult to attain, as they impose extremely strict measures within the farmed animals since we want to annihilate the intramural disease incidence. On the other hand, we also impose strict biosafety measures, to eliminate the imported disease carriers.

## 6.5 Analysis of a disease with two states

In this section we consider an epidemic spreading in a population of fixed size, a disease that has two states, the recoverable mild form and the strong, possibly lethal one. The population is thus subdivided into three classes: the susceptibles $S$, the weakly infected $W$, and the $V$ strongly affected ones. The

# Epidemic models

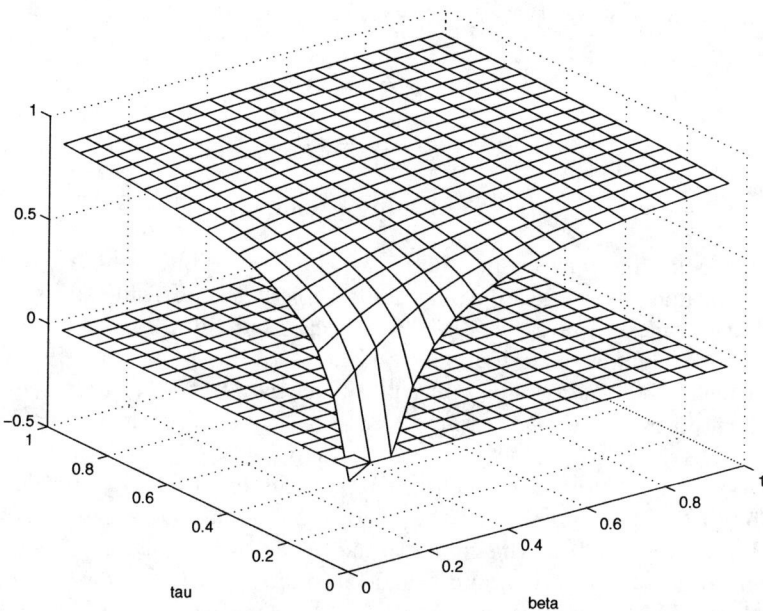

**FIGURE 6.2**: Plot of the surface $I^*(\beta, \tau)$ in the unit square. It becomes negative near the origin.

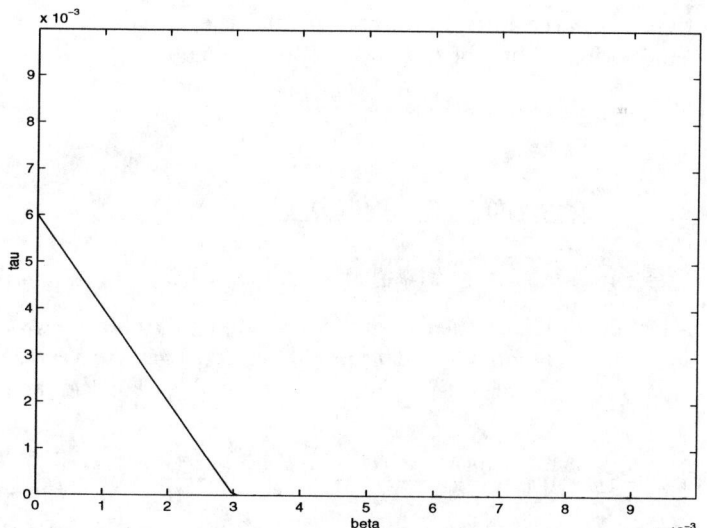

**FIGURE 6.3**: Contour line $i^* = 0$; the surface is negative in the triangle $\nabla$ near the origin; compare Figure 6.2 and notice the very small scale $O(10^{-3})$ on both axes.

governing equations are then

$$\frac{dS}{dT} = \rho S - \beta SW - (\gamma_w + \gamma_v)SV + \pi W + \nu V, \qquad (6.49)$$

$$\frac{dW}{dT} = -\lambda W + \beta SW + \gamma_w SV - \pi W,$$

$$\frac{dV}{dT} = \lambda W + \gamma_v SV - \nu V - \mu V, \quad \text{and} \quad N = S + W + V.$$

The first equation allows reproduction of the susceptibles, whose offsprings are born sound, at the net reproduction rate $\rho$. It then considers the disease incidence $\beta$ due to contact with weakly infected individuals, producing again only new weakly infected individuals, and to contact with the strongly affected ones, which, when "successful," may produce new infected either in the weak or virulent forms, at respective rates $\gamma_w$ and $\gamma_v$. Recovery from either infected state is in principle possible, although we suppose that the recovery rate $\pi$ from the strong infection is much lower than the one from the mild form of the disease, $\nu$. The dynamics of the infected subpopulations is completed by allowing a transition from the milder to the stronger one at rate $\lambda$. Finally, the virulent form may be lethal, and that is accounted for by the disease-related mortality rate $\mu$.

## Equilibria

The origin is clearly an equilibrium. To find the nontrivial ones, observe that the equilibrium values of susceptibles solve the quadratic

$$y(x) = \beta \gamma_v x^2 - (\gamma_v \lambda + \beta \mu + \gamma_v \pi + \beta \nu + \gamma_w \lambda)x + (\lambda + \pi)(\mu + \nu),$$

giving

$$S_{1,2} \equiv S_{+,-} = \frac{(\lambda + \pi)\gamma_v + \gamma_w \lambda + \beta(\mu + \nu) \pm \sqrt{\Delta}}{2\beta \gamma_v}, \qquad (6.50)$$

$$\Delta = ((\lambda + \pi)\gamma_v + \gamma_w \lambda)^2 + \beta^2(\mu + \nu)^2 - 2\beta(\mu + \nu)((\lambda + \pi)\gamma_v - \gamma_w \lambda) > 0,$$

where the last condition on the discriminant is imposed to have real solutions. In such a case it easily follows that $y(0) > 0$ and $y'(0) = -(\gamma_v \lambda + \beta \mu + \gamma_v \pi + \beta \nu + \gamma_w l) < 0$ imply that both $S_1$ and $S_2$ are feasible. For the infected, we find then

$$V_1 = \frac{\rho}{\mu} S_1, \quad W_1 = \frac{\rho(\lambda + \pi)(\mu + \nu)}{\beta \lambda \mu} - \frac{\rho((\lambda + \pi)\gamma_v + \gamma_w \lambda)}{\beta \lambda \mu} S_1, \quad (6.51)$$

$$V_2 = \frac{\rho}{\mu} S_2, \quad W_2 = \frac{\rho(\lambda + \pi)(\mu + \nu)}{\beta \lambda \mu} - \frac{\rho((\lambda + \pi)\gamma_v + \gamma_w \lambda)}{\beta \lambda \mu} S_2.$$

Feasibility then requires the following threshold condition for the susceptibles:

$$S \geq S_t = \frac{(\lambda + \pi)(\mu + \nu)}{(\lambda + \pi)\gamma_v + \gamma_w \lambda},$$

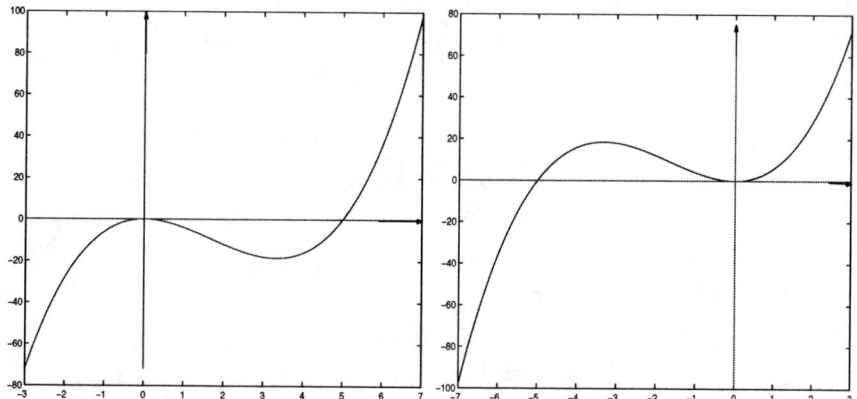

**FIGURE 6.4:** Possible forms of the function $\sigma^3 - C\sigma^2$, $C > 0$ on the left, $C < 0$ on the right.

which in turn shows that $S_1$ is infeasible, while $S_* \equiv S_2$ instead satisfies it. The feasible equilibria are then

$$E_0 = (0,0,0), \quad E_* = \left(S_*, \frac{\rho(\lambda+\pi)(\mu+\nu)}{\beta\lambda\mu} - \frac{\rho((\lambda+\pi)\gamma_v + \gamma_w\lambda)}{\beta\lambda\mu} S_*, \frac{\rho}{\mu} S_*\right).$$

## Stability

The Jacobian of the system is

$$J = J(S,W,V) = \begin{pmatrix} \rho - \beta W - (\gamma_w + \gamma_v)V & -\beta S + \pi & -(\gamma_w + \gamma_v)S + \nu \\ \beta W + \gamma_w V & -\lambda + \beta S - \pi & \gamma_w S \\ \gamma_v V & \lambda & \gamma_v S - \mu - \nu \end{pmatrix}.$$

Now, the eigenvalues at the origin are easily evaluated as $\sigma_1 = \rho$, $\sigma_2 = -\lambda - \pi$, $\sigma_3 = -\mu - \nu$ to show that it is inconditionally unstable.

The characteristic equation for $E_*$ can be written as $-\sigma^3 + C\sigma^2 + B\sigma + A = 0$,

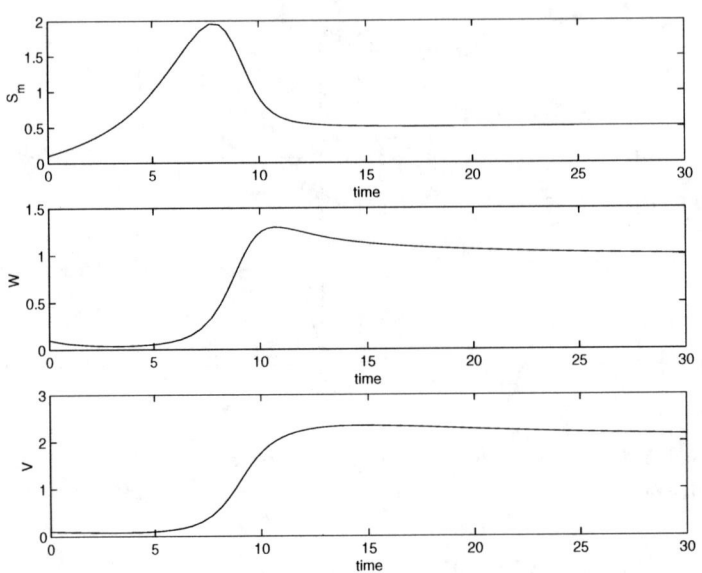

**FIGURE 6.5**: $\beta = 0.1$, $\gamma_w = 0.5$, $\gamma_v = 0.5$, $\lambda = 0.3$, $\mu = 0.1$, $\nu = 0.3$, $\pi = 0.3$, and $\rho = 0.4$, giving $A$, $B$, $C < 0$ but $A + BC \neq 0$.

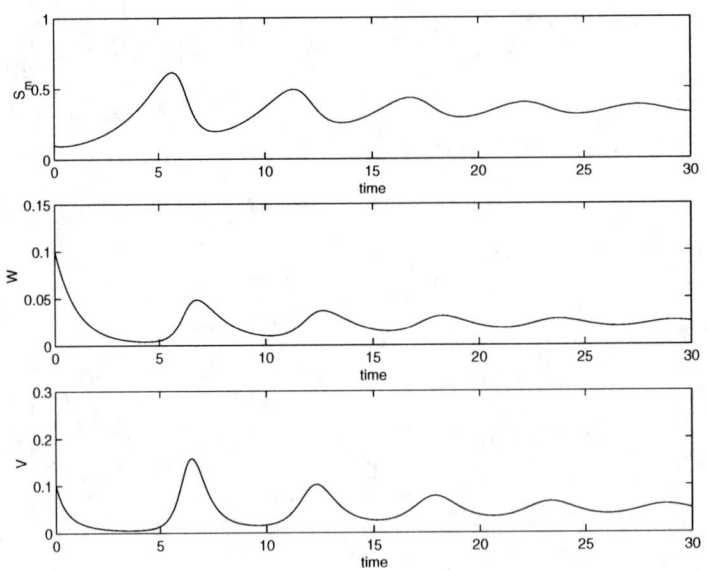

**FIGURE 6.6**: $\beta = 1.0$, $\gamma_w = 1.0$, $\gamma_v = 9.0$, $\lambda = 1.0$, $\mu = 3.5$, $\nu = 0.1$, $\pi = 0.1$, and $\rho = 0.5$.

with coefficients explicitly given by

$$A = -\rho S_2[\mu\beta + \gamma_w\lambda + \nu\beta - 2\beta S_2\gamma_v + \gamma_v\pi + \gamma_v\lambda],$$

$$B = \frac{1}{\lambda\mu}\left[-\rho\pi\mu^2 - \lambda^2\mu\nu - \rho\lambda\nu^2 - \rho\pi\nu^2 - \lambda\mu^2\pi - \rho\lambda^2\nu + \lambda\mu^2\beta S_2\right.$$
$$- \lambda\mu\pi\nu + \lambda^2\mu\gamma_w S_2 + l^2\mu\gamma_v S_2 - \lambda\mu\beta S_2^2\gamma_v + \lambda\mu\beta S_2\nu + \lambda\mu\pi\gamma_v S_2$$
$$- \rho\lambda\nu\pi - \rho\lambda\nu\mu - 2\rho\pi\mu\nu - \rho\pi\gamma_v^2 S_2^2 - \rho\lambda\gamma_v^2 S_2^2 - l^2\mu^2 + 2\rho\lambda\nu\gamma_v S_2$$
$$\left. + 2\rho\pi\mu\gamma_v S_2 + 2\rho\pi\nu\gamma_v S_2 - \rho\lambda\mu\beta S_2 + \rho\lambda\gamma_v S_2^2\beta - \rho\lambda\gamma_v S_2^2\gamma_w\right],$$

$$C = -\frac{1}{\lambda\mu}[l^2\mu + \rho\lambda\nu - \lambda\mu\gamma_v S_2 - \lambda\mu\beta S_2 + \lambda\mu\nu + \rho\pi\nu$$
$$- \rho\pi\gamma_v S_2 + \rho\pi\mu + \lambda\mu\pi + \lambda\mu^2].$$

We can seek the eigenvalues as intersections of the two functions $B\sigma + A$ and $\sigma^3 - C\sigma^2$, observing that the latter has one of the two forms in Figure 6.4. To impose stability, the case $C < 0$ is the most favorable one, as we need only to impose also $A, B < 0$ to have only roots with negative real parts.

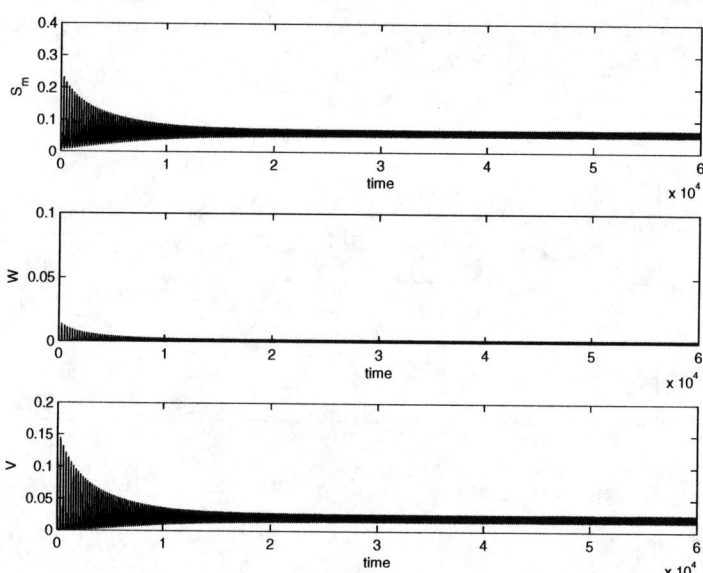

**FIGURE 6.7**: Limit cycle verified over a long time interval: $\beta = 0.003$, $\gamma_w = 0.2$, $\gamma_v = 0.9$, $\lambda = 0.1$, $\mu = 0.06$, $\nu = 0$, $\pi = 0.1$, and $\rho = 0.02$.

Instead of continuing the analysis in this direction, we concentrate on the pursuit of Hopf bifurcations. To this end, we seek the conditions for which the cubic characteristic equation factors into the product of two lower de-

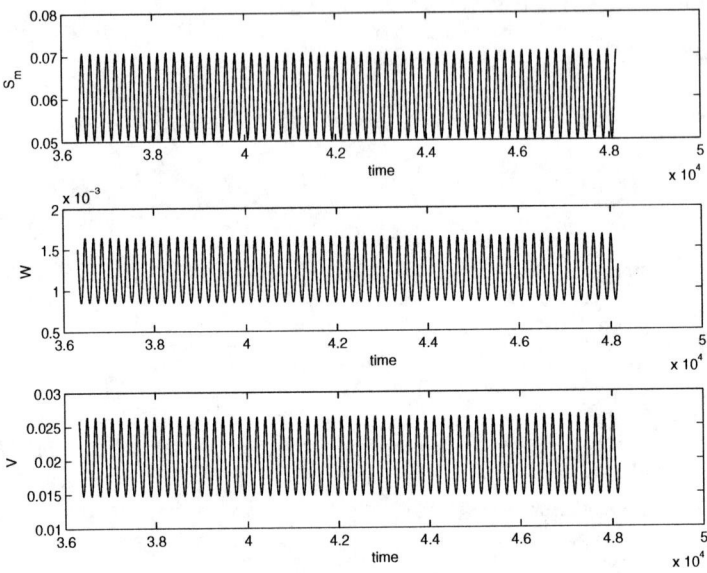

**FIGURE 6.8**: The system dynamics for the same parameters as in Figure 6.7 but in long-time simulations: one can readily see persistent oscillations.

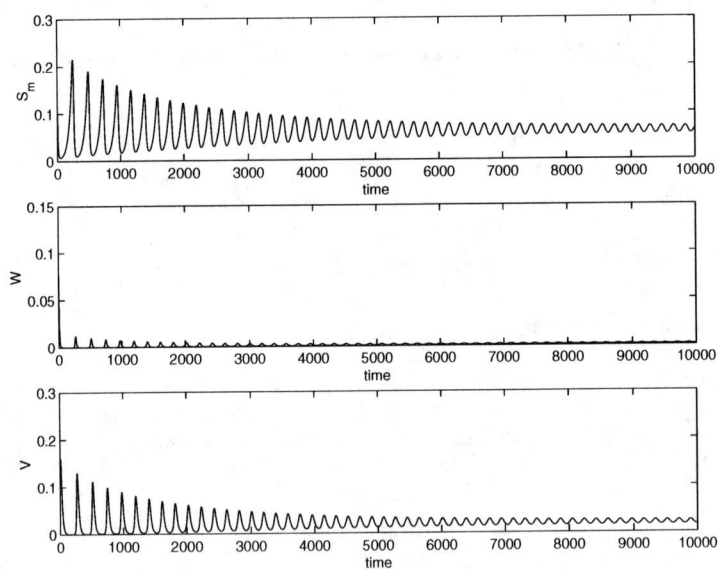

**FIGURE 6.9**: $\beta = 0.003$, $\gamma_w = 0.2$, $\gamma_v = 0.9$, $\lambda = 0.1$, $\mu = 0.06$, $\nu = 0.005$, $\pi = 0.1$, and $\rho = 0.02$.

gree polynomials, so that the following condition must hold in addition to $A, B, C < 0$:

$$\sigma^3 - C\sigma^2 - B\sigma - A = (\sigma^2 + \alpha_1^2)(\sigma - \alpha_2) = \sigma^3 - \alpha_2\sigma^2 + \alpha_1^2\sigma - \alpha_1^2\alpha_2,$$
$$A + BC = 0.$$

The analysis is then continued numerically. Specifically, for $\beta = 0.1$, $\gamma_w = 0.5$, $\gamma_v = 0.5$, $\lambda = 0.3$, $\mu = 0.1$, $\nu = 0.3$, $\pi = 0.3$, and $\rho = 0.4$, we find a stable equilibrium; see Figure 6.5.

The equilibrium can be slightly destabilized in the transient behavior; Figure 6.6 gives an example of this behavior.

To find persistent oscillations, we now set $\nu = 0$ and $\beta = 0.003$, $\gamma_w = 0.2$, $\gamma_v = 0.9$, $\lambda = 0.1$, $\mu = 0.06$, $\nu = 0$, $\pi = 0.1$, and $\rho = 0.02$. The results are in Figure 6.7, with a doublecheck on a smaller time range in Figure 6.8.

If we then change the value of $\nu = 0.001$, without changing the other parameter values, the oscillations are found to diminish in amplitude; see Figure 6.9.

# Chapter 7

# Ecoepidemic systems

In this chapter, we begin to consider ecoepidemic models. These are demographic models accounting for interactions among different populations of the types described in the preceding Chapter 4 in which a disease also spreads. To our knowledge, the first papers that appeared in this field are by Hadeler and Freedman (1989), Venturino (1992, 1994), and Beltrami and Carroll (1994). The name of the field is due to Joydev Chattopadhyay, following an earlier suggestion of the late Ovide Arino. Since those first investigations a number of papers have appeared in the literature. Among the more recent contributions to the field, we mention also Chattopadhyay and Arino (1999), Chattopadhyay et al. (2002), and Arino et al. (2004). In particular these ecoepidemiological models turned useful for the analysis of harmful algal blooms, see Chattopadhyay and Pal (2002) and Chattopadhyay et al. (2002; 2003; 2004). Ecoepidemic models involving more complex phenomena among the populations involved have been investigated with more sophisticated mathematical techniques in Venturino (2001; 2002; 2004; 2006; 2007), Haque and Venturino (2006a; 2006b; 2007), and Keller and Venturino (2007).

Epidemic models with vital dynamics for the population share the three dimensionality of the dynamical system that we have here (cf. Gao and Hethcote, 1992), but ultimately the trajectories will lie on a plane in the phase space, thus allowing the use of classical tools for their study. Unfortunately, this is not the case here.

The chapter is organized as follows. The ecoepidemic models considered here contain mass-action incidence, in the various demographic aspects of this situation. We start from predator prey models in which the disease affects first the predators (Section 7.1), then the prey (Section 7.2), and then the other two cases of competing (Section 7.3) and symbiotic communities (Section 7.4).

## 7.1 Prey–diseased-predator interactions

Assume first that the epidemics spreads only among predators and cannot be transmitted to the prey by interaction. This assumption is biologically sound for a wide range of diseases. We count only sound prey $U$, sound predators, $V$, and infected predators, $V_I$, and in this context the total number of predators is expressed by the variable $P = V + V_I$. By combining the classical predator prey model with the mass action SIS epidemic, with the disease only in the predators we obtain the following model, recalling that all the parameters are nonnegative:

$$\frac{dU}{dT} = U\left[a(1 - \frac{U}{K}) - cV - \eta V_I\right], \qquad (7.1)$$

$$\frac{dV}{dT} = V\left[d + eU - fP - \delta V_I\right] + (\nu + h)V_I,$$

$$\frac{dV_I}{dT} = V_I\left[k + gU - fP + \delta V - \nu - \mu\right].$$

Notice that here prey do not represent the only food source for the predators, since in their absence, and in the absence of the disease, the sound predators in the underlying classical demographic model do not die out, but tend to the equilibrium value $\frac{d}{f}$.

The first equation gives the dynamics of the prey, the last term accounting for the predation process due to infected predators. We split the effects produced by sound and infected predators, as their respective rates of predation may be different. Indeed, the parameter $\eta$ represents a prey loss due to hunting, but, contrary to our basic assumption, it could also mean the "instataneous" death of prey by disease contraction. It differs from $c$ since (a) $\eta < c$ models the situation in which sick predators are less able to catch prey or (b) $\eta > c$ would denote that the hunting abilities of sick predators may be unaffected, but prey surviving an attack may catch the disease and die of it. However, this violates our basic assumption, and then we would need to modify the model introducing also the infected prey. Situation (a) of reduced hunting efficiency is then assumed having in mind the large predators hunting using their speed. If disease affected, their running performance would be inferior to sound individuals. We could say that they are able to catch young or old prey at rate $\eta$, but the mature prey not as effectively as they could if they were sound. This then justifies condition (a).

The second and third differential equations express the predators' dynamics. In summing them we find

$$\frac{dP}{dT} = dV + eUV - fP^2 + hV_I + kV_I + gUV_I - \mu V_I, \qquad (7.2)$$

i.e., the demographics of this whole population, which depends on the subpopulations $V$ and $V_I$. In the second differential equation we find the basic quadratic model for the predators, and the interaction term due to the disease, involving the infected predators $V_I$. The last terms account for the disease recovery process, which togheter with the former disappear in the sum of (7.2), and the newborns coming from infected predators. The third differential equation contains a similar quadratic model, in this case formulated for infected predators (notice the $V_I$ term that premultiplies the bracket) and then the disease process with terms of reversed signs. Here also disease-related deaths are accounted for, expressed by the parameter $\mu$.

We further assume $g \neq e$, to allow for $g > e$, that infected predators get more "reward" from a caught prey, as food may be more valuable for them than for sound individuals.

Feeding on other sources takes place in both sound as well as diseased predators. This occurs at different net rates $d$, $h$, and $k$ and the offsprings of sound individuals are sound, this being expressed by the parameter $d$, while those of infected individuals may be sound, at rate $h$, or disease affected; this vertical disease transmission is expressed by the parameter $k$. Notice that in some instances this is a very reasonable assumption.

## Positively invariant sets

We begin to examine the long-term behavior of the model on a simplified version of (7.1), in that we take $\mu = 0$, i.e., biologically we state that the disease is not that virulent, and furthermore that infected individuals do not reproduce, $h = k \equiv 0$. An important preliminary question to answer before doing any further analysis is possibly to determine conditions for which trajectories will remain confined to a compact set, so that they will not "escape to infinity" i.e., that no species grows without limit.

We outline two theoretical tools for determining positively invariant sets, one geometric and the other analytic.

### Geometric approach

This method exploits the following property. Since the system is homogeneous, the axes and the coordinate planes are solutions. The fundamental existence and uniqueness theorem for ordinary differential equations ensures then that the trajectories cannot exit from the first orthant. The first tentative is to cut out these invariant sets in the three-dimensional phase space by using a suitable plane $\pi$ in the positive cone, i.e., the first orthant in the space, where $U$, $U$, $V \geq 0$. It will then be enough to show that the trajectories enter the tethrahedron defined by the coordinate planes and by this suitably defined plane. This means showing that the angle made by the "outward" normal to this plane with the tangent vectors to the trajectories on the mentioned plane $\pi$ is larger than $\frac{\pi}{2}$; see (7.4) below. The outward normal is the orthogonal line pointing away from the origin. This may be a usable method also applicable to other cases, for two reasons. Since it gives only sufficient conditions for the existence of such invariant sets, by working differently, at times we may indeed obtain different conditions under which bounded trajectories exist. Moreover, if needed, more complex surfaces might also be used, to possibly give alternative sets of sufficient conditions.

The construction starts from a generic plane in the $(U, V, V_I)$ phase space

$\tilde a U + \tilde b V + \tilde c V_I = \tilde d$ not crossing the origin, $\tilde d \neq 0$, and meeting the positive orthant $\tilde a, \tilde b, \tilde c > 0$. Dividing by $\tilde d$ we can cast it into the form

$$\pi: \quad AU + BV + CV_I = 1, \quad A, B, C > 0. \tag{7.3}$$

This plane has thus the outward unit normal $n = (A, B, C)$. Let the flow of the differential system be denoted by $x' = (U', V', V_I')$. This plane cuts out of the positive cone a compact positively invariant set if

$$n \cdot x'|_\pi \leq 0. \tag{7.4}$$

See the illustrative two-dimensional Figure 7.1 for a graphical representation, in which the plane $\pi$ is replaced by a straight line.

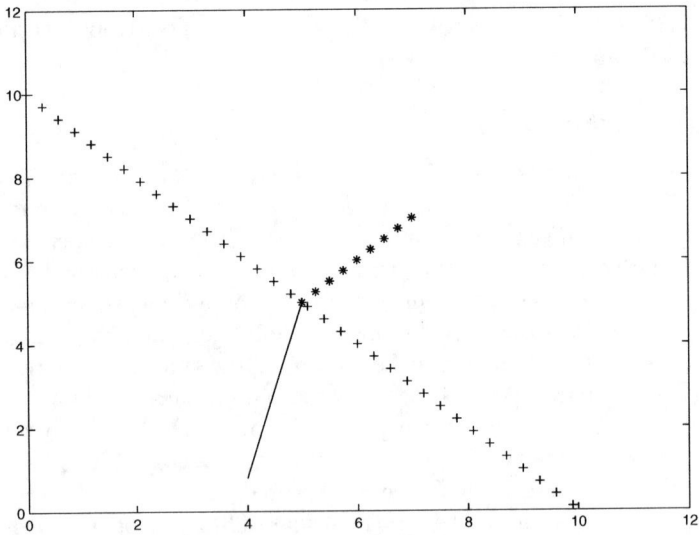

**FIGURE 7.1:** For the classical prey–predator dynamical system, the auxiliary line $\pi: U + V = 10$ is shown along with its gradient $\nabla \pi = (1, 1)$ (starred line) and the flow $\left(\frac{dU}{dT}, \frac{dV}{dT}\right)$ (continuous line). The angle of the latter vectors is obtuse, and if it is on the whole portion of the line $\pi$ in the first quadrant, the flow enters in the triangle, which is then the positively invariant set.

Use (7.3) to eliminate $V_I$ and obtain

$$AU\left[a - \frac{a}{K}U - cV - \frac{\eta}{C}(1 - AU - BV)\right] + \frac{B}{C}\nu(1 - AU - BV) \quad (7.5)$$

$$+ BV\left[d + eU - fV - \frac{1}{C}(\delta + f)(1 - AU - BV)\right]$$

$$+ (1 - AU - BV)\left(\delta V + gU - fV - \nu - \frac{f}{C}(1 - AU - BV)\right) \leq 0.$$

Rewrite (7.5) as follows:

$$AU\left[\left(a - \frac{\eta}{C} + \frac{2f}{C} + \frac{g}{A} + \nu\left(1 - \frac{B}{C}\right)\right)\right. \quad (7.6)$$

$$\left. + U\left(\frac{A}{C}(\eta - f) - \left(\frac{a}{K} + g\right)\right) + V\left(\frac{B}{C}(\eta - f) - (c + \delta - f)\right)\right]$$

$$+ BV\left[\left(d + \frac{f - \delta}{C} - \frac{f - \delta}{B} + \nu\left(1 - \frac{B}{C}\right)\right)\right.$$

$$\left. + U\left(e - g + \delta\frac{A}{C}\right) + V\delta\left(\frac{B}{C} - 1\right)\right] + \nu\left(\frac{B}{C} - 1\right) - \frac{f}{C} \leq 0.$$

If we impose all the above coefficients to be nonpositive, the inequality (7.6) follows. This amounts to satisfying the following system of inequalities, the relative equations of which represent surfaces in the $A$, $B$, $C$ space:

$$S_1: \quad (a + \nu)AC + gC + (2f - \eta)A - \nu BA \leq 0, \quad (7.7)$$

$$S_2: \quad A(\eta - f) - C\left(\frac{a}{K} + g\right) \leq 0, \quad (7.8)$$

$$S_3: \quad (f - c - \delta)C + B(\eta - f) \leq 0, \quad (7.9)$$

$$S_4: \quad (d + \nu)CB + (\delta - f)(C - B) - \nu B^2 \leq 0, \quad (7.10)$$

$$S_5: \quad (e - g)C + \delta A \leq 0, \quad (7.11)$$

$$S_6: \quad \delta(B - C) \leq 0, \quad (7.12)$$

$$S_7: \quad \nu(B - C) - f \leq 0. \quad (7.13)$$

The most complicated inequality to study is the one related to the quadric surface, i.e., (7.7). All the other ones are "cylinders" with axes parallel to one of the coordinate axes, for which it suffices to study the intersection with the plane of the remaining coordinates. Let $M^{(1)}$ be the solution set of the cylinders with axes parallel to the $A$-axis, $M^{(2)}$ be the respective solution set for the $B$-axis, and let us denote by $M^{(3)}$ the solution set of (7.7).

The solution set of $S_7$ is a half-space containing also the solution set of $S_6$. $S_3$ also gives a half-space, but feasibility requires three distinct subcases corresponding to the signs of the coefficients in the inequality $C \leq \frac{f-\eta}{f-c-\delta}B$. The last alternative arising, $c + \delta < f < \eta$, indeed gives an empty intersection with the first quadrant. We thus require either

a) $f > \max(\eta, c+\delta)$ and the slope in the $BC$ coordinate plane must be larger than one, i.e., $c+\delta > \eta$ must hold, giving in conclusion $f \geq c+\delta$. The intersection with the solutions of (7.12) and (7.13) is nonempty.

b) Alternatively, we impose $f < \min(\eta, c+\delta)$; the inequality then reverses the sign, and the intersection with the solution sets of (7.12) and (7.13) is always nonempty, given by the smaller of the two sets, i.e., $C \geq \max\left(1, \frac{f-\eta}{f-c-\delta}\right) B$.

c) Finally, if the coefficient is negative, $\eta < f < c+\delta$, the whole first quadrant of the $BC$ plane is a solution and no additional restrictions are required.

If nondegenerate, the conic (7.10) is a hyperbola, since the invariant given by the sum of the products of the coefficients of squared terms with the coefficient of the mixed product, i.e., $-\frac{1}{2}(d+\nu) - \nu \times 0$, is negative (e.g., see Bronshtein and Semendyayev, 1964, p. 254). It crosses the origin, with slope 1. Thus a possible nonempty intersection with the solution set of the former inequalities is possible only if its vertical asymptote lies to the right of the $C$-axis, i.e., the $B$ coordinate of its center is positive, $f > \delta$. Otherwise, a detailed analysis on the slopes of the asymptote and of $S_6$ shows that no crossings between the two lines are possible.

The condition $f > \delta$ found in this case, combined with the alternatives for (7.9), give respectively the possible modified alternatives

$$f \geq c+\delta > \eta; \quad \delta \leq f \leq \min(\eta, c+\delta); \quad \max(\eta, \delta) \leq f \leq c+\delta. \quad (7.14)$$

The solution set of inequalities (7.14) is thus the region $M^{(1)}$.

The cylinders with axes parallel to the $B$-axis are given by $S_5$ and $S_2$. $S_5$ gives half the space as a feasible solution if $g > e$, a cylinder with axes parallel to the $B$-axis, while $S_2$ has either the whole first quadrant as a solution for $\eta < f$, or an upper section for $\eta > f$. In the latter case, combining with $S_5$, we always find the nonempty intersection given by $C \geq \max\left(\frac{\delta}{g-e}, \frac{\eta-f}{g+\frac{a}{k}}\right) A$, the solution set for $S_5$ and $S_2$, i.e., $M^{(2)}$.

Applying together the above conditions, the two sets of cylinders $M^{(1)}$ and $M^{(2)}$ always give nonempty intersections, since they have mutually orthogonal axes. Thus a feasible solution set for the system of inequalities (7.8)–(7.13) exists.

The quadric (7.7) is in general a hyperbolic paraboloid. Indeed, its invariants are

$$\Delta \equiv \begin{vmatrix} 0 & -\frac{\nu}{2} & \frac{a+\nu}{2} & f-\frac{\eta}{2} \\ -\frac{\nu}{2} & 0 & 0 & 0 \\ \frac{a+\nu}{2} & 0 & 0 & \frac{g}{2} \\ f-\frac{\eta}{2} & 0 & \frac{g}{2} & 0 \end{vmatrix} = \left(-\frac{g\nu}{4}\right)^2 > 0,$$

$$\rho_3 \equiv rank \begin{pmatrix} 0 & -\frac{\nu}{2} & \frac{a+\nu}{2} \\ -\frac{\nu}{2} & 0 & 0 \\ \frac{a+\nu}{2} & 0 & 0 \end{pmatrix} = 2,$$

$$\rho_4 \equiv rank \begin{pmatrix} 0 & -\frac{\nu}{2} & \frac{a+\nu}{2} & f-\frac{\eta}{2} \\ -\frac{\nu}{2} & 0 & 0 & 0 \\ \frac{a+\nu}{2} & 0 & 0 & \frac{g}{2} \\ f-\frac{\eta}{2} & 0 & \frac{g}{2} & 0 \end{pmatrix} = 4.$$

Instead of a complete study of the quadric, we provide conditions to ensure that the solution set of (7.7) has a nontrivial intersection with the solution set $M^{(1)} \cap M^{(2)}$ of the former ones, (7.8)–(7.13), by investigating the restrictions of $S_1$ with planes parallel to the coordinate planes.

For $A = k > 0$, the inequality depends on a straight line, $C \leq k\frac{\nu B + (\eta - 2f)}{(a+\nu)k+g}$, whose ordinate at the origin is positive if $\eta \geq 2f$. This ensures a nonempty intersection with the solution set $M^{(1)}$. Should this ordinate at the origin instead be negative, an empty intersection with $M^{(1)}$ always exists, as the slope is always smaller than 1, $\frac{\nu k}{(a+\nu)k+g} < 1$. Combining with (7.14), for which the first and last conditions have to be neglected, we obtain the second equation in (7.15).

Restricting $S_1$ to $B = l \geq 0$ gives instead the inequality ruled by the hyperbola $C \leq A\frac{\eta - 2f + \nu l}{(a+\nu)A+g}$. A feasible solution in the first quadrant of the $AC$ plane exists only if $\eta + \nu l \geq 2f$. But the latter follows from $\eta \geq 2f$, which is true because it was derived above. The slope at the origin is $C'(A)|_{A=0} = \frac{1}{g}(\eta - 2f + \nu l)$. If it is larger than the biggest of the two slopes of $S_2$ and $S_5$, a nonempty intersection with $M^{(2)}$ can be found. Thus we need $\frac{1}{g}(\eta - 2f + \nu l) \geq \max\left(\frac{\delta}{g-e}, \frac{\frac{a}{K}+g}{\eta-f}\right)$. The fractions on the right are positive, since $g > e$ and $\eta > f$. In these conditions, a feasible solution for inequality (7.6) can then be found, thus giving the second equation in (7.15) by pushing down $l$ to its lower possible bound, $l = 0$.

In summary, for the SIS model with $\nu \neq 0$, $h = 0$, $k = 0$, $\mu = 0$, a compact positively invariant set exists if the next conditions are satisfied:

$$g > e, \quad \delta \leq f \leq \min\left(\frac{\eta}{2}, c+\delta\right), \tag{7.15}$$

$$\frac{1}{g}(\eta - 2f) \geq \max\left(\frac{\delta}{g-e}, \frac{\frac{a}{K}+g}{\eta-f}\right).$$

Notice that since $\nu$ does not appear in the inequalities just found, the sufficient conditions for the existence of a compact positively invariant set for the SI model with mass action incidence coincide with the above ones. Moreover, since (7.15) are a number of inequalities smaller than the number of parameters appearing in them, they can be satisfied very likely. But should they be violated, no feasible region found with this method would then exist, i.e., bounded by a plane. It may, however, be provided by the choice of another appropriate surface instead of the plane $\pi$, still satisfying (7.4). In any case, one should always remember that there are cases when the approach may fail to give a suitable set of parameters.

**Analytic approach**

The approach we are presenting here applies to the model (7.1) without the simplifying assumptions, so that in this paragraph, we can still take $h \neq 0$, $k \neq 0$, $\mu \neq 0$. Let us define the new function $\Psi = U + V + V_I$, i.e., the total population living in the environment. Using $\kappa \geq 0$, and adding $\kappa \Psi$ to the sum of the right-hand sides of (7.1) gives

$$\frac{d\Psi}{dT} + \kappa \Psi = \chi(U, V, V_I) , \qquad (7.16)$$

where we have, recalling that $P = V + V_I$,

$$\chi(U, V, V_I) \equiv (a + \kappa)U + (d + \kappa)V + (h + k + \kappa - \mu)V_I - \frac{a}{K}U^2$$
$$- U(cV + \eta V_I) + U(eV + gV_I) - f(V + V_I)P$$
$$= (a + \kappa)U - \frac{a}{K}U^2 + (d + \kappa)V + (h + k + \kappa - \mu)V_I$$
$$+ U[(e - c)V + (g - \eta)V_I] - fV^2 - fV_I^2 - 2fVV_I.$$

Assume $c > e$, $\eta > g$ to get an upper bound

$$\chi(U, V, V_I) \leq (a + \kappa)U - \frac{a}{K}U^2 + (d + \kappa)V \qquad (7.17)$$
$$+ (h + k + \kappa - \mu)V_I - fV^2 - fV_I^2$$
$$= (a + \kappa - \frac{a}{K}U)U + (d + \kappa - fV)V$$
$$+ (h + k + \kappa - \mu - fV_I)V_I .$$

The three parabolas in (7.17) are bounded above by the heights of their respective vertices, so that

$$(a + \kappa - \frac{a}{K}U)U \leq \frac{(a + K)^2}{4a} = H_1, \quad (d + \kappa - fV)V \leq \frac{(d + \kappa)^2}{4f} = H_2,$$

$$(h + k + \kappa - \mu - fV_I)V_I \leq \frac{(h + k + \kappa - \mu)^2}{4f} = H_3,$$

and it then follows

$$\frac{d\Psi}{dT} \leq -\kappa\Psi + (H_1 + H_2 + H_3) \equiv -\kappa\Psi + H.$$

Solving, from Gronwall's inequality, we find

$$\Psi(t) \leq \Psi(0)e^{-\kappa t} + \frac{H}{\kappa}(1 - e^{-\kappa t}) \leq \max\left(\Psi(0), \frac{H}{\kappa}\right). \quad (7.18)$$

More generally, we could try to study the nature of the function $\chi$ on the right-hand side of (7.16), i.e., what kind of quadric surface it is, and determine the largest point, i.e., the point farthest from the origin. Indeed, such a surface can be seen as a level surface of a 4-dimensional hypersurface, the maximum of which will correspond to its largest value.

### Equilibria analysis

We now study of the linearized system (7.1) with $h = k = \mu = 0$, near the equilibria, whose superscript in the notation emphasizes that they belong to the "mass"-action incidence model. We find the following points:

$$E_0^{(m)} \equiv (U_0, V_0, V_{I0}) \equiv (0,0,0),$$

with eigenvalues $a$, $d$, $-\nu$, which show it to be unstable, and

$$E_1^{(m)} \equiv (U_1, V_1, V_{I1}) \equiv \left(\frac{af - cd}{\frac{a}{K}f + ce}, \frac{ae + \frac{a}{K}d}{\frac{a}{K}f + ce}, 0\right),$$

with the feasibility condition $af \geq cd$. The matrix of the linearized system factors, to give the eigenvalue

$$\frac{(\delta - f)\left(ae + \frac{a}{K}d\right) - \nu\left(\frac{a}{K}f + ce\right) + g(af - cd)}{\frac{a}{K}f + ce};$$

the other eigenvalues are the roots of the quadratic polynomial $P(\mu) \equiv \mu^2 + \mu b_1 + b_2$, with

$$b_1 = \frac{\frac{a}{K}(fa - cd) + f\left(ae + \frac{a}{K}d\right)}{\frac{a}{K}f + ce}, \quad b_2 = \frac{(af - cd)\left(ae + \frac{a}{K}d\right)}{\frac{a}{K}f + ce}.$$

Because of the feasibility condition, both coefficients are nonnegative, $b_1, b_2 \geq 0$, implying that, if we exclude particular cases, both roots of the quadratic have a negative real part. The stability condition is thus given by the negativity of the first eigenvalue, i.e.,

$$gU_1 + (\delta - f)V_1 < \nu. \quad (7.19)$$

This equilibrium is a focus if the discriminant of the quadratic $P(\mu)$ is negative. Notice that for $\nu = 0$, i.e., for the SI case, if feasible, the equilibrium is unstable for $\delta \geq f$:

$$E_2^{(m)} \equiv (U_2, V_2, V_{I2}) \equiv \left(0, \frac{d}{f}, 0\right),$$

with eigenvalues

$$\frac{af - cd}{f}, \quad -d, \quad \frac{d\delta}{f} - d - \nu.$$

It is stable if $af < cd$ and $d\delta < f(\nu + d)$.

$E_3^{(m)} \equiv (U_3, V_3, V_{I3}) \equiv (K, 0, 0)$, with eigenvalues

$$-a, \quad \left(ae + \frac{a}{K}d\right)\frac{K}{a}, \quad \left(ga - \nu\frac{a}{K}\right)\frac{K}{a},$$

from which its instability follows;

$$E_{4,5}^{(m)} \equiv (U_{4,5}, V_{4,5}, V_{I4,5}) \equiv \left(0, V_{4,5}, [V_{4,5}(\delta - f) - \nu]\frac{1}{f}\right).$$

Here $V_{4,5}$ solves the quadratic equation $\phi(V) \equiv \delta^2 V^2 - (df + 2\nu\delta)V + \nu^2 = 0$. For feasibility, observe that since the discriminant reduces to $df(df + 4\nu\delta) \geq 0$ and $\phi(0) = \nu^2 \geq 0$, $\phi'(0) = -(df + 2\nu\delta) \leq 0$, $\phi''(0) = 2\delta^2 \geq 0$, the trinomial has two nonnegative roots. Thus $V_{4,5} \geq 0$ is ensured. For $\delta \leq f$, $V_{I4,5} \geq 0$ is impossible. Thus $\delta > f$ is necessary for feasibility, and $V_{I4,5} \geq 0$ gives a lower bound on $V_{4,5}$, i.e., $V_{4,5} \geq \frac{\nu}{\delta - f}$. As $\phi\left(\frac{\nu}{\delta - f}\right) = \frac{f\nu}{(\delta - f)^2}(f\nu - d(\delta - f))$, if $d(\delta - f) < f\nu$, for feasibility of both equilibria we need to impose $\phi'\left(\frac{\nu}{\delta - f}\right) = \frac{f}{\delta - f}(d(f - \delta) + 2\delta\nu) < 0$, but the two conditions combined give the contradiction $2f\nu < d(\delta - f) < f\nu$; hence both $E_{4,5}^{(m)}$ are infeasible. For $d(\delta - f) > f\nu$ we have $\phi\left(\frac{\nu}{\delta - f}\right) < 0$, so that only $E_5^{(m)}$ is feasible. The eigenvalues of $E_5^{(m)}$ are then given by

$$-\frac{V_5}{f\nu^2}\left[\delta^2 V_5(af + \eta\nu) - af(df + 2\delta\nu) + \nu^2(cf - \delta\eta)\right]$$

and by the roots of the trinomial $\theta(\lambda) = f\lambda^2 + \lambda\theta_1 + \theta_0$, $\theta_1 = \delta(\delta - f)V_{4,5} - (\nu\delta + 2\nu f + df)$, $\theta_0 = df((\delta - f)V_{4,5} + \nu) + 2\nu f(\nu - \delta V_{4,5})$. The negativity of the first one is then necessary, again entailing the following lower bound on $V_5$:

$$V_5 > \frac{af(df + 2\delta\nu) + \nu^2(\delta\eta - cf) + f\eta\nu(d + \nu)}{\delta^2(af + \eta\nu)} \equiv r^*.$$

This can be discussed as above in terms of the trinomial $\phi$. The conditions for the sign of the remaining ones are algebraically very much involved; only in some cases are we able to state some clearcut results. In particular, $\nu > \delta V_5$ is impossible, since $\phi\left(\frac{\nu}{\delta}\right) = -\frac{df\nu}{\delta} < 0$; we must then have $\nu < \delta V_5$, and this entails that $\theta(0) > 0$ follows if we require $d[V_5(\delta - f) + \nu] > \delta V_5 - \nu$. Since the parabola $\theta(\lambda)$ is convex, the case $\theta(0) < 0$ gives immediately the instability of $E_5^{(m)}$. We also obtain that $\frac{d\theta}{d\lambda}(0) < 0$ for $(\delta - f)(\delta V_5 - \nu) < (d + \nu)f$. Letting $p^* = \frac{f(\nu+d)}{\delta(\delta+f)} + \frac{\nu}{\delta}$, we have that the sign of $\frac{d\theta}{d\lambda}(0) > 0$ is equivalent to the inequality $V_5 > p^*$. To investigate the latter lower bound, we calculate

$$\phi(p^*) = -f \frac{-f\nu^2\delta + fd\nu\delta + d\nu\delta^2 + 2f^2\nu d + f^2 d^2}{(f+\delta)^2 \delta}.$$

Thus, if we require $d > \delta$, it follows $\phi(p^*) < 0$, i.e., the equilibrium is unstable. To impose $\phi(p^*) > 0$, i.e., $V_5 < p^*$, gives the much stronger requirement $fd\nu\delta + d\nu\delta^2 + 2f^2\nu d + f^2 d^2 < f\nu^2\delta$. In such a case, however, both roots of the trinomial $\theta(\lambda)$ are either real and negative, or complex conjugate, but with a negative real part. In this case, thus, $E_5^{(m)}$ would be stable.

We find then

$$E_{6,7}^{(m)} \equiv \left(\frac{\eta\nu + af - V_{6,7}(\eta\delta + cf - \eta f)}{\frac{a}{K}f + g\eta}, V_{I6,7},\right.$$
$$\left.\frac{(\delta - f)\frac{a}{K}V_{6,7} - \nu\frac{a}{K} + g(a - cV_{6,7})}{\frac{a}{K}f + g\eta}\right).$$

Here the values $V_{6,7}$ are the roots of the quadratic $\psi(V) \equiv \psi_2 V^2 + \psi_1 V + \psi_0 = 0$, where

$$\psi_2 = e\eta(\delta - f) + fg(\eta - c) + \delta^2 \frac{a}{K} + c(ef - g\delta),$$
$$\psi_1 = \delta ga - eaf - \frac{a}{K}df + afg - 2\delta\frac{a}{K}\nu - d\eta g - e\eta\nu + \nu gc,$$
$$\psi_0 = \nu^2 \frac{a}{K} - \nu ga.$$

For the SI model ($\nu = 0$), as $\psi_0 = 0$, $V_6 = 0$ and there is only one nontrivial equilibrium $V_7 = -\frac{\psi_1}{\psi_2}$. Feasibility for the SIS model, i.e., for the case $\nu \neq 0$, for $U_{6,7} \geq 0$, gives

$$V_{6,7} \leq \frac{af + \eta\nu}{(c-\eta)f + \delta\eta} \equiv V^* \quad \text{for} \quad \delta\eta + (c-\eta)f > 0. \tag{7.20}$$

No corresponding limitation exists for $\delta\eta + (c-\eta)f \leq 0$ and feasibility is always ensured. We have thus an upper bound on the size of the predator population. For $V_{I6,7} \geq 0$, we obtain instead the following bounds on the size of $V_{6,7}$:

$$V_{6,7} \geq \frac{\frac{a}{K}\nu - ag}{(\delta - f)\frac{a}{K} - cg}, \quad \text{for} \quad \frac{a}{K}\nu \geq ag, \quad \frac{a}{K}\delta \geq f\frac{a}{K} + cg, \quad (7.21)$$

$$V_{6,7} \leq \frac{ag - \frac{a}{K}\nu}{cg - (\delta - f)\frac{a}{K}}, \quad \text{for} \quad \frac{a}{K}\nu \leq ag, \quad \frac{a}{K}\delta \leq f\frac{a}{K} + cg.$$

For

$$\frac{a}{K}\nu \leq ag, \quad \frac{a}{K}\delta \geq \frac{a}{K}f + cg, \quad (7.22)$$

$V_{I6,7} \geq 0$ holds automatically, while for $\frac{a}{K}\nu \geq ag$, $\frac{a}{K}\delta \leq f\frac{a}{K} + cg$, it is impossible. To study the bounds (7.21) we consider the trinomial $\psi(V)$, from which we have a quadratic inequality. Concentrating again only on the discussion of sufficient conditions for feasibility, we assume (7.22) to hold, and consider (7.20). Since in this case $\psi(0) = \psi_0 = \nu\left(\frac{a}{K}\nu - ag\right) < 0$, there is exactly one positive root, $V_7$. To satisfy (7.20) it is sufficient to require then that $\psi(V^*) > 0$, a condition that reduces to

$$cf + (2\nu\eta + 2a\delta + \eta d)\frac{\nu c}{ad} + \frac{\eta^2 \nu \delta}{af} < \eta(f - \delta) + \frac{\eta^2 \nu}{d} + \frac{1}{ad}(a\delta + \nu\eta)^2 + \frac{\nu^2 c^2}{ad}. \quad (7.23)$$

Thus (7.22) and (7.23) ensure the feasibility of $E_6^{(m)}$. An alternative approach for feasibility consists in assuming both $U_{6,7}^{(m)}$ and $V_{I6,7}^{(m)}$ to be feasible, i.e., (7.22) and the inequality

$$\eta\delta + (c - \eta)f \leq 0. \quad (7.24)$$

Then no restrictions on the size of $V_{6,7}^{(m)}$ are necessary, and we need only to require the latter to be nonnegative. This can be achieved by observing that the quadratic form $\psi(V)$ possesses at least one positive root. Since $\psi(0) < 0$, it is enough to require $\psi'' > 0$, which gives

$$\psi_2 > 0. \quad (7.25)$$

We do not study the stability of this equilibrium point.

### 7.1.1 Some biological considerations

Our analysis has shown some sufficient conditions on the parameters ensuring the boundedness of the system trajectories; interestingly, they are independent of the parameters $a$, $d$. Thus net reproduction rates do not affect the existence of this invariant set.

Also, since the conditions do not depend on $\nu$, the positive invariant set exists for both the SIS ($\nu > 0$) and SI ($\nu = 0$) models. Geometrically, the net reproduction rates push the trajectories of the dynamical system away from the origin, but this effect in a sense disappears far away from the origin, a fact reflected by their absence from the sufficient conditions giving the invariant set.

Notice also that the condition $g > e$ is necessary, in this analysis, for the existence of the invariant set. For $g \le e$, indeed, condition (7.11) becomes impossible and thus the existence of the invariant set is not ensured. However, by rearranging the grouping of terms, another existence proof might be set up. The boundedness of the $f$ parameter in (7.15) shows that the "competition" term in the predators must not be too low, since it exceeds the infection rate of the disease, $\delta$, but also not too large, being limited by the returns in hunting $c$ and $\eta$ obtained by sound and diseased predators.

The points $E_0^m$, $E_1^m$, $E_2^m$, $E_3^m$ correspond to the equilibria of the underlying demographic model, and therefore are biologically important as being disease free. The origin remains an unstable equilibrium also in the ecoepidemic model. The same feasibility condition for the demographic equilibrium $E_1$, which is also globally asymptotically stable, also holds for $E_1^m$, but the latter is stable only if the additional condition (7.19) holds.

As for the prey–predator model, $E_2^m$ is stable when $E_1^m$ is infeasible, but in the ecoepidemic model another condition must be satisfied $\delta d < f(\nu + d)$, which for the SI model ($\nu = 0$) becomes $\delta < f$. $E_3^m$ is unstable as its underlying demographic counterpart.

In the ecoepidemic model there are also new equilibria not appearing in the classical prey–predator model. No prey survive at $E_{4,5}^m$. Notice that $E_5^m$ is feasible for $\delta > f$, i.e., when $E_2$ of the classical model is unstable. To make $E_5^m$ unstable, the level of the predator equilibrium population, $V_5$, should not exceed the ratio of recovery to contact rates, namely, $\frac{\nu}{\delta}$. But this is impossible for the SI model ($\nu = 0$). If $V_4$ and $V_5$ do not exceed the quantity $\frac{\nu}{\delta-f}$, both equilibria $E_{4,5}^m$ are infeasible.

Feasibility for equilibria $E_{6,7}^m$ is very difficult to be interpreted biologically. But combining the two inequalities (7.22) we find again $\delta > f$, so that once more feasibility is related to the instability of the prey–predator equilibrium $E_2$.

The disease is endemic at $E_5^m$, but the prey are also wiped out. If the goal is the eradication of an obnoxius pest, one could introduce a diseased predator in the ecosystem and aim for $E_5^m$. But the same result could be achieved in the classical prey–predator model by striving for equilibrium $E_2$. Thus the introduction of the disease seems to be useless.

Let us now attempt a partial answer to one biologically relevant question, namely, whether the introduction of a prey can eradicate a disease spreading in the predators. From $E_{4,5}^m$, in case of the SI model, $\nu = 0$, we find that $E_4^{(m)}$ migrates to the origin, which as we know is always an unstable equilibrium, but we also have $V_5^{(m)} = \frac{df}{\delta^2}$, from which it follows $V_{I,5}^{(m)} \ge 0$ for $\delta \ge f$. Thus

$E_5^m$ is feasible, with eigenvalues

$$\frac{1}{\delta^2}\left[\delta\left(a\delta - d\eta\right) + df\left(\eta - c\right)\right], \left[-f \pm \sqrt{f\left(5f - 4\delta\right)}\right]\frac{d}{2\delta}.$$

They have negative real parts if $d\eta > a\delta$, $c > \eta$, $4\delta > 5f$. Thus the introduction of the prey, although they in the end disappear, causes the predators to partly survive the epidemics unaffected, since part of the population survives, at $V_5^{(m)} > 0$. Comparing with the SI epidemics model, for which all individuals ultimately become infected, the presence of the prey acts as a "recovery" from the disease, i.e., what the SIS model prescribes. The prey thus allows the survival of a part of the predator population unaffected by the disease.

We conclude the Section by mentioning a result on the more general model allowing disease-related deaths. For small values of $\mu$ the behavior near the interior endemic equilibrium is similar to the one seen above, but increasing its value moderately, a bifurcation occurs; see Figures 7.2 and 7.3. Increasing it further however, the persistent oscillations disappear and the trajectories tend to a different (boundary) equilibrium.

## 7.2 Predator–diseased-prey interactions

We now consider the disease spreading among the prey. We count sound prey $U$, infected prey $U_I$, and sound predators $V$. Since the epidemics spreads only among the former, the system takes the form

$$\frac{dU}{dT} = U[a - bU - cV - \lambda U_I] + \gamma U_I, \qquad (7.26)$$

$$\frac{dU_I}{dT} = U_I[\lambda U - kV - \gamma],$$

$$\frac{dV}{dT} = V[d + eU - fV - hU_I].$$

The equations state that the disease can be overcome at rate $\gamma$. The sign of $h$ is understood to be undetermined, to describe interactions possibly leading to quick deaths of predators upon contact, $h > 0$, or predators hunting of infected animals occurs at a different rate $h < 0$ than the one for sound prey.

Let us examine the boundedness of the system trajectories first. Proceeding from the easier SI model, if $h > 0$, the set $[0, U^\circ] \times [0, \infty) \times [0, V^\circ]$, $U^\circ \geq a/b$, $d + eU^\circ - fV^\circ \leq 0$, is positively invariant in the $U$ $U_I$ $V$ space.

Indeed, it suffices to determine the sign of $\frac{dU}{dT}$ and $\frac{dV}{dT}$, deleting the terms $-\lambda U_I$ and $-hU_I$ from the equations (3.1). The given conditions ensure that $U' < 0$ on $U = U^\circ$, while the one for $V^\circ$ states that the point $(U^\circ, V^\circ)$ lies in the region $V' < 0$.

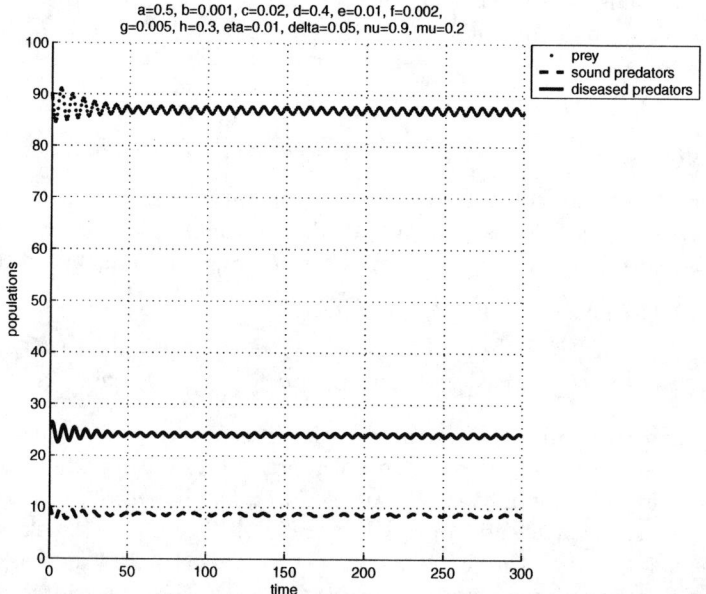

**FIGURE 7.2**: Persistent oscillations around the endemic equilibrium, for a moderate value of $\mu = 0.2$, top prey, middle diseased predator, bottom sound predators.

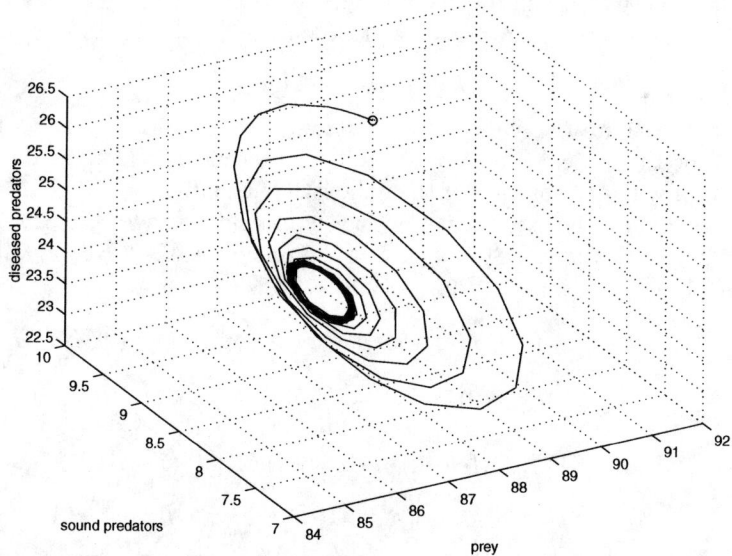

**FIGURE 7.3**: Phase space view of the above limit cycle around the endemic equilibrium, for a moderate value of $\mu = 0.2$.

For $h < 0$ and $b > k$, instead, an invariant set is given by the finite part of the positive cone cut out by the plane $\pi : AU + BU_I + CV = 1$, where $A, B, C > 0$ satisfy

$$aAB + \lambda(B - A) < 0, \quad \lambda A < B(c + \lambda), \quad hA < B(\lambda - e); \quad (7.27)$$
$$dCB - (kB + hC) < 0, \quad \lambda C < B(b - k), \quad hC < B(f - k).$$

Now we must also consider the term $-hU_I$, but working with the plane $\pi$ as above, imposing that its normal $n \equiv (A, B, C)$ makes an obtuse angle with the flow $x \equiv (U, U_I, V)$, we must then have $n \cdot x'|_\pi < 0$. This becomes

$$UA[(aAB + \lambda(B - A))(AB)^{-1} + V/B(\lambda C + (k - b)B)$$
$$+ U/B(\lambda A - (c + \lambda)B)] + CV[(dCB - (kB + hC))(CB)^{-1}$$
$$+ V/B(hC + (k - f)B) + U/B(hA + (e - \lambda)B)] < 0.$$

It is enough to require that each coefficient be negative, i.e., this amounts to (7.27). Geometrically they mean an intersection of a hyperbola with two straight lines. For the first equation in (7.27), the hyperbola has a vertical asymptote at $B = \lambda/a$ and slope at the origin $A'(0) = 1$. As the slope of the first line is larger, no other conditions must be imposed. For the second line, no other restriction is necessary in the cases $\lambda < e$ and $\lambda > e$, since in both these cases there will be a nonempty intersection with the solution set of the first inequality.

For the second equation in (7.27), instead, the pole for the hyperbola lies in the left half-plane, and its slope at the origin is $C'(0) = -k/h > 0$. The linear inequalities can be rewritten as follows:

$$C < (b - k)/\lambda \ B, \qquad C > (f - k)/h \ B.$$

We need to require the slope of the first line to be positive so as to find a solution for the system, i.e., $b > k$, while the second one always admits an intersection with the solution of the quadratic inequality.

We explore now the existence of positively invariant sets for the SIS model, exploiting the idea of the SI model. For the plane $\sigma : AU + BU_I + CV = 1$, we have here

$$n \cdot x'|_\pi \equiv U\frac{A}{B}[(aB - (\lambda + \gamma A)) + U((\lambda - \gamma)A - (b + \lambda)B)$$
$$+ V((k - c)B + (\lambda - \gamma)C)] + \frac{\gamma}{B}(A - 1)$$
$$+ V\frac{C}{B}[U((e - \lambda)B + (h - \gamma)A) + (\gamma(1 - A) + dB$$
$$+ (\gamma - h)) + V((k - f)B + C(h - \gamma))] - kV < 0.$$

Dropping $-\gamma A$ in the next to the last bracket gives the two systems of inequalities in the $BC$ parameter plane:

$$(k - c)B + (\lambda - \gamma)C < 0, \quad dCB + (2\gamma - h)C - kB < 0, \quad (7.28)$$
$$(k - f)B + (h - \gamma)C < 0,$$

and in the $AB$ parameter plane:

$$B < A(\lambda + \gamma A)/(\lambda + aA), \quad (\lambda - \gamma)A - (b + \lambda)B < 0, \qquad (7.29)$$
$$(e - \lambda)B + (h - \gamma)A < 0, \qquad A < 1.$$

Examining in detail all possible situations, we are led to the following set of inequalities, giving ranges for the parameters for which (7.28) and (7.29) have solutions. Namely, (7.28) is solvable for $h > 2\gamma$ if either

$$f > k > c, \quad \frac{\gamma - \lambda}{k - c} > \frac{h - \gamma}{f - k}, \quad \gamma > \lambda$$

or

$$k < \min(c, f)$$

hold. For $2\gamma > h > 0$ solutions exist by picking one set of conditions from either

$$f > k > c, \quad \gamma > \lambda, \quad \frac{\gamma - \lambda}{k - c} < \frac{h - \gamma}{f - k}$$

or

$$k < \min(c, f),$$

and a second set from either

$$k > c, \quad \gamma > \lambda, \quad \frac{2\gamma - h}{k} < \frac{\gamma - \lambda}{k - c}$$

or

$$k < c.$$

Finally, for $h < 0$ we need

$$k > f, \quad \frac{\gamma - h}{k - f} > \frac{2\gamma - h}{k},$$

and one set from either

$$k > c, \quad \gamma > \lambda, \quad \frac{2\gamma - h}{k} > \frac{\gamma - \lambda}{k - c}$$

or

$$k < c.$$

The system (7.29) possesses solutions if we take one set of conditions from either

$$e < \lambda, \quad \gamma < h, \quad \gamma - h < e - \lambda,$$

or

$$e > \lambda.$$

The parameter ranges giving solutions for the system (7.28) need to be paired then with those solving the system (7.29). In particular, notice that a set of easily understandable sufficient conditions are given by $h > 0$, $k < \min(c, f)$, $e > \lambda$. Thus the system trajectories are bounded in the case of the diseases prey having a strong impact on the predators, if the predation rate on infected prey is less than both the predators' logistic pressure term and the hunting rate on the sound prey, and the conversion rate of the predators exceeds the disease incidence.

For $h < 0$ some sufficient conditions for the existence of a compact positively invariant set are given, for instance, by $c > k > f$, $e > \lambda$, which are easily interpretable as suitable rates in the model, and the more complicated condition involving also the recovery rate of diseased individuals:

$$\gamma(k - f) < f(\gamma - h).$$

## Equilibria and their stability

The origin is easily found to be the first of the equilibria $E_i \equiv (U_i, U_{I,i}, V_i)$. $E_0 \equiv O$ has eigenvalues $a$, $-\gamma$, $d$, and thus it is unstable.

Then we have $E_1$ with only a nonzero component given by $V_1 = \frac{d}{f}$. Its eigenvalues are $-c\frac{d}{f} + a$, $-k\frac{d}{f} - \gamma$, $-d$. It is locally asymptotically stable then if $fa < cd$. Another equilibrium with only a nonzero component is $E_1$, with $U_2 = \frac{a}{b}$. The eigenvalues are $-a$, $\lambda\frac{a}{b} - \gamma$, $\frac{1}{b}(db + ea)$, thus is it always unstable. The next equilibrium is the one corresponding to the demographic model without disease. Letting $D = bf + ce$, it is $E_3$ with $U_3 = \frac{1}{D}(af - cd)$, $U_{I,3} = 0$, $V_3 = \frac{1}{D}(db + ea)$. It is feasible for $af \geq cd$. One eigenvalue can be explicitly evaluated as $[(af - dc)\lambda - k(db + ea)]/D - \gamma$; the other ones are the roots of the quadratic $T(\mu) \equiv \mu^2 + \mu\, b_1 + b_2$, with

$$b_1 = -\frac{t_1}{D}, \quad b_2 = -\frac{t_3}{D},$$
$$t_3 = (db + ea)(dc - fa),$$
$$t_1 = -fa(b + e) + bd(c - f).$$

Explicitly, they are given by

$$T_{1,2} = [2(ce + fb)]^{-1}[t_1 \pm t_2^{1/2}], \quad t_2 = t_1^2 + 4t_3(bf + ce).$$

The feasibility condition $U_3 > 0$ implies $t_3 < 0$, so that $t_2 < t_1^2$. Hence both roots of the quadratic have a negative real part. Stability is thus ensured by negativity of the first eigenvalue, $\lambda U_3 < kV_3 + \gamma$.

The equilibrium $E_4$ with all nonnegative components is analyzed only for the $SI$ model, $\gamma = 0$. Here

$$V_4 = \frac{\lambda}{k}U_4, \quad U_{I,4} = -\frac{Z}{\Delta}, \quad U_4 = \Psi\frac{k}{\Delta}, \quad \Psi \equiv \lambda d - ah,$$
$$\Delta \equiv \lambda(\lambda f - hc) - k(bh + \lambda e), \quad Z \equiv \lambda(dc - fa) + k(db + ea).$$

Feasibility gives either

$$\Psi > 0, \quad \Delta > 0, \quad Z < 0, \tag{7.30}$$

or

$$\Psi < 0, \quad \Delta < 0, \quad Z > 0. \tag{7.31}$$

The characteristic polynomial is $\Xi(\mu) = \mu^3 + a_1\mu^2 + a_2\mu + a_3$ with

$$a_1 = \frac{\Psi}{\Delta}(f\lambda + kb), \quad a_2 = \frac{\Psi}{\Delta^2}k\lambda[Z(h-\lambda) + \Psi D], \quad a_3 = -k\left(\frac{\Psi}{\Delta}\right)^2 \lambda Z.$$

The Routh–Hurwitz criterion gives the following results. If (7.30), then both $a_1 > 0$ and $a_3 > 0$ are consequences of the existence assumptions and the stability requirement follows if and only if $a_1 a_2 > a_3$, i.e., explicitly,

$$(f\lambda + bk)[D\Psi + Z(h - \lambda)] + Z\Delta > 0. \tag{7.32}$$

For the case $h < 0$, we can restate it, since the last two terms are equivalent to

$$Z[(f\lambda + bk)(h - \lambda) + \Delta] \equiv Z \ \lambda[h(f - c) - k(b + e)].$$

Then it suffices to ask $f > c$ for (7.32) to hold. If (7.31) are satisfied, then again $a_1 > 0$ follows from the feasibility, but $a_3 > 0$ contradicts the assumption $Z > 0$. Thus (7.31) never give stable equilibria.

Notice that the equilibrium $E_4$ for the SIS model splits to give rise to a pair of equilibria.

Finally, note that for the SI model $\gamma = 0$, we also have the following extra equilibrium, $E_* \equiv (U_*, U_{I,*}, V_*) \equiv (0, U_{I,*}, 0)$, with an arbitrary value for $U_{I,*}$. The eigenvalues are $a - \lambda U_{I,*}$, $0$, $d - hU_{I,*}$. The trajectories then lie on the plane $U_I = U_{I,*}$. The equilibrium is locally asymptotically stable if $U_{I,*}$ is large enough, i.e., for $\frac{a}{\lambda} < U_{I,*}$, $\frac{d}{h} < U_{I,*}$.

## Summary of findings

In this model, the instability of the equilibria $E_0$, $E_2$ is inherited from the corresponding equilibria of the underlying predator prey system. The other equilibrium $E_1$ on the coordinate planes of the $U\ U_I\ V$ space is stable if and only if the corresponding point of the underlying prey–predator model is infeasible. Also, $E_1$ is stable exactly as the underlying demographic equilibrium, and this occurs if and only if $E_3$ is infeasible. Notice that a general feature of ecoepidemic models is that the interior equilibrium of the classical model becomes a boundary equilibrium in the ecoepidemic model, since it lies on one of the the coordinate planes of the phase space. In this case it is the point $E_3$.

The model also possesses the nontrivial equilibrium $E_4$, in which the disease is endemic. The two species survive at a different level than prescribed by the underlying demographic model. For stability, the analysis gives the following sufficient condition. The crowding effect of predators should be larger than the predation rate on prey dynamics, at least for the case when the predators do not contract the disease. In addition, the level of both populations can be different than the prescribed values at the nontrivial equilibrium of the classical model by imposing that

$$\lambda \frac{\Psi}{\Delta} \neq \frac{db+ea}{D}, \qquad k\frac{\Psi}{\Delta} \neq \frac{af-cd}{D}.$$

These are two extra conditions, we already have another three given by (7.30) to ensure existence, and one more for stability, (7.32). Overall, then, there are six conditions to be satisfied, with nine parameters, as $\gamma = 0$ here. We can thus conjecture that for a suitable selection of the parameters of the model, the equilibrium is feasible, stable, and at a level not prescribed by the classical prey–predator model. Thus the disease in the prey acts as a controller for the size of the predators. In particular, for the case of pest control, the parameters could be chosen so that the prey equilibrium is at a smaller level than the one provided by the classical model.

## 7.3 Diseased competing species models

In the literature [e.g., see Begon and Bowers (1995a), Holt and Pickering (1986) and the bibliography cited therein] models of competition type have been treated and the role of multispecies interactions discussed, where the host density is itself considered as a dynamical variable, and an "ecological" perspective is envisaged. An investigation of two exponentially reproducing populations, each of them affected by the disease, can be found in Holt and Pickering (1986). Coexistence is governed by inter- and intra-specific infections. The numerical simulations do not reveal stable limit cycles. At the same time, the complications of the theoretical analysis of the stability are stressed. This fact will also be reflected in this section, in which the most complicated equilibria are essentially intractable, and therefore investigated mainly by numerical means.

In Holt (1977) a comparison with the Lotka-Volterra model with apparent competition among the hosts is discussed. The main inferences state that the prevalence of infection does not depend on alternative species being present; stable coexistence is impossible for a disease spreading only among the host species. In Begon and Bowers (1995b) two species share an infective agent, and limit cycles are discovered around the equilibrium whose stability is regulated by inter- and intra-specific infections. However, a whole range of possible complex behaviors is found. The system is also examined from an ecological viewpoint, i.e., allowing one of the hosts to be a pest that the pathogen is supposed to fight.

Closely related to the present investigation are the competing systems with a disease-free species together with another epidemics-affected host, which are considered in Begon and Bowers (1995a). The classical paper by Anderson and May (1986) considers two competitors, of which one is disease affected. The effect of the latter is to forbid reproduction of the infected individuals. The results highlight the possibility that the epidemics-affected superior competitor that if unaffected, would eliminate the other competitor, settles instead toward coexistence with it.

Two pathogens "competing" over the same host have been considered for instance in Dobson (1985) and Hochberg et al. (1990). Hochberg and Holt (1990) demonstrates that coexistence is allowed at an intermediate level between the efficient exploitation of the infected hosts and the less virulent exploitation of healthy ones.

We consider two species whose sizes at time $T$ are denoted by $P(T)$ and $Q(T)$. They may represent two predators feeding on each other, or two species in the same habitat using the same resources, for instance, goats and cattle grazing on the same meadows. We take the $Q$ species to be affected by an epidemic, which cannot be transmitted to the $P$ population, indicating by $Q_I$ the infected $Q$'s. The model with mass-action incidence and the possibility of disease recovery is

$$\frac{dP}{dT} = P\left[a - bP - cQ - \eta Q_I\right], \qquad (7.33)$$
$$\frac{dQ}{dT} = Q\left[d - eP - f(Q + Q_I) - \delta Q_I\right] + \nu Q_I,$$
$$\frac{dQ_I}{dT} = Q_I\left[\delta Q - gP - f(Q + Q_I) - \nu\right].$$

Each equation describes the logistic growth for the $P$ and the $Q$ species; it contains the negative cross-product interaction terms whose negative sign describs competition. The last equation gives the dynamics of the diseased population, which is less able to compete for resources, this being described by the different parameter $\eta$ in the first equation.

## Local stability analysis

Some biologically important questions that can be answered in part by the stability analysis concern the possibility that the action of a competitor in an environment in which an epidemic spreads among a population might eradicate the disease, if and how a disease can push to extinction one of the competing species, and under what circumstances the disease creates unstable dynamics.

The equilibria $E_i \equiv (P_i, Q_i, Q_{I,i})$ of the system are the following always feasible points: the origin $E_0$ with eigenvalues $a, -\nu, d$, giving instability; $E_1$

with only a nonzero component $P_1 = \frac{a}{b}$ and eigenvalues $-a$, $-\frac{1}{b}(ag + b\nu)$, $\frac{1}{b}(bd - ae)$, showing it is stable if $ae > bd$; $E_2$ again with only one nonzero component, $Q_2 = \frac{d}{f}$, with eigenvalues $-d$, $\frac{1}{f}(af - cd)$, $\frac{1}{f}(d\delta - df - f\nu)$ and stability conditions

$$cd > af, \qquad f(\nu + \delta) > d\delta. \tag{7.34}$$

These points are the corresponding phase space equilibria of the underlying demographic model, also with similar stability conditions. Only for $E_2$ does an extra condition on the epidemiological parameters arise (7.34) showing that the disease then affects the infected species dynamics. Considering the SI model $\nu = 0$, (7.34) reduces to $f > d$, i.e., it gives an extra *demographic* condition on the model. Thus, disease introduction may alter the stability of $E_2$ simply since $f < d$ destabilizes the otherwise stable equilibrium in the disease-free environment, for $cd > af$. It also helps the weaker species if it is introduced in the "winning" species, as the former would thrive in an ecosystem where normally it is wiped out by the adversary competitor.

The role of (7.34) is important also from the epidemics point of view; if $Q$ were the only species in the environment, by the epidemics action it would tend to $\left(\tilde{Q}_\nu, \tilde{Q}_{I_\nu}\right) = \left(\frac{\nu}{\delta}, N - \frac{\nu}{\delta}\right)$. If a competitor is introduced with interaction parameters satisfying (7.34), it would then remove the disease, although by being eliminated itself altogether. This allows control of the epidemics by a *demographic* device. If the action is done immediately after an epidemic outbreak, the environment could return to the previous conditions and the final level of the $Q$ species would be given by $Q_2$, independent of the epidemic parameters, another important point.

$E_3$ with nonvanishing components $P_3 = \frac{1}{D}(cd - af)$, $Q_3 = \frac{1}{D}(ae - bd)$, $D = ce - bf$ is again a counterpart of the demographic model, with the same feasibility condition, namely, $bf - ce$, $bd - ae$, $af - cd$ must all have the same sign. The three eigenvalues are

$$\lambda_3 = \frac{1}{D}\left[g(af - cd) + (f - \delta)(bd - ae)\right] - \nu,$$

and the two roots of the same quadratic polynomial of the underlying two-dimensional demographic model, implying the very same instability situation. This equilibrium remains also a saddle in the three-dimensional phase space. Notice further that imposing $f > \delta$ entails $\lambda_3 < 0$, so that in turn from (7.33) $\frac{dQ_I}{dT} < 0$. In these conditions the trajectories in the phase space tend then to the coordinate plane $Q_I = 0$, so that the final behavior of the model reduces to the classical competing system unaffected by the presence of the disease. Since $\nu$ appears negative in $\lambda_3$, the epidemic in general cannot influence the demographics, but again it can be removed by introducing a competitor. Although the situation is similar to the one discussed for $E_2$, in this case, however, the system settles to an equilibrium that also contains the competitor population, while the original population attains the level $Q_3 \neq Q_2$.

Other endemic equilibria in which the sound species disappear are $E_{4,5}$ with $Q_{I,4,5} = \frac{1}{f}((\delta - f) Q_{4,5} - \nu)$, where $Q_{4,5}$ solve $\delta^2 Q^2 - (df + 2\nu\delta) Q + \nu^2$. Its roots are always positive as geometrically they can be represented by the intersections of the parabola $y_1 = (\delta Q - \nu)^2$ with the straight line $y_2 = dfQ$. For feasibility we need to investigate only $Q_{I,4,5}$, ensured if $Q_{4,5} > \frac{\nu}{\delta - f}$. For $f > \delta$, the point $E_{4,5}$ is infeasible, and, as seen above, for $E_3$ the system approaches the underlying demographic model behavior. Take then $\delta > f$; to ensure feasibility for $E_{4,5}$, we require the value of $y_2$ at the vertex of the parabola to be larger than $\frac{\nu}{\delta - f}$, or $df(\delta - f) > \delta$.

For $\nu = 0$, we find in particular $E_4 \equiv E_0$ and $E_5 = \left(0, \frac{df}{\delta^2}, \frac{d}{\delta^2}(\delta - f)\right)$, feasible for $\delta > f$. Stability can easily be established, the eigenvalues being

$$\mu_1 = -\frac{1}{\delta^2}[df(c - \eta) - \delta(a\delta - d\eta)], \quad \mu_{2,3} = -\frac{d}{2\delta}\left[f \pm \sqrt{5f^2 - 4f\delta}\right].$$

From feasibility it follows that $5f^2 - 4f\delta < f^2$ and $Re(\mu_{2,3}) < 0$, so that stability is then governed by $\mu_1$. Sufficient conditions for stability are the alternative pairs of inequalities

$$cf > \delta\eta, a\delta^2 > df\eta; \qquad a\delta < d\eta, c > \eta. \tag{7.35}$$

At $E_5$ the $P$ species disappears and $Q$ survives at a lower level $df\delta^{-2}$ than the one exhibited by the classical, disease-free model, which would be given by $df^{-1}$. This contrasts with the classical SI epidemics model in which no sound individual is ultimately present. The use of an external competitor to control a disease results in a mistake, since the disease attains the nonzero value $d(\delta - f)\delta^{-2}$ and the competitors vanish. In case the incidence is very high, the endemic level of the epidemics is however low, $\approx d\delta^{-1}$, and in such a circumstance the introduction of the competitor might be beneficial.

There are two other equilibria $E_{6,7}$ with components

$$P_{6,7} = \frac{1}{bf - g\eta}[Q_{6,7}(f\eta - \eta\delta - cf) + \eta\nu + af],$$

$$Q_{I,6,7} = \frac{1}{bf - g\eta}[Q_{6,7}[b(\delta - f) + cg] - (ag + b\nu)],$$

where $Q_{6,7}$ solve the equation $p(Q) \equiv a_2 Q^2 + a_1 Q + a_0 = 0$ with coefficients

$$a_2 = e\eta(\delta - f) + cf(e - g) + g(f\eta - c\delta) - b\delta^2,$$
$$a_1 = a[f(g - e) + g\delta] + d(bf - g\eta) + \nu(-e\eta + 2b\delta + cg),$$
$$a_0 = -\nu(b\nu + ag).$$

Again consider $\delta > f$. If $a_2 > 0$ in view of $a_0 \leq 0$, there is always a positive root of the above quadratic. Imposing $e > g$, to ensure the positivity of $a_2$ we could require

$$e\eta(\delta - f) + cf(e - g) > g(c\delta - f\eta) + b\delta^2. \tag{7.36}$$

If instead $a_2 < 0$, two positive roots exist in case the discriminant is positive. Consider in particular the SI model for which $\nu = 0$ implies $a_0 = 0$; we find $E_7 \equiv E_0$, $Q_6 = -\frac{a_1}{a_2}$. We must require $a_1$ and $a_2$ to differ in sign, i.e.,

$$e\eta(\delta - f) + cf(e - g) + g(f\eta - c\delta) < b\delta^2, \tag{7.37}$$

together with sufficient conditions ensuring that $a_1 > 0$, such as $bf > g\eta$ and $g > e$. We finally need to ensure the positivity of the other populations; $P_6 > 0$ can be ensured by bounding the size of $Q_6$ while for $Q_{I,6}$ we need $Q_6$ to be bounded below:

$$-\frac{a_1}{a_2} < \frac{af}{cf + \eta(\delta - f)} \equiv \hat{Q}, \quad -\frac{a_1}{a_2} > \frac{ag}{b(\delta - f) + cg} \equiv \check{Q}.$$

From the former we have

$$\frac{-(bf - g\eta)(d\eta(f - \delta) + \delta^2 a) + 2caf^2(g - e) + dfc(bf - g\eta)}{-(e\eta(\delta - f) + cf(g - e) + g(f\eta - c\delta) - b\delta^2)(cf + \eta(\delta - f))} < 0,$$

with positive denominator, using (7.37). From the conditions on $a_1$, the above inequality gives

$$(bf - g\eta)\delta^2 a > (bf - g\eta)d\eta(\delta - f) + 2caf^2(g - e) + dfc(bf - g\eta). \tag{7.38}$$

For (7.36), we find that the denominator is positive, so that the numerator must be negative, giving

$$2caf^2(g - e) + dfc(bf - g\eta) > (bf - g\eta)(d\eta f + \delta^2 a - \delta\eta d). \tag{7.39}$$

For the lower bound, proceeding similarly, the inequality is rewritten as

$$\frac{(bf - g\eta)(bd - ae)(\delta - f) - g((bf - g\eta)(af - cd) - 2afc(g - e))}{(e\eta(\delta - f) + cf(g - e) + g(f\eta - c\delta) - b\delta^2)(b(\delta - f) + cg)} > 0.$$

Assuming (7.36) and $\delta > f$, the denominator is positive and finally we have

$$(bf - g\eta)(bd - ae)(\delta - f) > g((bf - g\eta)(af - cd) - 2afc(g - e)).$$

For (7.37), the opposite condition must be required. We do not analyze the stability. In this equilibrium, the three competing subpopulations coexist at the nonzero level. Thus, if the competitor has been introduced to eradicate the disease, the net result is a failure, since the disease persists endemically, and the same holds true by introducing a disease to fight a competitor, as the latter survives.

### Invariant sets for the trajectories

The results that follow show that it is enough to consider the region around the origin to understand the $\omega$-limit dynamics of the system. Since we model competing species, this boundedness result should be expected. To prove it, consider the box in the $PQQ_I$ phase space with opposite vertices at the origin and at $X^* = (P^*, Q^*, Q_I^*)$, where $P^* > \frac{a}{b}$, $Q_I^* > \frac{\delta}{f}$, $Q^* > \max\left(\frac{d}{f}, \frac{\nu}{f+\delta}\right)$. We calculate the angle of the system's flow with each face of $B$ through $X^*$. Consider the plane $\pi_Q : Q = Q^*$, with outward unit normal vector $n_Q$; similar notations hold for the other two variables. We have

$$n_Q \cdot (P', Q', Q_I')|_{\pi_{Q^*}} = Q^*(d - fQ^*) - ePQ^* - Q_I[(f+\delta)Q^* - \nu]$$
$$\leq -ePQ^* < 0,$$

thus the angle between the flow and the outward normal to the face of $B$ lying on $\pi_Q$ is obtuse, i.e., the trajectories enter into $B$ on this face. Similarly, taking the plane $\pi_P : P = P^*$, with outward unit normal $n_P$, we have again

$$n_P \cdot (P', Q', Q_I')|_{\pi_P} = P[a - bP - cQ - \eta Q_I]|_{\pi_P}$$
$$= P^*[a - bP^* - cQ - \eta Q_I] \leq P^*[-cQ - \eta Q_I] < 0.$$

An analogous inequality can also be computed for $n_{Q_I}$, i.e., $n_{Q_I} \cdot (P', Q', Q_I')|_{\pi_{Q_I}} < 0$. The trajectories thus enter into $B$ but cannot exit from it, as the existence and uniqueness theorem for dynamical systems prevents the crossing of the faces lying on the coordinate axes; notice indeed that (7.33) is homogeneous. In conclusion, $B$ is a positively invariant box.

### 7.3.1 Simulation discussion

Our simulations show the occurrence of persistent oscillations in various parameter ranges. Figure 7.4 shows them for the parameter values

$$a = 1.1, \quad b = 0.9, \quad c = 1.2, \quad d = 2, \quad e = 1.1,$$
$$f = 1, \quad g = 19, \quad \delta = 20, \quad \eta = 1, \quad \nu = 0.$$

A similar feature is obtained in Figure 7.5 where the oscillations arise by allowing disease recovery, i.e. $\nu \neq 0$, for the parameter values

$$a = 2.0, \quad b = 1.0, \quad c = 1.0, \quad d = 1.0, \quad e = 0.25,$$
$$f = 0.99, \quad g = 5, \quad \delta = 15, \quad \eta = 15, \quad \nu = 0.09.$$

These sustained oscillations are remarkable also from the ecological viewpoint, especially because they are *not induced by the presence of a periodically driven forcing* term. Observe, however, that the populations drop to very low levels, a fact that perhaps hints at an internal structural instability of the model, since small perturbations of the environmental conditions might drive the populations to extinction. When the disease-free equilibrium is a saddle, $\hat{E}_1 \equiv (2, 0)$, namely for

$$a = 2, \quad b = 1, \quad c = 1, \quad d = 1, \quad e = 1, \quad f = \frac{1}{4}, \tag{7.40}$$

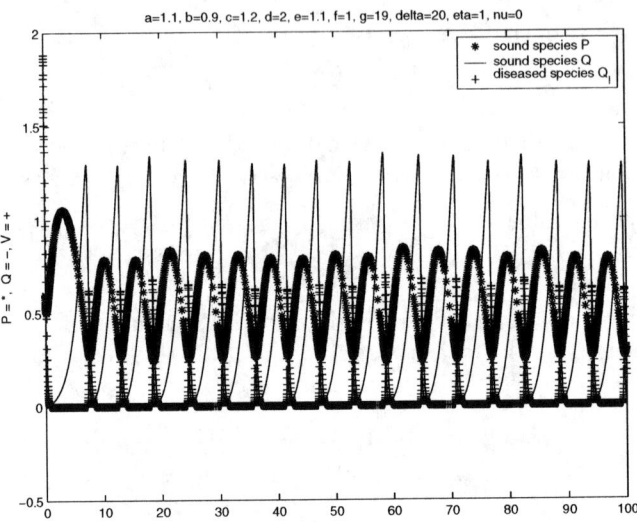

**FIGURE 7.4**: Diseased competing species: persistent oscillations for the three subpopulations.

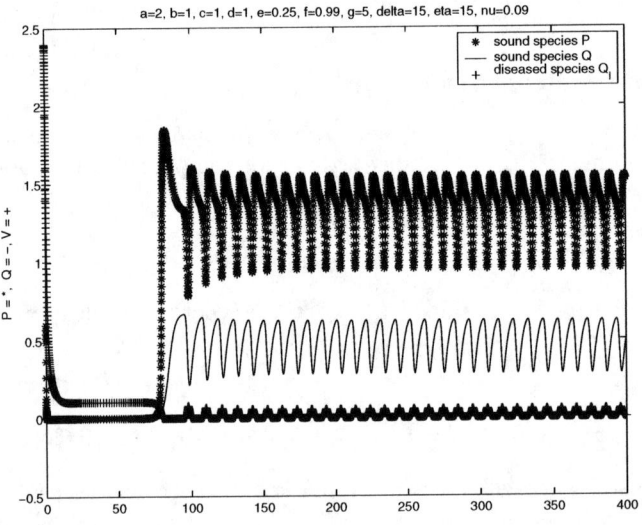

**FIGURE 7.5**: Diseased competing species: persistent oscillations for the three subpopulations after a transient phase, indicating the occurrence of a bifurcation depending on the disease recovery parameter $\nu$.

introducing the disease alters the system's behavior as follows. For $\eta = 30$, $\delta = 1$, $g = 0$, $\nu = 0$, the model settles to $\left(0, \frac{1}{4}, \frac{3}{4}\right)$. Taking $\delta = \frac{1}{2}$ the equilibrium becomes $(0, 1, 1)$, but for $\delta = \frac{1}{10}$ the equilibrium is $(1.9, 0, .003)$. Thus, from the equilibrium of the disease-free environment, in the diseased environment the system may settle to a point where either the sound species $Q$ or the infected one disappear *starting from the same initial conditions*, for different values of the incidence of the disease. Also the SIS model has similar consequences. For $\nu \geq \frac{1}{5}$, the equilibrium is $E_1 \equiv (2, 0, 0)$, but for $\nu = \frac{1}{10}$, the equilibrium becomes $E_5 \equiv (0, 1.37, .97)$ and for $\nu = \frac{1}{20}$, it is $E_5 \equiv (0, 1.19, .99)$. Thus, the introduction of a *recoverable* disease or a disease *with moderate incidence* in the succumbing competitor may lead to the *extinction of the predominant species*, provided it also deeply influences the competitor (i.e., if $\eta$ is large).

Conversely, a disease affecting the dominant species but with *low effect on the dominated species*, (i.e., for $\eta$ small), may have the effect of wiping the former out, in favor of the succumbing one, or to force the dominant one to a lower value, where the disease remains endemic, when it has a bigger impact on the dominated population (i.e., for larger $\eta$). Indeed, for the same demographic parameters (7.40), the initial condition $(P_0, Q_0) \equiv (1, 3)$ leads to the equilibrium $\hat{E}_2 \equiv (0, 4)$. Now for the ecoepidemic system with $\nu = 0, g = 0$ and $\delta = 1$, and initial condition $(1, 3, 1)$ we find the equilibrium $E_1 \equiv (2, 0, 0)$ for $\eta \leq .46$ and the equilibrium $E_5 \equiv (0, 1.64, .92)$, for larger values of $\eta$.

We notice, finally, that the epidemics can be eradicated for suitable choices of the parameter values, by introducing a competitor $P$ of the affected species. The final value attained by the sound subpopulation of the latter may however be at a higher or lower level than the one attained by the system without the species $P$.

## 7.4 Ecoepidemics models of symbiotic communities

Specific biological examples of symbiotic species affected by diseases that we can encompass with the models under investigation here are, for instance, the chestnut trees (*Castanea sativa*) in symbiosis with mushrooms (*Cantharellus cibarius*, *Boletus* spp., *Amanita* spp.) and affected by chestnut cancer (*Endothia parasitica*), the L-form bacteria that form nonpathogenic symbiosis with a wide range of plants, and the latter then become capable of resisting further bacterial pathogens (Walker et al., 2002). Also, very recently it has been discovered that the soil nematode *Caenorhabditis elegans* transfers the rhizobium species *Sinorhizobium melotiti* to the roots of the legume *Medicago truncatula* (Horiuchi et al., 2005).

Instead of providing a more or less complete analysis, as was done in the former sections, our study here concentrates mainly on two fundamental biological questions. Of interest is to provide an answer on whether and how the epidemics in one of the species affects also the dynamics of the "sound" population in a mutualistic community. Moreover, we consider if the introduction of a symbiotic population in an environment where an already infected species lives may fight the disease to the point of eradicating it. We will analyze the equilibria but also mainly provide simulation results obtaining some remarks that are not intuitive. Our findings indicate that for suitable parameter ranges, the model exhibits an interesting feature: The equilibrium shifts to a *higher* value, indicating that the disease seems to have a positive effect on the environment. This also agrees with recent field experiments revealing the same phenomenon.

We need here two variables for the symbiotic species $P$ and $Q$ and one more $Q_I$ for the infected individuals of population $Q$. By assumption, the disease will not be able to propagate to the other species. Notice that in the equation for the sound individuals there will now be two interaction terms, one of each sign, as the commensalism is beneficial for that population, but the negative sign accounts for the disease incidence. Also, encounters with a diseased individual may not result in the very same advantage gained by associations with a sound one. To model the benefit obtained by the sick individuals by interacting with the unaffected population, we may assume either that the gain may be larger for the former than for the latter, or, conversely, in particular if the disease impairs the commensalism or for instance the wandering in the environment. Notice also that the demographic aspect has to account for both sound and infected individuals in the population pressure terms hindering the growth of the species. More specifically, in both equations for the affected species the logistic terms must be given by products of the relevant populations with the totality of the population $Q + Q_I$.

We may even allow reproduction of sound $Q$'s begetting sound offsprings at rate $d$ and of diseased individuals giving birth to both sound and disease-affected newborns at rates $h$ and $\ell$, respectively, i.e.,

$$\frac{dP}{dT} = P(a - bP + cQ + kQ_I), \tag{7.41}$$

$$\frac{dQ}{dT} = Q(d + eP - f(Q + Q_I) - \delta Q_I) + (\nu + h)Q_I,$$

$$\frac{dQ_I}{dT} = Q_I[\ell + \delta Q + gP - f(Q + Q_I) - \nu].$$

**Equilibria**

Let the equilibria be denoted by $E_k \equiv (P_k, Q_k, Q_{I,k})$. The origin $E_0$ is unstable, as its eigenvalues are $a$, $d$, $\ell - \nu$, thus showing the same nature as its demographic counterpart. Among the boundary equilibria, we find first the ones with only one nonzero component, namely, $E_1$ and $E_2$ with $P_1 = \frac{a}{b}$ and eigenvalues $-a$, $\frac{ae}{b} + d$, $\ell - \nu + \frac{ag}{b}$, showing again instability, and with $Q_2 = \frac{d}{f}$ and eigenvalues $-d$, $a + \frac{cd}{f}$, $\frac{d\delta}{f} + \ell - d - \nu$ once again unstable. These instabilities once again arise from the corresponding results for the equilibria of the disease-free symbiotic model. For $E_3$, letting $D \equiv bf - ce$, the two nonzero components are $P_3 = \frac{1}{D}(af + cd)$, $Q_3 = \frac{1}{D}(ae + bd)$, i.e., it is a disease-free point. The sign of the first eigenvalue $\lambda_1^{[3]} \equiv \ell - \nu + (\delta - f)Q_3 + gP_3$ governs stability as the other two are the pair of roots of the characteristic equation of the underlying classical demographics equilibrium.

For stability, the condition $\ell + (\delta - f)Q_3 + gP_3 < \nu$ needs to be satisfied.

Let us remark that for a very virulent epidemic, i.e., for a large $\delta$, or for a very small recovery rate $\nu$, the stable equilibrium of the disease-free symbiotic system becomes unstable, and this is even more stringent for an SI infection. In such a case, indeed, a very large $f$ is needed to keep stability, which entails a large crowding effect on the epidemics-affected species. Conversely, if the latter is bounded above by disease incidence, $E_3$ is surely unstable.

There is then a pair of equilibria $E_{4,5}$ with only the affected species surviving with disease at the levels $Q_{I,4,5} = \frac{1}{f}[(\delta - f)Q_{4,5} + \ell - \nu]$, where $Q_{4,5}$ are the roots of

$$\Pi(Z) \equiv \delta^2 Z^2 - [f(\ell - d + h) + \delta(\ell - 2\nu - h)]Z + (\nu + h)(\nu - \ell) = 0. \quad (7.42)$$

For feasibility we must impose that $\delta Q_{4,5} + \ell - \nu > f Q_{4,5}$. One eigenvalue is immediate,

$$\lambda_1^{[4,5]} \equiv a + Q_{4,5}(c - k) + \frac{k}{f}(Q_{4,5}\delta + \ell - \nu)$$
$$> a + Q_{4,5}(c - k) + kQ_{4,5} = a + cQ_{4,5} > 0,$$

implying that $E_{4,5}$ are always unstable, if feasible.

The last two equilibria, $E_{6,7}$, in which the species coexist, have components

$$P_{6,7} \equiv \frac{1}{bf - gk}[Q_{6,7}(cf + k(\delta - f)) + af + k(\ell - \nu)]$$

$$Q_{I,6,7} \equiv \frac{1}{bf - gk}[Q_{6,7}(cg + b(\delta - f)) + ag + b(\ell - \nu)].$$

They will only be investigated numerically, as they arise from the solution of the quadratic equation $e_0 Z^2 + e_1 Z + e_2 = 0$, with

$$e_2 = (h + \nu)[b(\ell - \nu) + ag],$$
$$e_1 = \delta[b(2\nu - \ell + h) - ag] + f[b(d - h - \ell) + a(e - g)]$$
$$\quad + k[e(\ell - \nu) - dg] + cg(h + \nu).$$
$$e_0 = f(g - e)(c - k) + \delta(ek - \delta b - gc).$$

## Boundedness

To show that the trajectories ultimately lie in a compact set, let us define $E = \max(e, g)$, $D = \max(d, \ell + h)$, $K = \max(k, c)$. Summing the last two equations, and denoting the total disease-affected population by $R = Q + Q_I$, we find from (7.41)

$$\frac{dR}{dT} = dQ + (\ell + h)Q_I + (eQ + gQ_I)P - fR^2$$
$$\leq R[D + EP - fR] \equiv Rr(P, R),$$
$$\frac{dP}{dT} \leq P(a - bP + KR) \equiv Pp(P, R).$$

The setting of $p$ and $r$ to zero will give the loci in the $RP$ phase plane solutions of the inequalities obtained from the isoclines $\frac{dR}{dT} \leq 0$ and $\frac{dP}{dT} \leq 0$, respectively. The lines $p(P, R) = 0$ and $r(P, R) = 0$ intersect at $T \equiv (T_P, T_R) = (\frac{af+DK}{bf-EK}, \frac{aE+bD}{bf-EK})$, a feasible point whenever $bf > EK$. Assuming this condition, the quarter-circle centered at the origin and of radius given by $\rho \equiv \sqrt{T_P^2 + T_R^2}$, lying in the first quadrant, is easily seen to be a positively invariant set for (7.41), as the flow of the dynamical system there points inward. Hence all trajectories are bounded as ultimately they must enter into it.

### 7.4.1 Disease effects on the symbiotic system

Our simulations are aimed at observing the consequences entailed by the disease in the species $Q$ on the demographic equilibrium of the symbiotic system $P$-$Q$. In general, the equilibrium $E_3^{(d)}$ suffers from this change in the system, but there are parameter values for which interestingly the equilibrium shifts to a *higher* value. Thus the disease positively affects the environment.

We fix the parameters of the symbiotic system $a = 5$, $b = 1$, $c = 1$, $d = 5$, $e = 1$, $f = 2$, so that the equilibrium lies at $P_3 = 15$, $Q_3 = 10$. The effect of the modification of the remaining population parameters for the infected individuals, $g$, $k$, i.e., the return obtained from mutualism for both infected and disease unaffected species, $Q_I$ and $P$, and the infected reproduction rates resulting in sound and diseased offsprings, respectively, $h$, $l$, and the epidemic parameters $\delta$, $\nu$ can be investigated.

In Figure 7.6 we report one such typical simulation. When relevant the six graphs contain the horizontal lines indicating the symbiotic equilibria in absence of the disease. Columnwise and left to right, we have respectively the plots of the solutions of the system (7.41) for the disease-immune population $P$ and total population subject to disease, $R \equiv Q + Q_I$, for the disease-affected population $Q$, and the fraction infected versus sound individuals, $Q_I/Q$, and finally for the infected population $Q_I$, and the disease prevalence $Q_I/R$. A dramatic improvement in the equilibrium is found for the following choice of parameters $g = 2.4$, $h = 4.7$, $k = 0.8$, $\ell = 4.3$, $\delta = 13$, $\nu = 0.6$.

A large improvement of the equilibrium is found for the unaffected species $P$, which is more than one order of magnitude larger than $P_3$. While the $Q$'s are almost wiped out, an outsurge of $Q_I$ is observed, which drives the

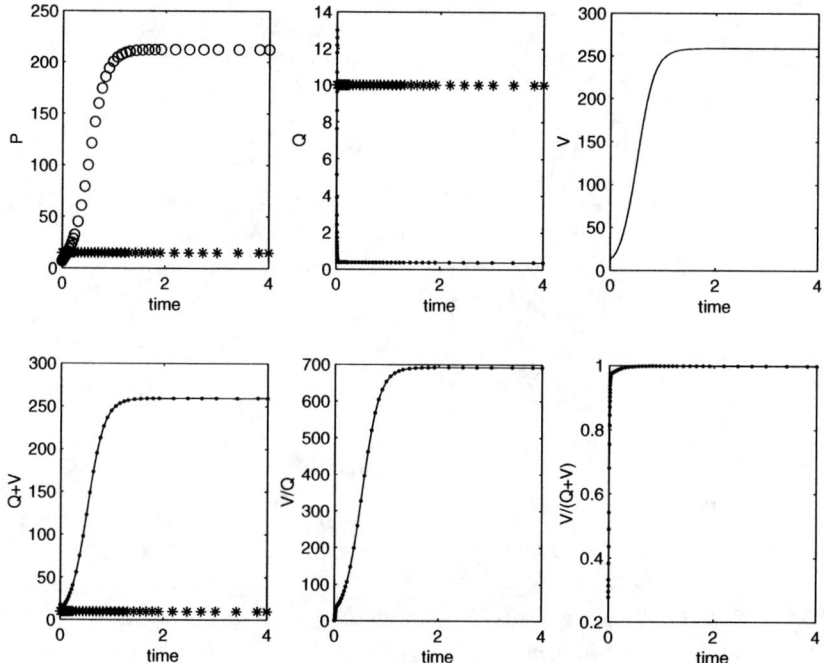

**FIGURE 7.6**: Disease effects on symbiotic system, for the ecoepidemic parameter values $g = 2.4$, $h = 4.7$, $k = 0.8$, $\ell = 4.3$, $\delta = 13$, $\nu = 0.6$ and demographic parameters $a = 5$, $b = 1$, $c = 1$, $d = 5$, $e = 1$, $f = 2$.

total population size again at a level more than one order of magnitude larger than $Q_3$. If one wants to harvest either one population, and the disease does not constitute a problem, this situation would be certainly an option to be seriously considered, to sustain the populations at much higher levels.

### 7.4.2 Disease control by use of a symbiotic species

Notice that disease eradication by introduction of a symbiotic species is not possible, as the best possible case for reducing the infected class $Q_I$, is to zero out the parameters expressing the return sound and infected obtain from the contact with the $P$ species, i.e. setting $g = 0$, $e = 0$. But in such case we have the following results.

Consider always the same epidemic model, for the parameters $d = 12$, $f = 2.25$, $\delta = 2.7$, $\nu = 2.5$, $h = 3$, $\ell = 10$, so that the equilibrium is $Q = 6.0576$, $Q_I = 4.5458$. Introduce then the symbiotic species with intrinsic reproduction rate and crowding coefficient $a = 10$, $b = 1$, i.e. at carrying capacity $P_3 = 10$. The remaining parameters express the gain $P$ have from interactions with sound and infectives of the other species and can only positively influence the

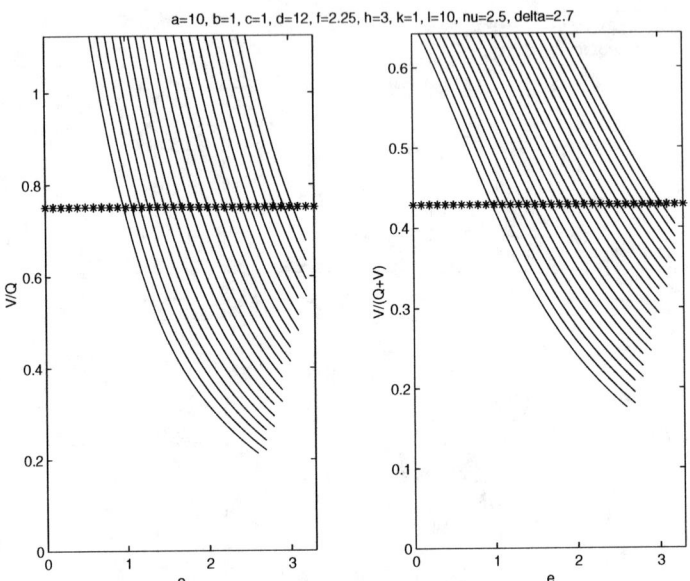

**FIGURE 7.7**: Disease control through parameters $e$ and $g$; each curve corresponds to values of $g$ in the range $[0, 0.6]$ with stepsize $0.03$

combined system. We take them as $c = 1$, $k = 1$. The new equilibrium is $P = 15.6268$, $Q = 1.9112$, $Q_I = 3.7156$, so that is not possible to obtain $Q_I = 0$, since the above choice provides the minimum return for the infected individuals $Q_I$. Thus it is not possible in general to eradicate the disease from the $Q$ species by introducing the symbiotic species $P$. But, on the contrary, it is possible to keep the ratio $Q_I/Q$ and the prevalence $Q_I/R$ small, by acting on the parameters $e$ and $g$, i.e., changing the return sound and infected get from the $P$ species. Figure 7.7 shows that the maximum possible reduction can be obtained by taking $g = 0$, $e \approx 2.6$. The horizontal line gives the ratio $Q_I/Q$ (left) and the prevalence $Q_I/R$ (right) for the original *SIS* equilibrium for the $Q$ species. The lines illustrate the behavior of the very same ratio at the equilibrium after introduction of the symbiotic species $P$. The lines from left to right correspond to increasing values for the parameter $g$ from $g = 0$ to $g = .6$ with stepsize $g = 0.03$. The lowest value of the curves gives the minimum prevalence, in the left plot.

What we just remarked seems in contradiction with the fact that the disease-free equilibrium $E_3$ may be stable. But the focus of this subsection are the consequences of introducing a new symbiotic species in the environment, *once the biologic parameters related to the disease-affected species $Q$ are fixed*. The eigenvalue $\lambda_1^{[3]}$ is only positively affected by $P_3$, so that it can only be destabilized, if before it was stable.

# Part III

# Spatiotemporal dynamics and pattern formation: deterministic approach

# Chapter 8

# Spatial aspect: diffusion as a paradigm

The analysis of the previous few chapters has largely ignored the spatial aspects of the population dynamics. The underlying assumptions are twofold. First, the results of the nonspatial analysis apply to the case of spatially homogeneous, "well-mixed" populations, which usually implies that the corresponding habitat is sufficiently small. Alternatively, the impact of spatial dimension(s) can possibly be ignored in a somewhat more exotic case when the individuals of a given species remain fixed in space at any time *and* in any generation.

Although these assumptions are not totally unrealistic, obviously, they do not always hold in the real world. Populations of ecological species do not remain fixed in space; their distribution changes continuously due to the impact of environmental factors, such as wind in case of airborne species, and/or due to self-motion of individuals. The properties of the motion can be significantly different depending on the particular ecological situation, i.e., on the behavioral traits characteristic for a given species and also on the temporal and spatial scale of the phenomenon under study. In this book we are not concerned with seasonal migrations, which are typical for many fish and bird species. Although migrations of that type may take place on a scale of thousands of kilometers, it often happens that a given flock of birds every spring returns to exactly the same pond or field, thus leaving the properties of the species spatial distribution essentially unchanged. Instead, we are interested in the dynamics that goes on a much smaller spatial scale but yet may eventually change the area of species dwelling as well as the features of the population spatial distribution inside.

A widely accepted and the most theoretically developed paradigm of the small-scale individual motion is the process called random walks. Having first appeared in physics, it was further developed and adapted for biological applications by Skellam (1951) and Okubo (1980).[1] One of its essential features is that the corresponding motion is isotropic in space (i.e., at each step the direction of motion is chosen randomly), and that has been a point of controversy and criticism. Indeed, it seems difficult to imagine that a mammal or

---

[1] The second edition of the Okubo book, cf. Okubo and Levin (2001), gives recent updates in this field and also contains an extended bibliography.

a bird or even an insect moves in a completely random manner, unless it has suffered from some sort of brain damage. However, this problem is easily resolved by choosing the adequate spatial scale of the random walk. The clue is that the random walk is just an abstract technique, which means that we are not at all obliged to associate each step in this highly theoretical procedure with the real "steps" or motion of a living being. Consider, for instance, the motion of a flying insect. Each of its short flights is apparently highly motivated and is by no means random, whether it is concerned with searching for food or avoiding predators or looking for a mating partner or something else. However, the "decision" for each new flight or "walk" is inevitably affected by a great variety of factors. Therefore, the direction of the next flight may be only slightly correlated with the direction of the previous one. Moreover, this loss of information is likely to accumulate with each new flight, so that after a certain number of steps the correlation will be lost completely (assuming, of course, that the external environmental conditions do not bring any preferred direction of motion). Now, what we should do is to coarse the scale of consideration combining several nonrandom flights together in order to produce one "step" for the random walk procedure. These heuristic speculations appear to be in a very good agreement with some data of field observations (cf. Root and P.M., 1984) showing that the motion of insects can indeed be very closely approximated as random on an appropriate spatial scale. Obviously, this way "to randomize" the motion is not restricted to insects and should be applicable to virtually any other species. Moreover, it has been shown recently that even the motion of humans can be very well approximated with random walks if considered on a relevant spatiotemporal scale (Brockmann et al., 2006).

The accepted assumption of isotropy of the individual motion does not fully define the random walk motion yet. Another very important property of the process is how the mean square displacement $<x^2>$ depends on time. A rather general case is given by the power law, $<x^2> \sim t^\sigma$, and then the three different cases are (i) Brownian motion with $\sigma = 1/2$, (ii) "subdiffusion" with $\sigma < 1/2$, and (iii) "superdiffusion" with $\sigma > 1/2$ (e.g., see Klafter et al., 1996). Although both case (ii) and especially case (iii) have been attracting considerable attention recently, cf. "Lévy flights," in this book we assume that the individual random walks are of standard Brownian type, which eventually results in the diffusion equation.

There are a few somewhat different theoretical routes to arrive at the "macroscopic" description of the population dynamics (e.g., by the diffusion equation) from the "microscopic" level of individual motion. A detailed description can be found elsewhere; for instance, an interested reader is advised to look into the classical works by Okubo (1980, 1986). Here we employ another way to derive the diffusion-type equations of population dynamics based on the "control volume" approach and making use of the available relation between the population flux and the corresponding population density.

Let us consider a population community consisting of $n$ species. At any moment of time $t$, the state of this community is described by the population densities $U_1, U_2, \ldots, U_n$. Let us consider a small "control volume" $V$ with the boundary given by a surface $S$; correspondingly, the population size of the $i$th species inside the volume $V$ is given as

$$N_i = \int_V U_i(\mathbf{R}, T) d\mathbf{R} , \qquad (8.1)$$

where $d\mathbf{R}$ is an infinitesimally small element of volume.

$N_i$ may change for two reasons, i.e., either due to the events of birth and death and/or due to the population fluxes through the boundary $S$. Let us begin with the latter. A cornerstone of the approach is then the equation for the fluxes. The simplest case is given by the Fick law when the flux of a given quantity is proportional to its gradient. In a general case, however, the flux of species $i$ may depend not only on its distribution but also on the spatial distribution of other species due to the phenomenon called crossdiffusion:

$$\mathbf{J}_i = -\sum_{j=1}^{n} D_{ij} \nabla U_j , \quad i = 1, \ldots, n, \qquad (8.2)$$

where $D_{ij}$ are nonnegative (cross)diffusion coefficients.[2]

The rate of increase/dicrease in the size of population $i$ per unit time is then given by integration of (8.2) over the surface enclosing the control volume $V$, i.e., by

$$\int_S \mathbf{J}_i \mathbf{ds} , \qquad (8.3)$$

where $\mathbf{ds}$ is an oriented element of surface $S$ directed outwards.

Equation (8.3) gives the rate of population change due to the motion of the individuals. In any real population, however, its individuals not only move but also, if considered on a time scale large enough, reproduce and die. Let the rate of the local population growth of species $i$ be described by a function $F_i(\mathbf{U})$ (which also takes into account the interspecific interactions such as predation, competition, etc.), where $\mathbf{U} = (U_1, U_2, \ldots, U_n)$. The total increase in $N_i$ over a time interval from $t_1$ to $t_2$ is then given by the following equation:

$$\Delta N_i = N_i(T_1) - N_i(T_0) = \int_{T_0}^{T_1} \left[ -\int_S \mathbf{J}_i \mathbf{ds} + \int_V F_i(\mathbf{U}) d\mathbf{R} \right] dT . \qquad (8.4)$$

Equation (8.4) allows us, in principle, to calculate the population size of a given species at any time $T$ provided that the initial species distribution is known. Note that it is obtained under rather unrestrictive assumptions and, as such, applies to a wide variety of real systems and situations. The reverse

---

[2] In a still more general case, $D_{ij}$ can be functions of time, space, and/or population density.

side of its generality, however, is that it is very difficult to study its properties other than through straightforward numerical simulations.

In order to arrive at an equation of population dynamics in a more convenient form, we have to make additional assumptions. Namely, we assume that the population densities are smooth functions so that there exists a continuous first derivative with respect to time and the second derivative with respect to space. The second term on the right-hand side of (8.4) can then be written as

$$-\int_S \mathbf{J}_i d\mathbf{s} = -\int_V \nabla \mathbf{J}_i d\mathbf{R} = \int_V \sum_{j=1}^n D_{ij} \nabla^2 U_j d\mathbf{R} \ . \tag{8.5}$$

Equation (8.4) takes the following form:

$$\int_V [U_i(\mathbf{R}, T_1) - U_i(\mathbf{R}, T_0)] \, d\mathbf{R} \tag{8.6}$$

$$= \int_{T_0}^{T_1} \left[ \int_V \sum_{j=1}^n D_{ij} \nabla^2 U_j d\mathbf{R} + \int_V F_i(\mathbf{U}) d\mathbf{R} \right] dT \ ,$$

or

$$\int_V \left( [U_i(\mathbf{R}, T_1) - U_i(\mathbf{R}, T_0)] - \int_{T_0}^{T_1} \left[ \sum_{j=1}^n D_{ij} \nabla^2 U_j + F_i(\mathbf{U}) \right] dT \right) d\mathbf{R} = 0 \ . \tag{8.7}$$

Let us note that Equation (8.7) is obtained for an unspecified volume $V$. Therefore, it can only hold if the expression under the integral equals to zero identically, i.e.,

$$[U_i(\mathbf{R}, T_1) - U_i(\mathbf{R}, T_0)] - \int_{T_0}^{T_1} \left[ \sum_{j=1}^n D_{ij} \nabla^2 U_j + F_i(\mathbf{U}) \right] dT = 0 \ . \tag{8.8}$$

Now, considering the limiting case $T_1 \to T_0$, from (8.8) we obtain the diffusion–reaction equation describing the dynamics of the $i$th species:

$$\frac{\partial U_i(\mathbf{R}, T)}{\partial T} = \sum_{j=1}^n D_{ij} \nabla^2 U_j + F_i(\mathbf{U}) \ . \tag{8.9}$$

Obviously, exactly the same chain of arguments can be repeated for any species from the given community so that, to describe the dynamics of an $n$-species system, we actually have a system of $n$ equations (8.9), $i = 1, 2, \ldots, n$.

Note that, in order to attain generality, the terms "volume" and "surface" have been used here in an abstract mathematical sense. Their heuristic meanings depend on the dimensionality of the specific system; for instance, in the case of a population dwelling on the Earth's surface (apparently two-dimensional), the "volume" actually means area and its boundary is a curve rather than a "surface."

The system (8.9) together with some of its extensions will be a focus of our analysis for the rest of this book. Its properties essentially depend on the properties of functions $F_i$, in particular, on the number of feasible steady states and their stability. For biological reasons, there must always exist at least one steady state, i.e., the extinction state $(0, \ldots, 0)$; however, conditions of existence/nonexistence of nontrivial equilibria as well as conditions of their stability vary very much from system to system. In fact, addressing these issues in terms of a biologically reasonable model is a core issue of mathematical biology, cf. Part II. Remarkably, as we will show in the following few chapters, some important features of the system spatiotemporal dynamics (cf. "biological turbulence") appear to depend more on the system behavior in the vicinity of a given nontrivial steady state rather than on the global structure of the phase space.

Let us also mention that, while crossdiffusion is a rather common phenomenon in some other natural sciences, e.g., in chemistry, in population dynamics it is rarely observed. Therefore, in order to avoid unnecessary generalizations, in the following we neglect its effect assuming that $D_{ij} = 0$ for $i \neq j$.

The properties of the (nonlinear) functions $F_i$ create a "skeleton" of the system dynamics; however, its spatiotemporal behavior also depends essentially on the initial conditions. Consider the case when, along with the extinction state, there exists a coexistence steady state $\bar{\mathbf{U}} = (\bar{U}_1, \bar{U}_2, \ldots, \bar{U}_n)$. In case the initial species distribution creates some sort of a boundary or front, i.e.,

$$U_i(X, Y, T) \to 0 \text{ for } X \to -\infty, \quad U_i(X, Y, T) \to \bar{U}_i \text{ for } X \to \infty \quad (8.10)$$

(i=1,...,n), then the generic solution of the system (8.9) describes a traveling population front propagating along or against axis $X$, cf. Kolmogorov et al. (1937) and Fisher (1937).

The traveling wave solutions of the diffusion–reaction system (8.9) are of high ecological relevance. They arise naturally in the problems related to exotic species spread, e.g., as a result of biological invasion (Shigesada and Kawasaki, 1997; Petrovskii and Li, 2006). In terms of pattern formation, they also provide the simplest pattern arising as a result of the system's self-organization. However, since they have been studied intensively in relation to other applications (such as, for instance, combustion and flame propagation; see Zeldovich et al., 1985; Volpert et al., 1994), as well as in a somewhat more abstract mathematical aspect (Aronson and Weinberger, 1975, 1978; Fife, 1979), here we will not focus on traveling population fronts per se. What makes them relevant to the scope of this book is that the propagation of a traveling front may trigger the formation of complex spatiotemporal patterns.

That will be considered in detail in Chapter 10. In the next chapter, however, we start with a more classical mechanism of pattern formation that is not related to front propagation.

# Chapter 9

# Instabilities and dissipative structures

In the previous chapter, we mentioned a traveling population front as the simplest theoretical example of a spatiotemporal structure. A front separates space into two regions with different properties, e.g., quantified by high and low values of population density, respectively. In each of these regions the population is distributed homogeneously. Therefore, the system actually exhibits spatial heterogeneity (in terms of large gradients of the population density) only within a narrow crossover region.

Front-like structures or patterns play an important role in ecology, but they of course do not exhaust all possible scenarios of spatiotemporal population dynamics. Spatial distribution of ecological populations is usually very inhomogeneous, so that patches of high population density in some way alternate with patches of low population density. Correspondingly, of high ecological relevance are scenarios of pattern formation where self-organized spatial heterogeneity would not be confined to only a certain domain, which is much smaller in size than the area actually available for the population dynamics, but could be observed, in principle, throughout the whole space.

In this chapter, we revisit some classical mechanisms of pattern formation in diffusion–reaction models of population dynamics, namely, those resulting from the Turing instability and the differential flow instability. The seminal work by Turing (1952) was eventually followed by a huge number of other publications reporting examples of the Turing instability in various systems along with some generalizations of the original mechanism. In particular, this topic is covered by many reviews, textbooks, and research monographs in mathematical biology, e.g., see Okubo (1980), Meinhardt (1982), Murray (1989), Ben-Jacob et al. (2000), Britton (2003), and Solé and Bascompte (2006). Thus, since the main goal of this book is to give an account about recent advances in this field, we address the Turing instability only briefly (yet with enough details to make our description comprehensible); an interested reader can find more information in the sources cited above.

Although the original study done by Turing was concerned with interacting chemical species, Segel and Jackson (1972) showed that the same ideas could potentially be applied to population dynamics as well. Correspondingly, here we also recall some more recent studies that used the Turing ideas in order to explain patterns in population communities – in particular, the patterns observed in semiarid vegetation systems; cf. Section 3.2.4. Finally, we conclude the chapter with a discussion (Section 3.2.5) of the Turing scenario relevance to ecological pattern formation in a general case.

## 9.1 Turing patterns

The basic idea of Turing's work (Turing, 1952) is very simple, although highly nontrivial: A steady state that is locally asymptotically stable in a nonspatial system can become unstable in the corresponding diffusive system. Relevant mathematical analysis shows that, on the onset of instability, the system first becomes unstable with respect to a spatially heterogeneous perturbation with a certain wavenumber. That leads to the formation of a regular spatial structure.

Remarkably, although the Turing instability manifests itself in a most wonderful way in nonlinear systems, where it results in the so-called "dissipative patterns" (Glansdorff and Prigogine, 1971), in a purely mathematical perspective, it is actually a property of a linear system. Indeed, let us consider the following system of linear diffusion–reaction equations:

$$\frac{\partial U(\mathbf{R},T)}{\partial T} = D_1 \nabla^2 U + a_{11} U + a_{12} V , \qquad (9.1)$$

$$\frac{\partial V(\mathbf{R},T)}{\partial T} = D_2 \nabla^2 V + a_{21} U + a_{22} V , \qquad (9.2)$$

where $U$ and $V$ are the state variables for the two interacting agents at the position $\mathbf{R} = (X,Y)$ and time $T$.

Apparently, $(0,0)$ is a steady state of the system (9.1)–(9.2) without space, i.e., without diffusion terms. This state is stable under the following conditions:

$$a_{11} + a_{22} < 0, \qquad (9.3)$$

$$a_{11}a_{22} - a_{12}a_{21} > 0 \qquad (9.4)$$

(which reflects the simple fact that both eigenvalues must have negative real parts), and it is unstable if at least one of these conditions is violated. In terms of the "full" system (9.1)–(9.2), conditions (9.3)–(9.4) mean that the spatially homogeneous steady state $u(\mathbf{r}) \equiv 0$, $v(\mathbf{r}) \equiv 0$ is stable with respect to spatially homogeneous perturbations.

We consider the case that the steady state $(0,0)$ is stable so that conditions (9.3)–(9.4) hold. Now, the question is whether the system remains stable in the case of an inhomogeneous perturbation.

Mathematical treatment of this issue is slightly different depending on whether the system (9.1)–(9.2) is considered in a bounded or in an unbounded domain. In the former case, any solution of the system (9.1)–(9.2) can be expanded into a Fourier series so that

$$U(\mathbf{R}, T) = \sum_{n,m=0}^{\infty} U_{nm}(\mathbf{R}, T) = \sum_{n,m=0}^{\infty} \alpha_{nm}(T) \sin \mathbf{kR} , \qquad (9.5)$$

$$V(\mathbf{R}, T) = \sum_{n,m=0}^{\infty} V_{nm}(\mathbf{R}, T) = \sum_{n,m=0}^{\infty} \beta_{nm}(T) \sin \mathbf{kR} \qquad (9.6)$$

(assuming the zero-function Dirichlet conditions at the domain boundaries), where $0 < x < L_x$ and $0 < y < L_y$, $L_x$ and $L_y$ giving the size of the system in the directions of $x$ and $y$, respectively, $\mathbf{k} = (k_n, k_m)$ and $k_n = \pi n/L_x$, $k_m = \pi m/L_y$ are the corresponding wavenumbers.

Since Equations (9.1)–(9.2) are linear, an interplay between different terms in (9.5)–(9.6) is impossible. Therefore, it is not necessary to work with the general solution (9.5)–(9.6) because it will decay with time if and only if $u_{nm}$ and $v_{nm}$ decay for any $n$ and $m$. (In more rigorous mathematical terms, it is a consequence of the fact that $\cos(k_n x + k_m y)$ and $\cos(k_p x + k_q y)$ are linearly independent for any $(n, m) \neq (p, q)$.) Thus, it is sufficient to consider the properties of a single pair $u_{nm}$, $v_{nm}$ for unspecified indices $n$ and $m$.

Having substituted $u_{nm}$ and $v_{nm}$ into (9.1)–(9.2), we obtain:

$$\frac{d\alpha_{nm}}{dT} = (a_{11} - D_1 \mathbf{k}^2) \alpha_{nm} + a_{12} \beta_{nm} , \qquad (9.7)$$

$$\frac{d\beta_{nm}}{dT} = a_{21} \alpha_{nm} + (a_{22} - D_2 \mathbf{k}^2) \beta_{nm} , \qquad (9.8)$$

where $\mathbf{k}^2 = k_n^2 + k_m^2$.

A general solution of (9.7)–(9.8) has the form $C_1 \exp(\lambda_1 t) + C_2 \exp(\lambda_2 t)$, where the constants $C_1$ and $C_2$ are determined by the initial conditions and the exponents $\lambda_{1,2}$ are the eigenvalues of the following matrix:

$$A = \begin{pmatrix} a_{11} - D_1 \mathbf{k}^2 & a_{12} \\ a_{21} & a_{22} - D_2 \mathbf{k}^2 \end{pmatrix} . \qquad (9.9)$$

Correspondingly, $\lambda_{1,2}$ arise as the solution of the following equation:

$$\lambda^2 - (\hat{a}_{11} + \hat{a}_{22}) \lambda + (\hat{a}_{11} \hat{a}_{22} - a_{12} a_{21}) = 0 , \qquad (9.10)$$

where the notations $\hat{a}_{11}$, $\hat{a}_{22}$ are introduced for convenience so that

$$\hat{a}_{11} = a_{11} - D_1 \mathbf{k}^2 , \qquad \hat{a}_{22} = a_{22} - D_2 \mathbf{k}^2 . \qquad (9.11)$$

In the case that the system (9.1)–(9.2) is considered in an unbounded domain, $-\infty < x, y < \infty$, Fourier series are no longer applicable; however, the solution can be expanded into a Fourier integral:

$$U(\mathbf{R}, T) = \frac{1}{2\pi} \int \tilde{U}(T, \mathbf{k}) e^{i\mathbf{k}\mathbf{R}} d\mathbf{k} ,$$

$$V(\mathbf{R}, T = \frac{1}{2\pi} \int \tilde{V}(T, \mathbf{k}) e^{i\mathbf{k}\mathbf{R}} d\mathbf{k} ,$$

(9.12)

where $\tilde{U}(T, \mathbf{k})$ and $\tilde{V}(T, \mathbf{k})$ are the Fourier transforms. Having substituted (9.12) into (9.1)–(9.2), we eventually arrive at the same Equation (9.10) for the linearized system's eigenvalues.

An inhomogeneous perturbation to the homogeneous steady state, cf. (9.5)–(9.6), decays with time if and only if all eigenvalues have negative real parts. Thus, the conditions of stability of the state $u(\mathbf{r}) \equiv 0$, $v(\mathbf{r}) \equiv 0$ with respect to a perturbation with a given wavenumber $k$ are given by

$$\hat{a}_{11} + \hat{a}_{22} < 0 \qquad (9.13)$$

and

$$\hat{a}_{11}\hat{a}_{22} - a_{12}a_{21} > 0 . \qquad (9.14)$$

A few important conclusions can be made simply from comparison between (9.3)–(9.4) and (9.13)–(9.14), without making any calculations. Indeed, it is readily seen that:

- If condition (9.3) holds then condition (9.13) holds as well. Thus, a change in stability, if any, can only be associated with (9.14).

- In case both $a_{11}$ and $a_{22}$ are negative, (9.14) follows from (9.4). Thus, a change in stability is possible only if $a_{11}$ and $a_{22}$ are of different sign, so that $a_{11}a_{22} < 0$.

- Therefore, $a_{12}$ and $a_{21}$ must be of different signs, $a_{12}a_{21} < 0$, otherwise condition (9.4) is violated, which contradicts our assumption that $(0,0)$ is locally stable.

Interestingly, the above conclusions do not yet tell much about the spatial aspect of the system, which is quantified by the diffusion coefficients. However, the impact of space becomes explicit when we consider inequality (9.14) in more details. From (9.14) and (9.11), we obtain:

$$D_1 D_2 k^4 - (D_1 a_{22} + D_2 a_{11}) k^2 \qquad (9.15)$$
$$+ (a_{11}a_{22} - a_{12}a_{21}) \equiv Q(\mathbf{k}^2) > 0 .$$

Inequality (9.15) is a necessary condition of stability. Correspondingly, the locally stable homogeneous steady state becomes unstable with respect to a perturbation with a given wavenumber **k** if (9.15) is violated, i.e., if $Q(\mathbf{k}^2) < 0$ (see Figure 9.1). By virtue of (9.4), it can only be possible if

$$D_1 a_{22} + D_2 a_{11} > 0 .\tag{9.16}$$

From (9.16), one can derive an important relation between the diffusion coefficients. Indeed, let us recall that $a_{11}$ and $a_{22}$ are of different signs. Assume that $a_{11} > 0$ and $a_{22} < 0$; then

$$a_{11} + a_{22} = a_{11} - |a_{22}| < 0$$

so that

$$a_{11} < |a_{22}| \quad \text{and} \quad \frac{a_{11}}{|a_{22}|} < 1 .\tag{9.17}$$

Therefore, from (9.16) we obtain:

$$\frac{D_1}{D_2} < \frac{a_{11}}{|a_{22}|} < 1 .\tag{9.18}$$

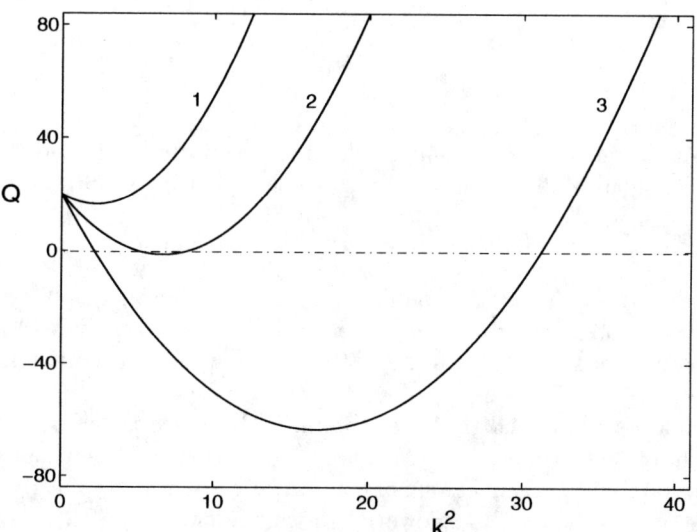

**FIGURE 9.1**: A sketch of the polynomial $Q(\mathbf{k}^2)$ for different values of $\epsilon = D_1/D_2$: curve 1 for $\epsilon = 0.65$, curve 2 for $\epsilon = 0.48$, curve 3 for $\epsilon = 0.3$. Coefficients $a_{ij}$ are calculated as in the Segel–Jackson system (9.35)–(9.36) with $\eta = 0.85$ and $a = 0.8$; correspondingly, $a_{11} = 16$, $a_{12} = -17$, $a_{21} = 20$, and $a_{22} = -20$.

It is readily seen that, in the opposite case $a_{11} < 0$ and $a_{22} > 0$, inequality (9.16) appears to be equivalent to $D_1/D_2 > 1$. Thus, we arrive at the following conclusion:

- Destabilization of the locally stable homogeneous steady state and the subsequent pattern formation are only possible if the diffusion coefficients are not equal.

Provided the inequality (9.16) holds, a sufficient condition for the instability onset is reached when $Q < 0$ in its minimum,

$$Q(\mathbf{k}_m^2) < 0, \tag{9.19}$$

where

$$\mathbf{k}_m^2 = \frac{D_1 a_{22} + D_2 a_{11}}{2 D_1 D_2} \tag{9.20}$$

(assuming that $D_1 D_2 \neq 0$) and the corresponding spatial period is then given as

$$l_m = 2\pi \left( \frac{2 D_1 D_2}{D_1 a_{22} + D_2 a_{11}} \right)^{1/2}. \tag{9.21}$$

Taking into account (9.15), from (9.19) and (9.20) we obtain:

$$(a_{11} a_{22} - a_{12} a_{21}) - \frac{(D_1 a_{22} + D_2 a_{11})^2}{4 D_1 D_2} < 0. \tag{9.22}$$

Since both terms on the left-hand side are positive, cf. (9.4) and (9.16), we can take a square root so that (9.22) takes the following form:

$$D_1 a_{22} + D_2 a_{11} > 2 (D_1 D_2)^{1/2} (a_{11} a_{22} - a_{12} a_{21})^{1/2}. \tag{9.23}$$

Note that if (9.23) holds, the condition (9.16) holds automatically.

Inequality (9.23) considered together with (9.3)–(9.4) defines the domain in the parameter space of the system (9.1)–(9.2), where the Turing instability can be observed.

It should be mentioned that the system (9.1)–(9.2) by itself does not have any clear biological meaning. Normally, a solution of a diffusion–reaction equation would have the meaning of concentration or density of a given agent, e.g., the population density. However, in the system (9.1)–(9.2), the domain $U \geq 0$, $V \geq 0$ is not an invariant manifold and the system trajectories can cross the axes $U = 0$ and $V = 0$. This means that the population density can become negative (or vice versa, a population can appear from thin air), which is, of course, biological nonsense. What is important, however, is that any realistic system can be reduced to (9.1)–(9.2) in case we consider its behavior in vicinity of a homogeneous steady state. In this case, $U$ and $V$ denote not the population densities but (small) transient perturbations to the steady state values $\bar{U}$ and $\bar{V}$, respectively.

Indeed, let us consider a two-species system in a general form:

$$\frac{\partial U}{\partial T} = D_1 \nabla^2 U + \phi(U, V), \qquad (9.24)$$

$$\frac{\partial V}{\partial T} = D_2 \nabla^2 V + \psi(U, V), \qquad (9.25)$$

where different ecological situations can be taken into account by choosing an appropriate form of the functions $\phi$ and $\psi$; cf. Chapter 2.

Now, assume that for given nonlinear feedbacks $\phi(U,V)$ and $\psi(U,V)$ there exists a "positive" coexistence state $(\bar{U},\bar{V})$, where $\bar{U} > 0$ and $\bar{V} > 0$. By means of introducing $\hat{U} = U - \bar{U}$ and $\hat{V} = V - \bar{V}$, where $\hat{U}$ and $\hat{V}$ are small, Equations (9.24)–(9.25) can be linearized so that we arrive at the system (9.1)–(9.2) where

$$a_{11} = \frac{\partial \phi}{\partial U}, \quad a_{12} = \frac{\partial \phi}{\partial V}, \quad a_{21} = \frac{\partial \psi}{\partial U}, \quad a_{22} = \frac{\partial \psi}{\partial V}, \qquad (9.26)$$

and the hats are omitted.

Considering (9.26) together with the necessary conditions (9.3)–(9.4) (see also the conclusions itemized earlier in this section), one can immediately identify some cases where the Turing instability is impossible. In order to give an instructive example, let us consider a Lotka–Volterra system of two competing species:

$$\frac{\partial U}{\partial T} = D_1 \nabla^2 U + r_1 U \left(1 - \frac{\alpha_{11}}{r_1} U\right) - \alpha_{12} UV, \qquad (9.27)$$

$$\frac{\partial V}{\partial T} = D_2 \nabla^2 V + r_2 V \left(1 - \frac{\alpha_{22}}{r_2} V\right) - \alpha_{21} UV. \qquad (9.28)$$

The null-isoclines are given by the following equations:

$$r_1 - \alpha_{11} U - \alpha_{12} V = 0, \qquad (9.29)$$
$$r_2 - \alpha_{21} U - \alpha_{22} V = 0. \qquad (9.30)$$

Obviously, under some restrictions on the coefficients (for details, see Shigesada and Kawasaki, 1997), there exists a coexistence steady state $(\bar{U},\bar{V})$. Calculating $a_{ij}$ according to (9.26) and taking into account (9.29)–(9.30), we then obtain

$$a_{11} = -\alpha_{11}\bar{U}, \quad a_{12} = -\alpha_{12}\bar{U},$$
$$a_{21} = -\alpha_{21}\bar{V}, \quad a_{22} = -\alpha_{22}\bar{V}.$$

Since all $\alpha_{ij}$ are positive (or, at least, nonnegative), all $a_{ij}$ appear to be negative (nonpositive) and the necessary conditions for the Turing instability are violated. Thus, we conclude that in a system of two competing species the Turing instability cannot occur.

On the contrary, in a prey–predator system, the Turing instability may become possible. To reveal possible constraints on the choice of parametrization, let us first consider the following generic equations:

$$\frac{\partial U}{\partial T} = D_1 \nabla^2 U + f(U)U - r(U)UV , \qquad (9.31)$$

$$\frac{\partial V}{\partial T} = D_2 \nabla^2 V + \kappa r(U)UV - g(V)V , \qquad (9.32)$$

where we use slightly different notations compare to Chapter 2.

Correspondingly, the matrix entries are given as

$$a_{11} = f'(\bar{U})\bar{U} - r'(\bar{U})\bar{U}\bar{V} , \qquad a_{12} = -r(\bar{U})\bar{U} ,$$
$$a_{21} = [r'(\bar{U})\bar{U} + r(\bar{U})]\kappa\bar{V} , \qquad a_{22} = -g'(\bar{V})\bar{V} ,$$

where prime denotes a derivative with respect to the argument.

One important observation can be made already in this general case: The Turing instability cannot occur if the predator mortality is density independent. Indeed, if $g(V) \equiv const$, then $a_{22} = 0$. Correspondingly, $a_{11}$ must be negative in order to ensure the local stability of the steady state, cf. (9.3), which means that condition (9.16) cannot be fulfilled. Thus, the formation of dissipative patterns in a prey–predator system is impossible unless the predator mortality is density dependant.[1] In particular, it means that the Turing instability is not possible in the classical Lotka–Volterra system.

More observations can be made in more specific situations. For instance, let us assume that $f(U)$ and $g(V)$ are a decreasing and an increasing function, respectively. (In biological terms, it may mean that the populations are not affected by the Allee effect.) Then it is readily seen that the Turing instability can only take place if density dependence in predation is taken into account. Indeed, consider that predation is described by a bilinear term, which means that $r(U) = const$ and $r'(U) = 0$. Then we immediately obtain that both $a_{11} < 0$ and $a_{22} < 0$, which makes the Turing instability impossible.

We want to emphasize here that relations (9.3)–(9.4) are only necessary conditions for the Turing instability and they alone cannot guarantee that the instability actually takes place. The full set of conditions also includes relation (9.23), and then one must prove that for a given choice of functional responses the corresponding domain in the parameter space is not empty. A biologically reasonable parametrization usually turns conditions (9.3)–(9.4) and (9.23) into a complicated system of inequalities that can only be solved numerically. There are not many cases that can be treated analytically; the one considered below was originally studied by Segel and Jackson (1972).

---

[1] The situation is different if predation is ratio dependent, e.g., see Alonso et al. (2002).

In order to make the problem analytically solvable, we have to keep it as simple as possible (and yet biologically sensible). Therefore, we consider a prey–predator system where predation is described by a bilinear term. Correspondingly, to fulfill the necessary conditions for the Turing instability, we have to introduce the Allee effect into the prey growth. This is, however, in agreement with numerous biological data showing that the Allee effect is a common phenomenon in population dynamics (Courchamp et al., 1999). Also, we take into account density dependence in the predator mortality, which may account for either the intra-specific competition or the impact of a top predator (in the latter case it is called a "closure term," cf. Steele and Henderson, 1992a). Having chosen a polynomial parametrization for $f(U)$ and $g(V)$, we arrive at the following system:

$$\frac{\partial U}{\partial T} = D_1 \nabla^2 U + (A_0 + A_1 U - A_2 U^2)U - r_0 UV , \qquad (9.33)$$

$$\frac{\partial V}{\partial T} = D_2 \nabla^2 V + \kappa r_0 UV - (B_0 + B_1 V)V , \qquad (9.34)$$

where $A_0$, $A_1$, $A_2$, $B_0$ and $B_1$ are nonnegative parameters.

Since here we are more interested in giving an example of a system prone to the Turing instability rather than in making a biologically meaningful prediction, in order to avoid tedious calculations we assume that $A_2 = B_0 = 0$. Introducing, for convenience, dimensionless variables as $t = A_0 T$, $x = (D_2/A_0)^{-1/2} X$, $y = (D_2/A_0)^{-1/2} Y$, $u = (r_0 \kappa / A_0) U$, and $v = (B_1/A_0)V$, from (9.33)–(9.34) we obtain:

$$\frac{\partial u}{\partial t} = \epsilon \nabla^2 u + (1 + au)u - \eta uv , \qquad (9.35)$$

$$\frac{\partial v}{\partial t} = \nabla^2 v + uv - v^2 , \qquad (9.36)$$

where $a = A_1/r_0 \kappa$, $\eta = r_0/B_1$, and $\epsilon = D_1/D_2$.

The null-isoclines of the system are given by

$$1 + au - \eta v = 0 , \qquad u - v = 0 , \qquad (9.37)$$

so that the steady-state density is

$$\bar{u} = \bar{v} = \Omega , \qquad (9.38)$$

where $\Omega = (\eta - a)^{-1}$. Obviously, the steady state must be situated in the first quadrant; thus, we obtain the following constraint on the parameter values:

$$\eta > a . \qquad (9.39)$$

For the matrix of the linearized system we have

$$a_{11} = a\Omega , \quad a_{12} = -\eta\Omega , \quad a_{21} = \Omega , \quad a_{22} = -\Omega . \qquad (9.40)$$

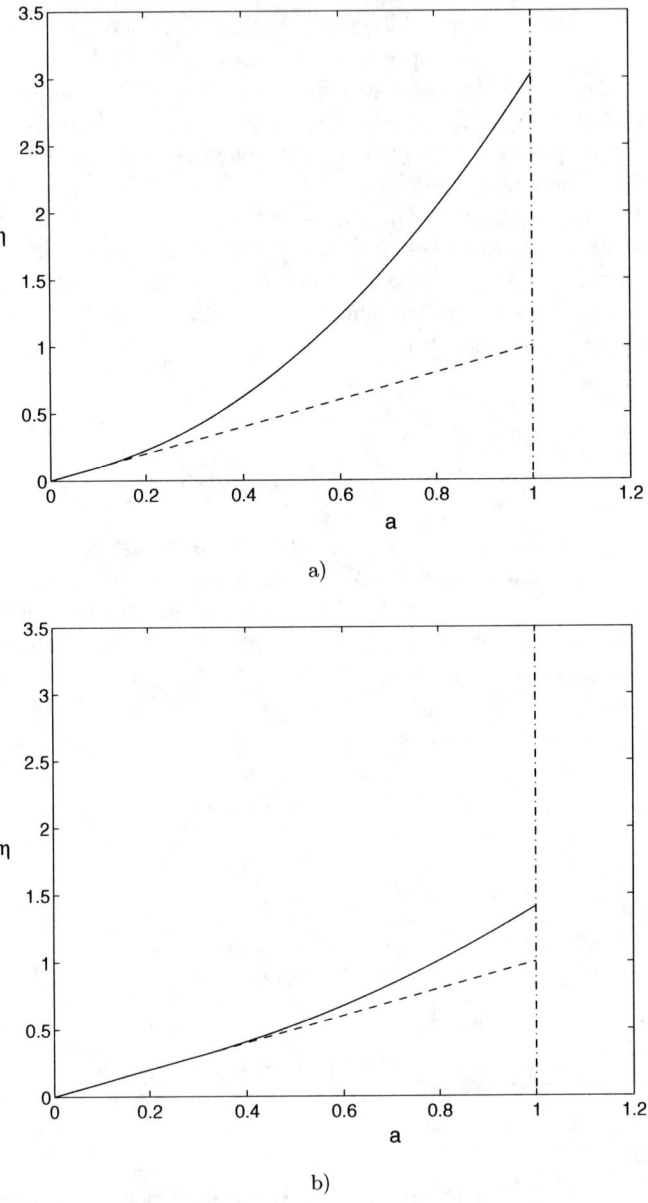

**FIGURE 9.2**: Parameter plane $(a, \eta)$ for the model (9.35)–(9.36) in the cases (a) $\epsilon = 0.1$ and (b) $\epsilon = 0.3$; other parameters are the same as in Figure 9.1. Turing instability can be observed for the parameter values inside the curvilinear triangle(s).

Takong (9.3) into account, we therefore obtain that

$$a < 1 . \tag{9.41}$$

Since $a_{11} > 0$ and $a_{22} < 0$, the condition (9.18) is applicable so that we obtain

$$\epsilon < a . \tag{9.42}$$

Finally, from (9.23) we arrive at

$$a - \epsilon > 2\sqrt{\epsilon(\eta - a)} , \tag{9.43}$$

from which we obtain that

$$\eta < \frac{(a - \epsilon)^2}{4\epsilon} + a . \tag{9.44}$$

The relations (9.39), (9.41), (9.42) and (9.44) taken together define the required domain in the parameter space; see Figure 9.2. It is readily seen that the diffusivity ratio is a controlling factor of crucial importance. While for $\epsilon \ll 1$ the parameter domain corresponding to the Turing instability can be of considerable size, for $\epsilon \approx 1$ it shrinks to a single point. This situation appears to be typical and does not depend much on the specific parametrization of the functional responses.

An interesting question is what the actual scenario of pattern formation may look like. The shape of the stationary structures achieved in the large-time limit is determined by the specific form of the problem nonlinearity (being described by a stationary solution of Equations (9.35)–(9.36), i.e., for $\partial u/\partial t = \partial v/\partial t = 0$) and therefore is unlikely to be affected by the choice of the initial conditions. However, the transient stage of the system dynamics obviously can be different depending on the presence/absence of "partial" perturbations with different $k$.

In order to give an instructive example, we consider the dynamics of the system (9.35)–(9.36) in the one-dimensional case and for the following initial conditions:

$$u(x, 0) = \bar{u} , \tag{9.45}$$

$$v(x, 0) = \bar{v} \text{ for } 0 < x < x_1 \text{ and } x_2 < x < L , \tag{9.46}$$

$$v(x, 0) = \bar{v} + v_0 \sin\left(\frac{2\pi(x - x_1)}{l_0}\right) \text{ for } x_1 < x < x_2 , \tag{9.47}$$

where the parameters have obvious meanings.

It can be expected that the system evolves differently depending on whether $l_0$ is "critical" (i.e., lies inside the critical range where the instability takes place) or not, i.e., whether $Q(2\pi/l_0)$ is negative or positive (see Figure 9.1).

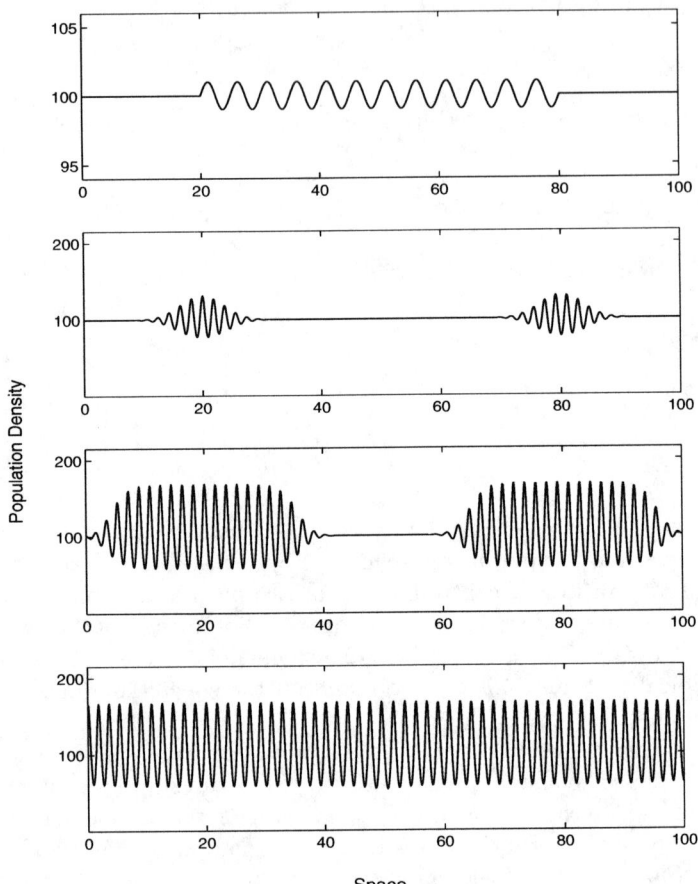

**FIGURE 9.3**: Pattern formation in the Segel–Jackson system (9.35)–(9.36) calculated numerically for the diffusivity ratio $\epsilon = 0.75$. Panels from top to bottom correspond to $t = 0$, $t = 5$, $t = 10$, and $t = 20$, respectively. Other parameters are $a = 0.95$ and $\eta = 0.96$, so that $l_m \approx 1.8$. The spatial period of the initial perturbation of the steady state is $l_0 = 5$.

Note that the function (9.46) is not monochromatic; due to its piecewise nature, its Fourier expansion apparently contains an infinite spectrum of wavenumbers. Thus, one can expect that, for parameter values allowing for the Turing instability, the patterns begin to emerge in the vicinity of points $x = x_1$ and $x = x_2$. As for the periodic initial perturbation in the middle of the domain, it will be "amplified" for critical values of $l_0$ and will decay otherwise. This heuristic expectation is confirmed by numerical simulations; see Figures 9.3 and 9.4. It is readily seen that the pattern first emerges in a certain subdomain, which appears to be different for $l_0$ being critical or noncritical, and eventually spreads over the whole area.

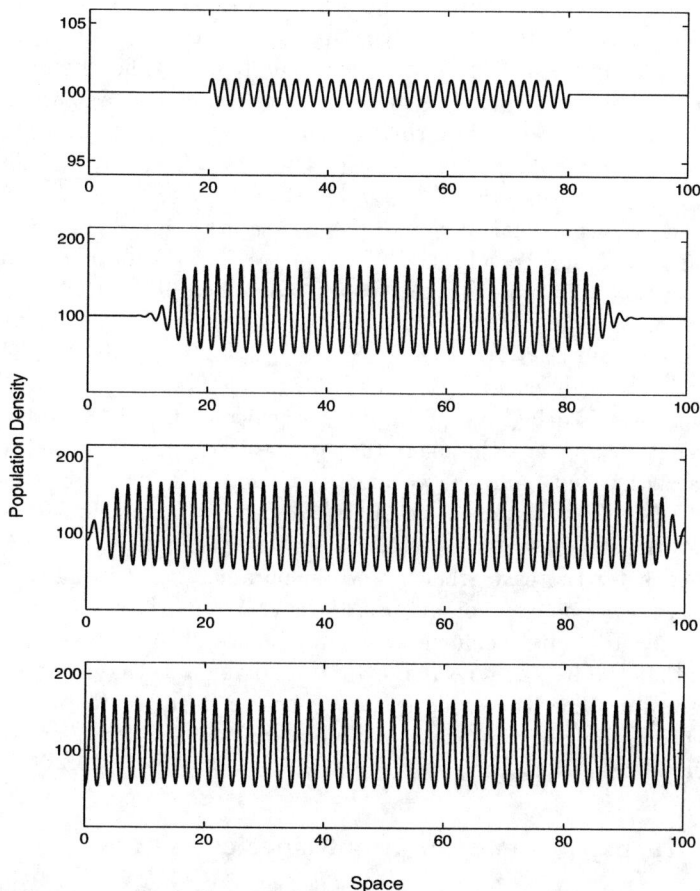

**FIGURE 9.4**: Pattern formation in the Segel–Jackson system (9.35)-(9.36). Panels from top to bottom correspond to $t = 0$, $t = 5$, $t = 10$, and $t = 20$, respectively. All parameters are the same as in Figure 9.3, except for the wavelength of the initial perturbation, which is now $l_0 = 2$.

It should also be mentioned that the Turing patterns are rather sensitive to the boundary conditions. While the initial conditions are likely to affect only the transient stage of pattern formation, the boundary conditions may affect both the transient stage and the large-time asymptotics. In particular, in a spatially bounded system, a stationary pattern can be formed only in case the boundary conditions are consistent with the intrinsic properties of the pattern *and* with the size and shape of the domain. A simple example of such consistency for the one-dimensional version of system (9.1)–(9.2) is given by the zero-flux Neumann conditions, provided $L/l_*$ is an integer, where $L$ is the length of the domain and $l_*$ is the intrinsic spatial period of the emerging pattern. The situation when the boundary conditions are not consistent with the intrinsic properties of the patterns is usually called the "boundary forcing." In this case, the system is more likely to evolve into a spatiotemporal pattern (e.g., periodical both in time and space) rather than into a stationary one; for more details and references, see Dillon et al. (1994) and Shoji and Iwasa (2003), and also Meinhardt (1982).

How can pattern formation due to the Turing instability take place in reality? Let us consider a certain natural system (e.g., a population community) that is in a spatially homogeneous steady stable equilibrium. Such an equilibrium normally appears as a result of balance between various deterministic "forces." However, the dynamics of any real system is actually a result of the interplay between deterministic and stochastic factors, and this steady state is continuously disturbed by spatially heterogeneous stochastic perturbations and fluctuations. Unless the magnitude of these perturbations is very large, which rarely happens, and the parameters are outside the Turing instability range, the perturbations decay with time and the system remains in a close vicinity of the steady state; therefore, no patterns emerge.

Now, consider the case when a certain parameter is gradually changing with time; for population communities it can be, for instance, either a result of adaptation to climate change or a consequence of natural selection. If it happens that the parameter value enters the domain in the parameter space of the given system where the system becomes unstable to heterogeneous perturbation, the corresponding spatial modes will be amplified, and that results in pattern formation.

### 9.1.1 Turing patterns in a multispecies system

In the above section, we were concerned with a dissipative instability in a system of two interactive species. A question that immediately arises is whether a similar scenario can be observed for a system with more than two species. The Turing instability is widely regarded to be responsible for morphogenesis (Turing, 1952; Meinhardt, 1982; Murray, 1989) and is also considered as a probable explanation for pattern formation in some ecological communities (Segel and Jackson, 1972; Levin and Segel, 1976; Alonso et al., 2002); however, natural systems are normally multispecific.

A closer look at the analysis done in the previous section shows that, from a mathematical point of view, a search for instability conditions is essentially an eigenvalue problem. Indeed, let us consider the following general model of an $N$-species community:

$$\frac{\partial U_i}{\partial t} = D_i \nabla^2 U_i + f_i(U_1, \ldots, U_N) , \qquad i = 1, \ldots, N , \qquad (9.48)$$

where notations have the same meanings as above.

It is readily seen that the corresponding linearized system in the vicinity of a spatially homogeneous steady state $(\bar{U}_1, \ldots, \bar{U}_N)$ has the form

$$\frac{du_i(t)}{dt} = \sum_{j=1}^{N} a_{ij} u_j(t), \qquad i = 1, \ldots, N, \qquad (9.49)$$

in case of a spatially homogeneous perturbation, and

$$\frac{d\tilde{u}_i(t;k)}{dt} = \sum_{j=1}^{N} \left(a_{ij} - k^2 D_i \delta_{ij}\right) \tilde{u}_j(t;k), \qquad i = 1, \ldots, N, \quad (9.50)$$

in case of a spatially inhomogeneous perturbation with the wavenumber $k$. Here $a_{ij} = \partial f_i / \partial U_j$ (where the derivatives are taken in the steady state), $\delta_{ij}$ is the Kronecker symbol, and $u_i(t)$ and $\tilde{u}_i(t;k)$ are the amplitudes of the perturbation and its Fourier transform, respectively. Obviously, for $k = 0$, system (9.50) coincides with (9.49).

According to preconditions of the Turing instability, the steady state $(\bar{U}_1, \ldots, \bar{U}_N)$ is assumed to be stable with respect to spatially homogeneous perturbations. Therefore, the Turing instability takes place when all eigenvalues of the system (9.49) have negative real parts but at least one eigenvalue of the system (9.50) has, at least for some $k$, a positive real part. Therefore, identification of the Turing instability is equivalent to identification of the corresponding eigenvalue properties for matrix $A(k)$ with $k = 0$ and $k > 0$.

For the sake of clarity, and in order to avoid tedious calculations, here we consider in some detail only a special case of a three-species system basing mainly on the paper by Qian and Murray (2003). A more general study of this problem, which addresses the case of a system with an arbitrary number of interacting species, can be found in Satnoianu et al. (2000).

Let us consider the following general model of a three-species community:

$$\frac{\partial U}{\partial T} = D_1 \nabla^2 U + \phi(U, V, W), \qquad (9.51)$$

$$\frac{\partial V}{\partial T} = D_2 \nabla^2 V + \psi(U, V, W), \qquad (9.52)$$

$$\frac{\partial W}{\partial T} = D_3 \nabla^2 W + \rho(U, V, W). \qquad (9.53)$$

The matrix of the linearized system is

$$A(k) = \begin{pmatrix} a_{11} - D_1 k^2 & a_{12} & a_{13} \\ a_{21} & a_{22} - D_2 k^2 & a_{23} \\ a_{31} & a_{32} & a_{33} - D_3 k^2 \end{pmatrix}. \qquad (9.54)$$

The eigenvalues are the solutions of the characteristic equation, i.e.,

$$\lambda^3 + p_2\lambda^2 + p_1\lambda + p_0 = 0 \tag{9.55}$$

where

$$p_2 = -\operatorname{tr} A(k), \qquad p_0 = -\det A(k) \tag{9.56}$$
$$p_1 = M_{11}(k) + M_{22}(k) + M_{33}(k),$$

and $M_{ij}$ are cofactors.

According to the Routh–Hurwitz criterium (see Section 4.4.3), all the eigenvalues have negative real parts if and only if the following conditions hold:

$$\text{(a)} \ p_2 > 0, \quad \text{(b)} \ p_0 > 0, \quad \text{(c)} \ q = p_0 - p_1 p_2 < 0. \tag{9.57}$$

Since the steady state is locally stable, the properties (a)–(c) hold for $k = 0$. What we need to do now is to understand how the values of $p_0$, $p_2$, and $q$ change with an increase in $k$, in particular, whether it may happen that for some $k$ one or more of the conditions (a)–(c) above get violated. The latter situation would mean instability with respect to a perturbation with a given wavenumber $k$.

By direct calculation, we obtain that

$$q = \operatorname{tr} A \cdot (M_{11} + M_{22} + M_{33}). \tag{9.58}$$

From (9.56), (9.57), and (9.58), we readily see that necessary conditions of the steady state stability are $\operatorname{tr} A < 0$ and $\det A < 0$. Since

$$\operatorname{tr} A(k) = \operatorname{tr} A(0) - \sum_{i=1}^{3} D_i k^2 \tag{9.59}$$

and $\operatorname{tr} A(0) < 0$, it is clear that condition (9.57a) cannot be violated for any $k > 0$.

In order to reveal how $\det A$ changes with an increase in $k$, we may notice that its value is affected by inhomogeneity only through the matrix diagonal elements. Therefore, we can consider $\det A$ as a function of three variables, $a_{11}$, $a_{22}$, and $a_{33}$, and apply a standard differential-based technique:

$$d(\det A) = \sum_{i=1}^{3} \frac{\partial(\det A)}{\partial a_{ii}} d(a_{ii}) \tag{9.60}$$
$$= M_{11} d(a_{11}) + M_{22} d(a_{22}) + M_{33} d(a_{33}),$$

where $d(a_{ii}) < 0$, $i = 1, 2, 3$. Recalling that $\det A(0) < 0$, in case all $M_{ii} > 0$ ($i = 1, 2, 3$) we obtain $d(\det A) < 0$, which means that an increase in $k$ cannot break condition (9.57b).

Similarly to (9.57b), for the condition (9.57c) we obtain

$$dq = \sum_{i=1}^{3} \frac{\partial q}{\partial a_{ii}} d(a_{ii}) = \left[M_{22} + M_{33} + (\text{tr}A)^2 - a_{11}\text{tr}A\right] d(a_{11}) \quad (9.61)$$
$$+ \left[M_{11} + M_{33} + (\text{tr}A)^2 - a_{22}\text{tr}A\right] d(a_{22})$$
$$+ \left[M_{11} + M_{22} + (\text{tr}A)^2 - a_{33}\text{tr}A\right] d(a_{33}) .$$

Assuming all $M_{ii} > 0$ in order to satisfy condition (9.57b), and since $\text{tr}A < 0$, we obtain that condition $a_{ii} < 0$ ($i = 1, 2, 3$) is sufficient to make $dq < 0$. Recalling that (9.57c) holds for $k = 0$, it means that it cannot be violated for any $k > 0$.

From (9.60) and (9.61), we arrive, respectively, at the following conditions of the steady-state "absolute" stability (i.e., stability for any $k$):

(i) All diagonal cofactors of matrix $A(k)$ must be positive;
(ii) All diagonal elements of matrix $A(k)$ must be negative.

The two above conditions taken together are sufficient to ensure absolute stability of a given steady state. It means that instability for some $k > 0$ can only be observed if at least one of them is violated. Thus, we arrive at the following necessary condition of Turing instability (Qian and Murray, 2003):

- The largest diagonal element of matrix $A(k)$ must be positive and/or the smallest diagonal cofactor of matrix $A(k)$ must be negative.

We want to emphasize that this is a necessary condition and it does not, by itself only, guarantee onset of the Turing instability. In order to obtain a sufficient condition of the Turing instability, we have to additionally introduce certain constraints on the values of the species diffusivity.

Although in the case of a three-species system it is hardly possible to obtain explicit analytical relations similar to (9.16) and (9.18), an instructive example can be built easily. By virtue of the Routh–Hurwitz criterium, instability takes place if and only if one of the conditions (9.57a–9.57c) is broken. Let us consider (9.57b). After some standard although rather tedious calculations, we obtain:

$$p_0(k) = D_1 D_2 D_3 k^6 - (D_1 D_2 a_{33} + D_2 D_3 a_{11} + D_1 D_3 a_{22}) k^4 \quad (9.62)$$
$$+ (D_1 M_{11} + D_2 M_{22} + D_3 M_{33}) k^2 - \det A .$$

Under conditions (i)–(ii) all terms on the right-hand side are positive and so $p_0(k) > 0$ for any $k$.

Now, let us assume, without any loss of generality, that the smallest cofactor is $M_{33}$, so that $M_{33} < 0$. The third term on the right-hand side of (9.62) then becomes negative if $D_1$ and $D_2$ are sufficiently small. Let $D_1/D_3 \ll 1$ and

$D_2/D_3 \ll 1$; then $p_0(k) < 0$ for some $k > k_0$, where

$$k_0 \approx \left( \frac{\det A}{D_3 M_{33}} \right)^{1/2}. \tag{9.63}$$

(Note that, for $k \to \infty$, $p_0(k)$ is always positive due to the contribution of the higher-order terms in (9.62), however small [but positive] the coefficients $D_1$ and $D_2$ can be.)

In a similar manner, let us assume that all diagonal cofactors are positive and the largest diagonal element is $a_{11}$, so that $a_{11} > 0$. Considering the case that $D_2 \sim D_3$ but $D_1/D_2 \ll 1$ and $D_1/D_3 \ll 1$, from (9.62) we obtain:

$$p_0(k) \approx -(D_2 D_3 a_{11}) k^4 + (D_2 M_{22} + D_3 M_{33}) k^2 - \det A. \tag{9.64}$$

Clearly, the right-hand side of (9.64) becomes negative for sufficiently large $k$.

The above condition of the Turing instability has been obtained in terms of diagonal elements of the community matrix and the corresponding cofactors. We should recall, however, that $M_{ii}$ obviously depend on $k$. Therefore, now we have to find out under what conditions the cofactors actually remain positive for any $k$ and under what conditions they may become negative.

Consider cofactor $M_{11}$:

$$M_{11} = \begin{vmatrix} a_{22} - D_2 k^2 & a_{23} \\ a_{32} & a_{33} - D_3 k^2 \end{vmatrix} = (a_{22} - D_2 k^2)(a_{33} - D_3 k^2) - a_{23} a_{32}.$$

Assuming all $a_{ii}$ are negative, $a_{23} a_{32} < 0$ is a sufficient condition for $M_{11}(k) > M_{11}(0) > 0 \ \forall k$. Similarly, it is readily seen that $a_{13} a_{31} < 0$ and $a_{12} a_{21} < 0$ are sufficient conditions for $M_{22}(k)$ and $M_{33}(k)$ being positive for any $k$.

Note that the latter conditions have a clear biological interpretation. If, for any two species $i$ and $j$, the two corresponding elements $a_{ij}$ and $a_{ji}$ of the community matrix are of different sign, it means that the interspecific interaction is asymmetrical. Growth of one of these species appears to be damaging for the other one; if one of the species wins, the other looses; etc. Immediate biological examples are given by prey–predator and host-parasite interactions. In a more general perspective, it is sometimes said that the species interacting in this way make an "activator–inhibitor" pair. Thus, condition (i) of absolute stability can be reformulated in a more general way as follows:

($i_1$) For each pair of species, their interaction must be of the activator–inhibitor type.

In the case that the three species described by system (9.51)–(9.53) correspond to three trophic levels, e.g., prey, predator, and top predator, the above condition means that there must be a certain "looping" in the species interaction so that the upper level can be negatively affected by the lower one by means of some direct interactions.

Correspondingly, a necessary condition of Turing instability is that at least one of the following inequalities should hold:

$$a_{12}a_{21} > 0, \quad a_{13}a_{31} > 0, \quad a_{23}a_{32} > 0. \tag{9.65}$$

The fact that the coefficients $a_{ij}$ and $a_{ji}$ have the same signs means that the species $i$ and $j$ interact in a "symmetric" way. The simplest example of such an interaction is given by competition; cf. Chapter 2. Comparing this conclusion to the conditions of Turing instability in a two-species system, we now arrive at a curious result: In a system of two competing species, Turing's instability is impossible, but it may become possible when the two species are affected by the same predator. Therefore, one might expect that the conditions of Turing instability are somewhat less restrictive in a multispecific community than they are in the two species case.

## 9.2 Differential flow instability

Pattern formation resulting from the Turing instability in a system of at least two interacting diffusive species is a phenomenon of great theoretical and practical interest. It is currently regarded as a generic process in morphogenesis, and it is also potentially important for ecological applications making a theoretical background for studying the heterogeneous distribution of ecological populations. However, it also has some inherent constraints; in particular, in a two-species system, the difference in species diffusivity must be sufficiently large. Apparently, this is not always the case in nature, and that gives a reason to look for some alternative scenarios of pattern formation.

It seems obvious that self-organized spatial patterning can only become possible when there is a relevant mechanism of species transport "integrating" the local dynamics at different positions into a single spatiotemporal system, which otherwise would be an ensemble of disconnected points or sites.[2] One mechanism of transport is diffusion, and that may result in diffusive instability. Note that diffusion must not necessarily be of the Brownian type; a possibility of pattern formation due to a diffusive instability in a more complicated system with fractional diffusion has been reported recently; see Gafiychuk and Datsko (2006). Another physically different mechanism is advection. Therefore, a question arises whether the existence of an advective flow in a system of interacting species may result in the instability of an otherwise stable homogeneous state.

---

[2] Here we are not interested in pattern formation in a disconnected system that may arise, for instance, as a result of specific initial conditions or inhomogeneous forcing.

Let us consider the following advection–reaction system:

$$\frac{\partial U(\mathbf{R},T)}{\partial T} = \phi(U,V) + \mathbf{w}_u \nabla U , \qquad (9.66)$$

$$\frac{\partial V(\mathbf{R},T)}{\partial T} = \psi(U,V) + \mathbf{w}_v \nabla V , \qquad (9.67)$$

where $U$ and $V$ are the population densities and $\mathbf{w}_u$ and $\mathbf{w}_v$ are constant velocities of advective transport of species $U$ and $V$, respectively.

In a general case, both velocities can be nonzero. However, by introducing the speed of relative motion, $\mathbf{w} = \mathbf{w}_u - \mathbf{w}_v$, and considering the system dynamics in a framework moving with speed $\mathbf{w}_v$, Equations (9.66)–(9.67) turn into

$$\frac{\partial U}{\partial T} = \phi(U,V) + \mathbf{w} \nabla U , \qquad (9.68)$$

$$\frac{\partial V}{\partial T} = \psi(U,V) \qquad (9.69)$$

without any loss of generality. A nontrivial impact of advection may be expected if $\mathbf{w} \neq 0$, in which case the system (9.68)–(9.69) is called a system with differential flow (Rovinsky and Menzinger, 1992).

Now, we assume that the corresponding nonspatial system, i.e., system (9.68)–(9.69) without the advection terms, possesses a stable steady state $(\bar{U},\bar{V})$ so that $\phi(\bar{U},\bar{V}) = \psi(\bar{U},\bar{V}) = 0$. Obviously, in the spatial system it corresponds to the homogeneous steady state $U \equiv \bar{U}$, $V \equiv \bar{V}$.

Since we are interested in stability, we focus on the system dynamics in the vicinity of $U \equiv \bar{U}$, $V \equiv \bar{V}$. Therefore, introducing new variables as $\hat{U} = U - \bar{U}$ and $\hat{V} = V - \bar{V}$, which we assume to be small, from (9.68)–(9.69) we arrive at the following linearized system:

$$\frac{\partial U}{\partial T} = a_{11}U + a_{12}V + \mathbf{w}\nabla U , \qquad (9.70)$$

$$\frac{\partial V}{\partial T} = a_{21}U + a_{22}V , \qquad (9.71)$$

omitting hats for simplicity. Local stability of the steady state $(\bar{U},\bar{V})$ implies that $\mathrm{tr}A = a_{11} + a_{22} < 0$ and $\det A = a_{11}a_{22} - a_{12}a_{21} > 0$.

Further analysis is similar to the one that we have previously applied to diffusive instability. For the sake of brevity, we restrict our consideration to the case of an unbounded space. Solution of the system (9.70)–(9.71) can be expanded into a Fourier integral; cf. Equation (9.12). Since the system (9.70)–(9.71) is linear, in order to study its stability it is sufficient to study the solution properties for a given $\mathbf{k}$:

$$U_k(\mathbf{R},T) = \tilde{U}(T,\mathbf{k})e^{i\mathbf{k}\mathbf{R}} , \qquad V_k(\mathbf{R},T) = \tilde{V}(T,\mathbf{k})e^{i\mathbf{k}\mathbf{R}} . \qquad (9.72)$$

Apparently, local stability of the system (9.70)–(9.71) means that the mode with $\mathbf{k} = 0$ is stable.

Substituting (9.72) into (9.70)–(9.71), we obtain:

$$\frac{d\tilde{U}_k(T)}{dT} = (a_{11} - i\mathbf{kw})\tilde{U}_k + a_{12}\tilde{V}_k , \qquad (9.73)$$

$$\frac{\tilde{V}_k(T)}{dT} = a_{21}\tilde{U}_k + a_{22}\tilde{V}_k , \qquad (9.74)$$

where $i^2 = -1$. Correspondingly, the eigenvalues $\lambda_{1,2}$ are the solutions of the following equation:

$$\begin{vmatrix} a_{11} + i\mathbf{kw} - \lambda & a_{12} \\ a_{21} & a_{22} - \lambda \end{vmatrix} = 0 , \qquad (9.75)$$

that is,

$$(a_{11} + i\mathbf{kw} - \lambda)(a_{22} - \lambda) - a_{12}a_{21} = 0 , \qquad (9.76)$$

so that

$$\lambda_{1,2}(\mathbf{k}) = \frac{1}{2}\left(\mathrm{tr}A + i\mathbf{kw} \pm \sqrt{\Gamma}\right) , \qquad (9.77)$$

where

$$\Gamma = \Delta - (\mathbf{kw})^2 + 2i\mathbf{kw}(a_{11} - a_{22}) , \qquad (9.78)$$

where $\Delta = (\mathrm{tr}A)^2 - 4\det A$.

For the equilibrium state to become unstable, there must exist an eigenvalue with a positive real part. After standard although rather tedious calculations, from (9.78) we obtain the following expression for the largest real part:

$$\mathrm{Re}\,\lambda(\mathbf{k}) = \frac{\mathrm{tr}A}{2} + \frac{1}{2\sqrt{2}} \cdot \qquad (9.79)$$

$$\cdot \left(\left[\left(\Delta - (\mathbf{kw})^2\right)^2 + 4(\mathbf{kw})^2(a_{11} - a_{22})^2\right]^{1/2} + \Delta - (\mathbf{kw})^2\right)^{1/2} .$$

Recall that $\mathrm{tr}A < 0$ because the steady state is assumed to be locally stable. However, the second term in (9.79) is apparently positive. The question then is, how actually large can it be for different $\mathbf{k}$?

For $\mathbf{k} = 0$ (i.e., for a spatially homogeneous perturbation), Equation (9.79) turns into

$$\mathrm{Re}\,\lambda(0) = \frac{1}{2}\left[\mathrm{tr}A + \left(\frac{\Delta + |\Delta|}{2}\right)^{1/2}\right] , \qquad (9.80)$$

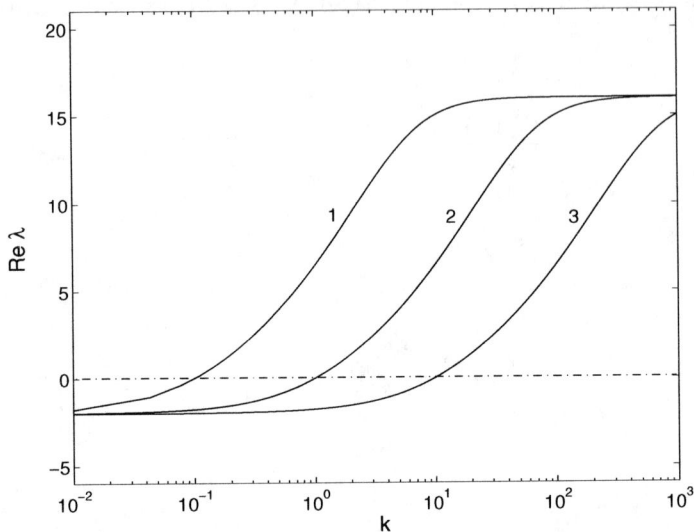

**FIGURE 9.5**: The real part of the eigenvalues vs wavenumber $k$ of an inhomogeneous perturbation of a locally stable homogeneous equilibrium state in a system with a differential advective flow, cf. (9.70)–(9.71); curve 1 for $w = 10$, curve 2 for $w = 1$, curve 3 for $w = 0.1$. Other parameters are $a_{11} = 16$, $a_{12} = -17$, $a_{21} = 20$, and $a_{22} = -20$.

which is negative by the assumption. However, in the short-wave limit $\mathbf{kw} \to \infty$, (9.79) takes the following form:

$$\text{Re } \lambda(\infty) = \frac{1}{2} \left( \text{tr}A + |a_{11} - a_{22}| \right). \tag{9.81}$$

The left-hand side of (9.79), considered as a function of $\mathbf{kw}$, may have a shallow minimum for intermediate values of its argument, but it will be increasing for large $\mathbf{kw}$. The outcome then depends on the relation between the matrix elements $a_{11}$ and $a_{22}$. Three different cases are possible:

$$\text{(a)} \quad a_{11} < 0, \ a_{22} > 0, \quad \text{(b)} \quad a_{11} > 0, \ a_{22} < 0, \tag{9.82}$$
$$\text{(c)} \quad a_{11} < 0, \ a_{22} < 0.$$

It is readily seen that, in case both $a_{11}$ and $a_{22}$ are negative, Re $\lambda(\infty)$ is negative as well and no instability can take place. However, in the cases (9.82a) and (9.82b), we obtain Re $\lambda(\infty) = a_{22}$ and Re $\lambda(\infty) = a_{11}$, respectively, which means that the spatially homogeneous steady state becomes unstable for sufficiently large $\mathbf{k}$. Thus, we arrive at the following necessary condition of differential flow instability:

- A change in stability is possible only if $a_{11}$ and $a_{22}$ are of different signs so that $a_{11}a_{22} < 0$.

Note that it coincides with one of the conditions of diffusive instability; cf. the lines below Equations (9.13)–(9.14).

Re $\lambda(\mathbf{k})$ as given by Equation (9.79) is shown in Figure 9.5 for the case $\mathbf{kw} = kw$. The system becomes unstable for $k > k_0$, where Re $\lambda(k_0) = 0$, the expression for the critical wavenumber $k_0$ being obtained from (9.79):

$$k_0 = -\frac{\mathrm{tr}\,A}{w}\left(-\frac{\det A}{a_{11}a_{22}}\right)^{1/2}. \tag{9.83}$$

In full agreement with the above analysis, the larger the speed $w$ of the differential flow, the smaller is $k_0$.

It should also be mentioned that the system response to perturbations with different $\mathbf{k}$ is apparently anisotropic. For the same value of $k$, the largest value of Re $\lambda$ is reached when $\mathbf{k}$ is parallel to $\mathbf{w}$, while in the direction perpendicular to $\mathbf{w}$ the spatially homogeneous steady state remains stable for any $k$. Consequently, Equation (9.83) gives the minimum possible critical value of $k$.

Thus, similar to the case of diffusive instability, the impact of a differential flow between two interacting species can make the spatially homogeneous steady state unstable to heterogeneous perturbations. The steady state remains stable with respect to long-wave (small $k$) spatial perturbations but will be destabilized by short-wave perturbations when the wavenumber exceeds a certain critical value.

There is one important distinction, though. In the diffusive-driven instability, the emerging positive eigenvalue is real; however, in the differential flow instability, the eigenvalue with a positive real part is apparently complex. Correspondingly, while the Turing patterns are standing, the existence of a non-zero imaginary part results in a traveling wave-type structure: It is readily seen from (9.72) that an emerging periodical pattern with an overcritical wavenumber $k$ will be traveling with a constant velocity

$$\mathbf{s} = -\frac{\mathbf{k}}{k^2}\,\mathrm{Im}\,\lambda(\mathbf{k}), \tag{9.84}$$

where

$$\mathrm{Im}\,\lambda(\mathbf{k}) = \frac{\mathbf{kw}}{2} + \frac{\mathrm{sign}(a_{11}-a_{22})}{2\sqrt{2}} \tag{9.85}$$

$$\cdot\left(\left[(\Delta-(\mathbf{kw})^2)^2 + 4(\mathbf{kw})^2(a_{11}-a_{22})^2\right]^{1/2} - \Delta + (\mathbf{kw})^2\right)^{1/2}.$$

Now let us recall that the whole analysis has been done in the framework moving with speed $\mathbf{w}_v$; cf. the lines above Equations (9.68)–(9.69). It means that, in the original framework, the emerging periodical structure will be traveling with velocity $\mathbf{w}_v + \mathbf{w}$; the instability conditions apparently remain the same regardless of the choice of framework.

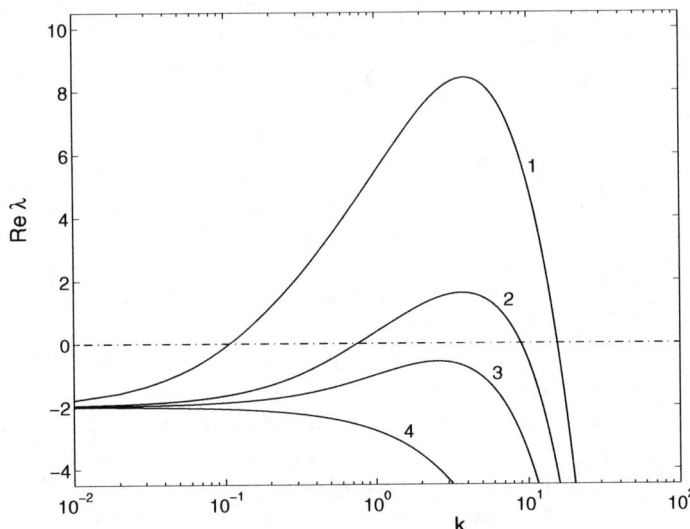

**FIGURE 9.6**: The largest real part of the eigenvalues vs wavenumber $k$ in a system with a differential flow and diffusion in the case that diffusivity is equal for both species. Here curve 1 for $w = 10$, curve 2 for $w = 2$, curve 3 for $w = 1$, curve 4 for $w = 0.1$; other parameters are the same as in Figure 9.5.

Note that the speed (9.85) of the traveling pattern is obviously different from $w$, its value depending both on the speed $w$ of the "external" flow and on the intrinsic system's kinetics quantified by the matrix $(a_{ij})$. This is a clear manifestation of the self-organized system's dynamics. The instability itself and the subsequent pattern formation are induced by the impact of the differential flow, but it cannot be reduced to it.

A drawback of the model system (9.66)–(9.67), or (9.68)–(9.69), is that the homogeneous steady state appears to be unstable with respect to modes with an infinitely large wavenumber, i.e., for any $k > k_0$ (see Figure 9.5 and Equation (9.83)). Since a perturbation of an arbitrary shape contains, in a general case, the whole spectrum of $k$, this means that even a very slow differential flow would destabilize the system. This does not seem realistic. Indeed, one can expect that perturbations from a short-wave part of the spectrum should be damped down by diffusion, which is always present in any physical, chemical, or biological system, even if its impact can be sometimes quite small compared to the impact of advection.

Therefore, although system (9.68)–(9.69) was very useful to demonstrate the principal possibility of steady-state destabilization by a differential flow,

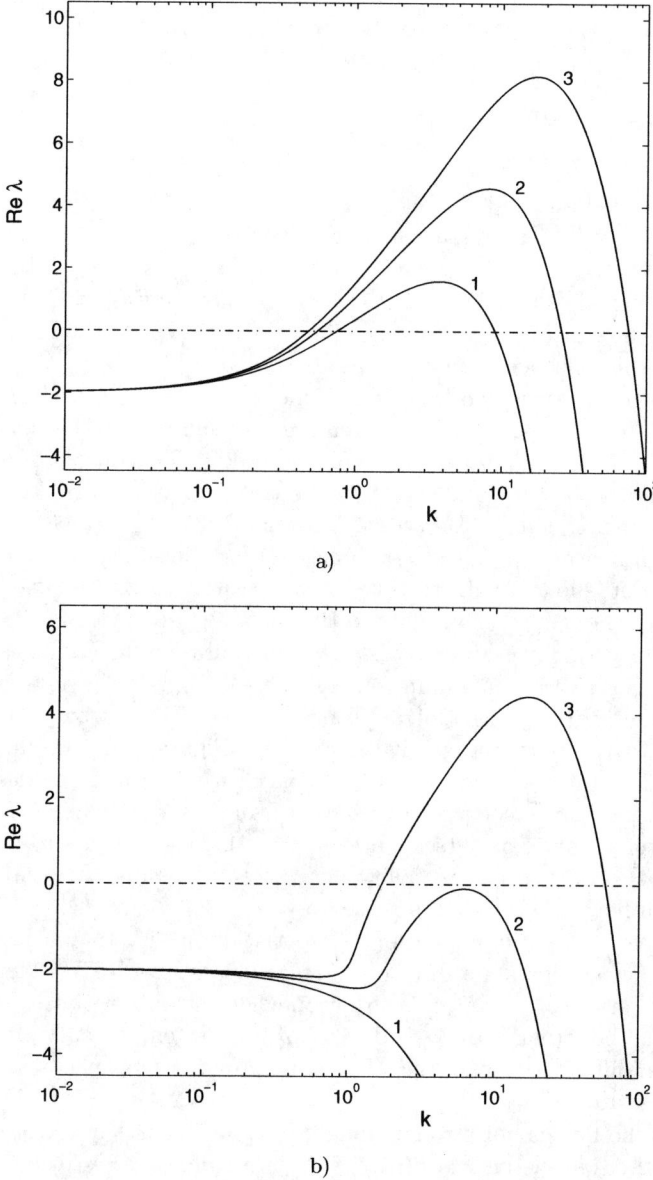

**FIGURE 9.7**: The largest real part of the eigenvalues vs wavenumber $k$ arising as a result of the combined effect of differential flow and differential diffusivity: (a) for $w = 2$ and (b) for $w = 0.1$. Here curve 1 for $\epsilon = D_1/D_2 = 1$, curve 2 for $\epsilon = 0.5$ and curve 3 for $\epsilon = 0.2$; other parameters as in Figure 9.5.

a more reasonable model should include diffusion for both species:

$$\frac{\partial U}{\partial T} = a_{11}U + a_{12}V + \mathbf{w}\nabla U + D_1\nabla^2 U \;, \tag{9.86}$$

$$\frac{\partial V}{\partial T} = a_{21}U + a_{22}V + D_2\nabla^2 V \;. \tag{9.87}$$

A complete model should also account for cross-diffusion, cf. Malchow (2000a); here we assume that its impact can be neglected.

Without advection, the system (9.86)–(9.87) was shown to be prone to diffusive instability (see Section 9.1) while without diffusion it is subject to differential flow instability. The question now arises what may be an outcome of the interplay between the two types of instability?

Focusing on the behavior of a $k$th mode, cf. (9.72), it is readily seen that Equations (9.73) to (9.79) remain essentially the same up to the change $a_{11} \to \tilde{a}_{11} = a_{11} - D_1\mathbf{k}^2$ and $a_{22} \to \tilde{a}_{22} = a_{22} - D_2\mathbf{k}^2$. The equation for Re $\lambda(\mathbf{k})$ now appears to be too cumbersome to be treated analytically; in particular, an explicit expression for the critical wavenumber is not available. However, it is straightforward to plot it as a function of $k$. Recalling that the system's stability limit is lowest with respect to perturbations with $\mathbf{k}$ being parallel to $\mathbf{w}$, we restrict our further insights to the case $\mathbf{kw} = kw$.

We begin with the case $D_1 = D_2$ when no diffusive instability is possible. Figure 9.6 shows the maximum real part of the eigenvalues as a function of $k$. Apparently, the impact of diffusion changes the situation considerably. Contrary to the purely advective system, cf. Figure 9.5, now the homogeneous steady state can become unstable only with respect to modes with the wavenumbers from a certain intermediate range, $k_{min} < k < k_{max}$. Moreover, for small values of $w$ (see curves 3 and 4), Re $\lambda$ appears to be negative for any $k$. Consequently, instability onset is only possible if the speed of the differential flow is large enough.

In case $D_1 \neq D_2$, the impact of differential diffusivity enlarges the parameter range where instability occurs. Figure 9.7a shows Re $\lambda(k)$ for different diffusivity ratios $\epsilon = D_1/D_2$ in a situation when the differential advective flow alone would be strong enough to destabilize the system; cf. curve 1. However, differential diffusivity makes the effect more prominent by means of increasing the relevant range of $k$.

In case the flow is not strong enough, instability onset is still possible due to the impact of differential diffusivity. Curve 1 in Figure 9.7b shows Re $\lambda(k)$ in the case when the differential flow instability is not possible. As soon as the diffusivity ratio becomes sufficiently large, stability of the homogeneous steady state gets broken; there emerges a range of $k$ where the eigenvalues' real part is positive (curve 3).

Thus, in this section we have shown that, in a system of two interacting species with diffusion and a differential advective flow, a locally stable homogeneous steady state may become unstable with respect to heterogeneous perturbation with a wavenumber form a certain intermediate range. As a result of instability onset, the system becomes capable of developing periodic spatial patterns of the traveling wave type.

A question may yet remain regarding the ecological significance of this generic mechanism. A relevant pair of species is given by a prey–predator system; however, the situation when the predator and prey populations move with respect to each other may seem rather exotic (but see Malchow and Shigesada, 1994). This seeming difficulty is immediately resolved if we realize that the interacting agents must not necessarily be biological species but can be treated, for instance, as a resource–consumer system. This apparently broadens the range of potential ecological applications of the above scenario; one appropriate example will be considered below.

## 9.3 Ecological example: semiarid vegetation patterns

In the previous sections of this chapter we revisited some classical theoretical approaches describing pattern formation in a system of two or more interacting species. A common feature of these approaches is that pattern formation becomes possible as a result of system instability with respect to a heterogeneous perturbation with a certain wavelength – or, more generally, wavelengths or wavenumbers from a certain range. For that reason, the above mechanisms, although apparently different in specific details, are often known under the same name of a "Turing-type" scenario. As a result of a Turing-type instability, the emerging patterns appear to be periodical. That brings a certain problem when one tries to link the Turing-type scenario(s) to the spatiotemporal patterns in ecosystems because the spatial distribution of ecological populations is usually highly irregular.

There is, however, at least one example of spatial patterning in ecosystems where the structures exhibit a remarkable degree of regularity. Vegetation in arid and semiarid regions of the world is often distributed in space not randomly but forms distinct regular patterns, normally in the form of stripes, where areas covered with bushes and trees alternate with areas/patches of grass or even bare ground; see Figure 9.8. Due to the apparent visual similarity of the structure to the pattern on the skin of a tiger, this vegetation pattern is often referred to as the "tiger bush." The phenomenon was originally reported by MacFadyen (1950a,b) and for some time it was regarded as an exotic feature of a particular land. The reason is that the variation of the population density across the stripes is not always easily observed on the spot, although in most cases it can be readily seen from the air. Thus, it was not before significant development of airborne methods of field observations, such as aerial photography, that it became clear that vegetation stripes in semiarid regions is a usual rather than an exotic phenomenon (Beard, 1967; White, 1969, 1970; Bernd, 1978; Greig-Smith, 1979); for more references also see Lefever and Lejeune (1997).

**FIGURE 9.8**: An example of tiger bush vegetation pattern in Niger (with permission from Lejeune et al., 1999). The dark gray patches correspond to high vegetation density, e.g., bushes and trees, and the white patches correspond to low vegetation density (grass or bare ground). The wavelength of the pattern is estimated to be about 70 m.

Apparent curiosity of this phenomenon stimulated several studies aimed at identifying the mechanism of the vegetation pattern formation and to explain its main properties, such as the typical wavelength of the pattern in the direction accross the stripes. Attempts were made to relate it to the spatial distribution of precipitation (Greig-Smith, 1979), to soil heterogeneity (Beard, 1967), and to the impact of anthropogenic activity (Wickens and Collier, 1971). These approaches, however, do not seem to catch the essence of the phenomenon simply because tiger bush is a much more general phenomenon. Indeed, it is often observed in areas where none of the above factors apply.

The generality of vegetation stripes in semiarid regions gives us a reason to assume that the underlying mechanism is intrinsic for this ecosystem type, which means that it arises from some of its inherent processes or interactions. A common feature for arid and semiarid ecosystems is that water is the main limiting factor, which may lead to a hypothesis that pattern formation might be a result of the interaction between vegetation (consumer) and water (resource). The principal source of water is rainfall; but what is the mechanism of its redistribution on the land surface? To apply a Turing-type scenario, the two interacting agents, i.e., plants and water, must exhibit different mobility. While mobility/diffusivity of a plant population originates in seed dispersal, in the case of water the relevant factors are not immediately clear.

An answer to the above question can be found in a more careful analysis of field observations. It was concluded by several authors that vegetation stripes are more likely to be found on a slope rather than on flat ground. On a sloping surface, water flows downhill. Consequently, we arrive at the following model of a system with a differential advective flow:

$$\frac{\partial U(X,Y,T)}{\partial T} = R - \Upsilon U - C(U,V) + W\frac{\partial U}{\partial X} + D_1 \nabla^2 U , \qquad (9.88)$$

$$\frac{\partial V(X,Y,T)}{\partial T} = \kappa C(U,V) - M(V) + D_2 \nabla^2 V , \qquad (9.89)$$

where $U$ is the amount of water per unit surface area, $V$ is the vegetation biomass, $R$ is the rate of water supply due to rains, $W$ is the speed of the water flow, $\Upsilon$ is the rate of water loss due to evaporation, function $C$ describes water consumption by plants, $\kappa$ is the consumption efficiency, $M$ is mortality, and $D_2$ is the plant diffusivity due to seed dispersal. The diffusive term in (9.89) takes into account the impact of the surface roughness on the water flow; the higher the roughness, the larger is $D_1$.

We mention here that parameter $R$ is not a momentary rate of water supply (due to the rainfall) but is a value averaged over a certain time interval covering at least one rain. Correspondingly, $U$ also gives an average value of surface water density; otherwise, $U$ would simply be zero for most of the time.

Apparently, the system (9.88)–(9.89) belongs to the type considered in the previous section and thus is capable of exhibiting pattern formation due to a Turing-type instability. The specific properties of the model depend on the choice of functions $C$ and $M$. As an illustrative example, cf. Klausmeier (1999), we assume that $C = eUV^2$ and $M(V) = M_0 V$, where $e$ and $M_0$ are coefficients. Introducing dimensionless variables as

$$t = \Upsilon T, \qquad x = \left(\frac{\Upsilon}{D_2}\right)^{1/2} X, \qquad y = \left(\frac{\Upsilon}{D_2}\right)^{1/2} Y,$$

$$u = \left(\frac{e}{\Upsilon}\right)^{1/2} \kappa U, \qquad v = \left(\frac{e}{\Upsilon}\right)^{1/2} V,$$

from (9.88)–(9.89) we obtain

$$\frac{\partial u(x,y,t)}{\partial t} = r - u - uv^2 + w\frac{\partial u}{\partial x} + \epsilon \nabla^2 u , \qquad (9.90)$$

$$\frac{\partial v(x,y,t)}{\partial t} = uv^2 - mv + \nabla^2 v , \qquad (9.91)$$

where $r = \kappa e^{1/2}\Upsilon^{-3/2}R$, $w = W(\Upsilon D_2)^{-1/2}$, $m = M_0/\Upsilon$, and $\epsilon = D_1/D_2$.

It should be mentioned here that the system (9.90)–(9.91) is mathematically identical to a chemical "substrate-autocatalyst" system, which has been studied extensively, e.g., see Rasmussen et al. (1996); Davidson (1998). Here we only recall as many of its properties as is necessary for the purposes of this section; more details can be obtained from the cited papers.

It is readily seen that the nonspatial system (9.90)–(9.91) may possess either one or three steady states.[3] Specifically, there is a semi-trivial "no vegetation" state $(r, 0)$ and two nontrivial coexistence states $(u_-, v_-)$ and $(u_+, v_+)$, where

$$v_\pm = \frac{1}{2m}\left(r \pm \sqrt{r^2 - 4m^2}\right), \qquad u_\pm = \frac{m}{v_\pm} . \qquad (9.92)$$

The semi-trivial steady state is feasible for any parameter value and is always linearly stable. The coexistence states exist only if $m < r/2$; for $m \geq r/2$, they merge and disappear in a saddle-node bifurcation. From the two coexistence states, the one with lower vegetation biomass, i.e., $(u_-, v_-)$, is always a saddle. As for $(u_+, v_+)$, it can be either a node or a focus and can be either stable or unstable. In particular, it is straightforward to see that, whatever the value of $r$, steady state $(u_+, v_+)$ is stable for sufficiently small $m$.

Moreover, Klausmeier (1999) showed that the steady state $(u_+, v_+)$ is feasible and stable for biologically reasonable parameter values. Therefore, what we have to do now is to reveal whether the corresponding spatially homogeneous steady state can be destabilized by a heterogeneous perturbation in a relevant parameter range and whether the properties of the emerging pattern are congenial with those observed in semiarid ecosystems.

Having applied the standard linearization technique, in the vicinity of $(u_+, v_+)$ the system (9.90)–(9.91) takes the following form:

$$\frac{\partial u}{\partial t} = \left(-\frac{rv_+}{m}\right) u - 2mv + w\frac{\partial u}{\partial x} + \epsilon \nabla^2 u , \qquad (9.93)$$

$$\frac{\partial v}{\partial t} = \left(\frac{rv_+}{m} - 1\right) u + mv + \nabla^2 v . \qquad (9.94)$$

Obviously, system (9.93)–(9.94) is exactly of the type that was considered in the previous section and thus all the results apply. In particular, instability onset is definitely to occur for sufficiently high speed of the flow and/or for sufficiently small diffusivity ratio.

Let us consider the case when the impact of diffusion can be neglected, which applies to plants with a short-range seed dispersal growing on a relatively steep slope with even ground surface. Then, the diffusion terms in

---

[3]There is also a case of two steady states that is not robust to parameter changes.

Equations (9.93)–(9.94) are missing and a simple estimate of the typical spatial wavelength is given by (9.83), which now turns into

$$k_0 = \frac{1}{w}\left(\frac{rv_+}{m} - m\right)\left(\frac{rv_+ - 2m}{rv_+}\right)^{1/2}. \qquad (9.95)$$

In order to calculate the corresponding wavelength of the pattern, we need to know the value of the ecological parameters. This is a subtle issue because the accuracy of ecological measurements is usually rather low. Moreover, parameter values can be of different orders of magnitude depending on whether the vegetation stripes consists of grass or bushes and trees, so that these two cases should be considered separately.

In the case of stripes of grass, using some typical parameter values according to Mauchamp et al. (1994) and Klausmeier (1999), one can obtain that $r = 1.9$ and $m = 0.45$. Consequently, it gives $v_+ \approx 4$ and $k_0 \approx 15.4w^{-1}$, so that the wavelength of the pattern is

$$l_0 = \frac{2\pi}{k_0} \approx 0.4w, \qquad (9.96)$$

which, in dimensional units, gives

$$L_0 \simeq 0.4\left(\frac{W}{\Upsilon}\right). \qquad (9.97)$$

For semiarid regions, the evaporation rate $\Upsilon$ is estimated to be 4 year$^{-1}$ (Klausmeier, 1999). On the other hand, $W$ apparently depends on the properties of a given landscape and can vary significantly. Using a tentative value $W = 200$ m·year$^{-1}$ we arrive at $L_0 \simeq 20$ m, which agrees very well with the typical measured values of grass stripe widths being between 1 and 40 m.

Similarly, for stripes of bushes and trees, one can obtain $r = 0.15$ and $m = 0.05$, so that $v_+ \approx 2.6$ and the wavelength of the pattern is $l_0 = 0.95w$ or

$$L_0 \simeq 0.95\left(\frac{W}{\Upsilon}\right). \qquad (9.98)$$

For the same values of $\Upsilon$ and $W$ as above, it gives $L_0 \simeq 50$ m.

We emphasize that, since the chosen value of $W$ is purely hypothetical, the above figures by themselves should not be taken too seriously. What is encouraging, however, is that this approach predicts, at least qualitatively, a correct relation between the width of the grassy and bushy stripes. Indeed, field data show that a typical stripe width in case of trees/bushes is a few times larger than that of grass.

Moreover, let us recall that the pattern formation scenario due to differential flow instability predicts that the emerging patterns are of the traveling wave type rather than stationary. Remarkably, field observations show that the vegetation stripes tend to move up the slope with approximately constant speed. Using Equation (9.84) in order to calculate the value of the traveling wave speed, we obtain that the stripes should move in the direction opposite to the water flow (i.e., up the slope) with the speed $v = 2.7$ m·year$^{-1}$ for grass and with $v = 0.65$ m·year$^{-1}$ for trees/bushes. Once again, although these values appear to be somewhat larger than those obtained in the field (which are 0.9 and 0.2, respectively), they give the correct predictions regarding the speed ratio.

In the case that diffusion cannot be neglected, an explicit relation for a typical wavenumber is not available and thus Equations (9.97) and (9.98) do not immediately apply. However, since the impact of diffusion tends to broaden the critical range of wavenumbers rather than shifting it (cf. Figure 9.7), and also taking into account the usual low accuracy of ecological data, these estimate are likely to remain valid, at least in the case when the diffusion coefficients are not very large.

### 9.3.1 Pattern formation due to nonlocal interactions

The above application of a conceptual advection–diffusion–reaction model to pattern formation in semiarid vegetation, also known as "tiger bush," leads to apparently encouraging results giving a reasonable estimate for the width of the vegetation band and for the speed with which the pattern gradually travels up the slope. Moreover, it gives a correct prediction of how these values change between the cases when the vegetation stripes are formed by grass and trees/bushes, respectively.

However, a closer look at published data of field observations reveals that, although the model relating the patterns to a differential flow instability agrees very well with some of them, there are some aspects that remain beyond its catch:

- A few studies reported that, in some cases, the stripes can be oriented not across the slope (i.e., parallel the contour lines) but along the slope (MacFadyen, 1950b; White, 1969). While the stripes of the first type tend to travel uphill, the stripes of the second type are stationary.

- On a flat surface, the vegetation pattern normally consists of patches rather than stripes. The first naive ideas were to relate this patchiness either to the impact of a topographic "noise" or to an inherent heterogeneity of the soil. However, a later study (see Couteron and Lejeune, 2001) proved that the patchy pattern has a clear periodical structure, which makes those explanations irrelevant.

A question then arises whether an alternative approach could be possible where all these features would appear as a result of the same dynamical mechanism. It should be noted here that a diffusion–reaction system (the system from the previous section without water flow, i.e., when applied to flat grounds) is still intrinsically capable of generating a spatiotemporal patchy structure; however, it takes place for parameter values inconsistent with those of a semiarid vegetation community (Klausmeier, 1999).

In order to outline what kind of a model it can be, let us have a somewhat deeper look at vegetation functioning. In an arid or semiarid region, the main controlling factor is the availability of water. An individual plant collects water from a certain area defined by its root system, so its success is apparently subject to competition with the neighboring plants. On the other hand, interaction with neighbors can have not only negative but also a positive effects: Obviously, the presence of other plants nearby results in a slower evaporation rate. A common sense supported also by biological data tells us that, in a situation where a single tree is likely to dry out and die, a group of trees has higher chances to survive.

Having based on these and other similar arguments, Lefever and Lejeune (1997) proposed the following model:

$$\frac{\partial U(\mathbf{R}, T)}{\partial T} = \Lambda_1[U(\mathbf{R}, T)] \cdot \Lambda_2[U(\mathbf{R}, T)] - \Lambda_3[U(\mathbf{R}, T)] ; \qquad (9.99)$$

see also Lejeune et al. (1999); Lefever and Lejeune (2000); Couteron and Lejeune (2001). Here factors $\Lambda_1[U]$ and $\Lambda_2[U]$ account for all collective effects enhancing and hampering reproduction, respectively (where the term "reproduction" includes all processes and stages between the production of seeds and appearance of new seedlings), and the last term stands for the plants' natural mortality.[4]

For the sake of simplicity, we assume that the natural mortality is a local process so that $\Lambda_3[U] \sim U$. On the contrary, along the lines of the above description, most the processes involved in reproduction have a clear spatial aspect and are essentially nonlocal. Therefore, $\Lambda_1[U]$ and $\Lambda_2[U]$ are functionals rather than functions, which we parameterize as follows:

$$\frac{\partial U(\mathbf{R}, T)}{\partial T} = \left( \alpha \int_{\infty}^{-\infty} w_1(|\mathbf{R} - \tilde{\mathbf{R}}|) U(\tilde{\mathbf{R}}, T)[1 + AU(\tilde{\mathbf{R}}, T)] d\tilde{\mathbf{R}} \right)$$
$$\cdot \left( 1 - \frac{1}{K} \int_{\infty}^{-\infty} w_2(|\mathbf{R} - \tilde{\mathbf{R}}|) U(\tilde{\mathbf{R}}, T) d\tilde{\mathbf{R}} \right) \qquad (9.100)$$
$$- MU ,$$

where $\alpha$ is the (linear) per capita growth rate, $M$ is the mortality rate, $K$ is the population carrying capacity in the limiting case $M \to 0$, and $A$ is a coefficient quantifying intensity of cooperative behavior. Due to their meanings, all the parameters are nonnegative. The dynamics takes place in a two-dimensional

---

[4]Similar ideas, but in a somewhat more general context, have recently been further developed by Genieys et al. (2006).

space so that $\mathbf{R} = (X, Y)$. The kernels $w_i(|\mathbf{R} - \tilde{\mathbf{R}}|)$ describe how the intensity of the plant-plant interaction changes along with the interplant distance $|\mathbf{R} - \tilde{\mathbf{R}}|$. The kernels must be normalized, i.e.,

$$\int_{-\infty}^{\infty} w_i(|\mathbf{R} - \tilde{\mathbf{R}}|) d\tilde{\mathbf{R}} = 1, \qquad i = 1, 2, \qquad (9.101)$$

where integration is taken over the whole space.[5]

At first glance, it may seem that, unlike diffusion–reaction equations, (9.100) does not describe any motion. Indeed, interplant competition occurs due to the interplay between the root systems, which are static. Positive cooperative effect occurs due to mutual shadowing, which is static as well. It is not exactly true, however. The point is that, besides these processes, (9.100) describes seed dissemination. The rate $\partial U(\mathbf{R}, T)/\partial T$ of the population growth at a given position $\mathbf{R}$ depends on the amount of seeds that land at this position. This is readily seen if we neglect competition, cooperation, and mortality (by setting $A = M = 0$ and $K \to \infty$) and keep only multiplication; Equation (9.100) then takes the form

$$\frac{\partial U(\mathbf{R}, T)}{\partial T} = \alpha \int_{\infty}^{-\infty} w_1(|\mathbf{R} - \tilde{\mathbf{R}}|) U(\tilde{\mathbf{R}}, T) d\tilde{\mathbf{R}}, \qquad (9.102)$$

where $\alpha$ is a product of the average number of seeds per adult plant and the survival rate, and the kernel $w_1$ gives the probability density of seed distribution (hence the normalization condition (9.101)). Equation (9.100) describes a diffusion-like processes; moreover, it can be shown that, on a certain spatial scale, it is actually reduced to a diffusion equation (cf. Petrovskii and Li, 2006).

Apparently, $w_i(|\mathbf{R} - \tilde{\mathbf{R}}|)$ should be a decreasing function of $|\mathbf{R} - \tilde{\mathbf{R}}|$ tending to zero for $|\mathbf{R} - \tilde{\mathbf{R}}| \to \infty$. The faster the decay in $w_i(|\mathbf{R} - \tilde{\mathbf{R}}|)$, the less prominent is the impact of nonlocality. In case $w_i(|\mathbf{R} - \tilde{\mathbf{R}}|) \sim \delta(\mathbf{R} - \tilde{\mathbf{R}})$, all processes are strictly local and (9.100) turns to a usual equation of nonspatial population dynamics:

$$\frac{dU}{dT} = \alpha U(1 + AU)\left(1 - \frac{U}{K}\right) - MU . \qquad (9.103)$$

Due to condition (9.101), the steady states of nonspatial Equation (9.103) correspond to the spatially homogeneous steady states of Equation (9.100); therefore, it is convenient to start with the nonspatial case.

In order to study (9.103), we first introduce dimensionless variables as $t = \alpha T$ and $u = U/K$, so that (9.103) turns into

$$\frac{du}{dt} = u(1 + au)(1 - u) - \mu u, \qquad (9.104)$$

---

[5]In equations (9.100), (9.101), and further on, the integrals are, actually, multiple integrals due to the dimensionality of space.

where $\mu = M/\alpha$ and $a = AK$. Correspondingly, the steady states are given by

$$\bar{u} = u_0 = 0 \quad \text{and}$$
$$\bar{u} = u_\pm = \frac{1}{2a}\left[(a-1) \pm \sqrt{(a-1)^2 + 4a(1-\mu)}\right]. \tag{9.105}$$

While the no-vegetation state $u = 0$ exists for any parameter values, feasibility of the nontrivial states $u_+$ and $u_-$ depends on the relation between $a$ and $\mu$. For $0 < a \leq 1$, the only positive state is $u_+$, and it only exists if $0 < \mu < 1$. For $a > 1$, the parameter range of $u_+$ existence extends to $0 < \mu < \mu_{max} = (a+1)^2/(4a)$. In the latter case, for $1 < \mu < \mu_{max}$, there also exists the other steady state $u_-$. It is readily seen, however, that the state $u = u_-$ is locally unstable and thus is not biologically relevant. The no-vegetation state $u = 0$ is stable for $1 < \mu < \mu_{max}$ and unstable for $0 < \mu < 1$. In its turn, the state $u = u_+$ is stable for all parameter values when it is feasible.

Coming back to the spatial model (9.100), our main goal is to check the stability of the homogeneous steady state $u = u_+$ with respect to heterogeneous perturbations and under what conditions the loss of stability, if any, can give rise to periodic spatial patterns. For that purpose, we now consider $u(x,t) = u_+ + \phi(x,t)$, where $\phi$ is small. From (9.100), after some standard transformations, we then obtain in the first order with respect to $\phi$:

$$\frac{\partial \phi(\mathbf{R},t)}{\partial t} = \int \left[\gamma_1 w_1(|\mathbf{R} - \tilde{\mathbf{R}}|) - \gamma_2 w_2(|\mathbf{R} - \tilde{\mathbf{R}}|)\right] \phi(\tilde{\mathbf{R}},t) d\tilde{\mathbf{R}}$$
$$- \mu\phi(\mathbf{R},t), \tag{9.106}$$

where

$$\gamma_1 = (1 + 2a\bar{u})(1 - \bar{u}) \quad \text{and} \quad \gamma_2 = \bar{u}(1 + a\bar{u}). \tag{9.107}$$

An arbitrary heterogeneous perturbation can be expanded into a Fourier integral:

$$\phi(\mathbf{R},t) = \frac{1}{2\pi} \int \Phi(t,\mathbf{k}) e^{i\mathbf{kR}} d\mathbf{k}. \tag{9.108}$$

Having substituted (9.108) to (9.106), we arrive at

$$\int \frac{d\Phi(t,\mathbf{k})}{dt} e^{i\mathbf{kR}} d\mathbf{k} = \int d\tilde{\mathbf{R}} \left[\gamma_1 w_1(|\mathbf{R} - \tilde{\mathbf{R}}|) - \gamma_2 w_2(|\mathbf{R} - \tilde{\mathbf{R}}|)\right] \cdot$$
$$\cdot \int \Phi(t,\mathbf{k}) e^{i\mathbf{k}\tilde{\mathbf{R}}} d\mathbf{k} - \mu \int \Phi(t,\mathbf{k}) e^{i\mathbf{kR}} d\mathbf{k}, \tag{9.109}$$

which is equivalent to

$$\int \mathcal{H} e^{i\mathbf{k}\mathbf{R}} d\mathbf{k} = 0 , \qquad (9.110)$$

where

$$\mathcal{H} = \frac{d\Phi(t,\mathbf{k})}{dt} + \mu \Phi(t,\mathbf{k}) \qquad (9.111)$$

$$- \Phi(t,\mathbf{k}) \int d\tilde{\mathbf{R}} \left[ \gamma_1 w_1(|\mathbf{R}-\tilde{\mathbf{R}}|) - \gamma_2 w_2(|\mathbf{R}-\tilde{\mathbf{R}}|) \right] e^{i\mathbf{k}(\tilde{\mathbf{R}}-\mathbf{R})} .$$

Apparently, in order to satisfy Equation (9.110), $\mathcal{H}$ should be equal to zero identically. Therefore, we obtain that $\Phi(t,\mathbf{k})$ should be a solution of the following equation:

$$\frac{d\Phi(t,\mathbf{k})}{dt} = \omega(\mathbf{k}) \Phi(t,\mathbf{k}) , \qquad (9.112)$$

where

$$\omega(\mathbf{k}) = -\mu + \int \left[ \gamma_1 w_1(|\mathbf{S}|) - \gamma_2 w_2(|\mathbf{S}|) \right] e^{i\mathbf{k}\mathbf{S}} d\mathbf{S} . \qquad (9.113)$$

Correspondingly, for any particular $k$, the solution of (9.112) is given by

$$\Phi(t,\mathbf{k}) = \Phi(0,\mathbf{k}) e^{\omega(\mathbf{k})t} , \qquad (9.114)$$

so that a small initial perturbation will decay if $\omega(\mathbf{k}) < 0$ and increase if $\omega(\mathbf{k}) > 0$. Obviously, in the former case the system is asymptotically stable with respect to an inhomogeneous perturbation with given $\mathbf{k}$; in the latter case it is unstable.

Equations (9.112) to (9.114) were obtained in a general case, i.e., without making any hypotheses about the functional form of $w_1$ and $w_2$ (except for their sufficiently fast decay to ensure the existence of the integrals). However, further analysis will not be instructive unless we make specific assumptions about the shape of the kernels. In the following, we assume that

$$w_i(S) = \frac{1}{L_i^2 \pi} \exp\left[-\left(\frac{S}{L_i}\right)^2\right] , \quad i = 1, 2, \qquad (9.115)$$

where parameters $L_1$ and $L_2$ give, respectively, characteristic distances where the enhancing and hampering cooperative processes are effective.

From (9.113) and (9.115), we obtain

$$\omega(\mathbf{k}) = -\mu + (1 + 2au_+)(1 - u_+) \exp\left(-\frac{k^2 L_1^2}{4}\right)$$

$$- u_+(1 + au_+) \exp\left(-\frac{k^2 L_2^2}{4}\right) . \qquad (9.116)$$

Instabilities and dissipative structures 241

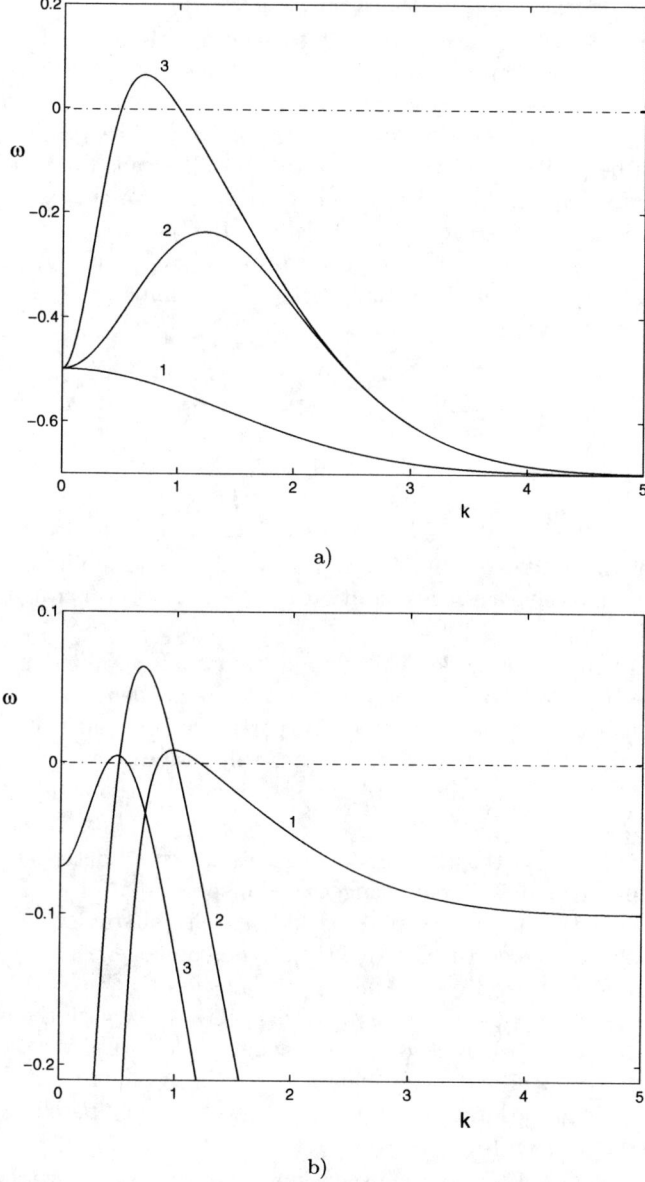

**FIGURE 9.9**: The stability diagram calculated according to (9.116) for (a) different values of $\nu = L_2/L_1$, curve 1 for $\nu = 1$, curve 2 for $\nu = 2$, and curve 3 for $\nu = 5$ (other parameters are $a = 0.8$, $\mu = 0.7$, $L_1 = 1$); and (b) for different values of $\mu$, curve 1 for $\mu = 0.1$, curve 2 for $\mu = 0.7$, and curve 3 for $\mu = 95$ (other parameters $a = 0.8$, $\nu = 5$, $L_1 = 1$).

Therefore, $\omega$ depends on $|\mathbf{k}|$ rather than on $\mathbf{k}$.

Expression (9.116) is rather difficult to study analytically; however, two things can be seen relatively easily. Both in the short-wave and long-wave limits $\omega(|\mathbf{k}|)$ appears to be negative, $\omega(|\mathbf{k}| \to \infty) \to -\mu < 0$, and $\omega(|\mathbf{k}| \to 0) \to -[(a-1)^2 + 4a(1-\mu)]^{1/2}u_+ < 0$, respectively. Analysis of conditions when $\omega$ can be positive for intermediate values of $|\mathbf{k}|$ leads to rather cumbersome and bulky equations, which we do not show here for the sake of brevity; details can be found in Lefever and Lejeune (1997). Briefly, function $\omega(|\mathbf{k}|$ appears to have a global maximum at a certain $|\mathbf{k}| = k_c$ and $\omega_{max}$ is positive provided $\nu = L_2/L_1$ is large enough. Indeed, assuming $\nu \gg 1$, for a small positive $|\mathbf{k}|$ we obtain

$$\omega(\mathbf{k}) \simeq -\mu + (1 + 2au_+)(1 - u_+)\exp\left(-\frac{\mathbf{k}^2 L_1^2}{4}\right) \quad (9.117)$$

$$\simeq a(1 - u_+)u_+ > 0$$

(taking into account that $\mu = (1 + au_+)(1 - u_+)$ and $0 < u_+ < 1$).

Note that the necessary condition $L_2/L_1 > 1$ of the instability onset is close in its meaning to the necessary condition $D_2/D_1 > 1$ of the Turing instability; cf. (9.16)–(9.18).

A typical behavior of $\omega(|\mathbf{k}|)$ is shown in Figure 9.9. While dependence on $\omega_{max}$ on $L_2/L_1$ is relatively straightforward, its dependence on other parameters is somewhat more complicated. In particular, dependence of $\omega_{max}$ on $\mu$ is non-monotonic, so that its maximum value (and, correspondingly, the widest range of $k$ where $\omega(|\mathbf{k}|) > 0$) is reached for an intermediate value of $\mu$; see Figure 9.9b.

These results prove that the model vegetation system described by Equation (9.100) is capable of generating periodic spatial patterns. It should be mentioned here that the above analysis addresses linear stability and, as such, is only valid when the perturbation of the original homogeneous steady state is small. Therefore, generally speaking, it does not tell much about the properties of the emerging patterns (except for their periodicity with the wavelength from a predicted range). Indeed, numerical simulations (Lefever and Lejeune, 1997; Lejeune et al., 1999) show that there can be two qualitatively different patterns arising in the large-time limit, namely, stripes and patches. Examples are shown in Figure 9.10.

Note that model (9.100) does not account for any spatial anisotropy. Therefore, it predicts that vegetation bands should not be strictly related to a sloping ground. Indeed, although on a flat surface a periodical patchy "spotted" pattern is more commonly observed, the tiger bush pattern has been occasionally observed as well (White, 1970). Remarkably, in the model (9.100), both pattern types arise as an outcome of the same dynamical mechanism. Orientation of the patterns was shown to be determined by the initial conditions (Lefever and Lejeune, 1997; Lejeune et al., 1999).

Instabilities and dissipative structures 243

**FIGURE 9.10:** Snapshots of population density distributions in space obtained from the model (9.100) with the kernels given by (9.115) for different parameter values. Black and white colors stand for high and low density, respectively. The left-hand panel obtained for $a = 1$, the right-hand panel obtained for $a = 0.8$; other parameters are $\mu = 0.95$ and $\nu = 10$. With permission from Lejeune et al. (1999).

Now, the next step is to understand how the system properties may change under the presence of spatial anisotropy. For that purpose, we assume that the kernels $w_1$ and $w_2$ are translated or "shifted" by fixed distances $\mathbf{R}_1$ and $\mathbf{R}_2$, respectively. Equation (9.100) now takes the following form:

$$\frac{\partial U(\mathbf{R},T)}{\partial T} = \left( \alpha \int_{\infty}^{-\infty} w_1(|(\mathbf{R}-\mathbf{R}_1)-\tilde{\mathbf{R}}|) U(\tilde{\mathbf{R}},T)[1+AU(\tilde{\mathbf{R}},T)]d\tilde{\mathbf{R}} \right)$$

(9.118)

$$\cdot \left( 1 - \frac{1}{K} \int_{\infty}^{-\infty} w_2(|(\mathbf{R}-\mathbf{R}_2)-\tilde{\mathbf{R}}|) U(\tilde{\mathbf{R}},T)d\tilde{\mathbf{R}} \right) - MU .$$

Generally speaking, in a real-world vegetation system there can be different sources of anisotropy, and they may act differently on different aspects of plants' reproduction. For instance, competition for water would be affected by anisotropy produced by a sloping surface while anisotropy in seed dissemination is obviously subject to wind direction, which is largely independent on landscape details. For simplicity, here we assume that there is only one preferred direction so that vectors $\mathbf{R}_1$ and $\mathbf{R}_2$ are parallel to each other. Considering $\mathbf{R}_1 = (0, R_1)$ and $\mathbf{R}_2 = (0, R_2)$ and using the same parametrization (9.115) for the kernels, after some rather tedious calculations, from (9.119) one can obtain[6] the following expression for $\omega(\mathbf{k})$:

$$\omega(\mathbf{k}) = -\mu + (1+2au_+)(1-u_+)\cos(R_1 k_y)\exp\left(-\frac{\mathbf{k}^2 L_1^2}{4}\right)$$
$$- u_+(1+au_+)\cos(R_2 k_y)\exp\left(-\frac{\mathbf{k}^2 L_2^2}{4}\right)$$
$$+ i\left[(1+2au_+)(1-u_+)\sin(R_1 k_y)\exp\left(-\frac{\mathbf{k}^2 L_1^2}{4}\right)\right.$$
$$\left. - u_+(1+au_+)\sin(R_2 k_y)\exp\left(-\frac{\mathbf{k}^2 L_2^2}{4}\right)\right]. \tag{9.119}$$

Therefore, the expression for $\omega(\mathbf{k})$ now contains both real and imaginary parts. While stability is still fully determined by the sign of Re $\omega$, the homogeneous steady state being unstable to a perturbation with given $\mathbf{k}$ if Re $\omega(\mathbf{k}) > 0$, the existence of a nonzero imaginary part opens a possibility of traveling wave patterns.

The principal properties of $\omega(\mathbf{k})$ given by (9.119) are best seen if we consider separately two particular cases, i.e., (i) when anisotropy affects only multiplication/cooperation ($R_1 \neq 0$, $R_2 = 0$), and (ii) when anisotropy affects only competition ($R_1 = 0$, $R_2 \neq 0$). The results are sketched in Figure 9.11. In the former case, the two domains in the $(k_x, k_y)$ plane where Re $\omega(\mathbf{k}) > 0$ are centered around the points $(\pm k_x^{(c)}, 0)$; in the latter case, they are situated around the points $(0, \pm k_y^{(c)})$, $k_x^{(c)}$ and $k_y^{(c)}$ being certain characteristic values.

Now, coming back to the semiarid vegetation patterns, we can associate the source of anisotropy with sloping ground so that the $y$-axis is directed along the gradient. This means that, in the cases (i) and (ii) (cf. left and right of Figure 9.11), the instability of the homogeneous state $u = u_+$ favors development of vegetation bands oriented down the slope and perpendicular to the slope, respectively. Thus, both stripe types appear as a result of essentially the same dynamical mechanism. Moreover, since in case (i) $k_y \approx 0$, Im $\omega(\mathbf{k}) \approx 0$ as well, which means that the downslope stripes are largely stationary. In case (ii), however, Im $\omega(\mathbf{k}) \neq 0$ and the vegetation stripes are traveling rather than stationary, which is in very good agreement with field observations.

---

[6]Subject to additional conditions $|R_i k_y| < \pi/2$ ($i = 1, 2$) in order to avoid model artifacts; cf. Lefever and Lejeune (1997).

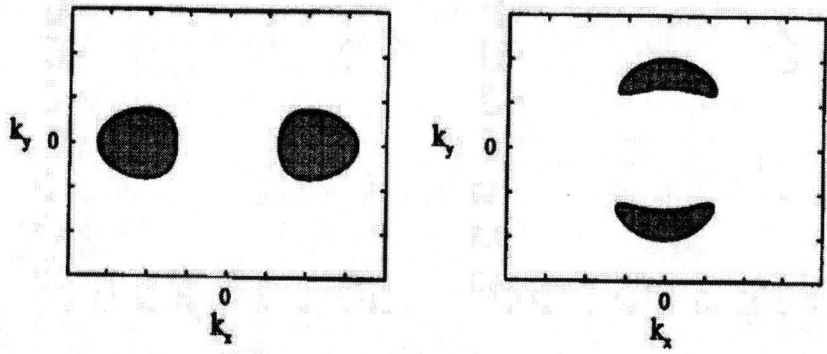

**FIGURE 9.11:** A sketch of the stability diagram of the model (9.119) in the $(k_x, k_y)$ plane obtained for $a = 1$ and $\mu = 0.9$. Anisotropy is assumed to act in the $y$ direction. Left-hand panel corresponds to the case when the anisotropy affects only multiplication/cooperation, $R_1 = 0.5$, $R_2 = 0$, and $\nu = 5$. Right-hand panel corresponds to the case when the anisotropy affects only competition, $R_1 = 0$, $R_2 = 0.5$, and $\nu = 3.33$. Grey color shows the instability domains where $\omega > 0$. With permissions from Lefever and Lejeune (1997).

## 9.4 Concluding remarks

During the last two decades, considerable progress has been made in further developing the Turing ideas and in identification of instabilities in various diffusion–reaction systems; see, for instance, Rovinsky and Menzinger (1992); Kapral and Showalter (1995); Andresen et al. (1999); Satnoianu et al. (2000, 2001). Although a vast majority of the results have been originally obtained for chemical systems, which implies different types of nonlinearities in the equations, there is no doubt that at least some of them are relevant to population dynamics (Malchow, 1993, 1995, 2000a,b). In particular, the pattern formation scenario due to the Turing instability was hypothetically considered as the basic mechanism underlying the commonly observed heterogeneous spatial distribution of marine plankton, the phenomenon also known as "plankton patchiness" (see Levin and Segel, 1976).

It should be mentioned that specific requirements of the "Turing type" instability (i.e., instability of a locally stable equilibrium to a heterogeneous perturbation with wavelengths from a certain finite range) are apparently different for different systems. Particular examples of these requirements are given by the condition that the diffusion coefficient ratio should be sufficiently large (sufficiently small) compared to unity in case of diffusive instability, or the speed of the differential flow should be large enough in case of advection-induced instability. There is one thing, however, that these seemingly different scenarios have in common: Onset of instability is not possible unless the interacting species or agents possess different "mobility," no matter whether the motion is random or ordered.

Another common feature, which stems from the very "definition" of the Turing-type instability, is that the emerging patterns are periodical in space. Meanwhile, although one real-world example of a spatially periodic ecological pattern was considered in the preceding section, in general, periodicity in the spatial distribution of ecological populations is rarely observed. In particular, the spatial distribution of plankton is remarkably irregular.[7] Therefore, the generality of the Turing scenario as a generic mechanism for ecological pattern formation remains highly controversial.

In this chapter, we focused on pattern formation in purely deterministic systems. Concerning real-world applications, a big issue is how the scenarios and the properties of emerging patterns can be altered by the presence of noise. There is a lot of evidence that noise can make Turing patterns less regular (cf. Malchow et al., 2004). This issue will be considered in detail in Part IV. Meanwhile, the reported impact of noise does not answer the question of whether a deterministic diffusion–reaction system is intrinsically capable of generating irregular patterns. Throughout the next few chapters, we revisit some recently discovered non-Turing mechanisms of pattern formation in deterministic models of population dynamics that result in a remarkably irregular spatial structure, qualitatively (and sometimes quantitatively) similar to those observed in nature.

---

[7]Not to mention the fact that the diffusivity of the phytoplankton (prey) and zooplankton (predator) is essentially the same because it originates in the same turbulent mixing.

# Chapter 10

## Patterns in the wake of invasion

Spatial distributions of ecological species are usually distinctly heterogeneous, but the main processes underlying these heterogeneities, as well as scenarios of ecological pattern formation, are often rather poorly understood. A conceptual model treating ecological patterns as the diffusion-induced "dissipative structures" in a system of interacting ecological species, such as a predator and its prey, was considered in the previous chapter. This approach, however, was shown to have severe limitations and, apart from a few specific cases, pattern formation due to a Turing-type diffusive instability can hardly be regarded as a common scenario.

A key to the understanding can probably be found in a historical perspective. Positions of ecological borders change with time (e.g., as a result of continuous climatic changes or, more recently, due to the impact of human activity) and the area inhabited by a given species a few thousands years ago might have been completely different from what it is today. Moreover, many of the species are known to be invaders and came from a distant land or even from another continent. These observations seem to give some reasons for making a hypothesis that the currently observed species distribution has a "historical dimension" and should be considered together with the corresponding processes of species spread, i.e., dispersal and/or migration. Note that, both in the case of moving ecological borders and in the case of invasion, species redistribution is associated with a traveling boundary. Therefore, in more mathematical terms the question can then be formulated as follows: Can it happen that spatiotemporal patterns are "triggered" by the propagation of a traveling population front, considering it as a basic scenario of biological invasion? Indeed, spatiotemporal population oscillations in the wake of invasion have been observed in some ecological data (Caughley, 1970; Jeltsch et al., 1992).

To the best of our knowledge, in the theoretical perspective, pattern formation caused by a propagating front was first discovered by Dunbar (1983), who observed promptly decaying oscillations behind the front. A somewhat more prominent effect was observed by Yachi et al. (1989) in numerical simulations in an ecoepidemic model describing the spread of rabies among a fox population. A few years later, the work was carried forward by Sherratt (1994a) and Sherratt et al. (1995), who showed that patterns arising in the wake can be chaotic. A curious phenomenon of "dynamical stabilization," which results in the convergence of system's trajectories to an unstable equilibrium and formation of a quasi-homogeneous species distribution in the wake of the front, was observed and studied in Petrovskii and Malchow (2000) and Malchow and Petrovskii (2002) using a bifurcation theory along with numerical simulations. Similar phenomena were also found in some chemical systems (Merkin et al., 1996; Merkin and Sadiq, 1996; Rasmussen et al., 1996; Davidson, 1998), which are mathematically similar to the models of population dynamics.

The chapter begins with an insight into a typical scenario of pattern formation in the wake of an invading predator, with special attention paid to the possibility of realistic irregular patterns. After that, we focus on the dynamical stabilization of an unstable equilibrium in the wake of invasion. Then we show that a similar succession of patterns can be observed in a system of competing species, thus indicating that the scenario is generic. Finally, we figure out how these results can be extended to ecoepidemic systems.

## 10.1 Invasion in a prey–predator system

The spatiotemporal dynamics of a prey–predator system is described by the following equations:

$$\frac{\partial U}{\partial T} = D\nabla^2 U + P(U) - E(U, V) , \qquad (10.1)$$

$$\frac{\partial V}{\partial T} = D\nabla^2 V + \kappa E(U, V) - MV \qquad (10.2)$$

where all notations have the usual meanings, cf. Chapter 2.

Contrary to the previous chapter, here we deliberately focus on the case when diffusivity is the same for both species in order to exclude the possibility of the Turing instability; therefore, all patterns to be observed should be ascribed to a non-Turing mechanism(s). Note that the assumption of equal diffusivity is not at all ecologically unrealistic. One immediate example is given by a marine ecosystem where diffusion takes place mainly due to turbulent mixing (see Okubo, 1980), which has practically the same impact on prey (e.g., phytoplankton) and predator (zooplankton and/or fish larvae). In terrestrial ecosystems, a predator's success is often reached more due to an optimal foraging strategy rather than due to a faster motion, and its diffusivity must not necessarily be higher than that of its prey.

For non-Turing pattern formation, effective analytical methods are yet to be developed, and results are usually obtained by means of computer simulations (but see Sherratt, 1994a). For that purpose, we need to decide about the specific functional form of prey growth and predation, i.e., about functions $P(u)$ and $E(u, v)$, respectively. Throughout this section, we assume that prey growth is logistic and predation is of Holling type II. Specifically, we use the following parametrization:

$$P(U) = -\frac{\alpha}{K} U(U-K), \qquad (10.3)$$

where $\alpha$ is the intrinsic growth rate, $K$ is the carrying capacity, and

$$E(U,V) = A\frac{UV}{U+H}, \qquad (10.4)$$

where $\gamma$ and $H$ are the per capita predation rate and the half-saturation prey density, respectively.

We want to emphasize, however, that, actually, the scenarios of pattern formation described below are proved to be robust to the choice of parametrization as long as the main properties of the functions remain qualitatively the same. For instance, instead of the algebraic Michaelis–Menten function (10.4), predator response to prey density could be described by the exponential Ivlev function (Sherratt et al., 1995; Petrovskii et al., 1998) or by a ratio-dependent function (Sherratt et al., 1995) without any significant change in the system dynamics.

We begin with the one-dimensional case so that $\nabla^2 = \partial^2/\partial X^2$. Introducing, for convenience, dimensionless variables as

$$t = \alpha T, \quad x = X\left(\frac{\alpha}{D}\right)^{1/2}, \quad u = \frac{U}{K} \quad \text{and} \quad v = \frac{AV}{\alpha K} \qquad (10.5)$$

(see Chapter 2 for more details), Equations (10.1)–(10.2) take the following forms:

$$\frac{\partial u}{\partial t} = \frac{\partial^2 u}{\partial x^2} + u(1-u) - \frac{uv}{u+h}, \qquad (10.6)$$

$$\frac{\partial v}{\partial t} = \frac{\partial^2 v}{\partial x^2} + k\frac{uv}{u+h} - mv, \qquad (10.7)$$

where $k = \kappa A/\alpha$, $m = M/\alpha$, and $h = H/K$ are dimensionless parameters. In agreement with ecological reality, we assume that the population system dwells on a finite domain of length $L$ so that $-L/2 < x < L/2$.

It seems reasonable to consider the corresponding nonspatial system first:

$$u_t = u(1-u) - \frac{uv}{u+h}, \qquad v_t = k\frac{uv}{u+h} - mv. \qquad (10.8)$$

In a general case there is no one-to-one correspondence between the properties of the spatial and nonspatial systems, the dynamics of the spatial system being much richer. Yet it is clear that the nonspatial system provides a certain "skeleton" for understanding more complicated dynamics of its spatial counterpart because of the evident relation between the steady states of Equations (10.8) and the spatially homogeneous stationary solutions of system (10.6)–(10.7).

Basic investigation of the system (10.8) can be done by standard linear stability analysis. It is not difficult to see that there are only three stationary points in the phase plane $(u, v)$, i.e., the extinction state $(0, 0)$, the prey-only state $(1, 0)$, and the coexistence state $(\bar{u}, \bar{v})$ where

$$\bar{u} = \frac{ph}{1-p}, \qquad \bar{v} = (1 - \bar{u})(h + \bar{u}) \qquad (10.9)$$

(denoting, for convenience, $p = m/k$). The type of stationary point depends on the eigenvalues that are the solutions of the equation

$$\lambda^2 - \lambda \operatorname{tr} A + \det A = 0, \qquad (10.10)$$

where $A$ is, as usual, the matrix of the linearized system. Obviously, in order to trace changes in the stability of a given steady state, it is sufficient to reveal any change of sign in $\operatorname{tr} A$, $\det A$, and $\Delta = (\operatorname{tr} A)^2 - 4 \det A$.

It is readily seen that $(0, 0)$ is a saddle for any value of the system parameters $k$, $m$, and $h$. The stationary point $(1, 0)$ is a saddle for

$$h < h_1(p) = \frac{1-p}{p} \qquad (10.11)$$

(see curve 1 in Figure 10.1), and only in this case is the coexistence state $(\bar{u}, \bar{v})$ situated in the biologically meaningful region $u \geq 0$, $v \geq 0$; otherwise, $(1, 0)$ is a stable node.

For $(\bar{u}, \bar{v})$ we obtain

$$\operatorname{tr} \bar{A} = \frac{p}{1-p}[(1-h) - p(1+h)], \quad \det \bar{A} = kp[1 - p(1+h)], \qquad (10.12)$$

so that the coexistence state changes its stability for

$$h = h_2(p) = \frac{1-p}{1+p}; \qquad (10.13)$$

cf. curve 2 in Figure 10.1. For all parameter values when the state $(\bar{u}, \bar{v})$ is unstable, i.e., below curve 2, it is surrounded by a stable limit cycle that appears through the Hopf bifurcation.

Figure 10.1 shows a sketch of the map in the parameter plane $(p, h)$ for a hypothetical value $k = 0.1$. Here curves 3 and 4 arise from the equation $\Delta = 0$ and show where the coexistence steady state changes its type from node to focus (or vice versa). Therefore, domain I above curve 1 corresponds to the case of $(\bar{u}, \bar{v})$ being a saddle-point, where the only attractor in the phase plane $(u, v)$ for these parameter values is the stable node $(1, 0)$. Domain II, between curves 1 and 3, corresponds to $(\bar{u}, \bar{v})$ being a stable node and domain III between curves 2 and 3 to $(\bar{u}, \bar{v})$ being a stable focus. Domains IV (between curves 2 and 4) and V (below curve 4) correspond to $(\bar{u}, \bar{v})$ being an unstable

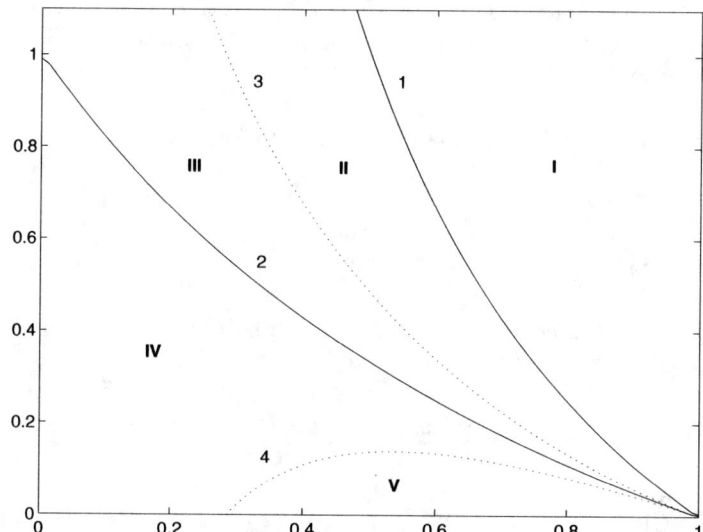

**FIGURE 10.1:** Map in parameter plane $(p, h)$ for $k = 0.1$; see comments in text. With permission from Petrovskii and Malchow (1999).

focus or unstable node, respectively, surrounded by a stable limit cycle that appears via Hopf bifurcation when crossing curve 2.

Note that, although Equations (10.6) and (10.7) contain three parameters, i.e., $k$, $m$, and $h$, tr$A$ actually depends on the ratio $p = m/k$ but not on $m$ and $k$ separately. As a result, it appears that variations in $k$ lead to only insignificant changes in the structure of the map in the $(p, h)$ plane. Indeed, curves 1 and 2 (defined by $h_1(p)$ and $h_2(p)$, respectively) are "universal" in the sense that their positions do not depend on $k$. As for curves 3 and 4, they show only a slight dependence on $k$ (for values $k \ll 1$, curves 3 and 4 approach curve 2, and for values $k \gg 1$ curve 3 gets close to curve 1 while curve 4 approaches axis $p$).

Having kept the properties of the nonspatial system in mind, now we proceed to the dynamics of the spatial system (10.6)–(10.7). However, firstly, we must complement the equations with boundary and initial conditions. Regarding the initial distribution of species, we consider an invasion of a predator into an area already inhabited by prey at the level of the carrying capacity:

$$u(x, 0) = 1 \text{ for any } x, \tag{10.14}$$

$$v(x, 0) = v_0 \text{ if } |x| < \frac{d_v}{2} \quad \text{and} \quad v(x, 0) = 0 \text{ if } |x| > \frac{d_v}{2}, \tag{10.15}$$

where $v_0$ is the predator density inside the initially invaded patch, $d_v$ being the patch diameter.

As for the boundary conditions, generally speaking, they should also be chosen according to the biological situation under study. For instance, a hostile environment outside of the domain would best be taken into account by setting the population densities at the boundary to zero, an "impermeable" boundary would be more adequately described by the no-flux conditions, etc. It should be noted, however, that, in case the time is not very large (so that the spreading predator population does not reach the domain boundary), the type of boundary condition does not really matter. In order to make it consistent with the initial distribution (10.14), the simulation results below are shown for the zero-flux conditions for both species.

Knowledge of the non-spatial dynamics seems to allow us to outline what the dynamics of the spatial system can be. For parameters from domain I we can expect that the predator goes extinct and there is no invasion. For parameters from domain II, we can expect the existence of a propagating front "switching" the system from the prey-only state $(1,0)$ to the stable coexistence state $(\bar{u},\bar{v})$. For parameters from domain III, we can probably expect some damped oscillations at the front. For domains IV and V, prediction is more difficult; due to the existence of the stable limit cycle, it seems reasonable to expect some sort of population oscillations.

These semi-intuitive predictions appear to be in a good agreement with results of computer experiments, at least for those parameter values when the coexistence state is stable. In particular, for domains I and II, it is predator extinction and invasion through propagation of monotonous population front, respectively.

For domain III, where the coexistence state is a stable focus, the situation becomes somewhat more interesting because the population front is no longer monotonous; see Figure 10.2. The papers by Dunbar should be mentioned here (Dunbar, 1983, 1984, 1986); he was the first to observe and study this phenomenon. The oscillations can be rather prominent at the front but decay promptly behind the front; on the whole, their amplitude becomes larger the closer the parameters are to curve 2, where the coexistence state $(\bar{u},\bar{v})$ looses its stability. Note that the oscillations do not change their shape and the pattern simply moves as a whole with a constant speed.

Oscillations at the propagating population front form an interesting pattern; however, its relation to the phenomenon of ecological patchiness may seem somewhat doubtful. The first concern is that the area occupied by these oscillations is rather small is size; thus, this pattern alone can hardly account for the common spatial heterogeneity of ecological populations. Second, the pattern has an apparently regular structure and is essentially stationary in the sense that the shape and relative positions of the peaks do not change with time. These properties are clearly different from those typically observed in nature.

Nevertheless, we want to emphasize that population oscillations at the front are not a mathematical artifact. Note that, due to oscillations' fast decay, the pattern consists of a distinct peak at the boundary separating the area

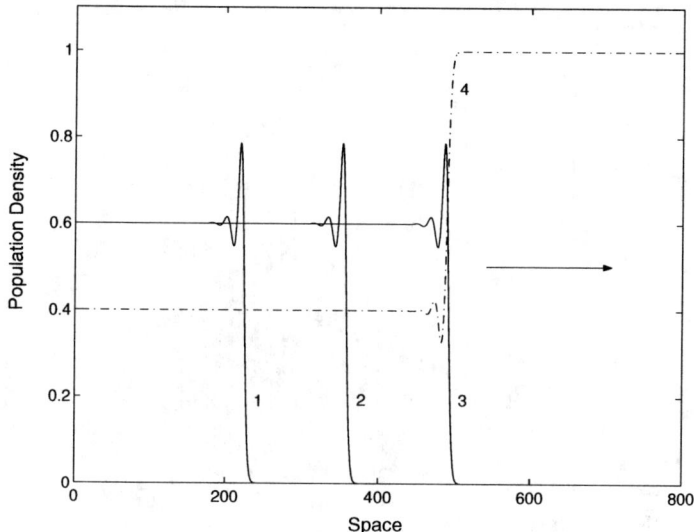

**FIGURE 10.2:** Predator density versus space shown at $t = 100$ (curve 1), $t = 200$ (curve 2) and $t = 300$ (curve 3) for parameters $k = 2.0$, $p = 0.4$, $h = 0.6$; the arrow indicates the direction of the predator invasion. The initial distribution is given by (10.14)–(10.15) with $d_v = 200$ and $v_0 = 0.5$. Since the problem is symmetrical with respect to the origin, only half of the domain is shown. The distribution of prey density has qualitatively similar features up to the apparent difference that $u \approx 1$ on the right-hand side of the traveling front; an example is given by the dashed-and-dotted curve 4 showing the prey distribution at $t = 300$.

inhabited by the given species from an empty space and a slightly heterogeneous population distribution behind. This means that the population density reaches its maximum at the boundary rather than inside. Remarkably, spatial patterns with similar properties have been observed for some insect species (Harrison, 1997; Brodmann et al., 1997) and they were attributed to a diffusion–reaction mechanism (cf. Hastings et al., 1997), although somewhat different from considered here.

Up to the minor distinctions mentioned above, for both domains II and III, a spatially homogeneous distribution of the populations arises behind the front with the population densities $u = \bar{u}$ and $v = \bar{v}$. The dynamics of the system becomes significantly different in the domains where the coexistence steady state is unstable, i.e., in domains IV and V below curve 2. In this case, there are population oscillations at the front as well, however, they are now much more distinct; see the top of Figure 10.3. Also, their rate of decay is lower compared to the case when $(\bar{u}, \bar{v})$ is a stable focus. Note that, in spite of the complicated spatial structure, the area occupied by periodic oscillations

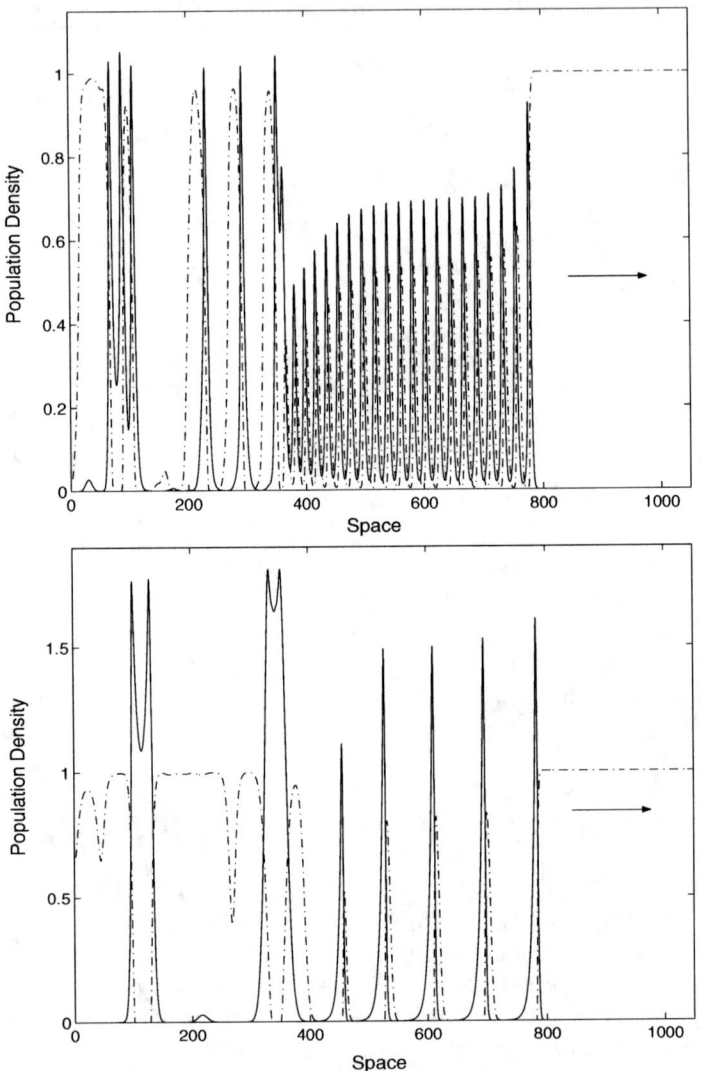

**FIGURE 10.3**: Densities of predator (solid curve) and prey (dashed-and-dotted curve) versus the coordinate calculated for (top) $k = 2.0$, $p = 0.5$, $h = 0.1$, $t = 380$ and (bottom) $k = 2.0$, $p = 0.2$, $h = 0.1$, $t = 290$. The initial conditions are the same as in Figure 10.2. In both cases, parameters correspond to oscillatory local kinetics. The arrows indicate the direction of the front propagation.

adjoining the front is a traveling wave, it moves as a whole with a constant speed and without changing its shape.

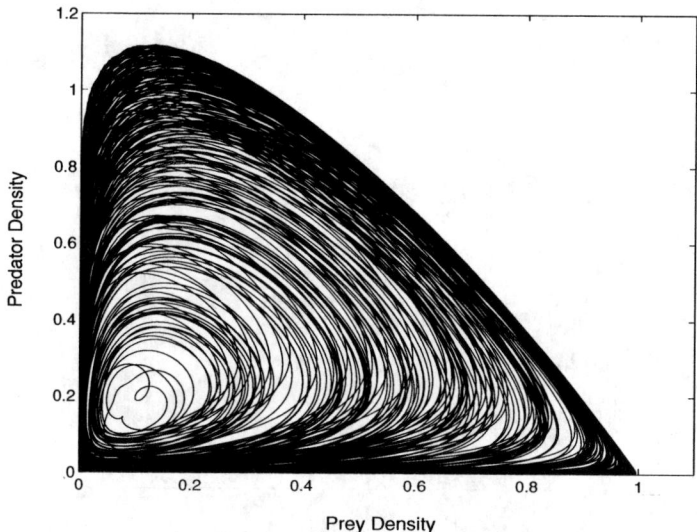

**FIGURE 10.4**: Phase plane of local population densities at a fixed point $x = 300$ after onset of irregular spatiotemporal oscillations. The trajectory is shown for the time interval between $t = 4000$ and $t = 8000$; parameters are the same as in Figure 10.3, top.

For parameter values when the limit cycle grows in size considerably and approaches the saddle-point $(0, 0)$, the spatial period of oscillations may become so large that the whole succession looks more like an ensemble of moving separated "patches" rather than a single pattern, cf. bottom of Figure 10.3 between $x = 500$ and $x = 900$, although it still retains its regular structure.

Remarkably, the regular population oscillations with decaying amplitude are not the only pattern induced by the propagating front. At the wake of the traveling wave, there also arise oscillations of a different sort; see the left-hand side of the upper panel in Figure 10.3. They are apparently aperiodic both in space and time: Figure 10.4 shows the phase plane of the local population densities obtained at a fixed position after irregular oscillations have developed. A closer look shows that the oscillations are chaotic (Sherratt et al., 1995), the corresponding value of the dominant Lyapunov exponent being estimated as $\lambda_D \sim 0.01$ (Medvinsky et al., 2002). Eventually, after the traveling wave leaves the domain, they occupy the whole area. That regime of spatiotemporal chaos appears to be self-sustained: Large-time computer simulations do not reveal any qualitative changes in the system dynamics after the onset of chaos.

We emphasize that the system (10.6)–(10.7) does not contain any prescribed spatial structure because none of the parameters depend on space. Therefore, although these chaotic spatiotemporal population oscillations are triggered by

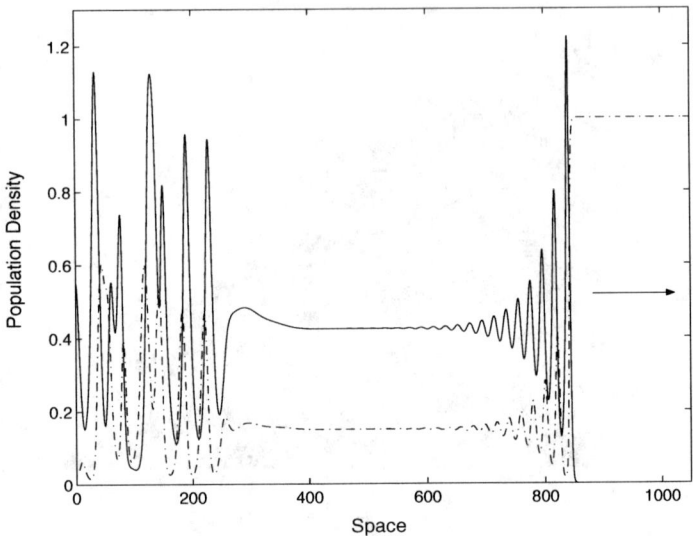

**FIGURE 10.5**: Densities of predator (solid curve) and prey (dashed-and-dotted curve) vs coordinate at $t = 400$ as obtained for parameters $k = 2.0$, $p = 0.3$, and $h = 0.35$ chosen from the region of dynamical stabilization. The initial conditions are given by (10.14)–(10.15) with $d_v = 200$ and $v_0 = 0.5$. Note that the plateau behind the oscillatory fronts corresponds to the unstable state $(\bar{u}, \bar{v})$.

the traveling front (and therefore may be linked, to some extent, to the initial conditions of special type – but see Chapter 11), they are self-organized in the sense that they are not induced by any sort of "environmental" heterogeneity.

Interestingly, for some parameter values, the onset of chaotic oscillations in the wake of the traveling population wave is preceded by a plateau emerging behind the strongly oscillating front; see Figure 10.5. The population densities inside this area of quasi-homogeneity correspond to those of the coexistence steady state, $u \approx \bar{u}$ and $v \approx \bar{v}$. This is a highly nontrivial result because, for these parameter values, the stationary point $(\bar{u}, \bar{v})$ is unstable in the corresponding nonspatial system. Thus, the "dynamical stabilization" of the unstable equilibrium is an essential spatiotemporal effect and cannot be reduced to the nonspatial dynamics. The plateau gradually grows with time and may reach a considerable length, sometimes nearly as large as the overall length of the domain; see Figure 10.6. Numerical simulations show that, in an unbounded domain, the whole structure would persist over an indefinitely long time, which indicates that the dynamical stabilization of the unstable equilibrium in the wake of invasion is not a transient regime but rather an "intermediate asymptotics" (Barenblatt and Zeldovich, 1971; Barenblatt, 1996). In the next section we provide a more detailed analysis of this phenomenon.

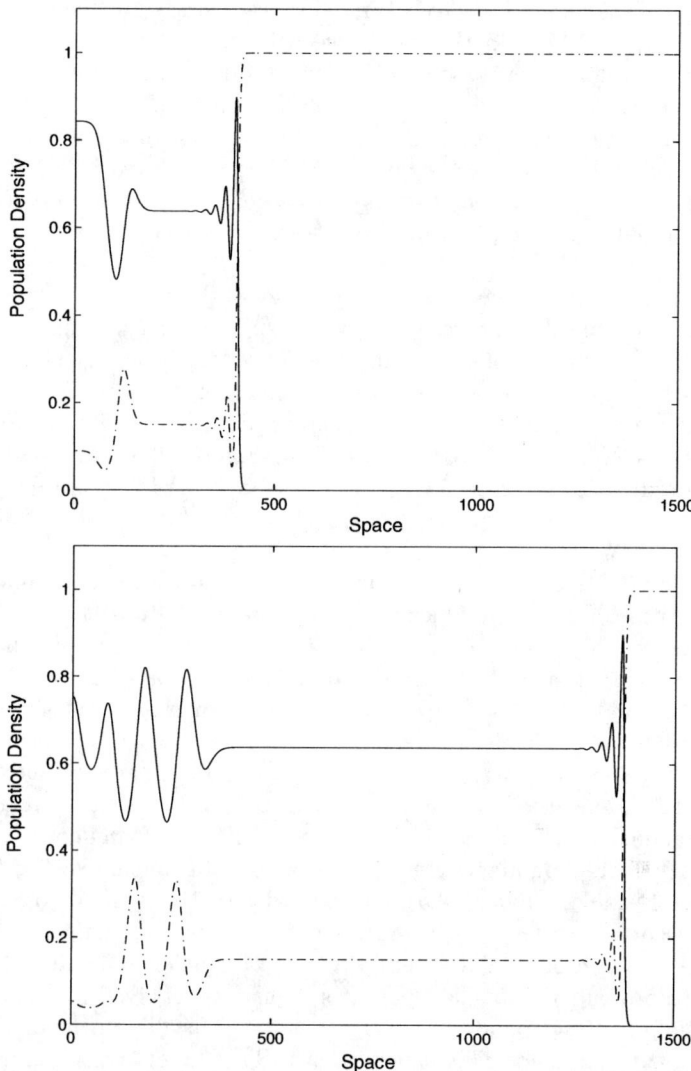

**FIGURE 10.6**: The density of predator (solid curve) and prey (dashed-and-dotted curve) calculated at $t = 350$ (top) and $t = 1400$ (bottom) for parameters $k = 0.5$, $p = 0.2$, and $h = 0.6$ and the initial conditions the same as above. The region of homogeneity in the middle corresponds to the locally unstable coexistence state.

It should also be mentioned that, in some cases, between the unstable plateau and the chaotic region there can be a region of regular spatiotemporal oscillations periodical in space and time (e.g., see the bottom of Figure 10.6).

Regular oscillations in the wake of invasion were studied in much detail by Sherratt (1994b, 1998); in this book, we will not specifically focus on this pattern for the following reasons. Regular periodical waves can be stable or unstable per se (Kopell and Howard, 1973; Sherratt, 1998); however, there is substantial numerical evidence that they are not persistent in the sense that they are finally displaced by the chaotic oscillations. Moreover, although the regular periodical oscillations can indeed be generated by the initial conditions of special type (e.g., finite), they are rarely observed in a more general case; see Chapter 11.

For parameters from domain V, the dynamics of the system is rather similar to that in domain IV. However, there is one difference: No dynamical stabilization of the unstable node $(\bar{u}, \bar{v})$ has been observed. The region of irregular oscillations begins right behind the front of the predator wave.

Now, our concern is to reveal to what extent the above results (in particular, the typical succession of patterns shown in Figure 10.5) can be extended to the two-dimensional case, i.e., to the system given by Equations (10.6)–(10.7) where $u = u(x, y, t)$, $v = v(x, y, t)$, and $\partial^2/\partial x^2$ is now changed to $\partial^2/\partial x^2 + \partial^2/\partial y^2$.

It should be mentioned here that numerical solution of the two-dimensional system brings considerable difficulty. The matter is that, as it is readily seen from the system dynamics in the one-dimensional case, the problem has a few distinctly different spatial and time scales. In particular, onset of chaos is a slow process and it does not happen until all preceding patterns emerge, such as oscillations at the invasion front and dynamical stabilization in the wake. However, the speed of the invasion front is relatively large and, by the time the chaotic oscillations develop, the front would propagate a significant distance. Correspondingly, the numerical domain must be large enough in order to avoid the impact of the boundary. As a result, a relevant domain size appears to be, in dimensionless units (10.5), on the order of $10^3$. On the other hand, the step-size $\Delta x$ of the numerical grid has to be sufficiently small in order to provide resolution of spatial heterogeneity on a smaller scale and to ensure sufficient approximation. The smallest spatial scale is given by the size of a single hump in the population density, which is typically on the order of 10; therefore, $\Delta x$ should normally be less than 1. That leads to a one-dimensional numerical grid with one thousand nodes, or even larger. In the corresponding two-dimensional system, it results in a grid as large as $10^6$ nodes. Although simulations on a grid as large as that are, in principle, within the capability of modern PCs, a single simulation run (i.e., for a given parameter set) takes many hours, sometimes a few days. That makes a detailed study virtually impossible. Thus, the results shown here should be regarded as illustrative rather than exhaustive.

Some snapshots of the system dynamics are shown in Figure 10.7, which is obtained for the same parameters as Figure 10.5. The initial conditions describe a spatially homogeneous distribution of prey at the carrying capacity, $u(x, y, 0) = 1$ for any $(x, y)$, and a finite distribution of predator inside a square domain centered around the origin, $v(x, y, 0) = v_0 = 0.2$ for $-10 < x, y < 10$ and $v(x, y, 0) = 0$ otherwise. Since the system apparently exhibits symmetry with regards to reflections $x \to -x$, $y \to -y$, the simulations are actually made only in the first quadrant of the system, imposing zero-flux conditions at the artificial boundaries $x = 0$ and $y = 0$.

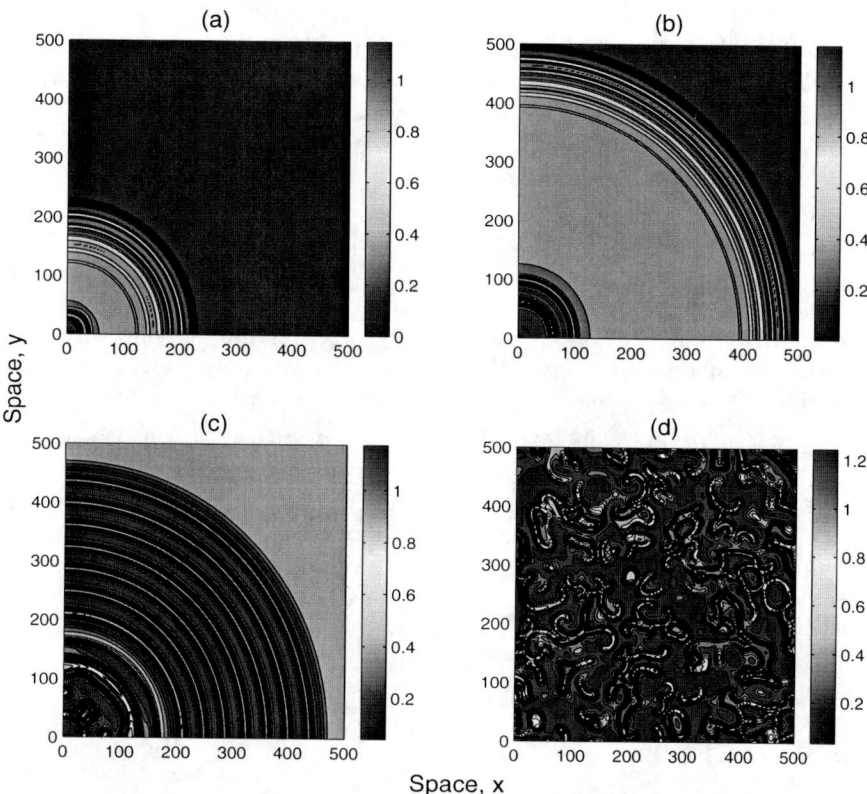

**FIGURE 10.7**: (See color insert.) Spatial distribution of the predator density in the two-dimensional case for (a) $t = 125$, (b) $t = 275$, (c) $t = 1000$, and (d) $t = 3500$. Parameters are the same as in Figure 10.5. Prey density shows qualitatively similar behavior up to the apparent difference that in front of the leading edge prey is at its carrying capacity.

It appears that the dynamics of the two-dimensional system follows more or less the same scenario as the corresponding one-dimensional system. At an early stage (see Figure 10.7a), a traveling population front is formed with decaying population oscillations in the wake. It is followed by dynamical stabilization of the unstable coexistence state so that the population densities converge to $\bar{u}$ and $\bar{v}$, respectively, and a (quasi-)homogeneous plateau is formed; cf. the area near the left-hand bottom corner of Figure 10.7a. As the leading front propagates further, the unstable plateau grows in size considerably (see Figure 10.7b); however, eventually it gives way to a periodical wavetrain. At the wake of the wavetrain, chaotic spatiotemporal oscillations start developing gradually (see the left-hand bottom corner of Figure 10.7c) and eventually occupy the whole domain. Chaotic oscillations are self-sustained and persistent, and, after the onset of chaos all over the domain, there are no other qualitative changes in the system dynamics.

Qualitatively, the emerging chaotic patterns (see Figure 10.7d) look surprisingly similar to those usually observed in spatial distributions of real ecological populations (for instance, in plankton distribution in the ocean; e.g., see Figure 11.13 in the next chapter). Therefore, a hypothesis can be made that it may be the prey–predator interaction that shapes, at least to a certain extent, the "geometry" of the ecological patchiness and the properties of the system's spatiotemporal dynamics in general. Indeed, prey–predator interactions are among the most common ones in ecosystems. Moreover, there is a widely accepted point of view that it is the prey–predator interaction(s) that are integrating separate species into a community by means of creating the mass flows along the trophic chain or web.

However, a visual resemblance between the patterns is not sufficient yet. In order to demonstrate that the patterns observed above are relevant to population dynamics in nature, at least, one has to show that pattern formation takes place on an appropriate spatial scale (recall that all the results have been obtained in dimensionless units). Also, the generality of the proposed scenario when patterns are generated by a propagating population front is disputable. Although a quick estimate of a typical patch size can be done relatively easy (based on the definition (10.5) of the dimensionless variables), we prefer to postpone detailed consideration of this issue until Chapter 11, where some more general scenarios of pattern formation in a conceptual prey–predator system will be revealed.

## 10.2 Dynamical stabilization of an unstable equilibrium

It is well known that a steady state that is stable in a nonspatial system may become unstable in the corresponding spatial system. Then, after the homogeneity is broken due to linear Turing instability (see Chapter 9), the nonlinear interactions between the components drive the system into the formation of standing spatial patterns (Nicolis and Prigogine, 1977). This is an irreversible process, i.e., the broken homogeneity is never restored unless the parameters of the system are changed considerably in order to violate the instability conditions.

Yet a kind of inverse process – in a somewhat more general sense – may occur. Namely, for some parameter values an "anti-Turing" phenomenon takes place: A locally unstable equilibrium of the system (10.8) can be made dynamically stable in the full diffusion–reaction system (10.6)–(10.7). In this case, for certain time- and length scales, formation of spatial patterns is suppressed and homogeneity is restored.

In the previous section we showed that, when the coexistence steady state $(\bar{u}, \bar{v})$ turns unstable and is surrounded by a stable limit cycle, the diffusive front may still switch the system to the state $u \equiv \bar{u}$, $v \equiv \bar{v}$. A typical wave profile is presented in Figure 10.5. One can see that, after prominent oscillations at the front of the wave, there comes the region where the concentrations $u(x,t)$ and $v(x,t)$ nearly reach their stationary (but unstable!) values $\bar{u}$, $\bar{v}$. Note that this curios phenomenon is not exotic in the sense that it can be observed for a wide range of parameter values (cf. Figures 10.5 and 10.6), and it takes place both in one and two spatial dimensions; see Figure 10.7. The unstable plateau exists for remarkably long time before it is finally displaced by the irregular spatiotemporal oscillations. Moreover, as it is readily seen from comparison between the top and bottom of Figure 10.6 (cf. also Figures 10.7a and 10.7b), the length of the plateau can grow with time.

A question remains as to the intrinsic properties of the dynamical system (10.1)–(10.2), or its particular case (10.6)–(10.7), that make stabilization of the locally unstable coexistence state possible. A related and somewhat more practical question is how to distinguish between the parameters when dynamical stabilization can and cannot occur. Below we address these issues by using two different approaches.

### 10.2.1 A bifurcation approach

The idea of the approach is rooted in the observation that the oscillating front preceding the formation of the unstable plateau propagates as a stationary traveling wave. It means that, by the corresponding change of variables, the original system of partial differential equations, cf. (10.6)–(10.7), can be reduced to a system of ordinary differential equations. A required traveling front solution connecting the prey-only state $(1, 0)$ in front of the front to the coexistence state $(\bar{u}, \bar{v})$ behind the front can then be regarded as a heteroclinic connection between the corresponding steady state in the phase space of the system; e.g. see Dunbar (1984).

Since there are parameter values where dynamical stabilization is observed and there are ones where it is not, it means that this heteroclinic connection appears as a result of a global bifurcation. The problem of distinguishing between the parameters can then be addressed by using methods of bifurcation theory (Kuznetsov, 1995). However, global bifurcations are difficult to study, and effective mathematical tools are not fully developed yet. Instead, we endeavor to relate the emerging heteroclinic trajectory to a local bifurcation in one of the two steady states. In particular, we are going to show that there is a parameter range where the type/stability of the coexistence state $(\bar{u}, \bar{v})$ makes the existence of the heteroclinic connection impossible while outside of that range it cannot be ruled out.

Let us consider a stationary traveling wave propagating with a certain speed $c_v$ and "switching" the system from the state $u = 1$, $v = 0$ to the state $u = \bar{u}$, $v = \bar{v}$. A relevant solution of the system (10.6)–(10.7) has the form $u(x,t) = u(\xi)$, $v(x,t) = v(\xi)$, where $\xi = x - c_v t$. Then, system (10.6)–(10.7) is reduced to

$$\frac{d^2 u}{d\xi^2} + c_v \frac{du}{d\xi} + u(1-u) - \frac{u}{u+h} = 0, \qquad (10.16)$$

$$\frac{d^2 v}{d\xi^2} + c_v \frac{dv}{d\xi} + k \frac{u}{u+h} v - mv = 0, \qquad (10.17)$$

which, having introduced auxiliary variables $p$ and $q$, turns into a system of four ordinary differential equations:

$$\frac{dp}{d\xi} = -c_v p + u(1-u) - \frac{u}{u+h}, \qquad (10.18)$$

$$\frac{du}{d\xi} = -p, \qquad (10.19)$$

$$\frac{dq}{d\xi} = -c_v q + \frac{ku}{u+h} - mv, \qquad (10.20)$$

$$\frac{dv}{d\xi} = -q, \qquad (10.21)$$

where the dot above a letter means the ordinary derivative with respect to $\xi$. Obviously, each homogeneous stationary state $(\bar{u}, \bar{v})$ of system (10.6)–(10.7) corresponds to a stationary point $(0, \bar{u}, 0, \bar{v})$ in a four-dimensional phase space $(p, u, q, v)$ of system (10.18)–(10.21).

Propagation of the front means that, while $\xi$ changes from $+\infty$ to $-\infty$, the corresponding trajectory in the phase space leaves the state $(0, 1, 0, 0)$ and eventually enters (in the limit $\xi \to -\infty$) the state $(0, \bar{u}, 0, \bar{v})$. It means that there must exist at least one trajectory entering $(0, \bar{u}, 0, \bar{v})$. In its turn, it means that, for the system (10.18)–(10.21) linearized in vicinity of $(0, \bar{u}, 0, \bar{v})$, there must exist at least one eigenvalue with $\text{Re}\lambda > 0$. (Note that, since $\xi$ is changing from $+\infty$ to $-\infty$ along the traveling wave profile, the stability of $(0, \bar{u}, 0, \bar{v})$ is associated with eigenvalues with positive real parts.)

It is readily seen that the eigenvalues $\lambda_{1-4}$ of the system (10.18)–(10.21) linearized in the vicinity of a given equilibrium point are solutions of the following equation:

$$z^2 + (\text{tr} A)z + \det A = 0, \qquad z = \lambda(c_v + \lambda), \qquad (10.22)$$

where $A$ is the matrix of the linearized nonspatial system (10.8). In particular, for the steady state $(0, \bar{u}, 0, \bar{v})$ the eigenvalues are given by the following expression:

$$\lambda_{1-4} = -0.5\left(c_v \pm \sqrt{(c_v^2 - 2\operatorname{tr}\bar{A}) \pm i\,2\sqrt{|\Delta|}}\right), \qquad (10.23)$$

where $\Delta = (\operatorname{tr}\bar{A})^2 - 4\det\bar{A}$ and $\bar{A}$ is the corresponding matrix. Since dynamical stabilization has been observed in numerical simulations only for $(\bar{u}, \bar{v})$ being an unstable focus, here we restrict our consideration to the case $\Delta < 0$.

From (10.23) we obtain

$$\operatorname{Re}\lambda = -0.5\,(c_v \pm \omega), \qquad (10.24)$$

where

$$\omega = \sqrt[4]{(c_v^2 - 2\operatorname{tr}\bar{A})^2 + 4|\Delta|}\,\cos\left(0.5\arctan\frac{2\sqrt{|\Delta|}}{c_v^2 - 2\operatorname{tr}\bar{A}}\right). \qquad (10.25)$$

Now, if $|\omega| < c_v$, all four eigenvalues have negative real parts. In this case, there can be no trajectory approaching the equilibrium point $(0, \bar{u}, 0, \bar{v})$ in the $\xi \to -\infty$ limit. However, for $|\omega| > c_v$, there are two eigenvalues with positive and two eigenvalues with negative real parts, and in this case such a trajectory can exist. Therefore, the heteroclinic connection and the corresponding change in the dynamical behavior of the system may emerge at the following critical relation:

$$|\omega| = c_v. \qquad (10.26)$$

We emphasize that relation (10.26) is obtained without any reference to the specific parametrization used in the systems (10.6)–(10.7) or (10.18)–(10.21). Therefore, one can expect that it remains valid for a more general class given by Equations (10.1)–(10.2), at least in the case that functions $E(u, v)$ and $P(u)$ have properties qualitatively similar to the Michaelis–Menten kinetics and the logistic growth, respectively. For any given parametrization of $E(u, v)$ and $P(u)$, relation $|\omega| > c_v$ gives a necessary condition of dynamical stabilization in the wake of the propagating front.

Taking into account (10.25), after some standard although tedious calculations, relation (10.26) takes a somewhat more specific form:

$$\frac{4\det\bar{A} - (\operatorname{tr}\bar{A})^2}{\operatorname{tr}\bar{A}} = 2c_v^2. \qquad (10.27)$$

Note that an exact expression for the speed of the predator wave $c_v$ is not available. However, there are many indications that the speed of the wave usually coincides with its minimum possible value (Murray, 1989), even if it is not always the case (Hosono, 1998).

The minimum speed value can be found by considering the solution properties in the vicinity of the steady state $(0, 1, 0, 0)$. Since $u$ and $v$ are population densities, they must be nonnegative. This means that the solution cannot be winding around $(0, 1, 0, 0)$. In its turn, it means that the eigenvalues of the linearized system may not be complex.

From equations (10.22) we obtain

$$\lambda = -0.5 \left[ c_v \pm (c_v^2 + 4z)^{1/2} \right] , \qquad (10.28)$$

where

$$z = -0.5 \left[ \mathrm{tr} A_1 \pm \left( [\mathrm{tr} A_1]^2 - 4 \det A_1 \right)^{1/2} \right] \qquad (10.29)$$

and $A_1$ is the matrix of system (10.8) linearized in the vicinity of $(1, 0)$. Since $(1, 0)$ is a saddle-point when the equilibrium $(\bar{u}, \bar{v})$ is situated inside the biologically meaningful region $u \geq 0$, $v \geq 0$, it holds $\det A_1 < 0$ and, therefore, $z_1 < 0 < z_2$. Thus, all the solutions of (10.28) are real if and only if

$$c_v^2 + 4z_1 \geq 0 , \qquad (10.30)$$

where $z_1$ corresponds to plus in Equation (10.29).

From (10.30), we obtain the lower bound for the spectrum of the possible speed values:

$$c_v \geq c_v^{min} = \left[ 2 \left( \mathrm{tr} A_1 + \left( [\mathrm{tr} A_1]^2 - 4 \det A_1 \right)^{1/2} \right) \right]^{1/2} \qquad (10.31)$$

(supposing that the wave propagates along axis $x$, i.e., $c_v > 0$).

Having assumed $c_v = c_v^{min}$, from (10.27) taken together with (10.31) and (10.12), we obtain an algebraic equation giving the critical relation between the problem parameters. Although the equation is rather bulky and cumbersome (we do not show it here for the sake of brevity), and its explicit analytical solution is hardly possible, it can be easily solved numerically. For each value of $p$ and $k$, it then gives the critical value of $h$. For a specific case $k = 0.1$, the results are shown in Figure 10.8 by curve 5. For the domain between curves 2 and 5, one can expect dynamical stabilization of the unstable equilibrium $(\bar{u}, \bar{v})$ in the wake of the propagating oscillating front, i.e., formation of a quasi-homogeneous unstable plateau with $u \approx \bar{u}$ and $v \approx \bar{v}$, which separates the invading front from the region of irregular spatiotemporal oscillations.

Note that the critical curve is situated in the parameter domain where $(\bar{u}, \bar{v})$ is an unstable focus. This is in agreement with the observation made from numerical experiments that dynamical stabilization in the system (10.6)–(10.7) is not observed if $(\bar{u}, \bar{v})$ is an unstable node.

It should be recalled here that the critical relation obtained in this way is not a criterion but rather a necessary condition; cf. the lines preceding Equation (10.26). A thorough mathematical study would have to prove that the trajectory entering $(0, \bar{u}, 0, \bar{v})$ actually started at $(0, 1, 0, 0)$. Moreover, strictly speaking, it is not an exact necessary condition either because of the

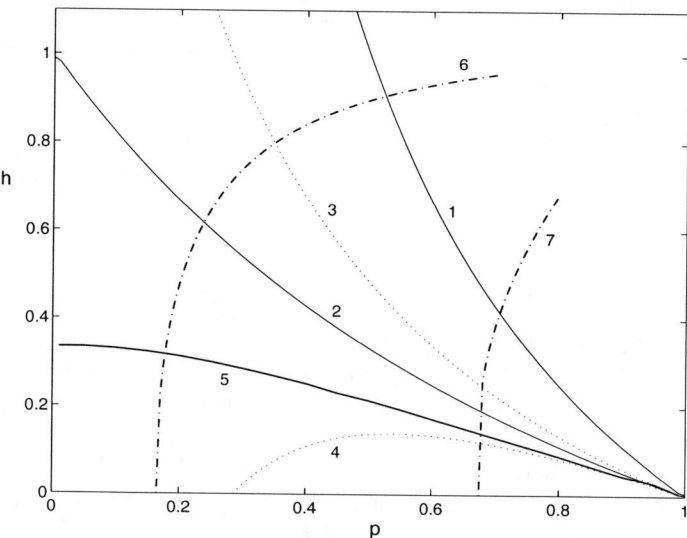

**FIGURE 10.8:** Analysis of dynamical stabilization – a map in the parameter plane of the system (10.6)–(10.7). Meanings of curves 1 to 4 are the same as in Figure 10.1. Curve 5 gives the critical relation (10.27) so that dynamical stabilization is only possible for parameters between curves 5 and 2. Curves 1 to 5 are obtained for $k = 0.1$. Curves 6 and 7 show the critical relation (10.46) obtained for $k = 0.1$ and $k = 1.2$, respectively.

above-made assumption $c_v = c_v^{min}$. Surprisingly, for values of $p$ not too close to 1 (see the next section for details), the obtained critical relation is in very good agreement with numerical experiments, so that curve 5 in Figure 10.8 indeed coincides with a boundary of the parameter domain where dynamical stabilization is observed (Petrovskii and Malchow, 2000).

The position of the parameter domain in the $(p, h)$ plane where dynamical stabilization takes place helps to put this phenomenon into a more general context of the global bifurcation structure of the system. Let us recall that a traveling wave with similar properties (decaying oscillations at the front with the population densities converging to the homogeneous state $u(x, t) \equiv \bar{u}$, $v(x, t) \equiv \bar{v}$; e.g. see Figure 10.2) is also observed when $(\bar{u}, \bar{v})$ is stable, i.e., for parameters from above curve 2 in Figure 10.8. Existence of the corresponding heteroclinic trajectory connecting $(0, 1, 0, 0)$ to $(0, \bar{u}, 0, \bar{v})$ was proved rigorously by Dunbar (1984). Since the dynamical stabilization domain is the one adjoining curve 2 from below, it becomes obvious that the traveling front with dynamical stabilization in the wake corresponds to the same trajectory. The coexistence state looses stability through the Hopf bifurcation when crossing curve 2 from above; however, the stability loss does not destroy the heteroclinic connection.

## 10.2.2 Comparison of wave speeds

The analysis done in the previous section, although providing a condition of dynamical stabilization, does not allow us to obtain any estimate regarding the plateau length. In this section, we endeavor to address the phenomenon of dynamical stabilization in a more heuristic way. We will show that some additional restrictions on parameter values, as well as an equation describing the plateau growth, may appear quite naturally by comparing the speed of relevant diffusive fronts.

The idea of the approach is as follows. We have already shown that stabilization of a locally unstable steady state occurs in the wake of a stationary traveling diffusive front propagating with a certain (constant) speed $c$. Furthermore, if $(\bar{u}, \bar{v})$ is unstable, the propagation of the diffusive fronts (no matter with or without dynamical stabilization in the wake) is normally followed by the onset of irregular spatiotemporal population oscillations. In the case that dynamical stabilization occurs, a remarkable thing is that there exists a distinct interface separating the unstable plateau from the region occupied by the oscillations; cf. Figures 10.5 and 10.6. A closer look (Petrovskii and Malchow, 2001b) shows that the interface propagates with a constant speed, which we denote as $w$.

Now, the spatial domain where dynamical stabilization takes place is bounded by two moving boundaries: the leading front propagating with a constant speed $c$ and the interface between the plateau and the region of irregular spatiotemporal oscillations propagating with a speed $w$. Therefore, the process is controlled by the relation between $c$ and $w$. In case $w < c$, obviously, the length of the domain grows with time as $(c-w)t$. Since the leading front behaves as a stationary traveling wave, its shape and "width" (i.e., the size of the region occupied by the regular damped oscillations) do not change with time. Correspondingly, an increase in the length of the domain locked between the two moving boundaries is only possible by means of an increase in the length of the unstable plateau. Thus, we obtain that

$$L_{plateau} = (c-w)t + L_0 , \qquad (10.32)$$

where $L_0$ is a constant.

In the opposite case, $w > c$, dynamical stabilization apparently cannot occur. The length of the plateau, even if it happens to appear as a result of specific initial conditions, should decrease with time until the region of irregular spatiotemporal oscillations would start immediately after the stationary traveling front. The embryo of the unstable plateau, if any, does not grow with time.

Thus, a simple necessary condition for the dynamical stabilization is $w < c$, so that the critical relation is given by

$$w = c . \qquad (10.33)$$

## Patterns in the wake of invasion

As it stands, however, relation (10.33) is not of much use because the actual values of speed are not known. The assumption that we are going to make at this point is that the diffusive fronts propagate with their minimum possible speed. Indeed, that is what usually happens in various diffusion–reaction systems (Murray, 1989), although a complete and thorough understanding of the reasons is still missing.

The value $c_{min}$ is given by Equation (10.31). Now we are going to obtain $w_{min}$. For that purpose, we consider the system far in front of the interface where the dynamics of the population densities $u(x,t)$ and $v(x,t)$ is described by small perturbations to the steady state $u = \bar{u}$, $v = \bar{v}$:

$$u(x,t) = \bar{u} + \Phi(x,t) , \quad v(x,t) = \bar{v} + \Psi(x,t) , \qquad (10.34)$$

so that $\Phi$ and $\Psi$ are the solutions of the linearized system

$$\frac{\partial \Phi}{\partial t} = \frac{\partial^2 \Phi}{\partial x^2} + a_{11}\Phi + a_{12}\Psi , \qquad (10.35)$$

$$\frac{\partial \Psi}{\partial t} = \frac{\partial^2 \Psi}{\partial x^2} + a_{21}\Phi + a_{22}\Psi , \qquad (10.36)$$

where $a_{ij}$ are the elements of the linearized system matrix.

Since an exponential function is a generic solution of a system of linear differential equations, and $\Phi$ and $\Psi$ must be, asymptotically, of the same order of magnitude, we look for a solution of (10.35)–(10.36) in the following form:

$$\Phi(x,t) = p(t)e^{-\nu x} , \quad \Psi(x,t) = q(t)e^{-\nu x} , \qquad (10.37)$$

where functions $p(t)$ and $q(t)$ are to be determined and $\nu$ is a coefficient.

Having substituted (10.37) into (10.35)–(10.36), we obtain the equations for $p(t)$ and $q(t)$:

$$\frac{dp(t)}{dt} = (\nu^2 + a_{11})p + a_{12}q, \quad \frac{dq(t)}{dt} = a_{21}p + (\nu^2 + a_{22})q. \qquad (10.38)$$

The general solution of the linear system (10.38) is

$$p(t) = B_1 e^{\omega_1 t} + B_2 e^{\omega_2 t} , \quad q(t) = C_1 e^{\omega_1 t} + C_2 e^{\omega_2 t} , \qquad (10.39)$$

where $B_1$, $B_2$, $C_1$, $C_2$ are constant coefficients and the exponents $\omega_{1,2}$ are the solutions of the following equation:

$$(a_{11} + \nu^2 - \omega)(a_{22} + \nu^2 - \omega) - a_{12}a_{21} = 0 . \qquad (10.40)$$

Hence,

$$\omega_{1,2} = \left(\nu^2 + \frac{\mathrm{tr}\bar{A}}{2}\right) \pm \frac{1}{2}\Delta^{1/2} , \qquad (10.41)$$

where $\Delta = (\mathrm{tr}\bar{A})^2 - 4\det\bar{A}$ and $\bar{A}$ is the matrix of the linearized system.

For simplicity, we restrict further consideration to the case of $(\bar{u},\bar{v})$ being an unstable focus. Respectively, $\Delta < 0$ and, considering (10.37) and (10.39) together, it is readily seen that the speed of the front is determined by the real part of the exponents. Since $\mathrm{Re}\,\omega = \nu^2 + (\mathrm{tr}\bar{A})/2$, Equations (10.37) take the form

$$\Phi(x,t) = \tilde{p}(t)\exp\left(-\nu\left[x - \left(\nu + \frac{\mathrm{tr}\bar{A}}{2\nu}\right)t\right]\right), \qquad (10.42)$$

$$\Psi(x,t) = \tilde{q}(t)\exp\left(-\nu\left[x - \left(\nu + \frac{\mathrm{tr}\bar{A}}{2\nu}\right)t\right]\right), \qquad (10.43)$$

where the functions $\tilde{p}(t)$ and $\tilde{q}(t)$ are periodical. Therefore, the speed of the front is given by

$$w = \nu + \frac{\mathrm{tr}\bar{A}}{2\nu}. \qquad (10.44)$$

The spectrum of possible values of the speed depends on the unknown parameter $\nu$. However, the spectrum (10.44) has a lower bound, which corresponds to $\nu_0 = [(\mathrm{tr}\bar{A})/2]^{1/2}$. Thus, the expression for the lower bound is

$$w_{min} = (2\,\mathrm{tr}\bar{A})^{1/2}. \qquad (10.45)$$

Note that (10.45) is found analytically without any additional assumptions and without any reference to the particular parametrization of the trophic responses in Equations (10.6)–(10.7)).

Having an expression for $w_{min}$ at hand, now we can return to the critical relation (10.33). Setting $c = c_{min}$ and $w = w_{min}$ and making use of (10.45) and (10.31), after a little algebra from (10.33) we finally obtain

$$\frac{p}{1-p}[(1-h) - p(1+h)] = -2k\left(p - \frac{1}{1+h}\right). \qquad (10.46)$$

The critical relation (10.46) is shown in Figure 10.8 by the dashed-and-dotted curves (obtained for different $k$). The parameter domain in the $(p,h)$ plane, where the relation between the speeds allows for dynamical stabilization to occur, is on the left of a given dashed-and-dotted curve and below the Hopf bifurcation curve 2.

Obviously, in order to make a prediction of dynamical stabilization more complete, the critical relation obtained from a comparison of speeds should be taken together with the one obtained through the bifurcation analysis; see the previous section. The remaining domain is not that large (between curves 2 and 5, on the left of curves 6 or 7). However, it should be mentioned that, while the position of curve 2 does not depend on $k$ and the position of curve 5 depends on $k$ only slightly, the position of the curve defined by (10.46) depends on $k$ significantly. With an increase in $k$, the critical curve (10.46) moves to the right; as an example, curve 7 shows its position for $k = 1.2$. As a whole, the parameter domain corresponding to dynamical stabilization tends to increase along with $k$.

## 10.3 Patterns in a competing species community

Predation is a very common type of species interaction but surely not the only possible one. Another ecologically relevant and important type is given by competition. In this section, we consider the dynamics of a community of competing species. Our goal is to find out what kind of spatiotemporal patterns can be observed in that system, and what conditions/restrictions are required for these patterns to emerge.

The first observation we want to make here is that, unlike the case of prey–predator systems, a system of two competing species is not capable of forming a pattern more interesting than a traveling monotonous population front (Namba and Mimura, 1980; Shigesada and Kawasaki, 1997). Therefore, in order to explore the possible scenarios of pattern formation in a competing community, we have to consider a somewhat more complicated model.

Apparently, the next level of complexity is a three-species community. Spatiotemporal dynamics of three competitive species is described by the following equations:

$$\frac{\partial U_1(X,T)}{\partial T} = D_1 \frac{\partial^2 U_1}{\partial X^2} + A_1(1 - r_{11}U_1 - r_{12}U_2 - r_{13}U_3)U_1 , \quad (10.47)$$

$$\frac{\partial U_2(X,T)}{\partial T} = D_2 \frac{\partial^2 U_2}{\partial X^2} + A_2(1 - r_{21}U_1 - r_{22}U_2 - r_{23}U_3)U_2 , \quad (10.48)$$

$$\frac{\partial U_3(X,T)}{\partial T} = D_3 \frac{\partial^2 U_3}{\partial X^2} + A_3(1 - r_{31}U_1 - r_{32}U_2 - r_{33}U_3)U_3 \quad (10.49)$$

(May, 1973; Hofbauer and Sigmund, 1988), where $U_1$, $U_2$, and $U_3$ are the species densities at time $T$ and position $X$; coefficient $A_i$ gives the intrinsic growth rate of the $i$th species; coefficients $r_{ij}$ describe intraspecific (for $i = j$) and interspecific (for $i \neq j$) competition, respectively. Coefficients $D_i$ describe the intensity of spatial mixing. Due to their biological meanings, all the parameters in (10.47)–(10.49) are nonnegative.

The system (10.47)–(10.49) is a system of nonlinear partial differential equations, for which effective analytical tools are largely lacking. Therefore, as in most parts of this book, an insight into the system properties is done via results of numerical simulations. On the other hand, Equations (10.47)–(10.49) contain a large number of parameters, which makes its detailed study virtually impossible. Therefore, our first task is to lessen the number of parameters.

One way to do this is to choose dimensionless variables. Namely, let us consider

$$u_i = r_{ii} U_i \ (i=1,2,3), \quad t = A_1 T, \quad x = \left(\frac{A_1}{D_1}\right)^{1/2} X. \tag{10.50}$$

Then, from (10.47)–(10.49) we obtain

$$\frac{\partial u_1}{\partial t} = \frac{\partial^2 u_1}{\partial x^2} + (1 - u_1 - r_{12} u_2 - r_{13} u_3) u_1, \tag{10.51}$$

$$\frac{\partial u_2}{\partial t} = \epsilon_2 \frac{\partial^2 u_2}{\partial x^2} + a_2 (1 - r_{21} u_1 - u_2 - r_{23} u_3) u_2, \tag{10.52}$$

$$\frac{\partial u_3}{\partial t} = \epsilon_3 \frac{\partial^2 u_3}{\partial x^2} + a_3 (1 - r_{31} u_1 - r_{32} u_2 - u_3) u_3, \tag{10.53}$$

where $\epsilon_2 = D_2/D_1$, $\epsilon_3 = D_3/D_1$, $a_2 = \epsilon_2/\epsilon_1$, and $a_3 = \epsilon_3/\epsilon_1$. In terms of original system (10.47)–(10.49), it means that some of the parameters in the equations can be set to unity without any loss of generality.

Furthermore, here we restrict our study to a special "cyclic" type of interspecific competition originally introduced by May and Leonard (1975) and assume that $r_{13} = r_{21} = r_{32} = \beta$ and $r_{23} = r_{31} = \gamma$. Thus we arrive at the following system:

$$\frac{\partial u_1}{\partial t} = \frac{\partial^2 u_1}{\partial x^2} + (1 - u_1 - \alpha u_2 - \beta u_3) u_1, \tag{10.54}$$

$$\frac{\partial u_2}{\partial t} = \epsilon_2 \frac{\partial^2 u_2}{\partial x^2} + a_2 (1 - \beta u_1 - u_2 - \gamma u_3) u_2, \tag{10.55}$$

$$\frac{\partial u_3}{\partial t} = \epsilon_3 \frac{\partial^2 u_2}{\partial x^2} + a_3 (1 - \gamma u_1 - \beta u_2 - u_3) u_3, \tag{10.56}$$

where we denote $\alpha = r_{11}$ to keep notations homogeneous. A fully cyclic competition would correspond to $\alpha = \gamma$; however, in order to make the system dynamics richer, we keep the possibility that $\alpha \neq \gamma$.

As we did in the case of prey–predator system, we begin with the corresponding nonspatial system:

$$\begin{aligned} \frac{du_1}{dt} &= (1 - u_1 - \alpha u_2 - \beta u_3) u_1, \\ \frac{du_2}{dt} &= a_2 (1 - \beta u_1 - u_2 - \gamma u_3) u_2, \\ \frac{du_3}{dt} &= a_3 (1 - \gamma u_1 - \beta u_2 - u_3) u_3. \end{aligned} \tag{10.57}$$

The system (10.57) has been studied in much detail by May and Leonard (1975) and Hofbauer and Sigmund (1988); here we only briefly recall those of its properties that are required for better understanding the spatiotemporal dynamics.

It is readily seen that system (10.57) can have as many as eight steady states. The number of steady states in the positive octant $\mathbf{R}_+^3 = \{(u_1, u_2, u_3) \mid u_i \geq 0,\ i = 1, 2, 3\}$ of the phase space can be different for different parameter values, and, in a general case, the structure of the phase space is rather complicated. However, assuming additionally that

$$\alpha = \gamma \quad \text{and} \quad \alpha + \beta > 2,\ \alpha > 1 > \beta, \tag{10.58}$$

it can be shown that the system (10.57) possesses exactly five stationary states. Namely, there are the trivial "no-species" state $(0, 0, 0)$ and three semi-trivial "one-species-only" states $(1, 0, 0)$, $(0, 1, 0)$, and $(0, 0, 1)$. There is also one coexistence steady state $(\bar{u}_1, \bar{u}_2, \bar{u}_3)$, the stationary value of the species concentrations being the solution of the following system:

$$\bar{u}_1 + \alpha \bar{u}_2 + \beta \bar{u}_3 = 1,\quad \beta \bar{u}_1 + \bar{u}_2 + \gamma \bar{u}_3 = 1,\quad \gamma \bar{u}_1 + \beta \bar{u}_2 + \bar{u}_3 = 1. \tag{10.59}$$

The trivial steady-state is always an unstable node and the semi-trivial states are saddle-points. Furthermore, under restrictions (10.58), the coexistence steady state is always a saddle-point. May and Leonard (1975) showed that, in this case, the only attractor in the phase space is a heteroclinic cycle consisting of the three "one-species-only" steady states and the trajectories connecting these states. In the $t \to +\infty$ limit, any trajectory in $\mathbf{R}_+^3$ ends at the heteroclinic cycle regardless of the initial conditions, which means that the system (10.57) is not permanent. From the ecological standpoint, it means that in a spatially homogeneous community at least one of the species inevitably goes extinct.

The properties of the system (10.57) change significantly when the restriction $\alpha = \gamma$ is relaxed. Assuming additionally that $\det R > 0$, where $R = (r_{ij})$, a test of the system permanence is

$$(\gamma - 1)^2 (\alpha - 1) < (1 - \beta)^3 \tag{10.60}$$

(for details, see Hofbauer and Sigmund, 1988), which evidently could not be satisfied under restrictions (10.58). For $\alpha \neq \gamma$, though, condition (10.60) may determine a nonempty set in the parameter space of the system. Taking into account that under condition $\det R > 0$ the solutions of the system (10.54)–(10.56) are uniformly bounded, permanence means the existence of an attractor belonging to the interior of $\mathbf{R}_+^3$. The nature of the attractor then depends on the stability of the coexistence state $(\bar{u}_1, \bar{u}_2, \bar{u}_3)$. Since it is the only steady state in $\mathbf{R}_+^3$, in the case that it is unstable, a stable limit cycle is expected to emerge. Following this chain of arguments, the existence of a stable limit cycle in the phase space of the system (10.57) has been proved by K. Kawasaki,[1] and the corresponding parameter range was identified. It should yet be mentioned that the system appears to be rather sensitive to a

---
[1] Personal communication.

variation of parameter values and the parameter range where the limit cycle exists is rather narrow; outside of that range the heteroclinic cycle is the attractor.

Now we proceed to the dynamics of the spatial system (10.54)–(10.56). Since in this chapter we are mostly interested in patterns generated by propagating population fronts, we use initial conditions described by functions of compact support. Specifically, the results shown below are obtained for

$$u_i(x,0) = u_{i0} \text{ if } |x| \leq \frac{d_i}{2}, \quad u_i(x,0) = 0 \text{ if } |x| > \frac{d_i}{2} \quad (10.61)$$

($i = 1, 2, 3$), where $u_{i0}$ and $d_i$ are the concentration of the $i$th species inside the originally invaded domain and the size (diameter) of the domain, respectively. To complete the mathematical formulation of the model, the boundary conditions are chosen to be zero-flux.

Some typical snapshots of the system dynamics are shown in Figures 10.9 to 10.11. (Note that the problem (10.54)–(10.56) with (10.61) is symmetrical with respect to the origin and only half of the domain is shown.) At an early stage, a succession of traveling population fronts is formed; see Figure 10.9. Species 3 wins the competition for open space and it spreads faster than the others. Eventually, however, it looses and is displaced by species 1, being actually put at the brink of extinction – note the very low population density of species 3 in the middle of the domain in Figure 10.9, bottom. Species 2 appears to be the weakest in the competition for space; it is the last to come into play and it spreads slower than the other two. Interestingly, species 2 takes some of the pressure from species 1 and saves it from extinction: In the wake of the last front, all three species coexist.

Each of the fronts has a complex shape formed by regular damped population oscillations. Comparison between species distributions obtained for different moments (e.g., cf. Figure 10.9, middle, and Figure 10.10, top) shows that each front propagates as a stationary traveling wave, so that the form of the oscillations does not change with time and the whole structure moves to the right with a constant speed. Apparently, this is a phenomenon of the same origin as the one observed in the prey–predator system; see Figure 10.2.

Also at a later stage, the system properties appear to be qualitatively similar to those of the prey–predator system when its local dynamics is oscillatory. In the wake of the fronts, after decay of the damped oscillations, dynamical stabilization of the unstable equilibrium $(\bar{u}_1, \bar{u}_2, \bar{u}_3)$ may take place; see Figure 10.9. The length of the unstable plateau can increase up to a very large value so that the species distribution becomes homogeneous nearly over the whole domain; cf. Figure 10.10a. In order to estimate the rate of the plateau growth and to identify the parameter range where dynamical stabilization can or cannot be observed, the method of speeds comparison may be applied; see the previous section. However, since the expressions for the fronts speed (more precisely, for the lower bounds) in the three-species system are rather

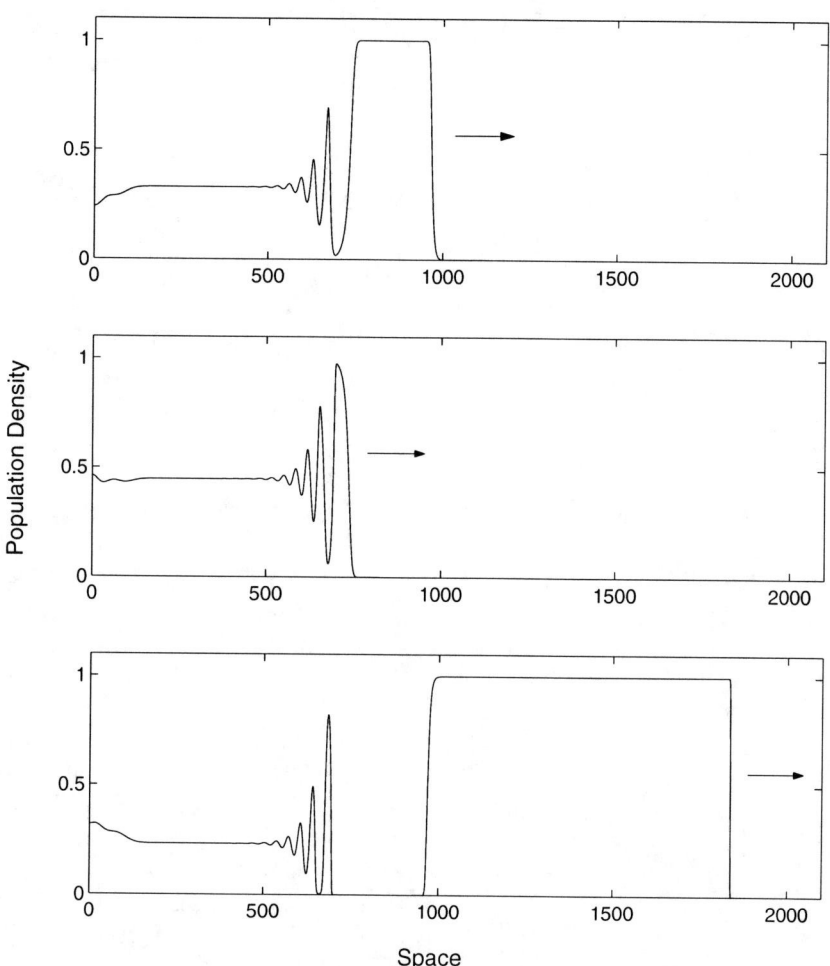

**FIGURE 10.9**: The spatial distribution of the species 1 (top), the species 2 (middle), and the species 3 (bottom) obtained at $t = 800$ for $\alpha = 1.08$, $\beta = 0.8$, $\gamma = 1.24$, $a_2 = 1.0$, $a_3 = 5.0$, $\epsilon_2 = 1.0$, $\epsilon_3 = 0.2$ and initial conditions (10.61) with $u_{10} = 1.1$, $u_{20} = 1.2$, $u_{30} = 1.3$, $d_{1,2,3} = 100$. For these parameters, the coexistence state is unstable and "surrounded" by a stable limit cycle. The arrows indicate the direction of fronts propagation. Since the problem is symmetrical with respect to the origin, only one half of the domain is shown, i.e. for $x \geq 0$.

complicated (Petrovskii et al., 2001), the critical relation appears to be very cumbersome.

Behind the unstable plateau, irregular spatiotemporal population oscillations emerge that eventually occupy the whole domain; compare the left-hand

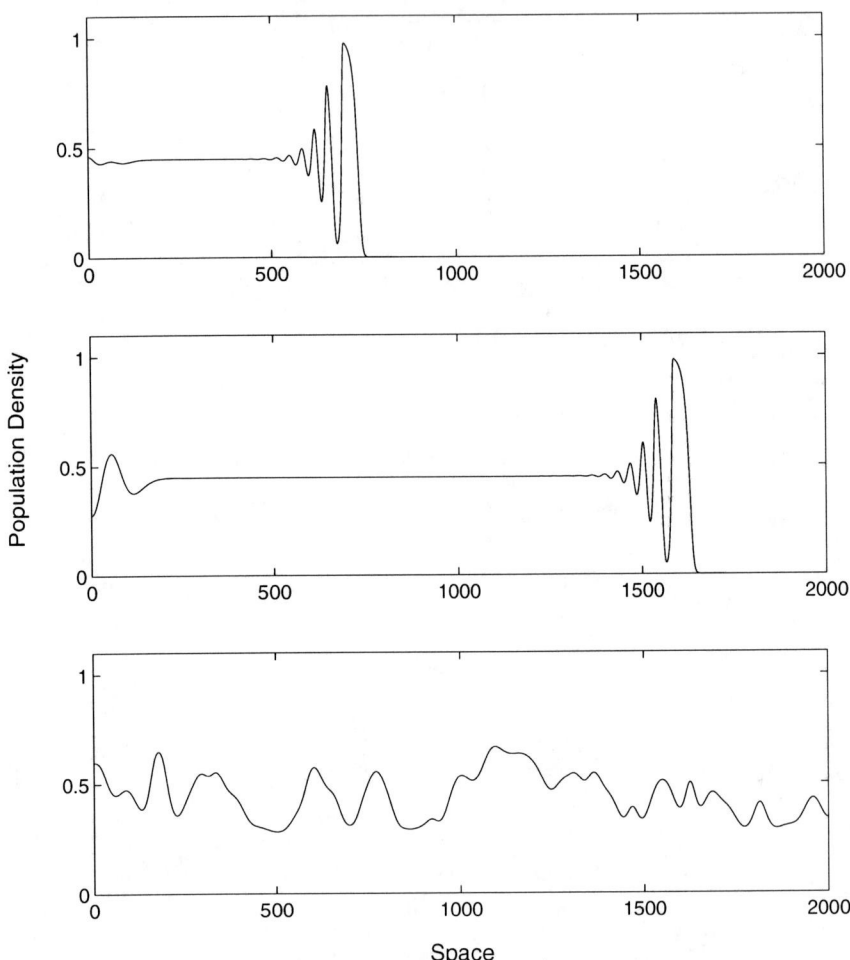

**FIGURE 10.10**: Snapshots of the spatial distributions of the species 2 for $t = 800$ (top), $t = 1800$ (middle), and $t = 16600$ (bottom). Parameters are the same as in Figure 10.9. Note that the plateau behind the oscillating front corresponds to the locally unstable equilibrium $(\bar{u}_1, \bar{u}_2, \bar{u}_3)$.

side of Figures 10.9 and 10.10a with Figure 10.10b. An important observation to be made here is that, after the oscillations invade over the domain, the temporal variations of the population densities also become remarkably irregular. A question naturally arises regarding the degree of this irregularity. This issue was addressed by Petrovskii et al. (2001) by means of checking the sensitivity of solutions to a small perturbation of the initial conditions. It was found that, while at the early stage of system evolution the perturbed and unperturbed solutions are very close to each other, starting from a certain moment the discrepancy begins to grow very fast (exponentially) so that the difference between the solutions soon becomes of the same order as the solutions themselves. This sort of behavior is a fingerprint of chaos (Nayfeh and Balachandran, 1995). Therefore, the dynamics of the system (10.54)–(10.56) after excitation of irregular spatiotemporal oscillations can be classified as chaotic. The dominant Lyapunov exponent is, however, rather small and is roughly estimated to be on the order of $10^{-4}$.

Numerical simulations show that the formation of chaotic spatiotemporal patterns at a later stage of the system dynamics is a typical phenomenon in the sense that it is robust to variation of the system parameters in a wide range, even if the variation results in a change of the phase space structure. Indeed, while the patterns in Figure 10.10 are obtained for stable limit cycle being the attractor, Figure 10.11 shows the patterns obtained in the case that the criterion (10.60) of permanence is broken and the hetroclinic cycle is an attractor. No unstable plateau is formed in this case, and the chaotic region starts immediately after the regular oscillations at the front.

It should also be mentioned that, as well as in the prey–predator system, the chaotic oscillations are self-sustained. Large-time numerical simulations show that, after the onset of chaos, the system dynamics does not undergo any qualitative changes.

Therefore, in this section we showed that, in a three competing species system where the local kinetics is given either by a heteroclinic attractor or by a stable limit cycle, population fronts are followed by the formation of an unstable plateau and/or by excitation of chaotic spatiotemporal population oscillations. These results provide an important extension to the analysis and inferences of the previous sections showing that the formation of complex spatiotemporal patterns is a more general phenomenon and should not be necessarily attributed to the prey–predator interactions. Even more importantly, they should not be attributed to the existence of a stable limit cycle either: Essentially the same type of system's dynamics is observed in the case of an heteroclinic attractor.

One ecologically important conjecture of this study must not be overlooked. Let us recall that, in the case of an heteroclinic attractor, in the nonspatial system the system's trajectory is gradually approaching the boundary of the first octant in the phase space. From the ecological standpoint, it would obviously correspond to species extinction because the system moves through a succession of states with very low population densities (Gilpin, 1972; May, 1972a). Moreover, a species would go extinct without any chance of re-colonization because, due to spatial homogeneity, its population density falls to a dangerously low value simultaneously all over the domain. However, the situation appears to be different in the spatial system: Due to homogeneity breakdown and the subsequent pattern formation, no extinction takes place and all three species coexist throughout the domain; cf. Figure 10.11. Therefore, the predictions obtained from spatial and nonspatial models are qualitatively different. This is a new feature, which we have not observed for the prey–predator system

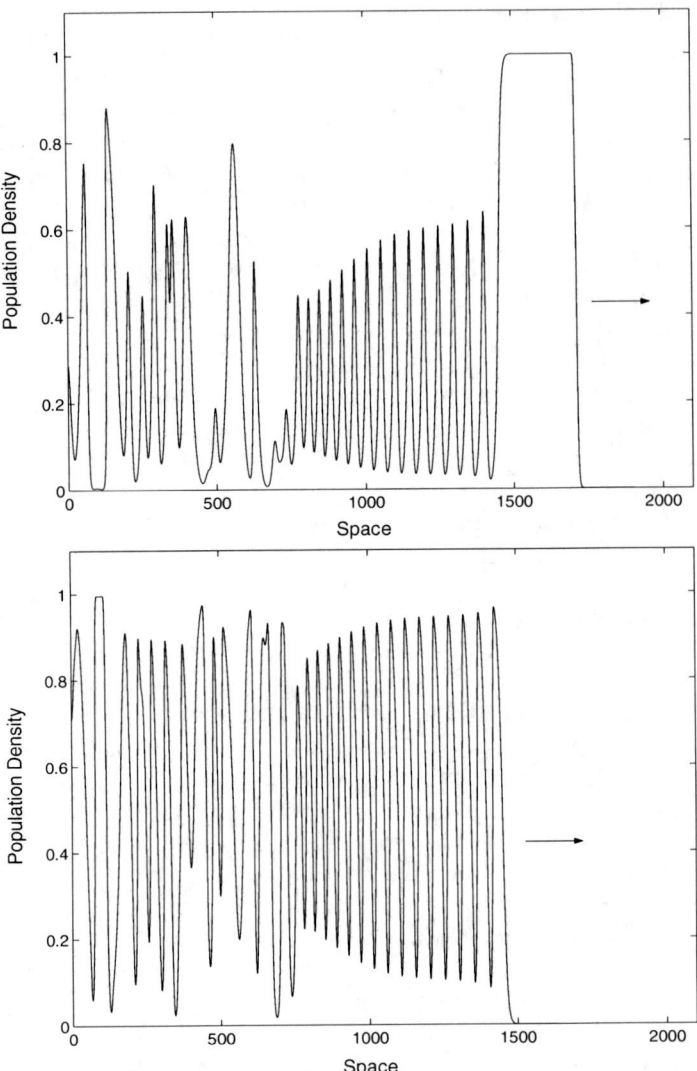

**FIGURE 10.11**: The spatial distributions of the species 1 (top) and the species 2 (bottom) obtained at $t = 3200$ for $\beta = 0.95$; other parameters are the same as in Figure 10.9. The system is not permanent and the heteroclinic cycle is the attractor.

considered in the previous sections. In Chapter 12, we will show that a similar situation may occur for a prey–predator system when the prey growth is hampered by the Allee effect.

## 10.4 Concluding remarks

We showed that, in a conceptual population community such as a prey–predator system or a system of three competing species, a traveling population front can generate a variety of patterns. The system dynamics is especially rich when the coexistence state is unstable. A typical succession of patterns then looks as follows. Damped oscillations at the front are followed by a plateau emerging as a result of dynamical stabilization of the unstable coexistence state. The plateau is followed by a periodic wavetrain. In the wake of the periodic wave, irregular spatiotemporal oscillations gradually develop. Remarkably, even in the case when the periodic wave is stable with respect to small local perturbations, it often appears to be unstable globally, so that the periodic wavetrain is gradually displaced by irregular oscillations. Having once appeared, the irregular oscillations eventually occupy the whole domain. Depending on parameter values, the plateau and/or the periodical pattern can be missing; however, the irregular spatiotemporal oscillations always arise in the wake of invasion.

Note that the main properties of the system dynamics appear to be robust to details of the phase space structure and, contrary to a widely accepted opinion (Kuramoto, 1984), should not be tightly related to the existence of a stable limit cycle; essentially the same pattern emerges if the attractor is given by a heteroclinic cycle. Moreover, an insight into mathematically similar models of chemical systems shows that damped oscillations at the propagating front, dynamical stabilization, and excitation of irregular oscillations in the wake can be observed as well even in the case that the attractor is just as simple as a stable node, provided the coexistence state is unstable (Malchow and Petrovskii, 2002).

After excitation of irregular spatiotemporal oscillations, the dynamics of the system becomes chaotic. It should be mentioned here that the term "chaos" appeared originally in relation to the temporal behavior of spatially homogeneous systems (Lorenz, 1963; Li and Yorke, 1975; Gilpin, 1979; Hastings and Powell, 1991). In our case, however, the phase space of the homogeneous system does not contain a strange attractor but only a (stable) limit cycle or heteroclinic cycle. Chaotic dynamics becomes possible as a result of homogeneity breaking and the formation of irregular spatial species distribution. It seems reasonable to distinguish between the situations when a given system may or may not exhibit chaos in the nonspatial case. In order to stress the importance of the spatial dimension of the system, to refer to the chaotic patterns considered in this chapter the term "spatiotemporal chaos" is more appropriate.

As a whole, the dynamics of simple few-species model systems continuous in space and time appears to be remarkably rich and capable of generating realistically looking complicated irregular patterns. We want to note that appreciation of this fact once became a real breakthrough in understanding population dynamics (Pascual, 1993; Sherratt et al., 1995). Until the mid-nineties, irregularities were normally associated with the discreteness of the dynamics either in time, cf. "populations with nonoverlapping generations" (May, 1974), or space (Hassell et al., 1991), rather than with intrinsic instabilities of continuous systems.

In this chapter, we focused on patterns generated by propagating population fronts. Excitation of chaotic oscillations in a prey–predator system used to be essentially attributed to the front wakes (Sherratt, 1994b, 1998) and, correspondingly, to the initial conditions of special type (e.g., finite) generating the fronts. In their influential paper, Sherratt, Lewis, and Fowler (1995) insisted that formation of chaotic patterns in the wake of invasion is a "fundamentally different mechanism" from, for instance, the one considered by Pascual (1993) where spatiotemporal chaos was triggered by a small constant-gradient heterogeneity in the prey growth rate. Later studies, however, showed that it is not so. In the next chapter, we will show that onset of spatiotemporal chaos and the corresponding formation of distinctly heterogeneous, patchy population distribution is a much more general phenomenon and must not be necessarily attributed to propagating fronts.

A question yet remaining open is how the system dynamics may change if the population growth is affected by the Allee effect. Indeed, even in the simplest case of a single-species system continuous in space and time, the system's properties are significantly different depending on whether the Allee effect is present or not (cf. Lewis and Kareiva, 1993). This issue will be addressed in details in Chapter 12. We will show there that the impact of the Allee effect makes the system dynamics even richer, and, along with the patterns already considered here, it results in a new phenomenon in population spread and the corresponding pattern formation.

Since one of the goals of this book is to "bridge the gaps" between ecological and epidemiological models, another question remains as to whether and to what extent the results of this chapter can be extended onto patterns in the dynamics of epidemics. Here the following observation is to be made. The PDE-based diffusion–reaction models of epidemic spread are similar to the population models in the sense that the spatiotemporal dynamics of a given infectious disease arises as an interplay between dispersal/diffusion and local interaction between different subpopulations or groups, e.g., susceptibles and infected. Consequently, as well as in the case of population models, the possible scenarios of pattern formation in the wake of epidemic spread depend very much on the properties of the local kinetics.

Aiming to give more specific examples, the simplest SI model of disease spread with the mass-action law for the disease transmission rate qualitatively coincides with a prey–predator system where predation is of Holling type I, provided only susceptible individuals are capable of reproduction. Therefore, since in this system the endemic/coexistence state is always stable, the spreading epidemic would form either a monotonous traveling front or a front with promptly damped oscillations in the wake; cf. Figure 10.2. Numerical simulations in the SI model agree with these predictions perfectly (Murray, 1989). Furthermore, it is readily seen that introduction of the Allee effect into the reproduction rate (for more details see Chapter 12) can destabilize the endemic state and thus makes possible the formation of complicated irregular spatiotemporal patterns in the wake (Petrovskii et al., 2005).

In general, however, epidemic models tend to have local kinetics more complicated than that of population dynamics. One reason is that the above assumption that only susceptibles contribute to population growth actually applies only to a narrow range of diseases (e.g., to those affecting the reproduction system). For most other diseases, contribution of infected to population multiplication cannot be neglected. Obviously, it immediately results in a much more elaborated reproduction term. Another reason can be found in the application of more complicated schemes of disease transmission, such as proportionate mixing or even a combination of the mass-action law and the proportionate mixing (cf. Fromont et al., 1998).

The outcome of the interplay between different factors is very difficult to predict, though. Having been taken separately, the Allee effect, the proportionate mixing, and the contribution of infected to reproduction tend to make the kinetics more complex and the system dynamics richer. Surprisingly, when they are all taken into account, the resultant model may appear to have simpler properties. The simplest SI model with the Allee effect is capable of local limit cycle oscillations and the corresponding formation of complex spatiotemporal patterns; however, an SI model accounting also for the other two factors exhibits nothing more complicated than traveling fronts (Hilker et al., 2007).

# Chapter 11

# Biological turbulence

Spatial distribution of ecological populations is very rarely homogeneous. On the contrary, patches of high population density quite often alternate with patches of low density or even with uninhabited areas. It can occur on different spatial scales and in various environment and ecosystem types. In population biology and spatial ecology, this phenomenon may come under a variety of names such as plant/animal grouping or aggregation, patchiness, etc. In some cases (e.g., see Section 9.3), the corresponding spatial structure exhibits prominent regularity, even if it might be blurred by environmental noise. Much more often, however, populations are distributed irregularly, without any detectable trace of order.

Remarkably, although environmental and/or landscape properties apparently create a certain "frame" for any population heterogeneity, the properties of observed spatial structures often appear to be uncorrelated or only weakly correlated with those of the environment. This fact has been used as a conceptual basis for considering patchiness as a separate phenomenon, appearing more due to biological interactions rather than being straightforwardly reducible to physical/chemical properties of the environmental settings; see Levin (1990, 1992) and Powell (1995). Although patchiness due to environmental forcing is a practically important and interesting phenomenon, its reasons and the corresponding mechanisms are relatively clear; therefore, in this chapter (as indeed in most parts of this book) we will focus more on its possible biological origin.

Apparently, the mechanisms of self-organized biological patchiness can be different depending, for instance, on how complex are the "social" aspects of the population dynamics of a given species. Perhaps the very first example of animal aggregation that comes into mind is the formation of flocks, herds, schools, swarms, etc. In particular, fish schooling has been paid a lot of attention for obvious commercial reasons. However, here we are more interested in patterns that arise irrespectively of whether the given species may or may not possess this type of behavioral response. It means that the relevant spatial scale should be large enough so that the population density would not be affected by the small-scale spatial variations due to flock/swarm formation. Spatial heterogeneity that appears largely as a result of schooling behavior lies beyond the scope of this book; a reader interested in this issue is advised to use extensive existing literature, e.g. see Radakov (1973); Steele (1977);

Okubo (1986); Huth and Wissel (1994); Reuter and Breckling (1994); Niwa (1996); Parrish et al. (1997); Stöcker (1999); Parrish and Edelstein-Keshet (1999); Saffre and Deneubourg (2002).

As was mentioned above, specific mechanisms and the corresponding theoretical background for understanding self-organized ecological patterns may depend significantly on the environmental properties, in particular, on the degree of its heterogeneity. In a strongly fragmented landscape, the species areal consists of an ensemble of sites or habitats. In case their typical size is small enough so that the population distribution inside each of them can be considered as homogeneous (up to inevitable small fluctuations of purely stochastic origin), the spatial structure is predetermined and the properties of the spatiotemporal dynamics can then be expressed in terms of synchronization between population fluctuations at different sites. A relevant mathematical model should be space-discrete and may be given by coupled systems of ODEs, with each system describing the population dynamics inside the corresponding habitat. We mention here the work by Jansen (1995, 2001), who showed that the local prey–predator oscillatory dynamics becomes desynchronized as soon as the inter-habitat coupling becomes sufficiently weak. As a result of weak coupling, the local dynamics of the subpopulations become chaotic even if the intrinsic dynamics of the corresponding local populations is periodic.

These and other similar results, although very important for understanding the population dynamics on a metapopulation scale, give little information with regards to what can be the mechanisms of self-organized patchiness in a relatively homogeneous environment (e.g., within a given habitat, provided its size is large enough). In that case, a mathematical description continuous in space and time seems to be the most adequate, which results in diffusion–reaction equations. In the previous chapter, using a prey–predator system as a paradigm, we showed that formation of spatially irregular chaotic patterns in such systems can be triggered by biological invasion, i.e., by propagation of a population front emerging from finite initial conditions. Indeed, earlier work on ecological pattern formation essentially attributed self-organized heterogeneous populations distribution[1] to the wakes of propagating fronts (Mark Lewis, personal communication; also Sherratt (1994a); Sherratt et al. (1995, 1997)). It appears, however, that pattern formation in a spatial prey–predator system is a more general, inherent phenomenon and must not necessarily be associated with initial conditions of a special type or any predefined environmental heterogeneity (Petrovskii and Malchow, 1999; Petrovskii and Malchow, 2001b). Rather, it reflects a certain intrinsic instability of a spatially explicit prey–predator system. Remarkably, however, contrary to corresponding non-spatial systems, where an instability may eventually lead to species extinction

---

[1] In an unstructured environment; phenomenon of self-organized patchiness triggered by (but not correlated with) an environmental heterogeneity of special type was first observed and studied in detail in influential papers by Pascual (1993) and Pascual and Caswell (1997).

(Rosenzweig, 1971; Gilpin, 1972), this instability of spatiotemporal dynamics appears to be beneficial for the community functioning enhancing global species persistence (Petrovskii et al., 2004).

The chapter is organized as follows. In Section 11.1, we give a detailed description of the self-organized patchiness – "biological turbulence" – focusing on the generic scenario of pattern formation and transition to spatiotemporal chaos in a prey–predator system. In Section 11.2, we make a stronger accent on the spatial aspects of the system dynamics after the onset of chaos and also give an extension of the main results onto other similar systems. Section 11.3 makes an insight into possible ecological consequences of the phenomenon. The last section provides a discussion of the main results.

## 11.1 Self-organized patchiness and the wave of chaos

The scenario of pattern formation resulting in a remarkably irregular patchy spatial structure and transition to chaos, which will be a focus of this section, appear to be generic for a wide class of diffusion–reaction systems where local kinetics is oscillatory. Here "oscillatory" is used in a wider sense and does not necessarily imply the existence of a stable limit cycle in the phase space of the corresponding nonspatial system; in fact, having an unstable focus may be enough. The case when the local kinetics given by a limit cycle can be treated in terms of the so-called $\lambda$-$\omega$ systems (Kopell and Howard, 1973) or its generalization, known as the Ginzburg–Landau equation (Kuramoto, 1984), which are widely regarded as adequate models of spatiotemporal chaos (but see the discussion of the issue in Section 11.4). A more general theory is largely missing, though, and many results are still obtained through numerical simulations in specific systems. Correspondingly, in this section we consider a prey–predator system as the simplest example of a population community with oscillatory kinetics. Some generalizations and extensions of the results onto other systems will be made in Section 11.2.

A general model of a time- and space-continuous system of two interacting species is given by the following equations:

$$\frac{\partial U(\mathbf{R},T)}{\partial T} = D\nabla^2 U + \phi(U,V) , \qquad (11.1)$$

$$\frac{\partial V(\mathbf{R},T)}{\partial T} = D\nabla^2 V + \psi(U,V) , \qquad (11.2)$$

where $U$ and $V$ are the species densities, i.e., presently, the population densities of prey and predator, respectively. Since here we are mostly interested in the possibility of non-Turing patterns, we assume that the diffusivity is the same for both species.

Assuming the logistic growth for prey and the Holling type II for predation, and introducing dimensionless variables in a standard way, the system (11.1)–(11.2) takes the following specific form:

$$\frac{\partial u(\mathbf{r},t)}{\partial t} = \nabla^2 u + u(1-u) - \frac{uv}{u+h}, \qquad (11.3)$$

$$\frac{\partial v(\mathbf{r},t)}{\partial t} = \nabla^2 v + k\left(\frac{uv}{u+h} - pv\right), \qquad (11.4)$$

cf. equations (10.3) to (10.7) of the previous chapter.

The properties of the corresponding nonspatial model were considered in detail in Section 10.1, where it was shown that the only coexistence steady state $(\bar{u}, \bar{v})$ is unstable for $h < (1-p)/(1+p)$ and stable (when feasible) otherwise. When it is unstable, it is surrounded by a stable limit cycle. Since a possibility of the Turing instability has been ruled out, one can hardly expect pattern formation in the case when $(\bar{u}, \bar{v})$ is stable. Therefore, in the computer simulations below we will focus on the parameter range where the steady state is unstable.

We start our insight into the spatiotemporal dynamics of the system (11.3)–(11.4) by considering the one-dimensional case so that $u = u(x,t)$, $v = v(x,t)$, $\nabla^2 = \partial^2/\partial x^2$, and $0 < x < L$, $L$ being the domain size.

Obviously, the spatiotemporal dynamics of the system depends to a large extent on the choice of initial conditions. Contrary to the previous chapter, here we are interested in a situation when, at the beginning, both populations are present over the whole area. In a real ecosystem, the details of the initial spatial distribution of the species can be caused by quite specific reasons. The simplest and, in some sense, most general form of the allocated initial distribution would be spatially homogeneous initial conditions. However, in that case, the distribution of the species would stay homogeneous for any time, and no spatial pattern can emerge. To get a nontrivial spatiotemporal dynamics, one has to perturb the homogeneous distribution.

We begin with a hypothetical "constant-gradient" distribution:

$$u(x,0) = \bar{u}, \qquad (11.5)$$
$$v(x,0) = \bar{v} + \epsilon x + \delta, \qquad (11.6)$$

where $\epsilon$ and $\delta$ are parameters. Since the system (11.3)–(11.4) is invariant with respect to transformation $x \to -x$, we reduce our study to the case $\epsilon > 0$ without any loss of generality.

It appears that the type of the system dynamics depends significantly on $\epsilon$ and $\delta$. In case $\epsilon$ is small and $\delta$ is positive, the initial conditions (11.5)–(11.6) evolve to a smooth nonmonotonic spatial distribution of species; see Figure 11.1. The spatial distributions gradually vary in time; the local temporal behavior of the dynamical variables $u$ and $v$ is strictly periodical following the limit cycle of the nonspatial system.

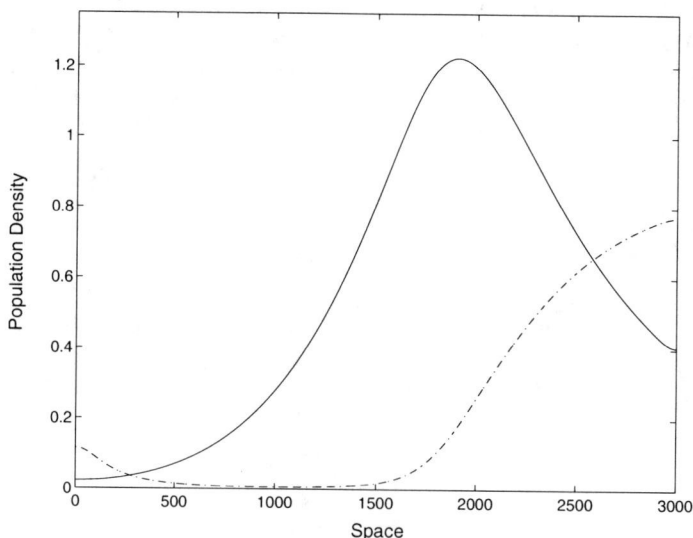

**FIGURE 11.1:** Population distribution over space at $t = 1200$ (solid curve for predator, dashed for prey) obtained for parameters $k = 2.0$, $p = 0.35$, $h = 0.3$ and the initial conditions (11.5)–(11.6) with $\epsilon = 10^{-5}$, and $\delta = 10^{-2}$).

A periodic temporal behavior together with a smooth spatial population distribution, like the one shown in Figure 11.1, is perhaps what is intuitively expected from the system (11.3)–(11.4). Indeed, in the field of theoretical ecology, a two-species diffusion–reaction system had long been regarded as too simple to be capable of producing anything more complicated than a regular pattern with regular dynamics, and that was one of the motivations that led researchers to build more complicated models.

However, for a slightly different set of parameters, the dynamics of the system undergoes principal changes; see Figure 11.2. In this case, the initial distribution (11.5)–(11.6) leads to the formation of a strongly irregular "jagged" dynamic pattern inside a subdomain of the system. The size of the region occupied by this pattern steadily grows with time (cf. top and bottom of Figure 11.2), so that finally irregular spatiotemporal oscillations invade over the whole domain.

Also, the temporal behavior of the densities $u$ and $v$ becomes completely different. Figure 11.3 shows the "local" phase plane of the system obtained at a fixed point $\bar{x}$ inside the region invaded by the irregular spatiotemporal oscillations, the trajectory now filling nearly the whole area inside the limit cycle.

This remarkable irregularity invokes a question of whether it is actually chaotic. In order to clarify the issue, sensitivity of the solutions to a small perturbation has been checked using different measures; for details, see Medvinsky et al. (2002). It was shown that, while during an initial stage of the system dynamics the difference between the perturbed and unperturbed solutions remains small, for a larger time it starts growing promptly so that shortly thereafter the difference becomes on the order of the solutions them-

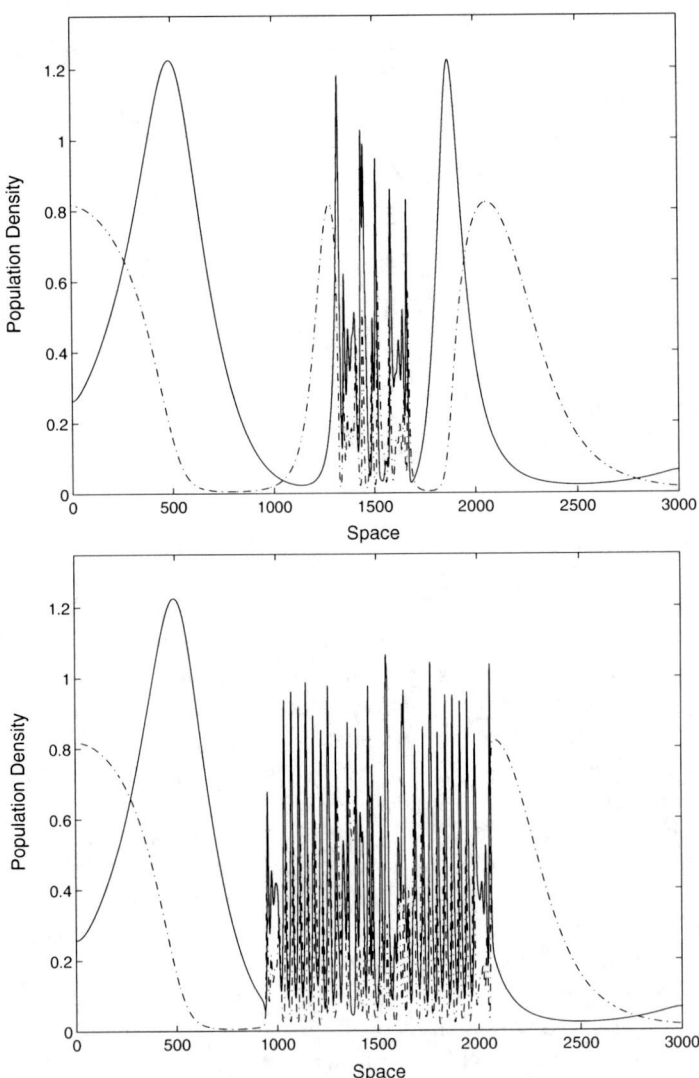

**FIGURE 11.2**: Population distribution over space at $t = 1000$ (top) and $t = 2000$ (bottom) obtained for the initial conditions (11.5)–(11.6) with $\epsilon = 10^{-5}$ and $\delta = -1.5 \cdot 10^{-2}$; other parameters are the same as in Figure 11.1). Solid line for predator, dashed for prey.

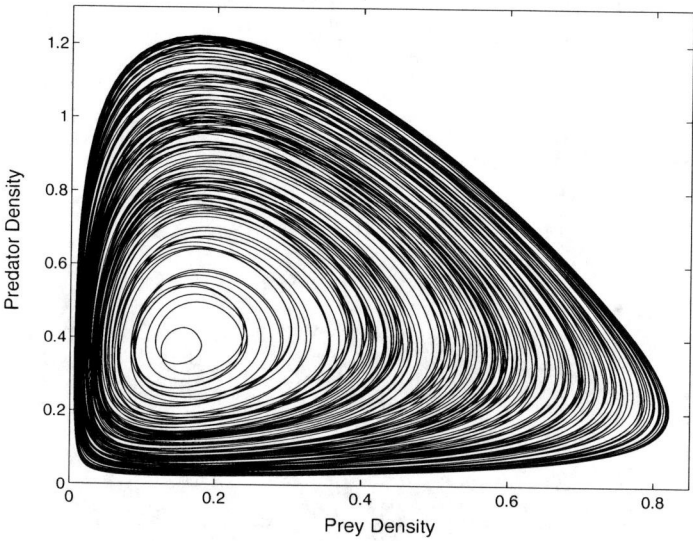

**FIGURE 11.3**: Local phase plane of the system (11.3)–(11.4) after formation of irregular sharp pattern; the trajectory is shown for time interval between $t = 3000$ and $t = 5000$. Parameters are the same as in Figure 11.2. The envelope of the domain filled with the system's trajectory coincides with the limit cycle.

selves; see Figure 11.4. This behavior obviously complies with the definition of chaos. Also, chaos in the dynamics of the jagged pattern can be demonstrated by considering the spectra of the corresponding time series (Petrovskii and Malchow, 1999). Note that chaos in the system (11.3)–(11.4) is essentially spatiotemporal because the nonspatial system can only exhibit a simpler dynamics as given by steady states and limit cycles.

A curious property of the system dynamics is that, for each moment of time, there exist distinct boundaries separating the regions occupied by different dynamic regimes, i.e., by jagged/chaotic and smooth/regular patterns. Both numerical results and analytical estimates (Petrovskii and Malchow, 2001b) show that these boundaries propagate with an approximately constant speed in opposite directions, so that the size of the region with chaotic dynamics is always growing until it occupies the whole domain. Thus, the chaotic regime arises as a result of the propagation of the "wave of chaos," i.e., of the moving interface between the two regions. The scenario is essentially spatiotemporal: The chaos prevails as a result of displacement of the regular regime by the chaotic regime.

An issue to address is what factors determine the position of the "chaotic embrio" when it first appears in the course of the system dynamics. A careful analysis of the computer simulation results show that, in case the initial

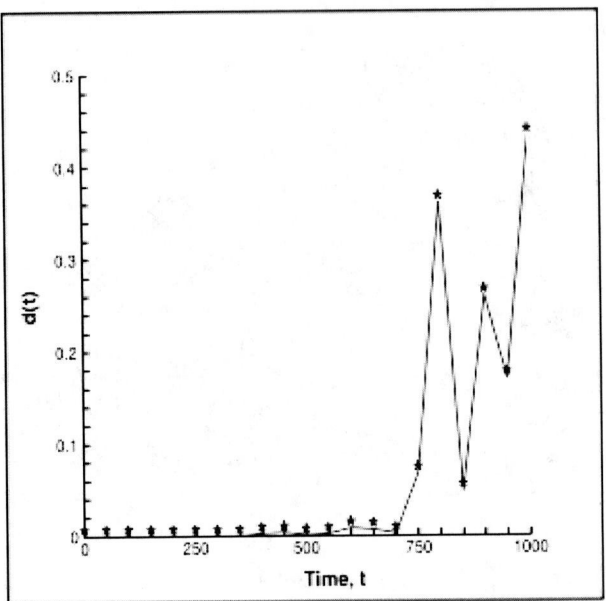

**FIGURE 11.4**: Distance $d(t)$ between perturbed and unperturbed solutions defined as $d(t) = \max_{(x)} |u_{pert}(x,t) - u_{unpt}(x,t)|$, $0 \le x \le L$. With permission from Medvinsky et al. (2002).

species distribution contains a point of "phase singularity" (cf. Kuramoto, 1984), it is in a vicinity of this point that the jagged chaotic pattern first develops. For the initial conditions given by (11.5)–(11.6), such a point $x_*$ is obviously determined as a solution of the following equation:

$$v(x,0) = \bar{v} + \epsilon x_* + \delta = \bar{v} \quad \text{so that} \quad x_* = -\frac{\delta}{\epsilon} \qquad (11.7)$$

(hence the empirical observation that, for $\epsilon > 0$ and $\delta > 0$, the system may sometimes stay in the regular regime; cf. the lines below Equations (11.5)–(11.6)).

The existence of a phase-singularity point is not a necessary condition of the onset of chaos, though. Interestingly, when the initial conditions do not possess such a point (e.g., when $\epsilon\delta > 0$), the chaotic pattern can develop anyway provided $\epsilon$ is not very small. However, in this case, the factors determining the position of the chaotic embrio remain obscure.

The dynamics with more general initial conditions (e.g., nonmonotonic) can be even more complicated, showing a phenomenon that may be called intermittency, when the domains occupied by regular and chaotic patterns alternate in space. Indeed, this sort of spatially intermittent dynamics should be expected if we take into account that nonmonotonic initial conditions may include more than one point of phase singularity and, consequently, may evolve to a few chaotic embrios. As a particular example, we consider the initial conditions in the following form:

$$u(x,0) = \bar{u} , \tag{11.8}$$

$$v(x,0) = \bar{v} + \epsilon x + \delta + \epsilon_1 \cos\left(\frac{2\pi x}{L}\right) . \tag{11.9}$$

In this case, a slightly perturbed homogeneous initial species distribution evolves to a pattern where two domains occupied by the jagged chaotic pattern are separated by regions with a smooth pattern; see Figure 11.2. As in the previous case, the size of the chaotic domains steadily grows, so that they eventually displace the regular dynamics and occupy the whole region.

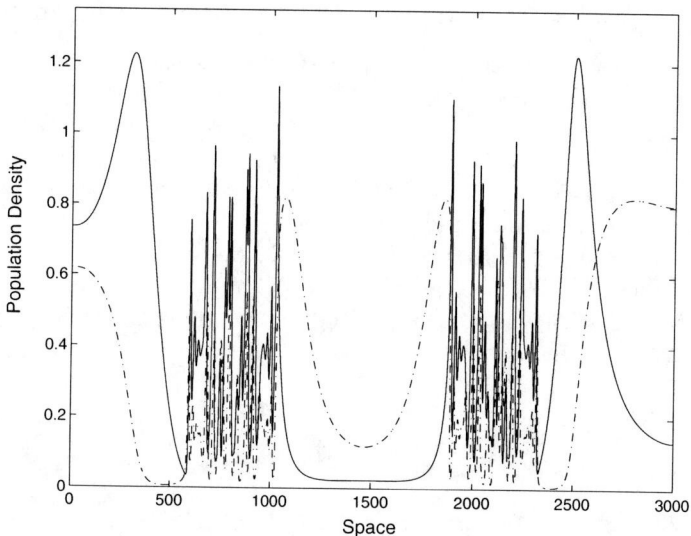

**FIGURE 11.5**: Intermittent population spatial distribution (solid curve for predator, dashed-and-dotted one for prey) obtained at $t = 450$ for slightly heterogeneous non-monotonic initial conditions (11.8)–(11.9) with $\epsilon_1 = 0.07$, $\epsilon = 10^{-5}$, $\delta = 0$, and $L = 3000$. Other parameters are the same as in Figures 11.1 to 11.3.

We want to emphasize, based on extensive evidence from numerous computer simulations, that the onset of chaos in the system (11.3)–(11.4) follows exactly the same scenario for any parameter value when the local kinetics of the system is oscillatory.[2] Therefore, the type of dynamics shown in Figures 11.2 to 11.5 should be regarded as typical rather than exotic.[3]

In addition, we note that apparent asymmetry between $u(x,0)$ and $v(x,0)$ in the initial conditions considered above, when $u(x,0)$ is constant and $v(x,0)$ is perturbed, is only a matter of choice and does not have any special meaning. Other forms of initial conditions, e.g., when $u(x,0)$ is perturbed and $v(x,0)$ is constant, or the distribution is perturbed for both species, lead to qualitatively the same dynamics of the system.

It should also be mentioned that the above results of computer simulations were very carefully checked and tested in order to exclude any significant numerical artifacts. In particular, sensitivity of the results to the size of the mesh steps was checked and they were chosen small enough in order to reach the required degree of approximation and accuracy. The numerical mesh used in the simulations was large enough (typically, with the number of nodes on the order of $10^4$) so that each peak in the chaotic pattern was approximated by a few dozens of nodes.

### 11.1.1 Stability diagram and the hierarchy of regimes

So far we have shown that, for a rather general class of initial conditions, the dynamics of the system (11.3)–(11.4) leads either to the formation of a time-periodical smooth spatial pattern or, via propagation of the "wave of chaos," to the formation of jagged chaotic spatiotemporal patterns. In this section, we will make a further insight into the conditions for the generation of the two regimes and into their persistence.

In order to address the first of these issues, we calculate the stability diagram, i.e., what can be the type of system dynamics in response to a perturbation of a spatially homogeneous initial distribution. For that purpose, we consider the following initial conditions:

$$u(x,0) = \bar{u} , \qquad (11.10)$$

$$v(x,0) = \bar{v} + A \sin\left[\frac{2\pi(x - x_0)}{S}\right] \quad \text{for} \quad x_0 \le x \le x_0 + S,$$

$$v(x,0) = \bar{v} \quad \text{for} \quad x \le x_0 \quad \text{or} \quad x \ge x_0 + S , \qquad (11.11)$$

and check the type of the system dynamics for the parameters $A$ and $S$ varying in a wide range.

The whole parameter plane $(S, A)$ appears to be divided into three regions: see Figure 11.6. If at least one of the parameters is small enough (region I),

---

[2] An interested reader is invited to undertake his/her own study into the system dynamics by using one of the computer codes provided on the attached CD.
[3] As yet another argument, recently, Upadhyay et al. (2007) observed the onset of spatiotemporal chaos through the propagation of the wave of chaos in a system different from (11.3)–(11.4).

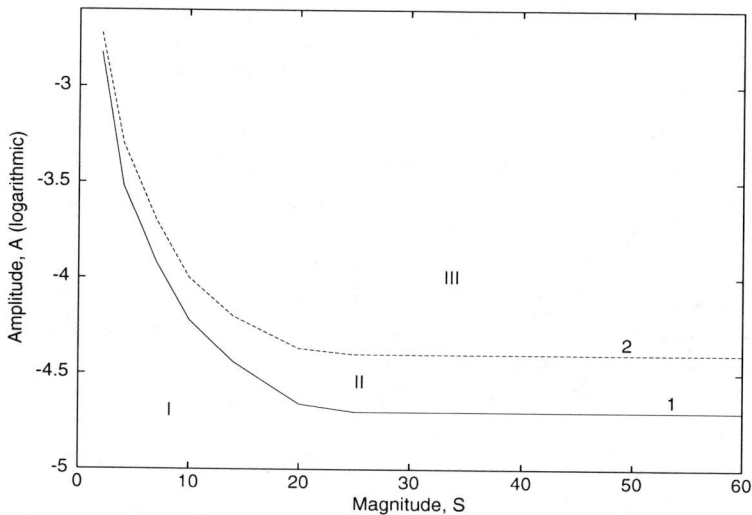

**FIGURE 11.6**: Stability diagram. A sketch of the map in the parametric $(S, A)$ plane (semilogarithmic) for the locally disturbed initial conditions (11.10)–(11.11) obtained for parameters $k = 2.0, p = 0.3$, and $h = 0.4$. Different domains correspond to the excitation of different regimes; see comments in the text. With permission from Petrovskii and Malchow (2001b).

the initial perturbation (11.10)–(11.11) of the stationary homogeneous spatial distribution does not lead to any pattern formation. In this case, the spatially homogeneous distribution is restored and the only consequence is that the local dynamics now becomes periodical corresponding to the limit cycle. If the values of amplitude $A$ and magnitude $S$ are larger (region II), then the initial perturbation leads to excitation of the smooth pattern; see Figure 11.1 as an example. Finally, for even larger values of $A$ and $S$ (region III), the initial condition (11.10)–(11.11) leads to excitation of the jagged chaotic pattern according to the scenario described above. These results are also summarized in the diagram in Figure 11.7; lines 1, 2, and 3 correspond to parameters from domains I, II, and III respectively. Note that the steady-state population densities for the parameters of Figure 11.6 are $\bar{u} = 0.171$ and $\bar{v} = 0.473$ whereas the minimum value of the perturbation amplitude $A$ sufficient to trigger the formation of chaotic pattern is typically quite small, i.e., on the order of $10^{-4}$. Numerical experiments show that, although particular figures can be somewhat different for a different set of parameters, the critical values of $A$ and $S$ remain of the same order.

These results lead to the conclusion that the formation of a jagged irregular dynamical spatial pattern is a rather typical, "natural" phenomenon for a locally oscillating, spatially explicit prey–predator community. The regular dynamics is stable with respect to perturbations of very small amplitude. However, a larger perturbation drives the system to spatiotemporal chaos. Note that, conceptually, it is in very good agreement with the conclusion made earlier regarding stability of periodic wakes (cf. Section 10.1): The regular (periodic) regime may be stable linearly but the basin of stability is very small, so that it is not persistent in computer simulations and would normally give way to chaos due to the impact of perturbation of various origins such as the impact of boundary conditions, imperfect numerical approximation, etc.

Figure 11.6 gives an account of system stability with respect to a small heterogeneous perturbation of a spatially homogeneous, locally unstable steady state. However, the situation when, prior to the perturbation, the system is in an unsteady state may look rather elaborate. Probably more natural would be an option when the initial spatially homogeneous population distribution would correspond to a stable limit cycle. Indeed, even if the homogeneity itself is preserved (cf. domain I in Figure 11.6), the perturbation drives the system from the equilibrium state to homogeneous oscillations. This case has been addressed as well, with similar conclusions about system stability, although it is more difficult to present the results in a compact form because the critical amplitude is different for different oscillation phases, i.e., for different positions of the system on the cycle. On the whole, the critical amplitude now appears to be larger than it was previously (i.e., for the initial conditions corresponding to the homogeneous unstable steady state), with the critical amplitude typically being between $10^{-3}$ and $10^{-2}$. Note that it is still well within a few per cent of the population density steady-state value, which confirms the above conclusion that the spatially homogeneous state is stable linearly[4] but is unstable "globally."

The next point is the persistence of the regimes. In order to address this issue, the system dynamics has been checked in long-term numerical simulations. Our results show that the smooth spatial pattern (see Figure 11.1) is not self-sustained. This regime of the system dynamics is, in fact, the process of a very slow relaxation (with a characteristic time $\tau_{phase} \sim 10^5$ to $10^6$) to the spatially homogeneous temporary periodical solution. The corresponding dynamics looks as follows. First, after a perturbation of the initial homogeneous state, the process of local relaxation to the stable limit cycle takes place (transition 2 in Figure 11.7). In each point $x$, the dynamical variables $u$ and $v$ approach the periodical solution of the homogeneous system. The local relaxation occurs almost simultaneously over the whole domain, and the time scale of this process is $\tau_{ampl} \sim 10^2$ to $10^3$, depending on the problem parameters. This stage can be regarded as the "amplitude relaxation"; as a result, the amplitude of the local oscillations becomes the same. However, the oscillations at different positions take place with different phases. Then, the process

---

[4]In the standard terminology of linear stability theory it would correspond to the "local stability"; however, here we reserve the word "local" for nonspatial systems.

of "phase relaxation" begins, cf. transition 4 in Figure 11.7, which is much slower but eventually leads, after time $\tau_{phase} \gg \tau_{ampl}$, to synchronization of the local oscillations all over the domain.

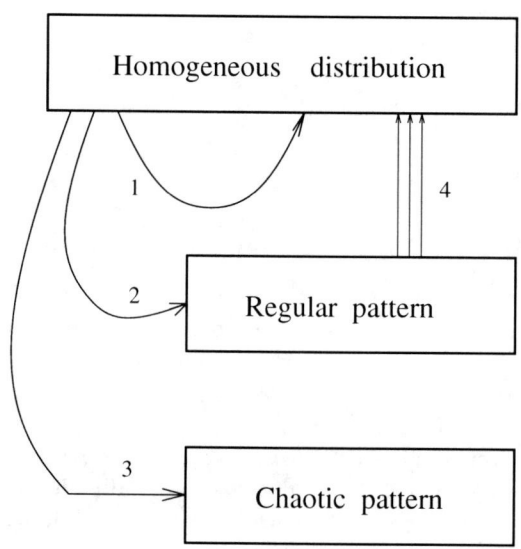

**FIGURE 11.7**: Hierarchy of regimes. The arrows show all possible transitions between different regimes observed in the course of the system dynamics. With permission from Petrovskii and Malchow (2001b).

Conversely, the regime of chaotic spatiotemporal oscillations is persistent. In this case, there are also two different time scales. The first scale, $\tau_{emb}$, corresponds to the formation of the chaotic embryo(s); this time is roughly estimated to be of the same order as $\tau_{ampl}$. The second time scale, $\tau_{dis}$, corresponds to the growth of the chaotic domain(s) and to the displacement of the regular pattern by chaos through the propagation of the "wave of chaos"; cf. Figure 11.2. The specific value of $\tau_{dis}$ apparently depends on the domain length $L$ and also on other parameters through the speed of the wave propagation (Petrovskii and Malchow, 2001b). After the chaotic phase occupies the whole domain, the dynamics of the system does not undergo any further quantitative changes. In particular, the long-time numerical simulations (up to $t \sim 10^6$) show that the form and size of the chaotic attractors both in a local $(u, v)$ plane (i.e., at a fixed $\bar{x}$) and in the plane $(<u>, <v>)$ of the spatially averaged values remain unchanged.

Therefore, based on the results of extensive numerical experiments, we have shown that the parameter range where spatial homogeneity in the system (11.3)–(11.4) is either maintained or gives way to a regular spatiotemporal pattern is quite narrow. For the parameters outside of this range, the system is driven to spatiotemporal chaos that is self-sustained and persistent.

We also mention that, having once emerged, the chaotic pattern appears to be robust with respect to moderate variations of the problem parameters $k$, $p$, and $h$. Interestingly, however, chaos may be suppressed when the parameters get sufficiently close to the Hopf bifurcation point, even if the local kinetics still remains oscillatory. In Section 11.2, we will address this issue in detail and show that it happens when the correlation length of the system exceeds the domain length.

### 11.1.2 Patchiness in a two-dimensional case

Thinking about the possible ecological implications of the above results, a question of immediate interest is to what extent they can be extended onto the case of two spatial dimensions and how the self-organized patchiness, if any, may look in a two-dimensional prey–predator system. Indeed, the dynamics of ecological populations rarely takes place along a line (but see Lubina and Levin, 1988); however, it is well known that the system's properties can be rather different depending on the number of spatial dimensions.

In order to address this issue, we consider the full two-dimensional version of Equations (11.3)–(11.4), where now $\nabla^2 = \partial^2/\partial^2 x + \partial^2/\partial^2 y$, $0 < x < L_x$, and $0 < y < L_y$. At the domain boundary, zero-flux conditions are imposed. The equations are solved numerically. As well as in the one-dimensional case, the type of system dynamics depends on the choice of the initial conditions. For a purely homogeneous population initial distribution, the system stays homogeneous forever. For a slightly perturbed homogeneous initial distribution (the shape of the perturbation can be different; cf. (11.5)–(11.6) and (11.10)–(11.11)), a smooth pattern arises, which is not persistent and gradually converges to a homogeneous distribution. For a greater initial perturbation, the system evolves to the formation of an irregular patchy spatial pattern. A typical example is shown below.

We consider the initial conditions describing a dome-shaped distribution of prey placed into a domain with a constant-gradient predator distribution:

$$u(x,y,0) = \bar{u} + \epsilon_1(x - 0.2L_x)(x - 0.8L_x) \quad (11.12)$$
$$+ \epsilon_2(y - 0.3L_y)(y - 0.7L_y),$$

$$v(x,y,0) = \bar{v} + \epsilon_3(x - 0.5L_x) + \epsilon_4(y - 0.45L_y), \quad (11.13)$$

where $\epsilon_1 = -2 \cdot 10^{-7}$, $\epsilon_2 = -6 \cdot 10^{-7}$, $\epsilon_3 = -3 \cdot 10^{-5}$, and $\epsilon_4 = -6 \cdot 10^{-5}$ are parameters.

Figure 11.8 shows snapshots of the prey density distribution over space at $t = 0, 150, 350$ and $1500$ (top to bottom) for the same parameter values as in Figures 11.1 to 11.6, i.e., for $k = 2.0$, $p = 0.3$, $h = 0.4$. Obviously, the

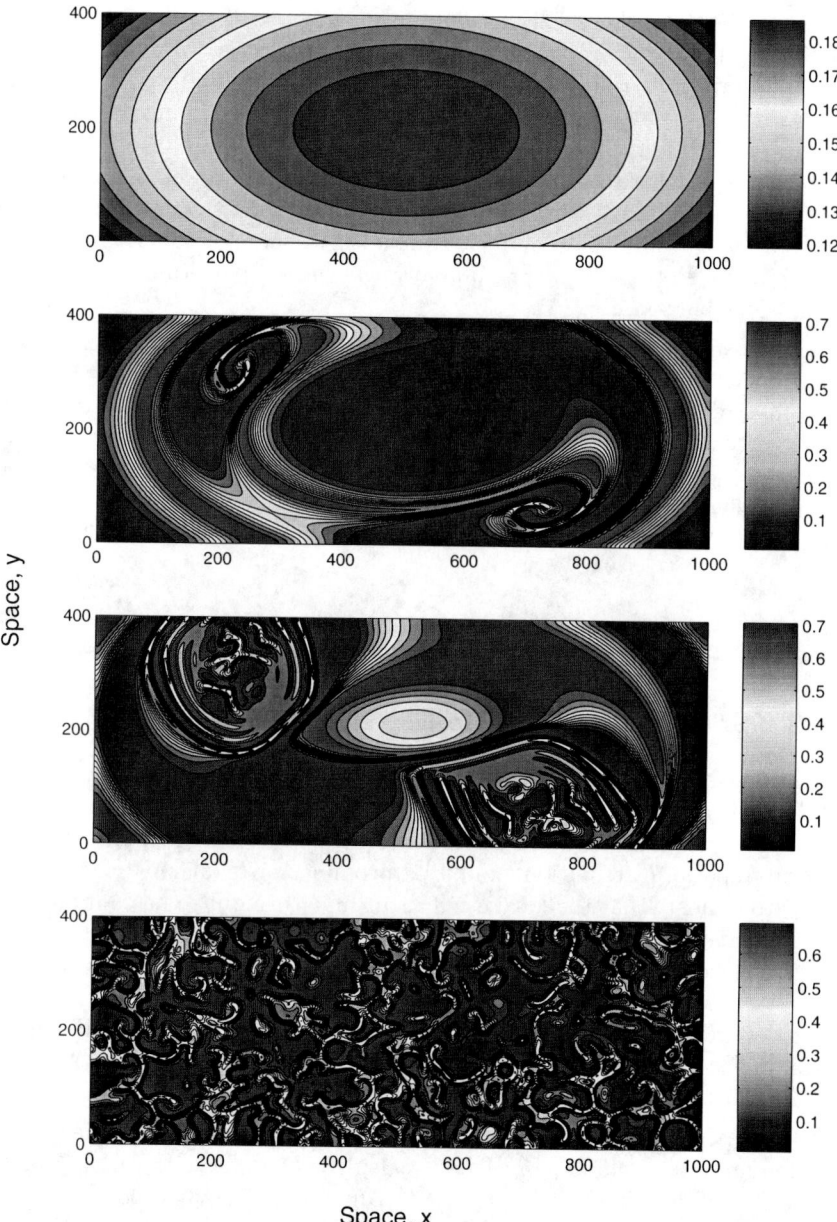

**FIGURE 11.8**: (See color insert.) Snapshots of the prey distribution over two-dimensional space for (top to bottom) $t = 0$, $t = 150$, $t = 350$, and $t = 1500$. Parameters are given in the text; see Equations (11.12)–(11.13) and below. Predator density shows qualitatively similar behavior except for very early stages of the system dynamics when the impact of initial conditions is essential.

system dynamics follows a scenario of pattern formation that is qualitatively similar to the one observed in the one-dimensional case. At an early stage, the population distribution is smooth and regular, forming spiral-like spatial structures. The chaotic embrios then appear in the vicinity of the spiral centers, which are readily seen to correspond to the points of the phase singularity $(x_*, y_*)$, where $u(x_*, y_*) = u_*$ and $v(x_*, y_*) = v_*$. A circular boundary separates the domains with jagged, patchy chaotic structures from the rest of the system where the dynamics is smooth in space and periodic in time. The domains grow gradually, and, finally, the chaotic pattern invades over the whole available space.

We want to emphasize that, although Figure 11.8 shows the results of a single simulation run accomplished for a given parameter set, the succession of patterns like "spirals → chaotic embrios → spatiotemporal chaos" is typical for a two-dimensional diffusion–reaction system with oscillatory kinetics. More examples (obtained for different initial conditions, different parameter values, and by different numerical methods) can be found in Medvinsky et al. (2002) and Garvie (2007); for an earlier reference see also Kuramoto (1984).

## 11.2 Spatial structure and spatial correlations

In the case of spatiotemporal chaos, the population densities fluctuate or oscillate with time in an irregular, stochastic-like manner. Furthermore, they also show a qualitatively similar behavior in space. The fact that the populations' spatial distribution exhibits a prominent irregular patchy structure seems to indicate that the local oscillations may turn out to be "out of phase" when the distance between any two positions becomes sufficiently large. This spatial aspect of the population dynamics is usually quantified in terms of the correlation length: Population fluctuations in points $\mathbf{r}_1$ and $\mathbf{r}_2$ can be regarded as independent (uncorrelated) in the case $|\mathbf{r}_1 - \mathbf{r}_2| > L_{corr}$. Within the distance $L_{corr}$, the population oscillations are correlated due to the diffusive coupling between neighboring points.

Clearly, the correlation length is an important characteristic of the system's spatiotemporal dynamics. In order to calculate $L_{corr}$, a variety of approaches can be used (Pascual and Levin, 1999; Durrett and Levin, 2000); for a discussion of related issues also see Rand and Wilson (1995). Perhaps the most common way is to derive the value of $L_{corr}$ from the properties of the spatial correlation function of the population fluctuations. It should be noted here that the state of a prey–predator community is naturally described by the two dynamical variables $u$ and $v$, i.e., the densities of prey and predator, respectively. Thus, for a general situation, one has to consider two autocorrelation functions as well as the cross-correlations. However, since both variables show a qualitatively similar behavior, here we restrict our analysis to the autocorrelation function for prey density $u$.

An immediate application of the standard definition leads to a problem. According to the usual approach, in case a dynamical variable $\psi$ is a function of a variable $\tau$ (which may have the meaning of time or position), the autocorrelation function is defined by the following expression:

$$F(\xi) = \lim_{Z \to \infty} \frac{1}{Z} \int_0^Z \psi(\tau + \xi)\psi(\tau) d\tau . \qquad (11.14)$$

In the problem under consideration, the prey density depends on two variables, position and time. Thus, applying definition (11.14) straightforwardly, we arrive at

$$\tilde{F}(\xi, t) = \lim_{Z \to \infty} \frac{1}{Z} \int_0^Z u(x + \xi, t) u(x, t) dx . \qquad (11.15)$$

Equation (11.15) has a few evident drawbacks. First, the autocorrelation function calculated according to (11.15) depends not only on the distance $\xi$ but also on time. The situation when the properties of $\tilde{F}$, considered as a function of $\xi$, are explicitly time dependent appears rather exotic and makes the interpretation of the results highly difficult. On the other hand, since the problem is essentially nonstationary, it seems reasonable that a proper definition of the autocorrelation function should take into account both spatial and temporal aspects of the system dynamics. Another problem is that, in order to obtain reliable results in computer simulations, the value of $Z$ in Equation (11.15) must be chosen sufficiently large. In practice, this means that the numerical grid must consist of at least a few dozen thousands of nodes (more likely, even tens of thousands), which would bring a considerable technical difficulty.

In order to overcome these problems, we use a modified definition of the autocorrelation function where the averaging over space is changed to the averaging over time:

$$K(\xi) = \lim_{T \to \infty} \frac{1}{T} \int_0^T u(x_0 + \xi, t) u(x_0, t) dt . \qquad (11.16)$$

Note that Equation (11.16) includes the standard definition as a particular case provided the system exhibits ergodic behavior. Let us also mention here that, although the value of $K$ formally depends on the parameter $x_0$, the results of the numerical simulations do not show any dependence on $x_0$. This fact has a clear interpretation indicating the "statistical homogeneity" of the system dynamics.

The autocorrelation function $K(\xi)$ calculated according to (11.16) in the case when the system is in the regime of spatiotemporal chaos is shown in Figure 11.9. Its properties (i.e., a fast decay for small $\xi$ followed by some

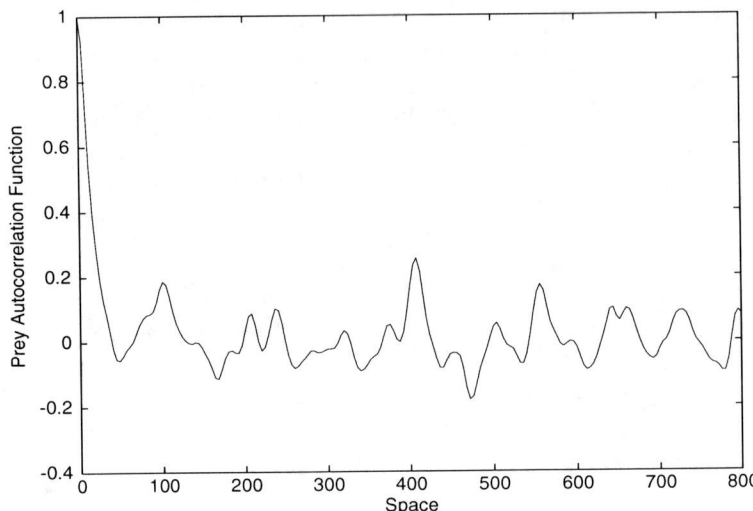

**FIGURE 11.9**: Autocorrelation function (11.16) calculated for the system (11.3)–(11.4) after onset of chaos. Parameters are the same as in Figure 11.2; averaging is done over time interval from $t = 4000$ to $t = 8000$. With permission from Petrovskii and Malchow (2001b).

irregular oscillations of relatively small amplitude) are typical for chaotic dynamics (Nayfeh and Balachandran, 1995). It should be mentioned here that the irregular oscillations of finite (nonzero) amplitude in $K(\xi)$ are the consequence of the finiteness of the averaging interval $T$; results of numerical experiments show that their amplitude tends to zero for increasing $T$. The dynamics of the system can now be characterized by the correlation length, its value being given by the first minimum of the autocorrelation function (Nayfeh and Balachandran, 1995; Abarbanel, 1996). For the parameters of Figure 11.9, $L_{corr} \approx 30$.

Existence of a finite correlation length means that the whole domain appears to be dynamically split into a number of subdomains, $N \simeq L/L_{corr}$ and $N \simeq (L/L_{corr})^2$ for one- and two-dimensional systems, respectively, their temporal behavior being virtually independent of each other. This self-organized splitting explains how the onset of chaos becomes possible in a time- and space-continuous prey–predator system: While the corresponding nonspatial dynamical system is of dimension 2 (in the sense of the dynamical systems theory), which makes chaos impossible, in the patchy regime its dimension grows with the number of subdomains and, therefore, can be much larger.

(It should be mentioned that, in case of the regular dynamics, the spatiotemporal behavior of the system (11.3)–(11.4) is highly correlated over the whole domain. Equation (11.16) then typically results in a monotonous, slowly decreasing curve. Since the regime of smooth patterns is a process of slow relaxation to the homogeneous spatial distribution, the autocorrelation function gradually changes with time so that the degree of correlation between different points increases. Correspondingly, the actual shape of the autocorrelation function depends significantly on both the initial conditions and the averaging time; for that reason, we do not show it here. In any case, however, in the large-time limit, temporal oscillations become synchronized throughout the system and $K(\xi)$ approaches 1 for any finite $\xi$.)

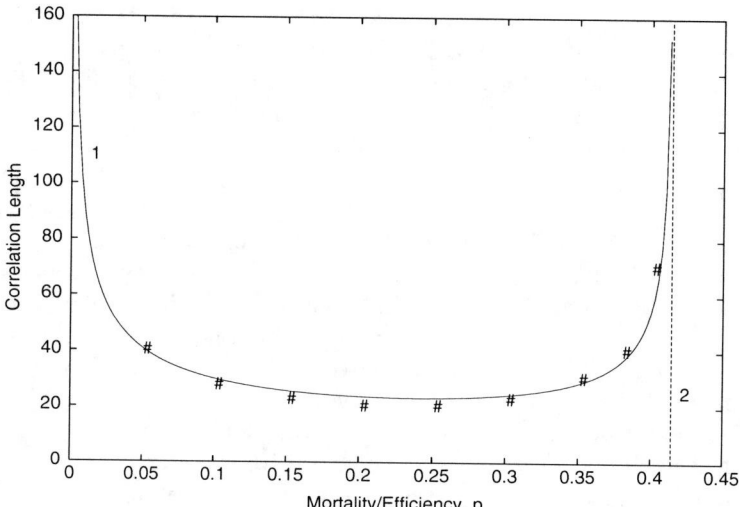

**FIGURE 11.10**: The value of the correlation length (#) in the chaotic prey–predator system (11.3)–(11.4) calculated for different parameter values along the line $h = p + 0.1$. Curve 1 shows the analytical prediction, see (11.26)–(11.27), and vertical line 2 shows the position of the Hopf-bifurcation point. With permission from Petrovskii and Malchow (2001a).

The next point is how $L_{corr}$ may depend on parameter values. A visual comparison between the spatial population distributions obtained for different parameters shows that they become "more patchy" when the parameters move away from the Hopf-bifurcation curve. In order to address this issue, the correlation length was calculated for the parameters varying along a certain line in the $(p, h)$ parameter plane. The results are shown in Figure 11.10. Indeed, it is readily seen that, although $L_{corr}$ changes only slightly in the intermediate parameter range, it tends to increase significantly when approaching the bifurcation point. This behavior also helps to explain why chaos is always suppressed in a close vicinity of the Hopf bifurcation; cf. the last paragraph in Section 11.1.1. As the system's parameters get closer to the bifurcation point, for any fixed length $L$ of the domain it appears that $L < L_{corr}$. This means that the oscillations get synchronized all over the domain and the system becomes dynamically equivalent to the corresponding homogeneous system where chaos is impossible.

So far as, we have shown that the onset of chaos may depend critically on the ratio $L/L_{corr}$. In Section 11.3, we will show that de-synchronization of population oscillations for $|\mathbf{r}_1 - \mathbf{r}_2| > L_{corr}$ due to transition to self-organized patchiness – onset of "biological turbulence" – has also crucial ecological implications. Altogether, it appeals for a method that would provide a fast and reliable estimate of the correlation length in a given system. Note that, theoretically speaking, its value can always be derived from the properties of a (exponentially) decaying correlation function. In turn, the spatial correlation function can be calculated either based on spatiotemporal data of field observations (cf. Ranta et al., 1997) or from a relevant model of the population dynamics. However, sufficiently detailed data are rarely available and obtaining $L_{corr}$ from a model usually takes a lot of computer simulations for each parameter set. Moreover, numerical data alone usually provide only limited information about the dependence of $L_{corr}$ on the parameters.

In the following section, we consider an alternative way to estimate the value of the correlation length. Using a general model of population dynamics and considering the stability of the population homogeneous distribution at equilibrium, we obtain a simple analytical formula for a certain "intrinsic length," which distinguishes between subcritical perturbations (preserving homogeneity) and supercritical ones (driving the system to pattern formation). We will then show that this "intrinsic length" appears to coincide with $L_{corr}$ up to a numerical coefficient on the order of unity.

### 11.2.1 Intrinsic lengths and scaling

We begin with a general case of an $n$-species system. The mathematical model is given by the following equations:

$$\frac{\partial U_i(\mathbf{R}, T)}{\partial T} = D_i \nabla^2 U_i(\mathbf{R}, T) + f_i(\mathbf{U}), \qquad i = 1, \ldots, n \qquad (11.17)$$

(cf. Chapter 2), where $\mathbf{U} = (U_1, U_2, \ldots, U_n)$, $U_i$ is the density of the $i$th species, $\mathbf{R}$ is the position in space, $T$ is the time, and the nonlinear functions $f_i$ describe the local "kinetics" of the system as given by multiplication, mortality, predation, etc.

We assume that the form of the functions $f_i$ provides the existence of at least one coexistence steady state $\bar{\mathbf{U}} = (\bar{U}_1, \bar{U}_2, \ldots, \bar{U}_n)$, so that

$$f_1(\bar{\mathbf{U}}) = f_2(\bar{\mathbf{U}}) = \ldots = f_n(\bar{\mathbf{U}}) = 0 , \qquad (11.18)$$

where $\bar{U}_i > 0$, $i = 1, \ldots, n$. In order to induce a nontrivial local dynamics of the system, we assume below that $n \geq 2$. Obviously, each steady state defined by Equation (11.18) corresponds to the homogeneous stationary state of the system (11.17).

A coexistence state can be of a different type, either stable or unstable. We begin with the purely temporal stability of the spatially homogeneous system. Applying the standard linear analysis to the system (11.17) without diffusion, i.e., to

$$\frac{dU_i(t)}{dt} = f_i(\mathbf{U}), \quad i = 1, \ldots, n, \qquad (11.19)$$

stability of the given steady state means that all the eigenvalues of the system (11.19) linearized in the vicinity of $\bar{\mathbf{U}}$ have negative real parts. Denoting $U_i(t) - \bar{U}_i = \epsilon_i(t)$, from system (11.19) we obtain

$$\frac{d\epsilon_i(t)}{dt} = a_{i1}\epsilon_1 + a_{i2}\epsilon_2 + \ldots + a_{in}\epsilon_n, \quad i = 1, \ldots, n, \qquad (11.20)$$

where $a_{ij} = \partial f_i(\bar{\mathbf{U}})/\partial U_j$ are the elements of the Jacobian of the kinetic system at the steady state. Then, the eigenvalues $\lambda_i$ ($i = 1, \ldots, n$) are given by the following equation:

$$\det(A - \lambda E) = 0, \qquad (11.21)$$

where $A = (a_{ij})$ and $E$ is the identity matrix.

Since the coefficients $a_{ij}$ evidently depend on a number of parameters (such as the population growth and mortality rates, predation rates, etc.), the eigenvalues also depend on the parameter values. Let us consider the case when a given steady state can change its stability depending on the values of parameters. Loss of stability means that the real parts of some eigenvalues become positive when a point in the parameter space crosses a certain critical hypersurface. An example is given by the Hopf bifurcation. Denoting the maximum real part of the eigenvalues as $\bar{\lambda}$, the system becomes (linearly) unstable when $\bar{\lambda}$ becomes positive.

Note that stability of the system (11.19) means stability of the stationary homogeneous state $U_i(x,t) = \bar{U}_i$, $i = 1, \ldots, n$ of the system (11.17) with respect to small spatially homogeneous perturbations. However, the situation becomes somewhat more complicated when we consider spatially heterogeneous perturbations. Now, $U_i(x,t) = \bar{U}_i + \epsilon_i(x,t)$, and, substituting it into (11.17), we obtain

$$\frac{\partial \epsilon_i(\mathbf{r},t)}{\partial t} = D_i \nabla^2 \epsilon_i(\mathbf{r},t) + a_{i1}\epsilon_1 + a_{i2}\epsilon_2 + \ldots + a_{in}\epsilon_n, \qquad (11.22)$$
$$i = 1, \ldots, n.$$

Considering a perturbation with a certain wavenumber $|\mathbf{k}|$, i.e., $\epsilon_i = C_i e^{\nu t}$

$\cos \mathbf{kr}$, $i = 1, \ldots, n$, from (11.22) we arrive at the following system:

$$\left(a_{11} - \nu - D_1 \mathbf{k}^2\right) C_1 + a_{12} C_2 + \ldots + a_{1n} C_n = 0,$$
$$a_{21} C_1 + \left(a_{22} - \nu - D_2 \mathbf{k}^2\right) C_2 + \ldots + a_{2n} C_n = 0,$$
$$\vdots$$
$$a_{n1} C_1 + a_{n2} C_2 + \ldots + \left(a_{nn} - \nu - D_n \mathbf{k}^2\right) C_n = 0.$$

A nontrivial solution exists if and only if

$$\det\left[(A - \mathbf{k}^2 B) - \nu E\right] = 0, \qquad (11.23)$$

where $B = (b_{ij})$, $b_{ij} = D_i \delta_{ij}$ and $\delta_{ij}$ is the Kronecker symbol. Denoting, for convenience, $\bar{\nu} = \max_{(i)} \operatorname{Re} \nu_i$ ($i = 1, \ldots, n$), a stationary homogeneous distribution is linearly stable with respect to the heterogeneous perturbation with given wavelength $l = 2\pi/|\mathbf{k}|$ for $\bar{\nu}(\mathbf{k}) < \mathbf{0}$ and unstable for $\bar{\nu}(\mathbf{k}) > \mathbf{0}$.

Assuming here, for simplicity,[5] that $D_1 = \ldots = D_n = D$, it is readily seen that $B = DE$ and Equation (11.23) coincides with (11.21) provided $\lambda = \nu + D\mathbf{k}^2$. As an immediate consequence, it means that the maximum real parts of the eigenvalues of the homogeneous and heterogeneous problems are related by means of the following equation:

$$\bar{\nu} = \bar{\lambda} - D\mathbf{k}^2. \qquad (11.24)$$

Relation (11.24) has a clear meaning. When the steady state $\bar{\mathbf{U}}$ becomes locally unstable due to a change in parameter values, i.e., $\bar{\lambda}$ becomes positive, it still remains stable in the distributed system with respect to heterogeneous short-wave perturbations, i.e., for

$$|\mathbf{k}| > k_0 = \left(\bar{\lambda}/D\right)^{1/2}. \qquad (11.25)$$

From (11.25), we arrive at the following formula for the critical wavelength:

$$l_0 = 2\pi \left(D/\bar{\lambda}\right)^{1/2}. \qquad (11.26)$$

The value of $l_0$ may be regarded as a certain "intrinsic length" of system (11.17). It gives the upper limit of the system stability with respect to small spatially heterogeneous perturbations, i.e., the perturbations with wavelengths less than $l_0$ are damped by diffusion. Note that, in order to obtain (11.26), we did not make any specific assumptions about the local kinetics of the system, e.g., about the type of inter-specific interactions.

In case a perturbation exceeds the stability limits, the system may be driven into spatiotemporal chaos. Although the dynamics of the general system

---

[5] An extension of this analysis onto the case of unequal diffusion coefficients can be found in Petrovskii et al. (2003).

(11.17) remains to be investigated, existence of chaotic patterns was previously shown for a few particular cases such as a prey–predator system and a three competing species system. In the chaotic regime, the spatial properties of system (11.17) can be quantified by the correlation length $L_{corr}$ which has a meaning similar to the "intrinsic length" $l_0$, i.e., it gives the maximum distance between two positions in space where the diffusive coupling is still essential. Thus, the effect of diffusion is quantified by $l_0$ at the early stage of the system dynamics and by $L_{corr}$ at its later stage, after transients have disappeared. The assumption we are going to make now is that some properties of the early stage dynamics may stay important at the large-time stage as well.[6] Then one can expect that the length $l_0$ remains intrinsic for the system after the onset of chaos. Since both $l_0$ and $L_{corr}$ are related to the same process, i.e., diffusive coupling, we then make a somewhat stronger hypothesis about the existence of a "scaling law:"

$$\frac{L_{corr}}{l_0} = c^* , \qquad (11.27)$$

where $c^*$ is a "structural" coefficient of the order of unity, its value depending on the type of the interspecific interactions but not on particular parameter values once the system is specified.

The above speculations can hardly be accepted as a proof, though. Therefore, in order to test hypothesis (11.27), now we are going to consider a few particular cases of the system (11.17) where dependence of the maximum real part of the eigenvalues $\bar{\lambda}$ on system parameters can be followed explicitly.

*Prey–predator system.* For a prey–predator system described by Equations (11.3)–(11.4), the real part of the eigenvalues (in dimensionless units) of the system linearized in the vicinity of $(\bar{u}, \bar{v})$ is given by the following equation:

$$\bar{\lambda} = \frac{\mathrm{tr}A}{2} \quad \text{where} \quad \mathrm{tr}A = \frac{p}{1-p}[(1-p) - h(1+p)], \qquad (11.28)$$

where $A$ is the matrix of the linearized system. Figures 11.10 and 11.11 show $L_{corr}$ obtained from numerical data (as the position of the first minimum of the spatial autocorrelation function) and $l_0$ calculated analytically from (11.26) and (11.28). It is readily seen that the ratio $L_{corr}/l_0$ remains constant (up to small fluctuations most likely caused by the computational error) if $\bar{\lambda}$ is sufficiently small.

*Three competing species system.* Another example of a biological system showing the formation of chaotic spatiotemporal patterns is given by a community of three competitive species; see Section 10.3. For simplicity, here we

---

[6] Although this assumption may seem exotic, a recent paper by Neubert et al. (2002) reveals another link between the early stage dynamics and the large-time dynamics of a diffusion–reaction system.

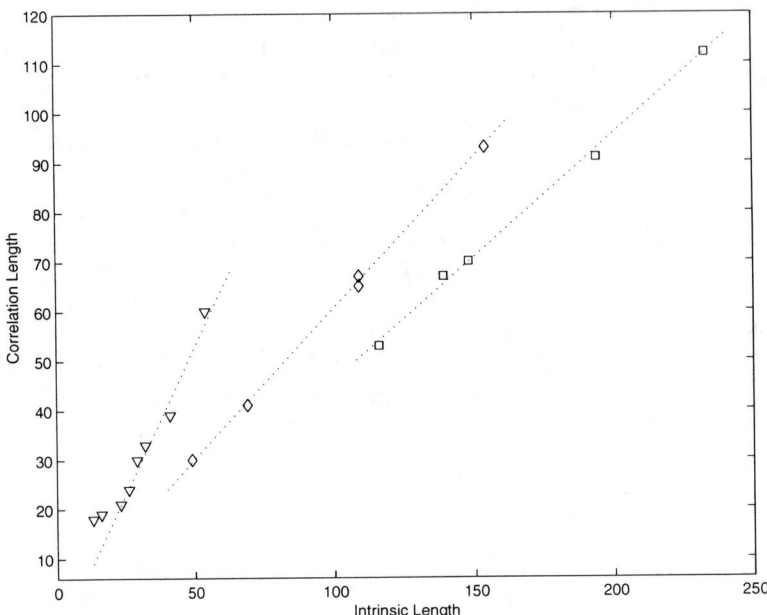

**FIGURE 11.11**: Numerically calculated correlation length $L_{corr}$ is shown vs intrinsic length $l_0$ for different systems: triangles for the prey–predator system, diamonds for the three competitive species system, and squares for the substrate-autocatalyst system. Obviously, in all three cases, $L_{corr}/l_0 = const$, as predicted by the scaling hypothesis (11.27). With permission from Petrovskii et al. (2003).

restrict our analysis to the special case of cyclic competition. The maximum real part of the eigenvalues of the linearized dimensionless system is then given by the following formula:

$$\bar{\lambda} = \frac{1}{1+\alpha+\beta}\left(-1 + \frac{\alpha+\beta}{2}\right). \tag{11.29}$$

When $\alpha + \beta > 2$, the steady state becomes locally unstable and perturbations of the homogeneous initial state can drive the system into spatiotemporal chaos (Petrovskii et al., 2001). Note that the structure of the phase space of system (10.54)–(10.56) is different compared to the prey–predator system: In the case of cyclic competition, no stable limit cycle exists and the attractor is given by a heteroclinic trajectory connecting the three unstable "one-species-only" states.

Table 11.1 and Figure 11.11 show the correlation length $L_{corr}$ calculated for different parameters $\alpha$ and $\beta$ as well as the corresponding value of $l_0$ given by (11.26) and (11.29). Apparently, both quantities exhibit essentially the same

**TABLE 11.1:** Relation between the (dimensionless) values of the correlation length $L_{corr}$ and the intrinsic length $l_0$ in the three competitive species community. With permission from Petrovskii et al. (2003).

| parameters | $\lambda$ | $l_0$ | $L_{corr}$ | $L_{corr}/l_0$ |
|---|---|---|---|---|
| $\alpha = 1.20,\ \beta = 0.90$ | 0.0161 | 49 | $30 \pm 2$ | 0.61 |
| $\alpha = 1.30,\ \beta = 0.80$ | 0.0161 | 49 | $30 \pm 2$ | 0.61 |
| $\alpha = 1.15,\ \beta = 0.90$ | 0.0082 | 69 | $41 \pm 2$ | 0.59 |
| $\alpha = 1.22,\ \beta = 0.80$ | 0.0033 | 109 | $65 \pm 2$ | 0.60 |
| $\alpha = 1.07,\ \beta = 0.95$ | 0.0033 | 109 | $67 \pm 2$ | 0.61 |
| $\alpha = 1.21,\ \beta = 0.80$ | 0.0017 | 154 | $93 \pm 5$ | 0.60 |

type of dependence on the parameters, so that their ratio stays constant. In particular, the scaling hypothesis (11.27) succeeds in predicting that the correlation length depends on the sum $\alpha + \beta$ rather than on $\alpha$ and $\beta$ separately, and that the value of $L_{corr}$ increases when $\alpha + \beta$ tends to 2.

*A two-species chemical system.* The two cases considered above, although essentially different in the type of interspecific interactions, still exhibit a certain mathematical similarity. Namely, in both cases there is only one stationary coexistence state. Meanwhile, it is well known that the existence of another steady state, particularly, the existence of a saddle-point, can change the dynamics significantly (Rai and Schaffer, 2001). A question thus arises whether the validity of Equation (11.27) is restricted to a specific structure of the phase space of the system, or it may have a wider application.

In order to address this issue, now we consider an example of different origin, i.e., the Gray–Scott model of an autocatalytic reaction in an open one-dimensional flow reactor:

$$\frac{\partial u}{\partial t} = \frac{\partial^2 u}{\partial x^2} + F(1-u) - uv^2, \qquad (11.30)$$

$$\frac{\partial v}{\partial t} = \frac{\partial^2 v}{\partial x^2} + uv^2 - (F+k)v \qquad (11.31)$$

(Gray and Scott, 1990) with properly chosen dimensionless variables (cf. Pearson, 1993). Now $u$ and $v$ are the concentrations of the substrate and the autocatalyst, respectively, $F$ is the flow rate, and $k$ is the effective rate constant of the decay of the autocatalyst. Although the system (11.30)–(11.31) does not immediately apply to a community of biological species, it provides a convenient test to check the generality of our scaling hypothesis (11.27).

One can readily see that, under condition $d = 1 - 4(F+k)^2/F > 0$, there are three spatially homogeneous steady states in the system (11.30)–(11.31): "substrate-only" state $(1,0)$, "substrate-dominated" state $(u_s, v_s)$

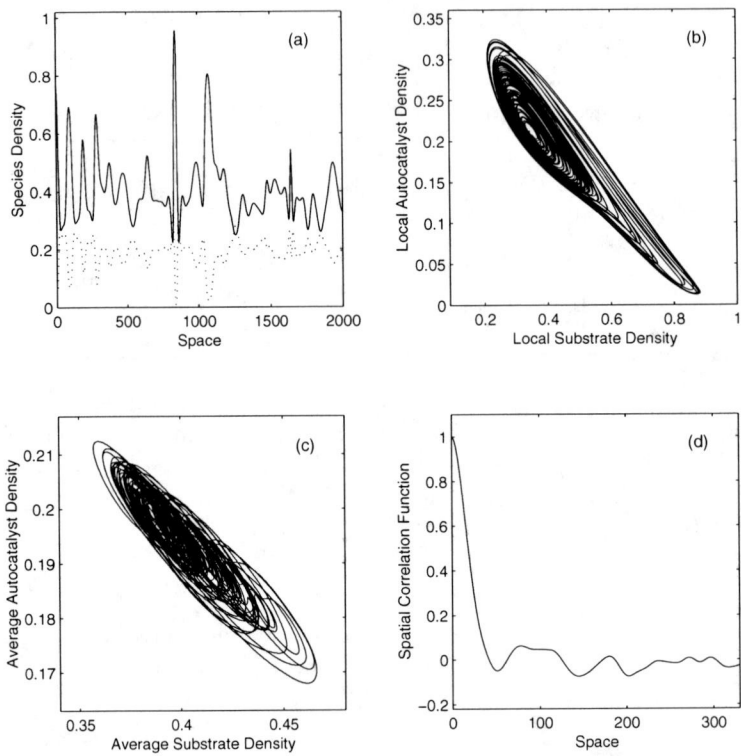

**FIGURE 11.12**: Spatiotemporal chaos in the substrate-autocatalyst system (11.30)–(11.31). With permission from Petrovskii et al. (2003).

and "autocatalyst-dominated" state $(u_a, v_a)$, where

$$u_s = \frac{1+\sqrt{d}}{2}, \qquad v_s = \left(\frac{F}{F+k}\right)\frac{1-\sqrt{d}}{2}, \qquad (11.32)$$

$$u_a = \frac{1-\sqrt{d}}{2}, \qquad v_a = \left(\frac{F}{F+k}\right)\frac{1+\sqrt{d}}{2}. \qquad (11.33)$$

When crossing the critical curve $d = 0$ in the $(k, F)$ plane towards smaller values of $k$, the two nontrivial states appear through a saddle-node bifurcation, the autocatalyst-dominated state being an unstable node.

The substrate-only state is always a stable node and the substrate-dominated state is always a saddle-point. A change in the local dynamics can be associated with the change of the type of the autocatalyst-dominated state, first of all, with the change of its stability. It is straightforward to see that the maximum real part of the eigenvalues at the autocatalyst-dominated state is given by the following equation:

$$\bar{\lambda} = \frac{1}{2}\left(k - v_a^2\right) \ . \tag{11.34}$$

The change of stability takes place when $\bar{\lambda} = 0$, which, after a little algebra, takes the following form:

$$\frac{k - F}{k + F} = \sqrt{d} \ . \tag{11.35}$$

The Hopf bifurcation that takes place when crossing the curve (11.35) is predicted to be supercritical for $k < k_{cr}$ (where $k_{cr}$ is estimated as about 0.035), and only in this case does a stable limit cycle appear (Rasmussen et al., 1996). Otherwise, no limit cycle arises and any trajectory starting in the vicinity of the autocatalyst-dominated state, after a number of twists of increasing amplitude, is finally attracted to the substrate-only state. Therefore, the structure of the local phase plane of the Gray–Scott model is essentially different from the prey–predator system (11.3)–(11.4).

For those parameter values when the autocatalyst-dominated state is unstable, the system (11.30)–(11.31) can exhibit spatiotemporal chaos (Merkin et al., 1996; Davidson, 1998). In the chaotic regime, the spatial and temporal behavior of the reacting species is qualitatively similar to the behavior of the biological species; see Figure 11.12. Although our main interest is focused on the dynamics of biological communities, the model (11.30)–(11.31) can be used for testing the scaling hypothesis (11.27). Figure 11.11 shows the value of $L_{corr}$ obtained in numerical simulations as well as the intrinsic length $l_0$. One can see that the ratio $L_{corr}/l_0$ stays constant, which "proves" the relation (11.27).

Thus, the results of our numerical simulations of chaotic spatiotemporal dynamics in different systems show that the hypothesis (11.27) about the relation between the intrinsic length $l_0$ and the correlation length $L_{corr}$ of the system is valid in a wide variety of cases; see Figure 11.11. We also want to mention that relation (11.27) remains valid for populations with different diffusivity, although the analytical expression for $l_0$ is less elegant in that case (Petrovskii et al., 2003).

An immediate consequence of the above conclusion is that the following equation can be used in order to estimate the value of $L_{corr}$:

$$L_{corr} = c^* l_0 \ , \tag{11.36}$$

where $l_0$ is given by Equation (11.26) and $c^*$ is a constant on the order of unity, e.g., $c^* \approx 1$ for the prey–predator system, $c^* \approx 0.6$ for the system of three competitive species, and $c^* \approx 0.47$ for the autocatalytic chemical system.

When the parameters approach the bifurcation point where the corresponding steady state becomes stable, $\bar{\lambda}$ tends to zero. Correspondingly, $l_0$ tends to infinity and (11.36) predicts that $L_{corr}$ grows unboundedly. This is in good agreement with our numerical results; see Figure 11.11. There is, however, a certain discrepancy between $L_{corr}$ and $l_0$ that is not described by the linear relation (11.27) further away from the bifurcation point, i.e., in the parameter range where $\bar{\lambda}$ becomes sufficiently large (correspondingly, $L_{corr}$ sufficiently small). It may indicate that $c^*$ is, in fact, a function of the system parameters with scaling properties so that $c^* \approx const > 0$ for $\bar{\lambda} \ll 1$.

## 11.3 Ecological implications

The importance of space in ecology has been increasingly recognized over the last two decades. When considered in space and time, ecological interactions may exhibit considerably different properties compared to the corresponding nonspatial systems. We have already demonstrated that a spatially homogeneous distribution of interacting species often appears to be unstable to heterogeneous perturbations and is prone to generating spatiotemporal patterns. In case a system's kinetics is oscillatory,[7] these patterns are remarkably irregular in space and chaotic in time. In order to study this phenomenon of "biological turbulence," a prey–predator system was used as a paradigm. However, self-organized patterning is clearly a more general phenomenon and not restricted to prey–predator interactions; in particular, extension of the main results has been made onto a system of competing species. Moreover, the prey–predator interactions are among the most common ones in population biology and are present in virtually every population community. Therefore, we expect that the mechanism of self-organized patchiness considered in the previous sections is likely to have a variety of profound implications for population dynamics and ecology. In this section, we will give two examples where the properties of biological turbulence enhance our understanding of ecological dynamics and may provide a solution for some long-standing ecological problems.

### 11.3.1 Plankton patchiness on a biological scale

Probably the most notorious example of ecological patterning, which has been studied intensively for several decades, is the heterogeneity of aquatic populations. In particular, the horizontal spatial distribution of plankton in oceans, seas, and large lakes is remarkably inhomogeneous. This phenomenon, which is often referred to as "plankton patchiness," has fundamental implications for almost every aspect of marine ecosystems organization (Valiela,

---

[7] Not necessarily along a stable limit cycle but in a broader sense; see the comment at the beginning of Section 11.1.

1995). Correspondingly, a number of attempts have been made to explain the plankton patterns by means of reducing them to the heterogeneity of the temperature field (Denman, 1976), to specific ocean hydrodynamics (Wyatt, 1973; Abbott and Zion, 1985), or turbulent mixing (Platt, 1972; Abraham, 1998; Neufeld, 2001; Reigada et al., 2003).

However, the problem as a whole seems to be much more complicated. Prominent spatial heterogeneity of plankton distribution occurs on different scales ranging from a few centimeters to thousands of kilometers. Analysis of field data (Powell et al., 1975; Weber et al., 1986) leads to the conclusion that the plankton patchiness on different scales is caused by the processes of different nature (Levin, 1990; Powell, 1995). On a scale of dozens of meters and less, plankton spatial distribution is indeed mainly controlled by turbulence (cf. Platt, 1972; Powell et al., 1975). On a scale of hundreds of kilometers and more, the heterogeneity of plankton distribution follows the heterogeneity of the temperature field (Weber et al., 1986). However, on the intermediate "biological scale," i.e., from hundreds of meters to dozens of kilometers, plankton patchy distribution is significantly affected by the interplay between the biological processes (such as population growth, predation, etc.) and the species dispersal, mostly due to turbulent mixing; see Levin and Segel (1976); Steele and Henderson (1992b); Powell (1995); Pascual and Caswell (1997); also see recent reviews by Medvinsky et al. (2002) and Martin (2003) for further references.

Remarkably, there is an opinion (Valiela, 1995) that the phenomenon of plankton patchiness is a manifestation of marine ecosystem's spatiotemporal self-organization, which is beneficial for the community as a whole. However, what particular mechanisms are behind this self-organization? Here we hypothesize that the dynamics of a plankton system is to a large extent controlled by the trophic interaction between phyto- and zooplankton acting as prey and predator, respectively. Consequently, we can apply the results obtained above for the prey–predator system and associate the chaotic spatiotemporal patterns with the plankton patterns on the biological scale. Indeed, there is a striking visual coincidence between the irregular spatial patterns generated in a prey–predator system (cf. Figure 11.8) and the patterns of spatial plankton distribution usually observed in the ocean; see Figure 11.13. It should also be mentioned that, although conclusive evidence is still lacking, there is a growing number of indications of chaos in the dynamics of aquatic populations (Medvinsky et al., 2002).

A visual coincidence can hardly be regarded as an acceptable proof, though. However, our hypothesis of the plankton patchiness origin can be tested in a more quantitative way. Recall that the biological scale has lower and upper bounds. The lower bound is called the KISS length; it corresponds to the spatial scale where the distractive impact of turbulent eddies on plankton patches is balanced by their growth due to population multiplication. The value of the KISS length (being on the order of a hundred meters) has been estimated consistently from a few different approaches (Kierstead and

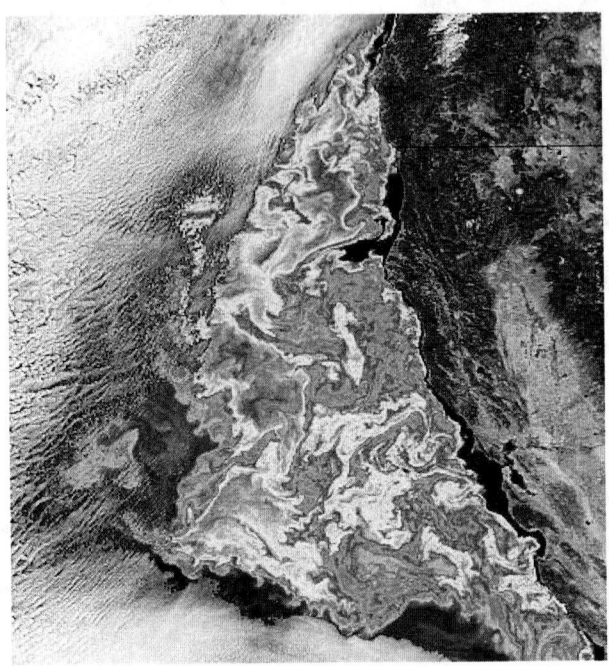

**FIGURE 11.13**: (See color insert.) Chlorophyll distribution off California coast. Orange and yellow patches correspond to high phytoplankton concentration. By courtesy of NASA; from http://oceancolor.gsfc.nasa.gov/FEATURE/gallery.html.

Slobodkin, 1953; Powell et al., 1975; Denman and Platt, 1976; Steele, 1978; Powell and Okubo, 1994). However, the origin of the upper bound remains poorly understood. Although suggestions have been made that it may reflect the impact of zooplankton on the system dynamics (cf. Steele, 1978), a particular mechanism has never been identified. Now, if we assume that plankton patchiness is a manifestation of the biological turbulence resulting from phytoplankton-zooplankton (prey–predator) interactions, the value that has a meaning similar to the upper bound of the biological scale is the correlation length.

In order to calculate $L_{corr}$, we use Equations (11.36), (11.26), and (11.28). Referring to the parameter estimates provided by different authors (Nisbet et al., 1991; Truscott and Brindley, 1994; Sherratt, 2001), we choose $\alpha = 1.0$ day$^{-1}$, $\gamma = 0.7$ day$^{-1}$, $\mu = 0.05$ day$^{-1}$, $\kappa = 0.15$, and $H/K = 0.3$ as typical values. The turbulent diffusivity is roughly estimated as $D = 10^5$ cm$^2$sec$^{-1}$ (Nihoul 1980). Using these values, we arrive at $L_{corr} \approx 30$ km (cf. Petrovskii et al., 2003), which is in a very good agreement with the estimates of the upper bound of the biological scale derived from field observations (Weber et al., 1986; Levin, 1990). Note that, since the (dimensionless) value of $L_{corr}$ remains of the same order within a certain parameter range (see Figure 11.10), the predicted value of about 30 km is robust and the agreement between the data and the theory is not violated by a reasonably small variation of parameter values.

The above agreement is encouraging; however, we also want to mention that a marine ecosystem is an extremely complex object where a wide range of complicated physical processes interplay with equally complicated biological ones. Therefore, it would hardly be realistic to relate plankton patchiness to a single reason. The "biological turbulence" arising due to the coupling between spatial mixing and local phyto-zooplankton oscillations is likely to manifest itself under conditions of statistically homogeneous and isotropic turbulent ocean flows. In case the ocean hydrodynamical flows exhibit some sort of anisotropy, other mechanisms may apply, creating an alternative or complementary framework for the plankton patchiness. In particular, shear flows at the edges of large ocean currents may also result in pattern formation on the scale of dozens of kilometers (Biktashev et al., 1999).

Moreover, the horizontal spatial mixing in plankton systems in oceans, seas, and large lakes takes place mainly due to the turbulent eddy diffusivity. Contrary to the usual Brownian diffusion, the intensity of the turbulent diffusion typically increases with the scale of the process because of the impact of larger eddies (Okubo, 1971). This can be quantified in terms of turbulence spectrum, i.e., velocity variance per unit wavenumber. While a conservative passive tracer such as water temperature or salinity has the spatial spectrum coinciding with that of the turbulent flow, the spectra of phyto- and zooplankton patchiness are different on the biological scale (Weber et al., 1986), presumably due to the impact of the local population growth (Levin, 1990). Having considered a phyto-zooplankton system near its steady equilibrium, Vilar et al. (2003) showed theoretically that, by means of including details of the turbulent spectrum into the model, the spectral properties of Equations (11.1)–(11.2) can indeed be made very similar to those observed in field data.

Finally, we mention that the whole concept of turbulent "diffusion" is a sort of mean-field approach containing very little information about the inherent stochastic fluctuations of the turbulent flows. Therefore, solutions of the diffusion–reaction models should be regarded as population densities averaged over the corresponding statistical ensemble rather than snapshots of actual plankton distributions. In order to make an insight into the transient spatial structures associated with turbulent pulsations, one should go beyond the mean-field approximation (Abraham, 1998), e.g., considering turbulent transport as "stochastic advection" (Neufeld, 2001).

## 11.3.2 Self-organized patchiness, desynchronization, and the paradox of enrichment

The paradox of enrichment (Rosenzweig, 1971), when an increase in the nutrient input into a prey–predator system destabilizes the community functioning and may even lead to extinction of the species, has been a challenge for a few generations of ecologists; cf. Gilpin (1972); May (1972a); Luckinbill (1974); Brauer and Soudack (1978); Jansen (1995); Abrams and Waters (1996); Bohannan and Lenski (1997); Nisbet et al. (1997); Genkai-Kato and Yamamura (1999); Holyoak (2000); Petrovskii et al. (2004); Jensen and Ginzburg (2005); and Morozov et al. (2007).

Theoretical arguments behind the paradox are simple, clear, and apply to a rather general case. Let us consider the following prey–predator system:

$$\frac{dU}{dt} = f(U)U - g(U)V, \qquad \frac{dV}{dt} = \kappa g(U)V - MV, \qquad (11.37)$$

where all terms have their usual meanings. For biological reasons, $g(U)$ is a monotonously increasing function; additionally, we assume that $g(U)$ increases linearly for small $U$, i.e., $g(U) = const \cdot U + o(U)$ (cf. Holling type II). We also assume that prey growth is not damped by the Allee effect; therefore, $f(U)$ is a monotonously decreasing function turning to zero for $U = K$, where $K$ is the carrying capacity. These assumptions are not necessary conditions for the paradox of enrichment to occur, but they make analysis easier and more straightforward.

The isocline of the second equation in (11.37) is a vertical straight line:

$$U = \bar{U} \quad \text{where} \quad g(\bar{U}) = \frac{M}{\kappa}, \qquad (11.38)$$

while the isocline of the first equation in (11.37) is given by

$$V(U) = \frac{f(U)U}{g(U)}. \qquad (11.39)$$

Under the above assumptions, the ratio $f(U)/g(U)$ is a monotonously decreasing function; correspondingly, function $f(U)U/g(U)$ has a single hump.

The coexistence steady state of the system is given by the intersection of the isoclines. Moreover, it is readily seen that the steady state is stable when the intersection takes place on the right of the hump (i.e., where function $f(U)U/g(U)$ decreases) and it is unstable on the left of the hump (where $f(U)U/g(U)$ increases). For the parameter values when it is unstable, it is surrounded by a stable limit cycle (see Figure 11.14, top), which appears through the Hopf bifurcation when the steady state passes the hump from right to left.

Now, how do the system properties change in response to enrichment? System eutrophication is thought[8] to increase the carrying capacity $K$. Since $K$

---

[8] It may also increase the linear per capita growth $r$, but see the comments after Equations (11.40)–(11.41).

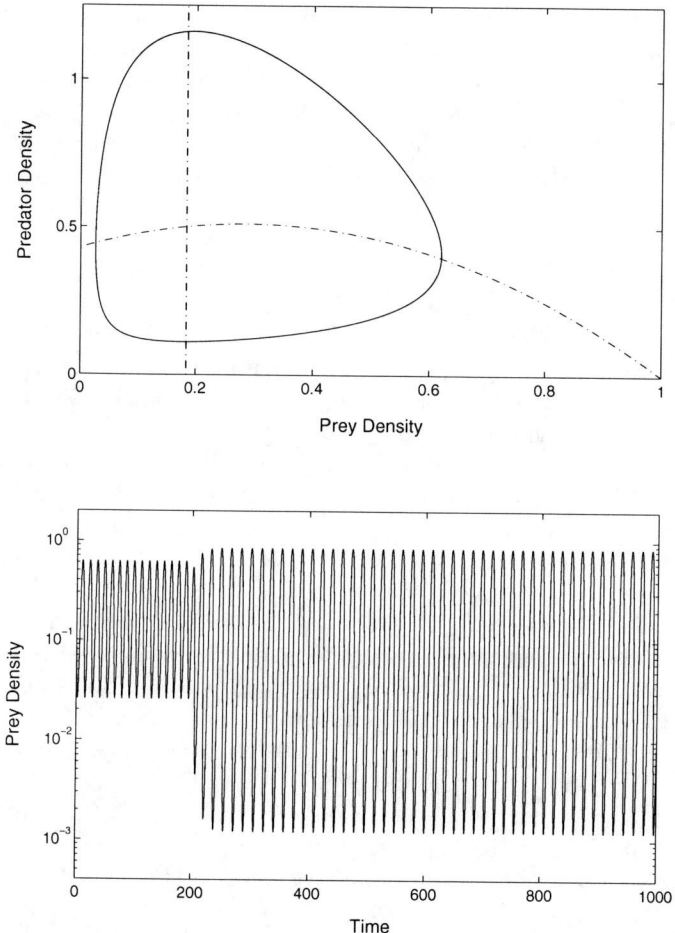

**FIGURE 11.14**: A sketch of the system's response to enrichment: (a) A typical phase plane of a prey–predator system. The dashed-and-dotted lines show the null-isoclines of the system, and the solid curve shows a stable limit cycle. The curves are obtained for the nonspatial version of (11.40)–(11.41) with parameters $k = 2$, $m = 0.6$, and $h = 0.43$. (b) Periodical oscillations of the prey density; enrichment of the system (when $K$ increases instantaneously by about 50 per cent at $t = 200$) results in oscillations of considerably larger amplitude.

is a natural scaling factor for the prey density, an increase in $K$ obviously "pulls" the whole plot of the first isocline to the right; in particular, the position of the hump moves to the right as well. Then, for any fixed position of the vertical isocline, a sufficiently large increase in $K$ will inevitably destabilize the steady state. As a result, the population densities start oscillating in time. The dynamics that may then lead to species extinction corresponds to the case when the trajectory in the phase space of the system comes close to the boundary of the biologically meaningful domain $U \geq 0$, $V \geq 0$. Remarkably, a further increase in $K$ leads to a fast increase in the size of the limit cycle, so that the minimum value of the population density reached in the course of oscillations eventually tends to zero (May, 1972b; Gilpin, 1972); see Figure 11.14, bottom. In terms of the dynamics of real ecosystems, as the minimum value of the population density decreases, population extinction becomes more probable due to stochastic environmental perturbations (Goel and Richter-Dyn, 1974; Lande, 1993). Moreover, the situations when the minimum population size falls to a value $\ll 1$ should be regarded as actual extinction even without any stochasticity.

The apparent contradiction between the intuitively expected positive impact of increasing nutrient input and its actual destabilizing effect inspired a number of modifications of the original prey–predator model. For instance, it was shown that enrichment of a prey–predator community does not necessarily diminish the minimum value of oscillating population densities in the cases of either the existence of invulnerable individuals within the prey population (Abrams and Waters, 1996) or in the presence of an alternative "unpalatable" prey (Genkai-Kato and Yamamura, 1999). Another way to increase the system's stability is to include ratio-dependence into the predation term (cf. Jensen and Ginzburg, 2005). However, these modifications have left open the question of whether the simplest one-predator–one-prey system is intrinsically unstable with respect to eutrophication. Although this kind of system response to eutrophication is not commonly seen in nature (McCauley and Murdoch, 1990), the self-regulating mechanisms of the system are not always clear. Furthermore, extinction of a prey–predator community following system eutrophication has been seen in some biological data (Luckinbill, 1974; Bohannan and Lenski, 1997).

The theoretical results mentioned above were obtained under the assumption that the interacting populations were homogeneous in space. A salient point of contemporary ecology has become the growing understanding that the dynamics of any biological community takes place not only in time but also in space, and that the properties of a spatiotemporal system can be essentially different from those of its nonspatial counterpart. The impact of space on the persistence of enriched prey–predator systems was indeed seen in laboratory experiments (Luckinbill, 1974), although the corresponding mechanisms remained unclear. More recently, it has been shown both in experimental studies (Holyoak, 2000) and theoretically (Jansen, 1995, 2001) that the existence of a predefined patchy spatial structure makes a prey–predator system less prone to extinction. In a spatially structured metapopulation, the temporal variations of the density of different subpopulations can become asynchronous and the events of local extinction can be compensated for due to re-colonization from other sites (Allen et al., 1993).

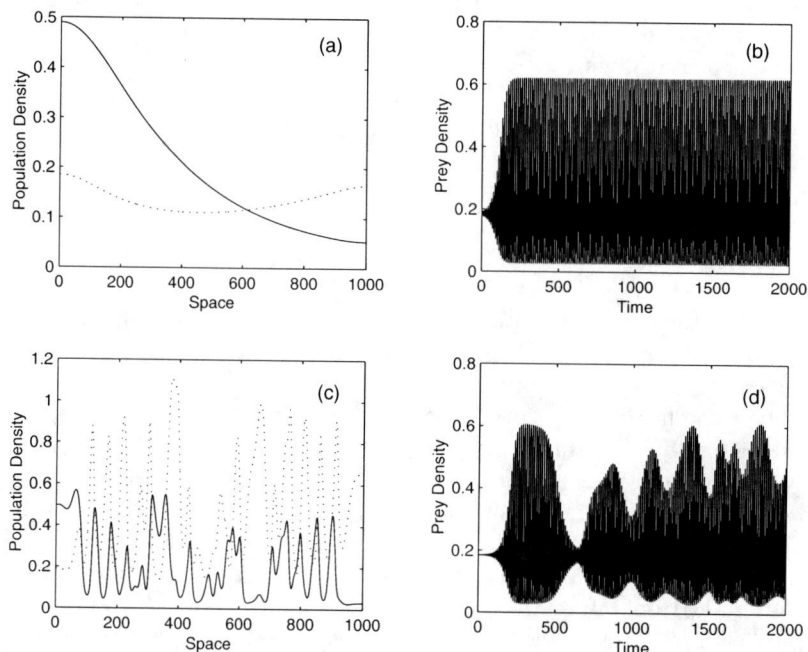

**FIGURE 11.15**: Spatial (a, c) and temporal (b, d) variations of the population densities in case of regular (a, b) and chaotic (c, d) dynamics as described by Equations (11.40)–(11.41). Snapshots of the species distribution in (a) and (c) (solid line for prey, dotted for predator) are taken at moment $t = 2000$. The corresponding initial conditions were chosen in the form of constant-gradient perturbation of the coexistence steady state, i.e., $u(x,0) = \bar{u}$, $v(x,0) = \bar{v} + \nu x + \delta$, with parameter values $\nu = 10^{-5}$ and (a) $\delta = 0.01$ and (c) $\delta = -0.005$. Other parameters are $k = 2.0$, $m = 0.6$, and $h = 0.43$ (with permission from Petrovskii et al., 2004).

Our goal here is to investigate the impact of enrichment on persistence / extinction of the species in a model prey–predator system taking into account the existence of the two different types of dynamics; see Figure 11.15. The idea is that, in the case that the temporal variations of the species are synchronized or strongly correlated over the whole space, as it takes place in the regular regime, a fall of the population density to a dangerously small value takes place almost simultaneously throughout the system and thus the species may go extinct without any chance of subsequent re-colonization (Allen et al., 1993). On the contrary, if the temporal variations are desynchronized, as it takes place in the chaotic regime, a local extinction does not necessarily lead to global extinction because the empty sites can be re-colonized via "diffusion" of the individuals from other sites. A related but somewhat more general question is whether these two types of dynamics are robust with respect to the system's enrichment.

In order to make a further insight into this problem, we use the prey–predator model from the preceding sections:

$$\frac{\partial u}{\partial t} = \frac{\partial^2 u}{\partial x^2} + u(1-u) - \frac{u}{u+h}v , \qquad (11.40)$$

$$\frac{\partial v}{\partial t} = \frac{\partial^2 v}{\partial x^2} + k\frac{u}{u+h}v - mv . \qquad (11.41)$$

Note that the system (11.40)–(11.41) is in dimensionless form where $h$, $k$, and $m$ are expressed through the original parameters as $h = H/K$, $k = \kappa A/\alpha$, and $m = M/\alpha$; see Section 10.1 for details. Enrichment of the system leads to an increase in the prey carrying capacity $K$ and thus to a decrease in $h$. A reasonable biological alternative would be an increase in the prey linear growth rate $\alpha$ and thus a decrease in $k$ and $m$. However, in Section 10.1 we showed that variations in $k$ do not have any significant impact on the stability of the steady state. In its turn, a decrease in $m$ has an impact similar to that of a decrease in $h$ (a point in the parameter plane of the system moves further away from the Hopf bifurcation curve; see Figure 10.1). Therefore, for the sake of simplicity, here we restrict our consideration to the case when enrichment only affects $h$.

In order to distinguish between different types of system dynamics, we use the spatially averaged population density as a convenient "measure" describing correlations between the temporal variations in population density at different positions in space. Indeed, in case temporal variations are strongly correlated throughout the domain (cf. Figure 11.15a), the amplitude of the variation of the average density is close to the amplitude of the local variations, which tends to grow with the system eutrophication. However, if the population oscillations are not correlated throughout the domain (cf. Figure 11.15c), the amplitude of the variation of the average density is significantly smaller than that of the local density.

We begin with the case when eutrophication occurs instantaneously so that the value of $h$ changes from an initial value $h_0$ to a new value $h_1 = h_0 - \Delta h$ at a certain moment $t_0$. We assume that, prior to eutrophication, the system is in the regime of regular spatiotemporal oscillations and use the spatial distribution shown in Figure 11.15a as the initial condition. The system dynamics is then studied through extensive numerical simulations, with the values of $r$, $h_0$, and $\Delta h$ varying in a wide range.

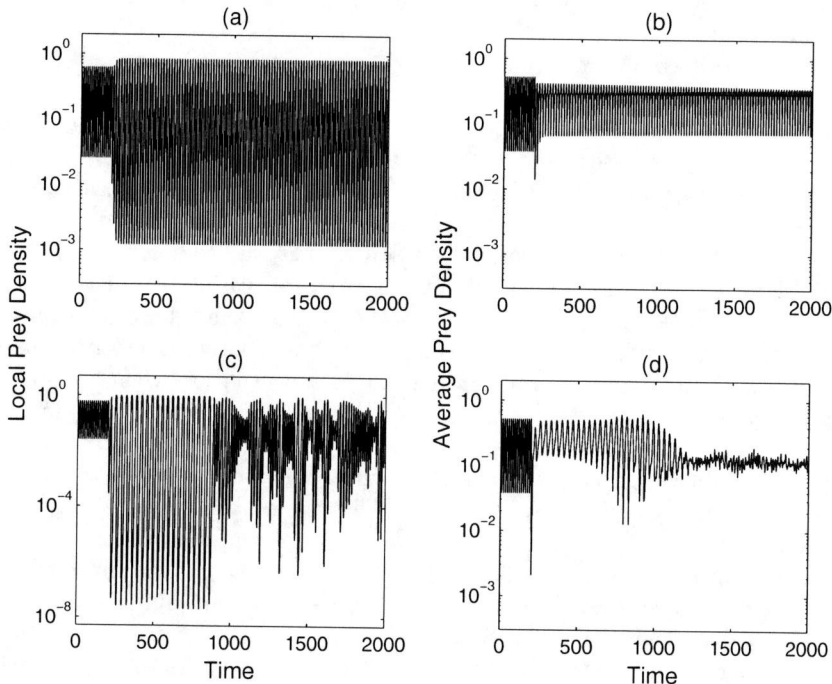

**FIGURE 11.16**: Temporal variation of the prey density (semilogarithmic plot) appearing as a response to an instantaneous (at $t = 200$) system's enrichment of different magnitude; (a-b) for $\Delta h = 0.13$ and (c-d) for $\Delta h = 0.27$. While regular dynamics persist for a smaller enrichment magnitude, a larger enrichment drives the system into spatiotemporal chaos; see comments in the text. In both cases, $h_0 = 0.43$; other parameters are the same as in Figure 11.15. For the initial conditions we used the regular "smooth" distribution of species shown in Figure 11.15a.

Some typical results are presented in Figure 11.16. Only temporal variations of the prey density are shown; the density of predator exhibits qualitatively similar behavior. Figures 11.16a and 11.16b show the system response after $h$ changed from $h_0 = 0.43$ to $h_1 = 0.3$ at the moment $t_0 = 200$. The amplitude of the local population oscillations increases considerably, so that the prey density periodically falls to a very small value. Interestingly, however, the minimum value of oscillating average prey density appears to be somewhat larger than it was before eutrophication: As a result of enrichment, the population fluctuations over the domain become less correlated. According to the above arguments, it makes global species extinction less probable. For comparison, enrichment in the corresponding nonspatial system would inevitably lead to oscillations of larger amplitude with a smaller minimum value of the population density; therefore, the impact of space has a clear positive effect on species survival.

Remarkably, the positive impact of space becomes much more prominent for eutrophication of a larger magnitude, i.e., for larger values of $\Delta h$. Figures 11.16c and 11.16d show the species temporal variations when at $t_0 = 200$ the value of $h$ decreases from $h_0 = 0.43$ to $h_1 = 0.16$. The system dynamics now changes from regular to chaotic, the properties of the species spatial distribution undergoing the corresponding changes (cf. Figures 11.15a and 11.15c). Although the local population density may fall to a very small value (see Figure 11.16c), the amplitude of the temporal variation of the average density decreases significantly; this reflects the fact that the oscillations of the species density at different positions in space become desynchronized. Desynchronized dynamics practically excludes the situation when the species density falls to a small value simultaneously at each position in space. In fact, the minimum value of the average population density now appears to be considerably larger than it was before eutrophication. Thus, following the above arguments, enrichment has made the probability of species extinction smaller, not larger.

Besides instantaneous changes, we also consider the case when eutrophication takes place gradually, during a certain finite time interval $\Delta t$. For simplicity, we assume that the value of $h$ changes from $h_0$ to $h_1$ linearly with time between $t_0$ and $t_1 = t_0 + \Delta t$. Typical results are shown in Figure 11.17. It is readily seen that, in this case, the type of system response depends also on the value of $\Delta t$. Simulations show that, while for small $\Delta t$ the system response is the same as for instantaneous enrichment (i.e., transition to chaos for sufficiently large $\Delta h$; cf. Figures 11.16d and 11.17, top), for sufficiently large $\Delta t$ the dynamics of the system remains regular; see Figure 11.17, bottom. Contrary to the case of instantaneous enrichment, the minimum value of the average population density now decreases drastically. That leads to a rather counter-intuitive conclusion that a fast eutrophication may appear to be less dangerous for the community functioning than a slow one.

The above results about the role of space and self-organized patchiness in resolving the paradox of enrichment are sketched in Figure 11.18, with the curves showing the "probability" (in a loose sense) of population extinction. Here curve 1 corresponds to the spatially homogeneous cases, i.e., to the classical paradox of enrichment. A monotonous increase in the probability of global extinction along with the magnitude of enrichment follows from the fact that the minimum average population density decreases when $\Delta h$ grows. The probability of extinction becomes dangerously high (above the horizontal dotted line) when the magnitude of enrichment exceeds a certain critical value $\Delta h_1$.

Curve 2 shows how the situation changes in a spatially heterogeneous system in the case of a fast (instantaneous) enrichment. Prior to eutrophication, the system is assumed to be in the regime of regular oscillations. A sudden increase in $K$ (decrease in $h$) makes the population oscillation less correlated

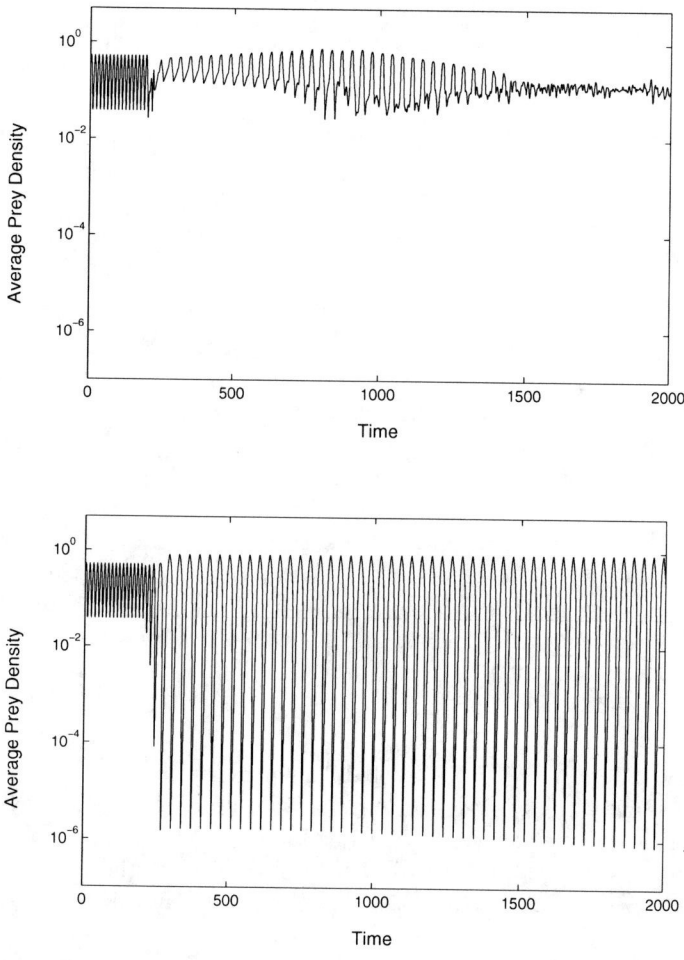

**FIGURE 11.17**: Temporal variation of spatially averaged prey density (semilogarithmic plot) in case of a gradual enrichment ($h_0 = 0.43$, $\Delta h = 0.27$) of different durations; top for $\Delta t = 5$ and bottom for $\Delta t = 50$. Other parameters and the initial conditions are the same as in Figure 11.16.

and thus increases the minimum average population density; hence the probability of global extinction decreases. A sufficiently large drop in $h$ drives the system into spatiotemporal chaos with an even larger value of minimum average density and, correspondingly, smaller probability of extinction.

Curve 3 outlines a typical response of a heterogeneous system to a slow (gradual) enrichment. In this case, a decrease in $h$ is accompanied by a decrease in the minimum average density; therefore, the larger the enrichment

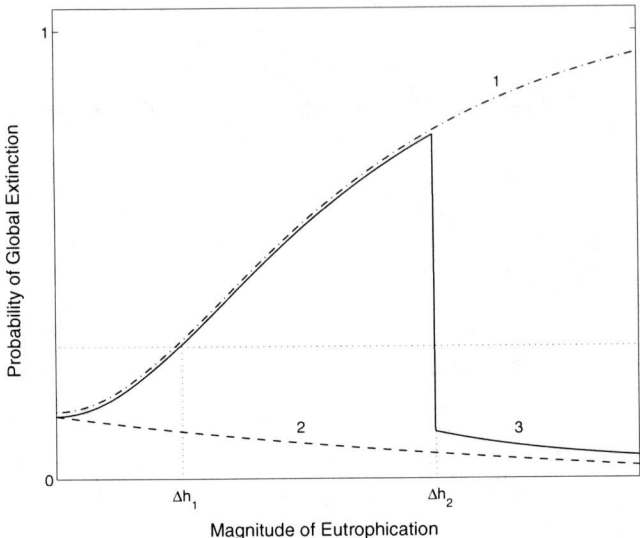

**FIGURE 11.18**: A sketch of the extinction probability dependence on the magnitude of a system's enrichment for spatially homogeneous (curve 1) and spatially heterogeneous (curves 2 and 3) systems; curves 2 and 3 show the system response for fast (instantaneous) and slow enrichments, respectively.

magnitude $\Delta h$, the higher the probability of extinction. However, this type of response only takes place for $\Delta h$ not too large. When $\Delta h$ exceeds the second critical value $\Delta h_2$ (which depends on the enrichment duration $\Delta t$), the regular dynamics changes to chaos. As a result, the local population oscillations become desynchronized and the probability of global extinction drops significantly.

Apart from the fact that the impact of space denounces the negative effect of eutrophication on global species persistence, the above results clearly show that enrichment can drive the system to spatiotemporal chaos. An issue of interest is then to understand how this type of system response can be modified by a variation in the length $L$ of the spatial domain. Since the system possesses an intrinsic length $L_{corr}$, one can expect that the properties of the system dynamics can be somewhat different for different values of the ratio $L/L_{corr}$, i.e., for different $L$ provided all other parameters are fixed. In order to address this issue, we have repeated simulations for various $L$. The results are summarized in Table 11.2. It is readily seen that regular dynamics in small habitats tends to be more stable to enrichment than it is in large ones. This is in a good agreement with more general conclusions about possible transitions to chaos in the spatially explicit prey–predator system; see Section 11.1.1 and also the paragraphs preceding Section 11.2.1. Consequently, since transition

**TABLE 11.2:** Chaos versus order. Pluses mark the cases when enrichment drives the system to spatiotemporal chaos, and minuses stand for the cases when the dynamics remains regu- lar; $L$ is the size of the domain. With permission from Petrovskii et al. (2004).

| $L$, dimensionless | $\Delta h$, dimensionless | | | | | |
|---|---|---|---|---|---|---|
| | 20 | 22 | 23 | 25 | 28 | 30 |
| 300 | − | − | − | − | − | + |
| 500 | − | − | − | − | + | + |
| 1000 | − | − | − | + | + | + |
| 1500 | − | − | + | + | + | + |
| 2000 | − | + | + | + | + | + |

to chaos is shown to decrease the probability of species extinction, it means that ecological communities dwelling in a small habitat are likely to be more endangered by enrichment than those living in large habitats.

The above analysis corresponds to the case when prior to eutrophication the system dynamics is regular. Now, what would be the difference if the system had already been in the regime of spatiotemporal chaos? In the previous sections we have shown that chaos is robust to parameter perturbations, unless either the steady state becomes stable and the local population oscillations are totally suppressed or the corresponding point in the parameter space comes very close to the Hopf-bifurcation curve so that $L_{corr} > L$. This is clearly not the case with enrichment when $h$ is decreasing, the system goes further away from the Hopf bifurcation, and the local oscillations become more prominent. Therefore, intuitively, it is rather obvious that a chaotic prey–predator system is unlikely to be sensitive to enrichment. Indeed, computer simulations show that the regime of spatiotemporal chaos is stable with respect to a decrease in $h$. In this case, although eutrophication of the system increases the amplitude of local population oscillations, it does not increase the danger of global species extinction in any significant way because the dynamics of the system remains chaotic and the population oscillations throughout the domain remain uncorrelated.

## 11.4  Concluding remarks

In this chapter, we have shown that the onset of "biological turbulence" – the formation of a prominent irregular spatial structure with the population oscillations desynchronized in space and chaotic in time – is an inherent property of a community of interacting populations. It should not be necessarily attributed to any environmental forcing (such as any predefined spatial heterogeneity), any special type of the initial conditions, or any particular type of interspecific interactions. The only necessary conditions that we were able to identify are that the system should possess a coexistence steady state and the corresponding system's kinetics must be oscillatory in a "broad sense," i.e., the coexistence state should be an unstable focus. The regular dynamics then appears to be unstable with respect to supercritical (but still very small) perturbations that drive the system into spatiotemporal chaos.

Note that, although conclusive evidence of chaos in the dynamics of real ecological populations is still lacking, chaotic population dynamics under laboratory conditions have indeed been observed (Costantino et al., 1997; Dennis et al., 2001; Becks et al., 2005; Cushing et al., 2003). The factors that may prevent observation of chaos in ecological data are not well understood yet. One plausible explanation is that the stationary ecological time series appropriate for the analysis simply cannot be made long enough due to the transient nature of the environmental conditions (Hastings, 2001, 2004). Nevertheless, here we have shown that spatiotemporal dynamics of a population community is intrinsically chaotic. This is likely to have crucial ecological implications (Scheffer, 1991b; Hastings et al., 1993): Chaos means sensitivity to the initial condition, and that makes an accurate long-term forecast of ecological dynamics principally impossible. For instance, we can calculate an average/expected size of a single population patch and the expected time of its existence, but we are not able to predict when and where it will actually appear; "we can indicate the range of things that can happen, but cannot predict when they will happen" (Scheffer, 1991b).

A more specific implication of our results concerns the choice of an adequate model to study chaotic dynamics. Namely, many theoretical results concerning spatiotemporal chaos in a system of interacting chemical or biological species have been obtained in terms of so-called $\lambda$-$\omega$ systems (the complex-variable analogue of the $\lambda$-$\omega$ system is called the Ginzburg–Landau equation); see Kopell and Howard (1973); Kuramoto (1984); Bohr et al. (1998), also Sherratt (2001) and the references therein. Briefly, a $\lambda$-$\omega$ system arises as the first order approximation of a general diffusion–reaction system considering the size of the limit cycle as a small parameter. An advantage of this approach is that $\lambda$-$\omega$ systems often appear to be mathematically much simpler than the original diffusion–reaction systems, especially in the case when the original system consists of many species. The $\lambda$-$\omega$ systems are formally valid for parameter values near the Hopf bifurcation, but there is also a strong opinion that they provide an adequate description of the corresponding diffusion–reaction systems in a much wider parameter range. Our results, however, seem to indicate that spatiotemporal chaos in the $\lambda$-$\omega$ systems is just a mathematical artifact without any clear relation to reality. It follows from Equations (11.26) and (11.36), as well as from simulations (e.g., see Figures 11.10 and 11.11), that the value of $L_{corr}$ grows unboundedly in vicinity of the Hopf bifurcation; therefore, in any finite spatial system (which is always the case for real biological or chemical systems) its size $L$ will be less than the correlation length. That makes the onset of chaos impossible exactly in the parameter range where the $\lambda$-$\omega$ systems are formally valid. This inconsistency might explain why the approach based on the $\lambda$-$\omega$ systems fails to describe the type of population dynamics correctly (cf. Sherratt, 2001), predicting spatiotemporal chaos for those cases where field observations report regular patterns.

# Chapter 12

# Patchy invasion

In the previous few chapters, we have demonstrated that self-organized patchiness uncorrelated with the environment is an inherent property of spatiotemporal population dynamics. Scenarios of pattern formation as well as the properties of the emerging heterogeneous population distribution can be rather different, though. In a system of nonlinearly interacting populations with unequal diffusivities, the Turing instability can destabilize a locally stable coexistence state and lead to the formation of a stationary, regular, spatially periodic structure. Conversely, destabilization of a population system with locally oscillatory kinetics normally results, even in the case of equal diffusion coefficients, in spatiotemporal chaos ("biological turbulence"), i.e., a spatially irregular patchy structure with the population densities fluctuating in time chaotically.

Whether intrinsic population cycles (e.g., due to the prey–predator interactions) are really common in nature remains a controversial issue (cf. White, 2001, 2004). Throughout this book, we strongly support the opinion that population oscillations are typical in population dynamics, even if they sometimes may arise as a response of a given community to some unfavourable, destabilizing factors (Gilpin, 1972; May, 1972a, 1973). The self-organized irregular patchiness then becomes a simple consequence of local oscillations as the corresponding spatially extended system is driven to spatiotemporal chaos by virtually any heterogeneous perturbation. For a few biologically reasonable situations, we have shown that the onset of biological turbulence follows the propagation of a certain traveling front – "the wave of chaos" – which separates the domains filled with chaotic population patches from the rest of the system where dynamics is regular.

Curiously, there can be another scenario when a prominent irregular patchy structure is formed straightforwardly, i.e., is not preceded by the propagation of a front. This scenario is usually associated with the existence of the Allee effect in the population growth (Petrovskii et al., 2002, 2005; Morozov et al., 2006). In this chapter, we revisit this phenomenon of "patchy invasion," identify biological situations when it is likely to occur, and make an insight into its basic properties. In particular, we will show that the patchy invasion may appear as a result of an overpressure on the invasive species in the course of biological control efforts. We will also show that the patchy invasion is invasion "at the edge of extinction," so that it takes place when no other scenario can apply and the invasive species would go extinct otherwise. Finally, we will have a look at some field data and show that this scenario of species spread and the corresponding pattern formation agree very well with what is often observed in nature.

## 12.1 Allee effect, biological control, and one-dimensional patterns of species invasion

Most of the analysis in previous chapters was done under assumptions that the population per capita growth rate $f(U)$ reaches its maximum when the population density $U$ tends to zero and that it decreases monotonically when the population density increases (turning to zero for $U = K$, where $K$ is the population carrying capacity). In mathematical terms, it means that the growth rate $P(U)$ should be a convex function. The simplest parametrization for the growth rate $P(U)$ is then the square polynomial, which is known as the logistic growth.

However, this is not always the case with real ecological populations. There has been a growing number of examples (Courchamp et al., 1999; Stephens and Sutherland, 1999), along with relevant theoretical arguments (Dennis, 1989; Stephens et al., 1999), where the per capita growth rate reaches its maximum value for an intermediate (nonzero) value of the population density. This is called the Allee effect (Allee, 1938) and the corresponding population dynamics is sometimes called the Allee dynamics. Special attention has been paid to the case of a "strong" Allee effect (when the population growth becomes negative if the population density goes below a certain threshold density $B$; cf. Owen and Lewis, 2001; Wang and Kot, 2001), in particular, because the deterministic population dynamics becomes more realistic in this case, excluding a possibility of population persistence once the population density has fallen to a very low value. Moreover, the Allee effect was shown to affect virtually all aspects of species interactions in space and time (Berryman, 1981; Amarasekare, 1998; Gyllenberg et al., 1999; Taylor and Hastings, 2005; Courchamp et al., 2008).

Throughout this section, we will focus on the spatiotemporal dynamics of the prey–predator system described by the following equations:

$$\frac{\partial U(X,Y,T)}{\partial T} = D_1 \left( \frac{\partial^2 U}{\partial X^2} + \frac{\partial^2 U}{\partial Y^2} \right) + P(U) - E(U,V) , \quad (12.1)$$

$$\frac{\partial V(X,Y,T)}{\partial T} = D_2 \left( \frac{\partial^2 V}{\partial X^2} + \frac{\partial^2 V}{\partial Y^2} \right) + \kappa E(U,V) - MV , \quad (12.2)$$

where all terms have their usual meanings.

We assume that predation is of Holling type II and use the standard functional form; see (2.5a) or (10.4). Regarding the prey growth, we now assume that it is damped by the strong Allee effect. Mathematically, the population growth under the impact of the Allee effect should be described by a function that is concave in a vicinity of $U = 0$ but becomes convex for larger $U$. For mathematical analysis and simulations, different parametrization can be used; perhaps the simplest and the most common one is the cubic polynomial:

$$P(U) = \tilde{\alpha} U(U - B)(K - U) , \qquad (12.3)$$

where $\tilde{\alpha}$ is a coefficient proportional to the maximum per capita growth rate; see Lewis and Kareiva (1993).

For technical reasons, here we introduce dimensionless variables differently from the two previous chapters, that is, as

$$u = \frac{U}{K}, \quad v = \frac{V}{\kappa K}, \quad x = X\left(\frac{a}{D_1}\right)^{1/2}, \quad y = Y\left(\frac{a}{D_1}\right)^{1/2}, \quad \text{and} \quad t = aT,$$

where $a = A\kappa K/H$, and $A$ and $H$ are the parameters quantifying predation. Then, from (12.1)–(12.2), we obtain

$$\frac{\partial u(x,y,t)}{\partial t} = \left(\frac{\partial^2 u}{\partial x^2} + \frac{\partial^2 u}{\partial y^2}\right) + \gamma u(u - \beta)(1 - u) - \frac{uv}{1 + \alpha u} , \qquad (12.4)$$

$$\frac{\partial v(x,y,t)}{\partial t} = \epsilon\left(\frac{\partial^2 v}{\partial x^2} + \frac{\partial^2 u}{\partial y^2}\right) + \frac{uv}{1 + \alpha u} - \delta v , \qquad (12.5)$$

where $\alpha = K/H$, $\beta = B/K$, $\gamma = \tilde{\alpha} HK/(A\kappa)$, $\delta = M/a$, and $\epsilon = D_2/D_1$. For the most of this chapter, we fix $\epsilon = 1$; the effect of differential diffusivity will be briefly addressed in Section 12.4.

In this section, we will focus on the one-dimensional dynamics; correspondingly, system (12.4)–(12.5) is reduced to

$$\frac{\partial u(x,t)}{\partial t} = \frac{\partial^2 u}{\partial x^2} + \gamma u(u - \beta)(1 - u) - \frac{uv}{1 + \alpha u} , \qquad (12.6)$$

$$\frac{\partial v(x,t)}{\partial t} = \epsilon\frac{\partial^2 v}{\partial x^2} + \frac{uv}{1 + \alpha u} - \delta v . \qquad (12.7)$$

Before proceeding to the spatiotemporal dynamics of the system (12.6)–(12.7), it is worth having a brief look at its nonspatial counterpart:

$$\frac{du}{dt} = \gamma u(u - \beta)(1 - u) - \frac{uv}{1 + \alpha u} , \quad \frac{dv}{dt} = \frac{uv}{1 + \alpha u} - \delta v . \qquad (12.8)$$

The isoclines of the system (12.8) are given by

$$\text{(a)} \quad v = \gamma(1 + \alpha u)(u - \beta)(1 - u) \quad \text{and} \quad \text{(b)} \quad u = \frac{\delta}{1 - \alpha\delta} . \qquad (12.9)$$

In the biologically meaningful first quadrant ($u \geq 0$, $v \geq 0$) of the phase plane, the isocline (12.9a) is dome-shaped and the stability of the coexistence steady state (given by the isocline intersection) is different depending on the position of the vertical line (12.9b). Considering $\delta$ as a controlling parameter, which will be justified later, for the values of delta large enough (but not exceeding $1/\alpha$ when the steady state disappears from the first quadrant) the intersection point is on the right of the hump and the steady state is stable. Let us note, however, that, contrary to the case of the prey–predator system with the logistic growth for prey, the extinction steady state $(0,0)$ is now stable for any parameter values. Whether the system trajectories go to the extinction state or to the coexistence state depends on their position with respect to the separatrix (see Figure 12.1), while the other two steady states $(\beta, 0)$ and $(1, 0)$ are the saddle-points.

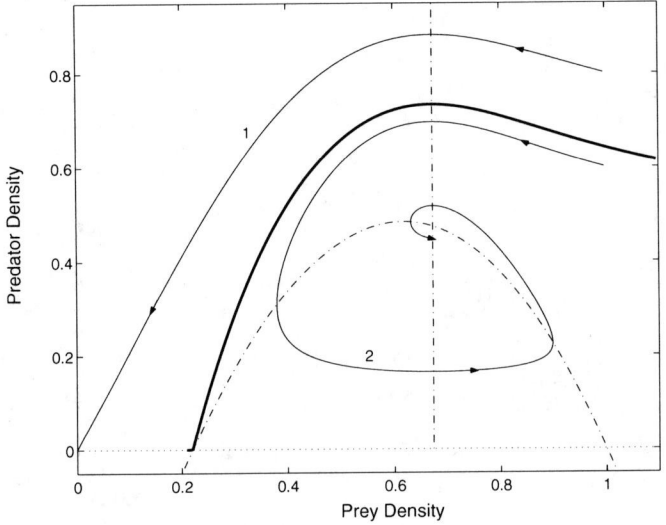

**FIGURE 12.1**: Phase plane of the system (12.8) for $\delta_{Hopf} < \delta < 1/\alpha$. Curves 1 and 2 show typical system trajectories while the thick curve shows the separatrix; see more comments in the text. Parameters are $\alpha = 0.1$, $\beta = 0.22$, $\gamma = 3$, and $\delta = 0.63$.

As $\delta$ becomes smaller, the intersection point passes the hump for $\delta = \delta_{Hopf}$ and the steady state loses its stability through the Hopf bifurcation. The attractors of the system are then the stable limit cycle and the extinction state, the two fields of trajectories being separated by the separatrix; see Figure 12.2.

The limit cycle does not persist for long, though. Along with a further decrease in $\delta$, it promptly grows in size. When $\delta$ reaches a certain critical

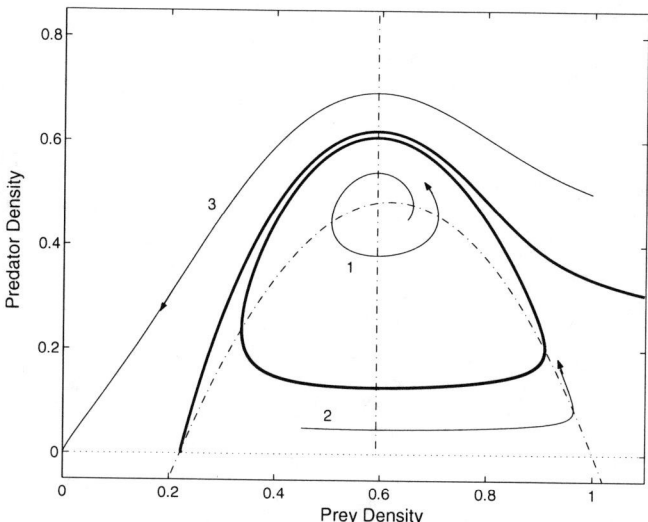

**FIGURE 12.2**: Phase plane of the system (12.8) for $\delta_{het} < \delta < \delta_{Hopf}$. Curves 1, 2, and 3 show typical system trajectories while the thick curves show the limit cycle and the separatrix; see more comments in the text. Here $\delta = 0.56$, and other parameters are the same as in Figure 12.1.

value $\delta_{het}$, it touches the saddle-points $(\beta, 0)$ and $(1, 0)$ and a heteroclinic connection is created. For $\delta < \delta_{het}$, the limit cycle disappears, the coexistence state is unstable, and the only attractor in the system is the extinction state; see Figure 12.3. The system kinetics then becomes "excitable": Depending on how far the initial point (given by the initial conditions of the system (12.8)) is from the extinction state, the system either returns to $(0, 0)$ straightforwardly (cf. curve 1) or after a long excursion over the phase plane (cf. curve 2). Which of the two trajectories actually takes place apparently depends on whether the initial point lies on the left or on the right of the separatrix connecting the coexistence steady state to the saddle $(\beta, 0)$; the two situations can be interpreted as "undercritical" and "supercritical" perturbations, respectively.

Therefore, the properties of the nonspatial system (12.8) accounting for the Allee effect are significantly different from those in case of the logistic growth; cf. Figure 11.14. Correspondingly, one can expect that the properties of the spatiotemporal dynamics can be rather different as well, and probably more complicated, because the structure of the phase plane and the succession of bifurcations have now become more complicated. For convenience, below we will refer to the phase portraits shown in Figures 12.1, 12.2, and 12.3 as, respectively, kinetics of type I (stable coexistence state), type II (unstable coexistence state, stable limit cycle), and type III (excitable kinetics).

Now, we are nearly ready to proceed to an analysis of the spatiotemporal system (12.6)–(12.7). However, one point that yet needs to be clarified is

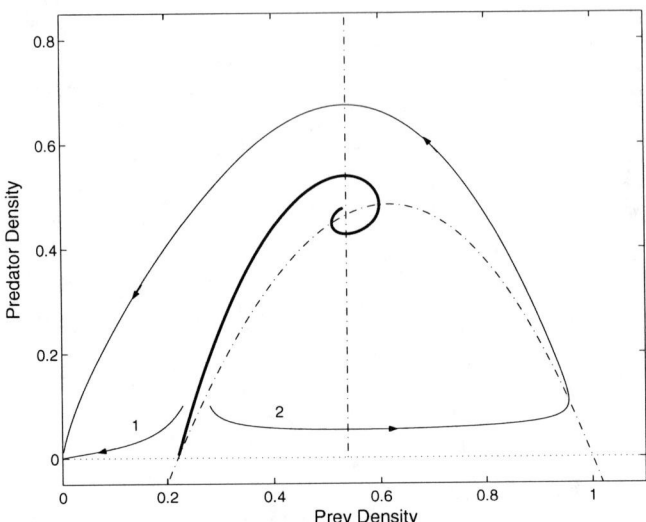

**FIGURE 12.3:** Phase plane of the system (12.8) for $\delta > \delta_{het}$. The thick curve shows the separatrix while curves 1 and 2, respectively, show typical system trajectories for "undercritical" and "supercritical" perturbations of the extinction steady state; see more comments in the text. Here $\delta = 0.51$, and other parameters are the same as in Figure 12.1.

the choice of the initial conditions. In order to do that properly, one has to take into account relevant biological arguments. Invasion of exotic species usually starts with the species introduction when a number of individuals of a given species are brought locally into a new ecosystem. Considering this species as a prey, it means that the initial condition for Equation (12.6) is most naturally described by a function of compact support. In particular, in numerical simulations we used the following initial population distribution:

$$u(x,0) = u_0 \quad \text{for} \quad -\Delta_u < x < \Delta_u, \quad \text{otherwise } u(x,0) = 0 , \quad (12.10)$$

where $u_0$ is the initial prey density and $\Delta_u$ gives the radius of the initially infested domain.

Regarding the spread of invasive pests, a lot of attention has been paid recently to a possibility of biological control which, contrary to more traditional control measures through application of chemical pesticides, is expected to be more effective and ecologically friendly. In particular, predation has been identified as a factor that can potentially slow down or even block the invasive species spread (Fagan and Bishop, 2000; Owen and Lewis, 2001; Fagan et al., 2002; Petrovskii and Malchow, 2005). Let us note here that, in order to be fully relevant to the issue of control, the biological factor itself must be controllable.[1] From that perspective, the use of predation for invasive species control seems to be adequate and convenient because the magnitude of predation can be relatively easily regulated through predator mortality (e.g., by means of additional harvesting or hunting); higher mortality would correspond to lower predation. For that reason, in our insight into the patterns of species spread below, we will use the dimensionless mortality $\delta$ as the main controlling parameter.

In practice, the idea to use predation as a tool in order to affect the species spread implies that, soon enough after introduction of the given exotic species, a small population of a relevant predatory species is introduced locally into the region already inhabited by its prey. Thus, the corresponding initial condition for Equation (12.5) is as follows:

$$v(x,0) = v_0 \quad \text{for} \quad -\Delta_v < x < \Delta_v, \quad \text{otherwise } v(x,0) = 0 \;, \quad (12.11)$$

where $v_0$ is the initial predator density and $\Delta_v$ is the radius of the area where the predator is introduced.

It should also be mentioned that the initial conditions (12.10)–(12.11) are somewhat idealized and in reality the form of the species initial distribution can of course be much more complicated. However, results of computer simulations show that the type of the system dynamics depends much more on the radius of the initially inhabited domain and on the population density inside rather than on the details of the population density profile.

### 12.1.1 Patterns of species spread

The system (12.6)–(12.7) is a system of nonlinear PDEs and thus is difficult to study analytically; however, it is relatively easy to reveal its main properties by means of numerical simulations.

Before proceeding to the description of possible invasion scenarios, the issue of a system's dynamics dependence on the initial conditions should be clarified somewhat further. In a space- and time-continuous prey–predator system (described by diffusion–reaction equations (12.1)–(12.2)) with logistic growth for prey, any infinitesimal initial distribution of prey results in its successful invasion. In case the prey growth is affected by the strong Allee effect, however, the system exhibits certain criticality due to the existence of the threshold population density: A successful invasion of prey cannot happen unless the initial density is large enough; otherwise, the species goes extinct (Lewis and Kareiva, 1993; Petrovskii and Li, 2006). Therefore, in order to exclude this somewhat trivial case, in computer experiments $u_0$ and/or $\Delta_u$ should always be chosen sufficiently large.

---

[1] For instance, the spread of airborne species is obviously affected by the direction and strength of the wind, but, since these two features are totally out of our control, we cannot use it for invasive species management.

All regimes of the system dynamics can be classified into three groups; see Figure 12.4. These groups correspond to extinction, invasion on a geographical scale when the species keep spreading until they reach the domain boundaries, and regional persistence when the species invade locally and spread over a certain area but do not go farther.

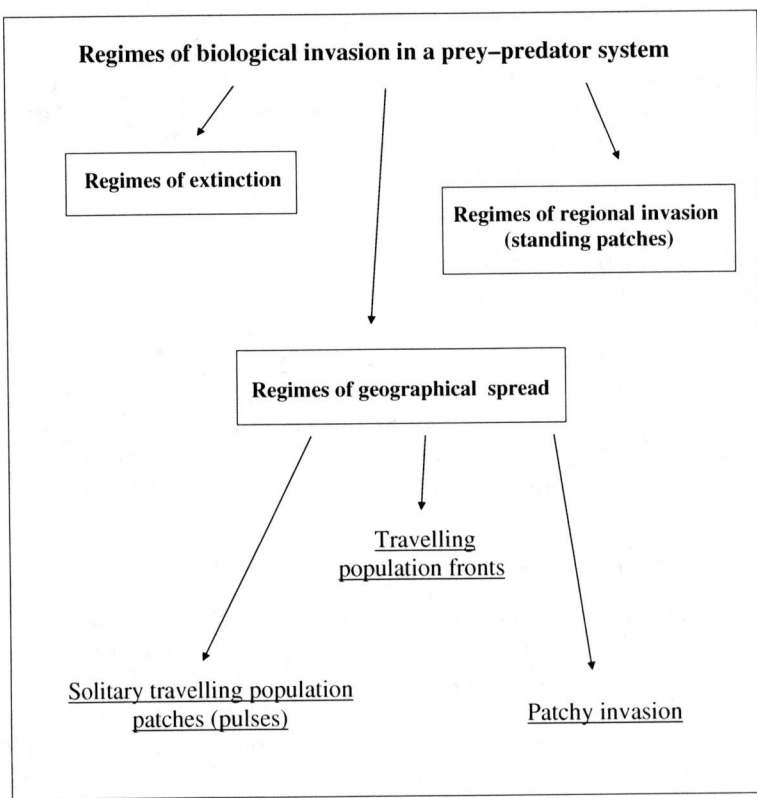

**FIGURE 12.4:** A diagram of possible invasion scenarios in a prey–predator system with the strong Allee effect.

We begin with the regimes describing the unbounded spread of the invasive species (geographical invasion). There are three qualitatively different scenarios of the species spread.

According to the first scenario, the species is spreading over space through propagation of a traveling population front. In front of the front the species is absent, and behind the front it is present at a considerable density. Depending on parameter values, in the wake of the front there can arise either a station-

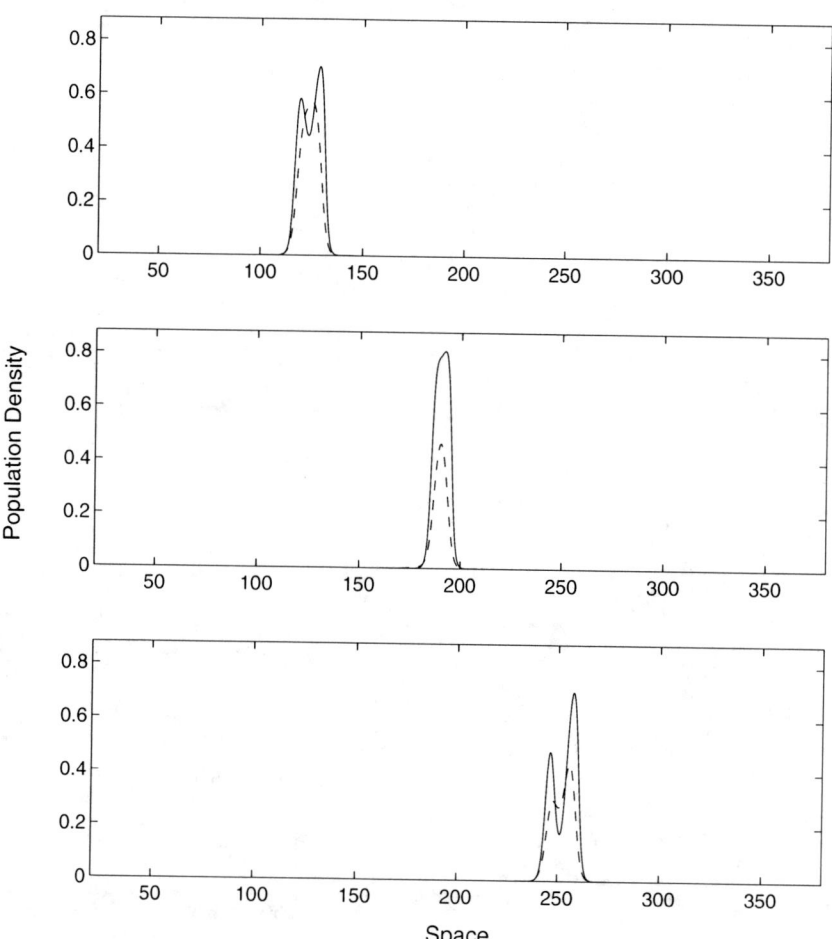

**FIGURE 12.5**: Solitary traveling patches of population density (solid curves for prey, dashed for predator) at $t = 350$, 600 and 850 (top to bottom, respectively) as predicted by the model (12.6)–(12.7). Parameters are $\alpha = 0.5$, $\beta = 0.28$, $\gamma = 3$, and $\delta = 0.425$. The initial conditions are given by (12.10)–(12.11) with $u_0 = 1$, $v_0 = 0.2$, $\Delta_u = 40$, and $\Delta_v = 20$. Since the problem is symmetrical with respect to the origin, only a half of the domain is shown.

ary spatially homogeneous species distribution (local kinetics of type I) or spatiotemporal population oscillations (local kinetics of either type II or III), which can be either regular or, more typically, chaotic. For some parameter values, the propagating front is linked to the domain with chaotic oscillations by means of a quasi-homogeneous plateau where the values of population

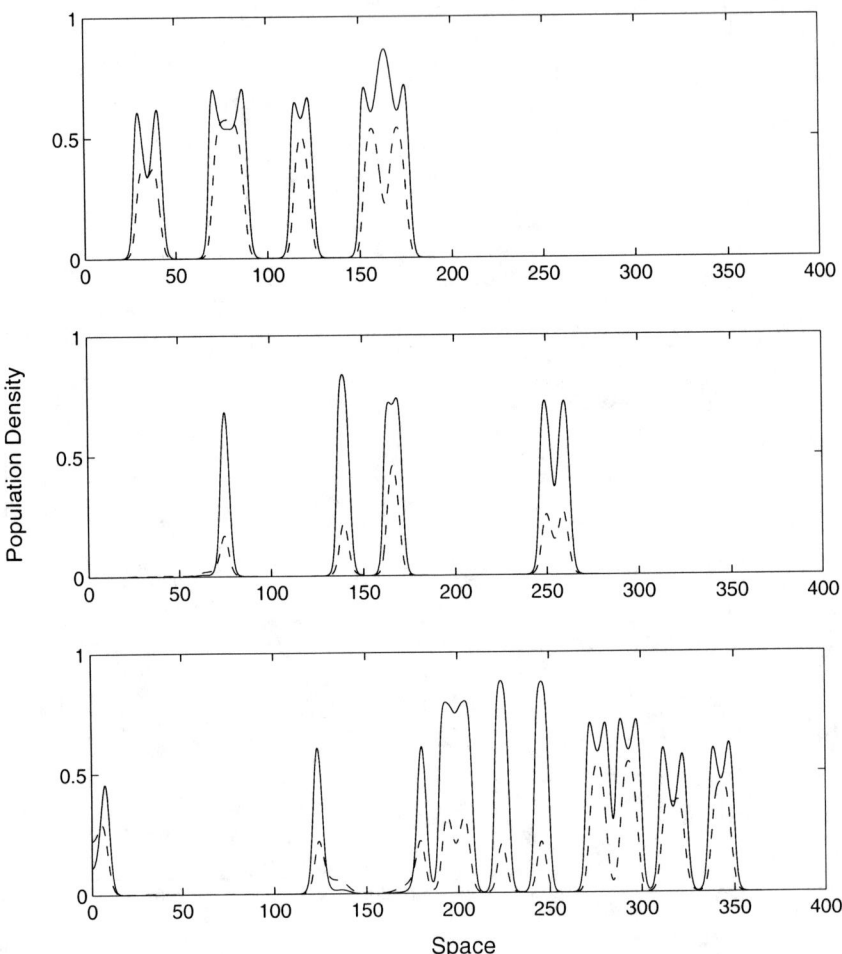

**FIGURE 12.6**: Species invade over space through irregular dynamics of separate patches. The snapshots of the population density (solid curves for prey, dashed for predator) is shown at $t = 600$, $1000$, and $1400$ (top to bottom, respectively). Parameters are $\alpha = 0.05$, $\beta = 0.28$, $\gamma = 3$, and $\delta = 0.53$, and the initial conditions are the same as in the previous figure.

density correspond to the locally unstable equilibrium; cf. the phenomenon of "dynamical stabilization" considered in Chapter 10. Apparently, the pattern of spread through the propagation of a population front is of high ecological relevance (cf. Shigesada and Kawasaki, 1997). However, as a whole, this scenario is very similar to what has already been observed and studied for the system without Allee effect; therefore, we do not go into more detail here.

According to the second scenario, the species spread over space via propagation of a solitary moving patch, or traveling "pulse"; cf. Figure 12.5. In this case, local kinetics is of type III; indeed, traveling pulses are often regarded as a "fingerprint" of excitable kinetics in a spatiotemporal system's dynamics (e.g., see Lindner et al., 2004, and references therein). Depending on parameter values, the traveling population pulse can be either stationary (when its shape does not change with time) or nonstationary (when its shape oscillates with time); in both cases, the pulse propagates with a constant speed.

Note that, in this case, the invasive species is absent both in front of the pulse and in its wake; the latter apparently means that invasion has failed. This fact has a curious ecological interpretation, i.e., there is species spread on a geographical scale but there is no invasion. In Sections 12.2 and 12.4, we will discuss this observation further and show that it may have important implications for biological control strategies and invasive species management.

From the point of ecological pattern formation, however, it is the third scenario that is probably the most interesting. In this case, invasion takes place through the formation and motion of separate patches and/or groups of patches; see Figure 12.6. However, the patch motion is now much more complicated than the simple locomotion in the case of traveling pulses. There is no traveling wave. The patches interact with each other, they merge and split, some of the patches or even groups of patches can disappear, new patches are formed, they can produce new groups of patches, etc. The invaded area grows and eventually the patches occupy the whole domain. Remarkably, the size of the domain occupied by the moving patches does always not grow monotonically: The boundary of the invaded domain can be set back when the leading group of patches goes extinct in the course of the system dynamics (Petrovskii et al., 2005).

### Regimes of regional invasion

When predation becomes high enough (specific value of $\delta$ depends on other parameters), unbounded species spread can no longer take place. However, invasion may still take place "regionally": For certain parameter values, evolution of the initial species distribution leads to the formation of quasistationary patches; see Figure 12.7. In this case, at early stage of the system dynamics (for $t \simeq 100$), two symmetric dome-shaped patches are formed. Their position then remains fixed, although their shape can be either stationary or oscillating. Interestingly, in the latter case, in spite of the fact that the spatial structure is pretty simple and regular, temporal fluctuations of the population density can exhibit rather complicated dynamics such as $n$-periodic limit cycles and even chaos (Morozov et al., 2004).

### Regimes of anomalous extinction

When the magnitude of predation is very large, simulations show that species extinction becomes inevitable, which apparently signifies the final success of the biological control strategy. Dynamically, it may happen in a rather

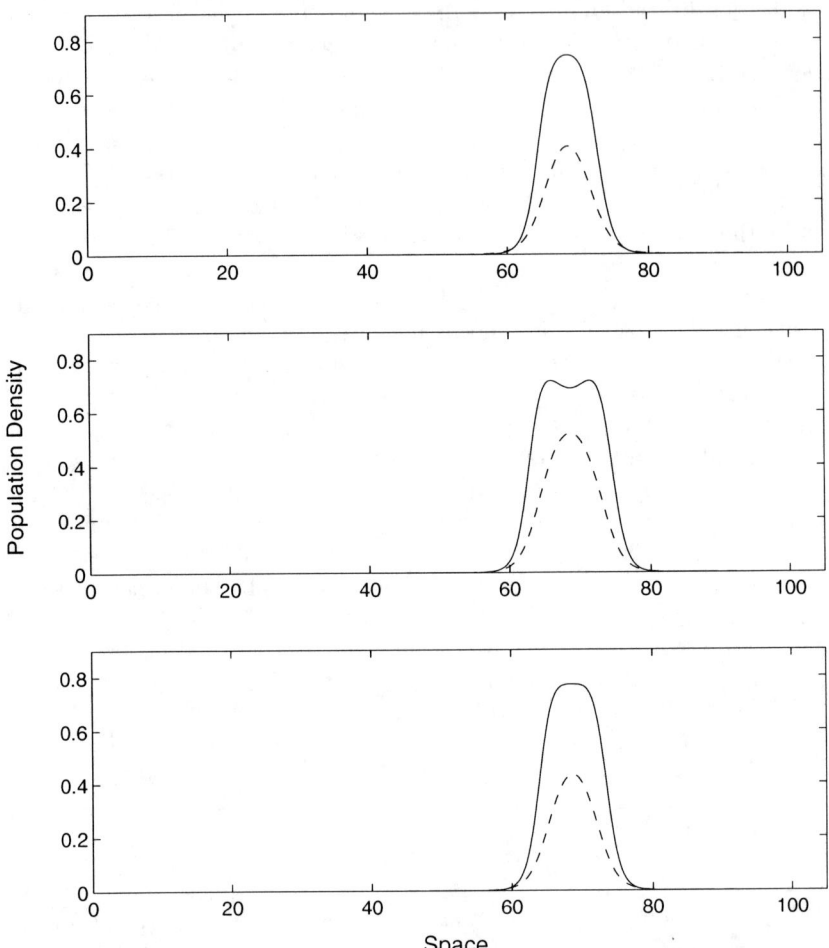

**FIGURE 12.7**: Regional invasion due to the formation of standing patches. The snapshots of population density (solid curves for prey, dashed for predator) are shown at $t = 200$, 750, and 2000 (top to bottom, respectively). Parameters are $\alpha = 0.5$, $\beta = 0.32$, $\gamma = 3$, and $\delta = 0.455$.

different manner, though. In the above, we have already mentioned that an introduced species affected by the strong Allee effect will go extinct if its initial population size is not large enough. In this case, the population size decreases exponentially while the population stays localized inside about the same domain where it had originally been introduced. This scenario of species extinction seems to agree perfectly with what is intuitively expected.

Due to the interplay between the Allee effect and predation, however, species extinction can also follow more exotic dynamical scenarios. There

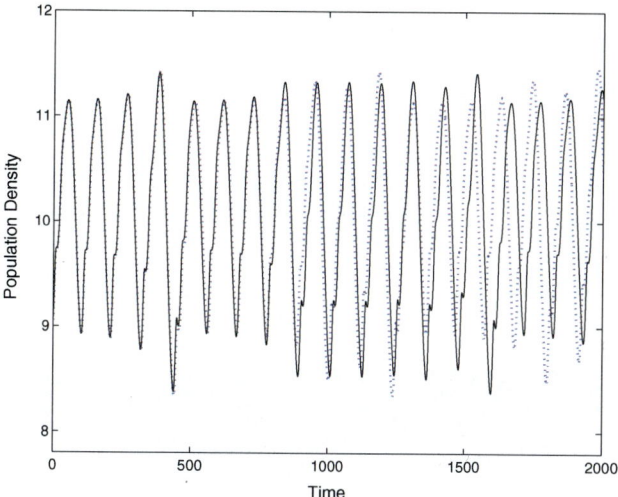

**Figure 4.6**

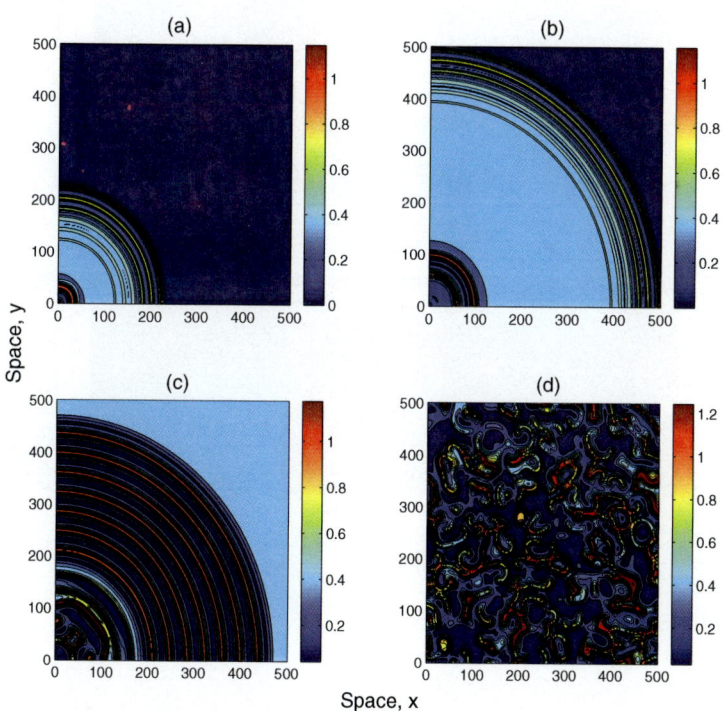

**Figure 10.7**

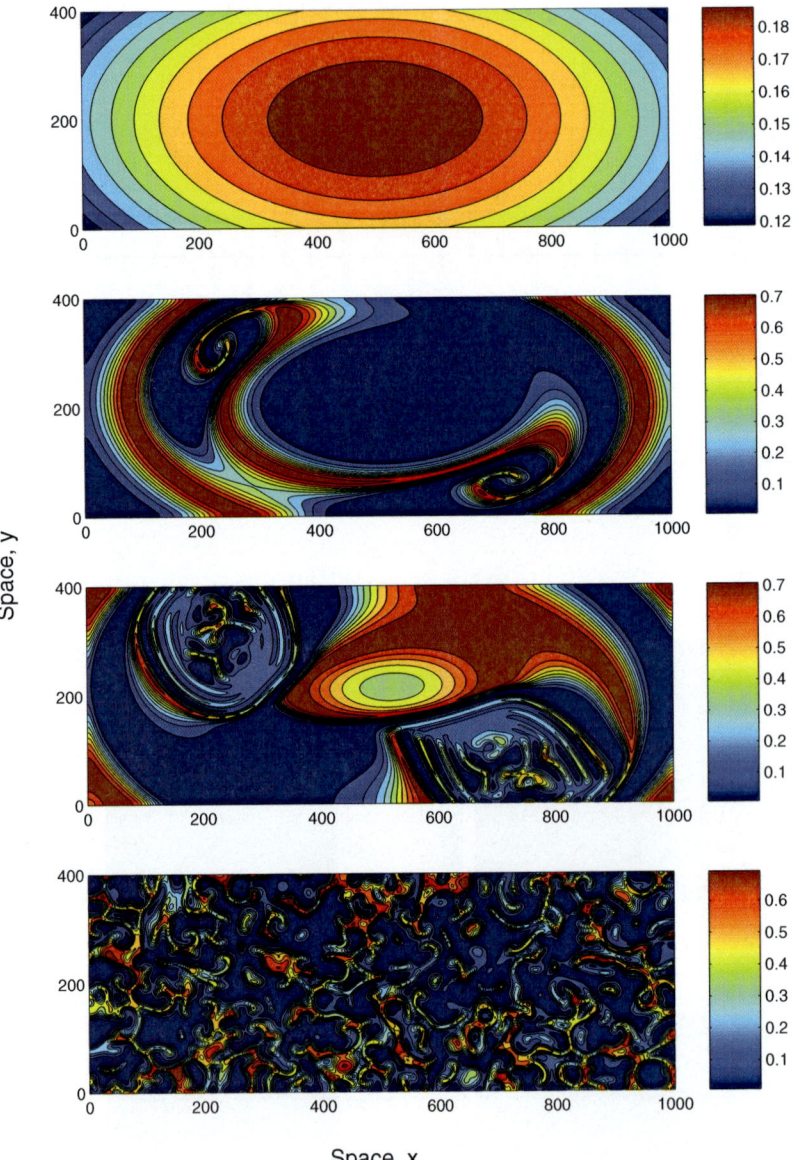

**Figure 11.8**

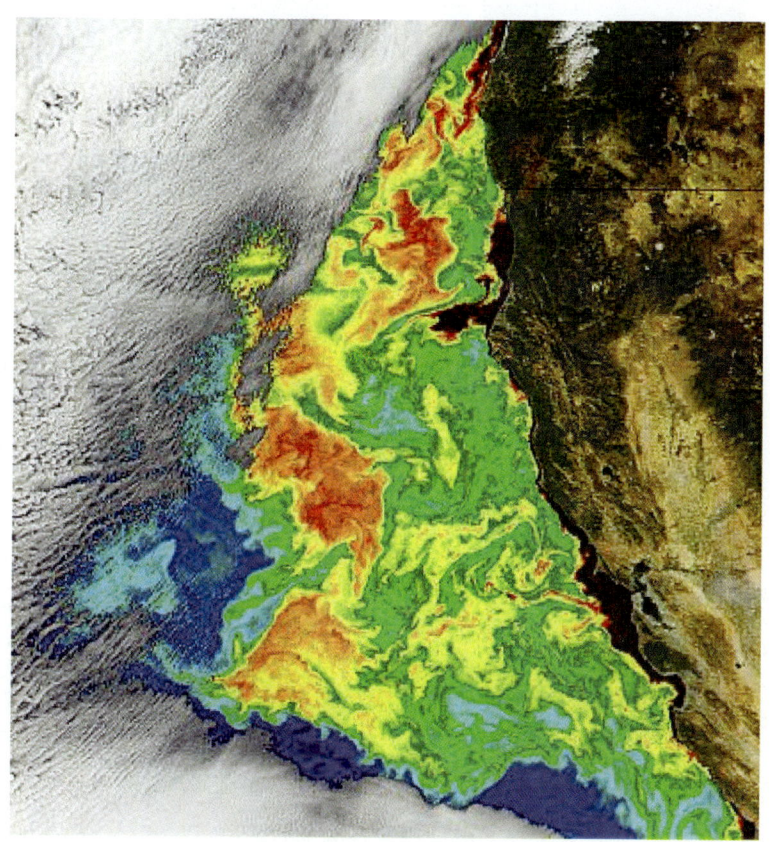

**Figure 11.13**

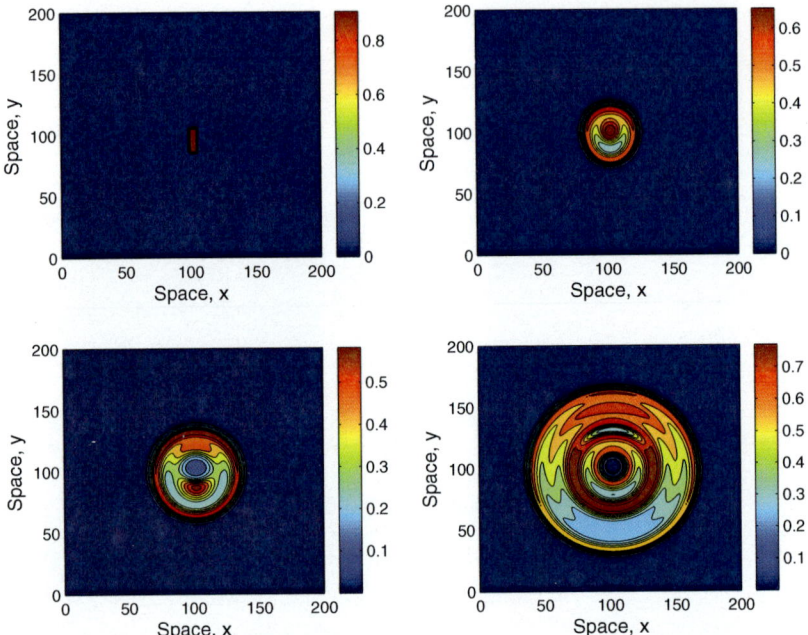

**Figure 12.16**

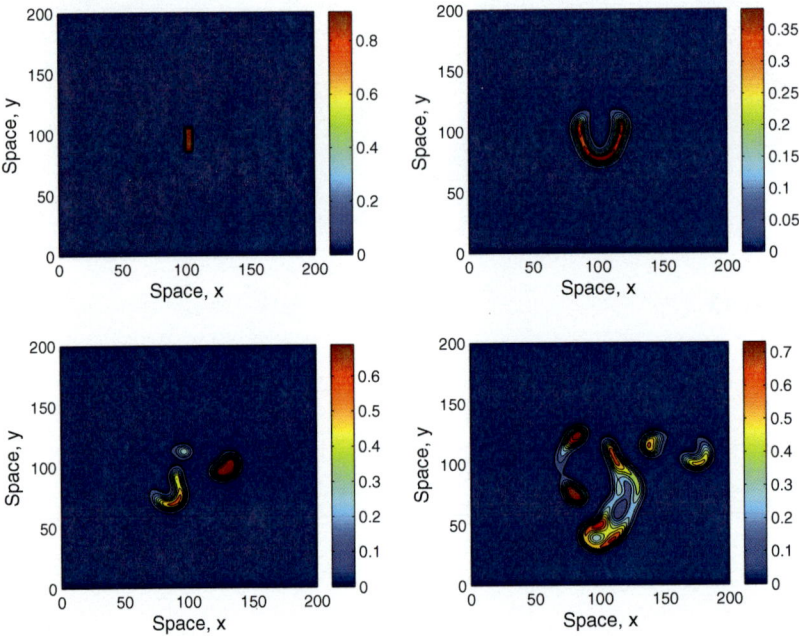

**Figure 12.17**

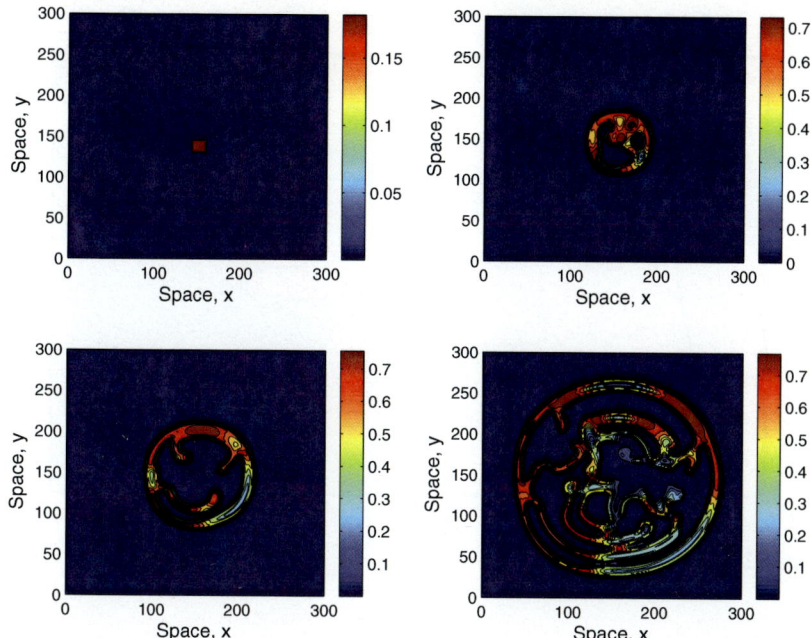

**Figure 12.18**

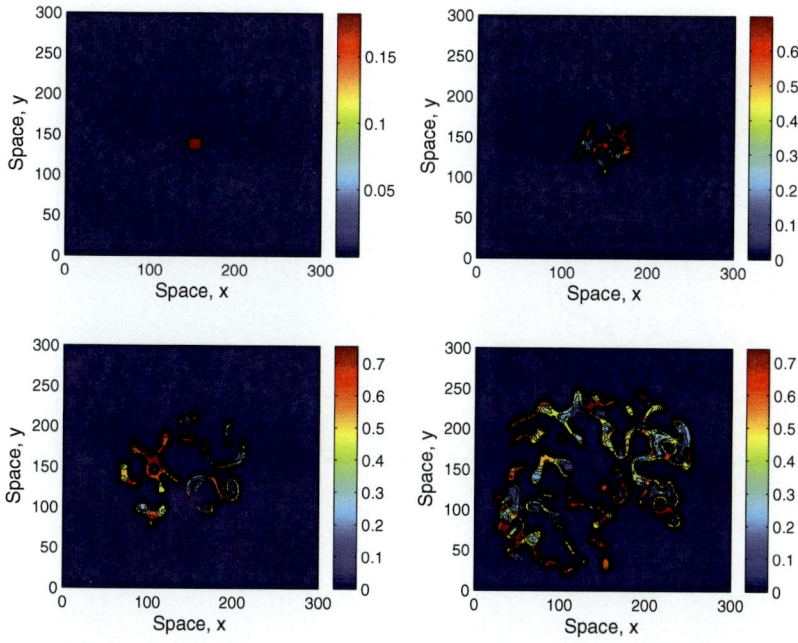

**Figure 12.19**

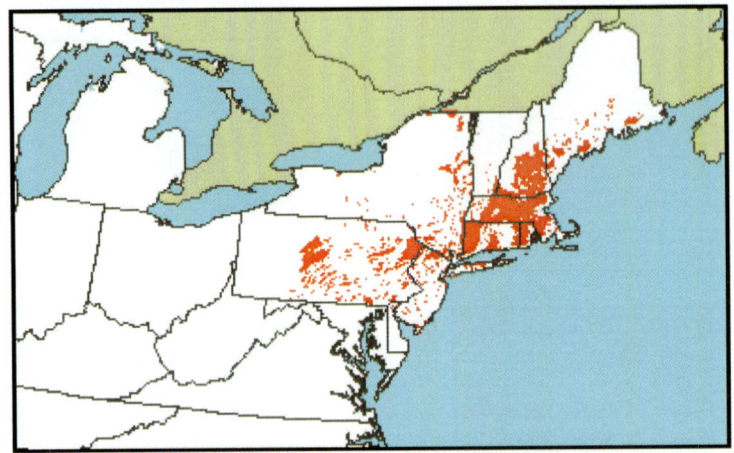

**Figure 12.21a**

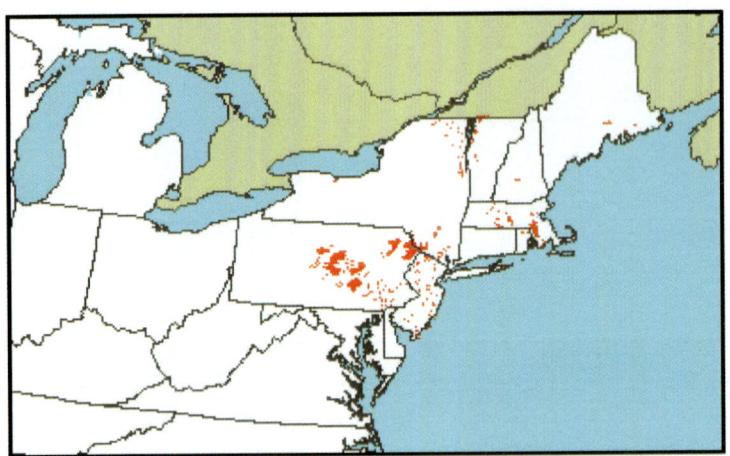

**Figure 12.21b**

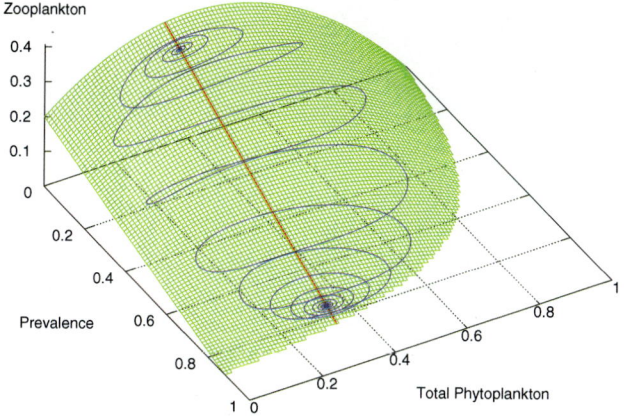

**Figure 15.1**

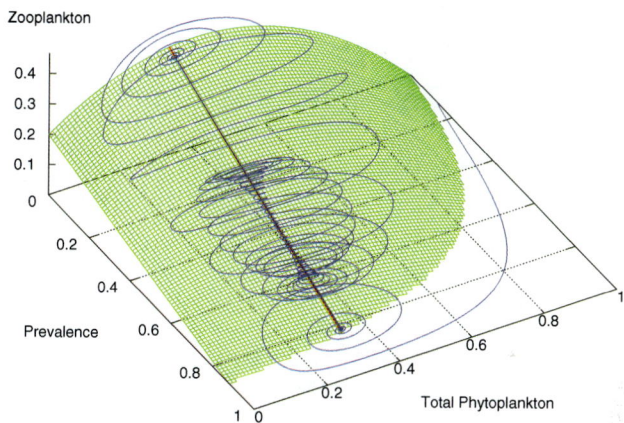

**Figure 15.2a**

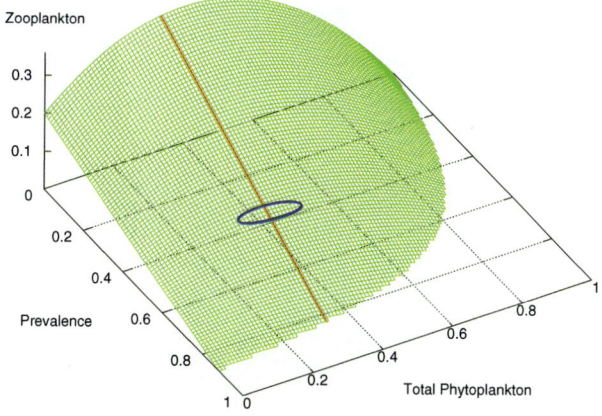

**Figure 15.2b**

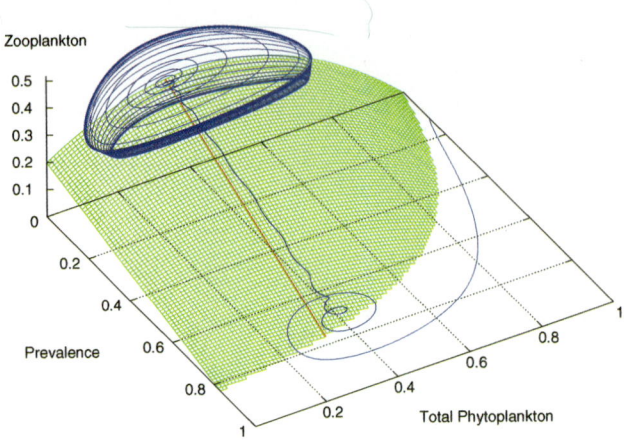

**Figure 15.3a**

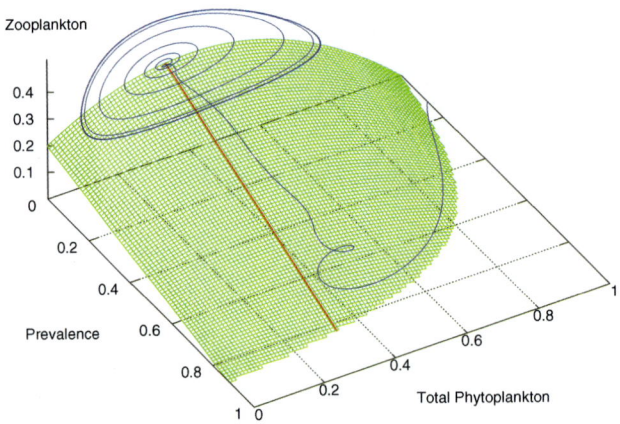

**Figure 15.3b**

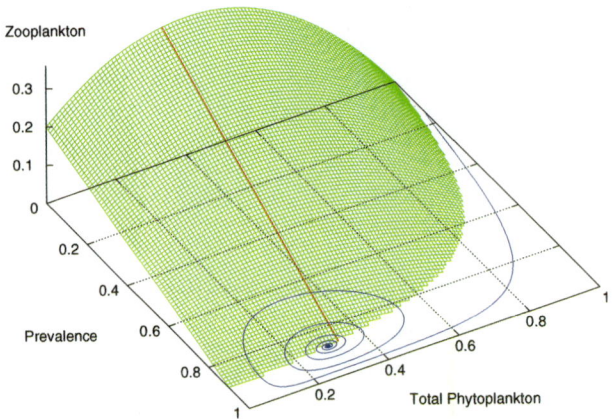

**Figure 15.4**

can be long-term transients that mimic some of the regimes of the species' geographical spread. In order to distinguish them from the trivial extinction scenario mentioned above, we refer to them as to "anomalous" extinctions.

Specifically, depending on parameter values, there can be two regimes when species extinction is either preceded by formation of a distinct long-living spatiotemporal pattern or by a long-distance population spread. In the first case, the initial conditions evolve into an ensemble of patches allocated over the domain. The patches interact with each other in a complicated manner similar to the patch dynamics shown in Figure 12.6. However, the regime is not self-sustainable and, finally, the species go extinct. In the second case, a moving patch is formed which propagates with approximately constant speed over distances much larger compared to the radius of the initial species distribution before the prey is caught by the predator; the pulse then decelerates, stops, and both species go extinct promptly.

We want to emphasize that, in both of these cases, the invasive population persists during a remarkably long time before the actual population decay takes place (typically, on the order of a hundred times longer than it would be in the case of the ordinary extinction) and/or it spreads over large distances. During that period, the system dynamics can be very similar to the corresponding regimes of geographical spread; see Figures 12.6 and 12.5, respectively.

Therefore, we have shown that the impact of the Allee effect (which, in a certain parameter range, makes the system kinetics excitable) brings to life two essentially new regimes of pattern formation that have not been observed in other systems. These regimes are (i) standing or traveling solitary patches of population density and (ii) the "patchy invasion," which is not associated with any traveling wave propagation. As a whole, the spatiotemporal dynamics of the prey–predator system appears to be much richer and much more complex under the impact of the Allee effect; see Figure 12.4. For comparison, in the corresponding system without the Allee effect (cf. Equations (10.6)–(10.7)), the only scenario is pattern formation in the wake of the traveling population front.

We note that the variety of dynamical regimes shown in Figure 12.4 can be presented in a more quantitative way as maps in the parameter space of the system, with different domains corresponding to different regimes. Since biological invasion per se is largely beyond the scope of this book, we are not showing them here; an interested reader can find more information in Petrovskii et al. (2005). What is important to mention is that, although a particular succession of regimes may be somewhat different depending on other parameter values, there are features of those maps that seem to be universal. Namely, the system dynamics always leads to species extinction for small $\delta$ (high predation) and to successful invasion through the propagation of a traveling front for large $\delta$ (low predation), while less trivial scenarios such as solitary patches and patchy invasion occur for intermediate values.

By now, we have focused on species' biological invasion and biological control in a system with one spatial dimension. This is a natural step in understanding the system spatiotemporal dynamics. Also, an advantage of an one-dimensional system is that numerical simulations are much faster than they would be in the corresponding multidimensional case; that makes it possible to study the problem thoroughly (cf. Petrovskii et al., 2005). However, in reality, biological invasion and the corresponding ecological pattern formation normally take place in two dimensions. The question thus arises to what extent the scenarios discussed in this section can be extended onto the dynamics of two-dimensional systems. This issue will be addressed below.

## 12.2 Invasion and control in the two-dimensional case

Now, we are going to make am insight into the patterns of biological invasion and corresponding species spread in the two-dimensional case, which is ecologically more realistic. As in the previous section, our main focus will be on pattern formation arising due to the interplay between the Allee effect and predation, the latter being regarded as a factor of biological control. The mathematical model is now given by the "full" system (12.4)–(12.5).

Since both the invasive species and its predator are introduced locally, relevant initial conditions for system (12.4)–(12.5) should be described by functions of compact support. Specifically, in this section we consider them in the form of rectangular patches:

$$u(x,y,0) = u_0 \quad \text{for} \quad |x - x_u| \leq \frac{\Delta_{ux}}{2} \quad \text{and} \quad |y - y_u| \leq \frac{\Delta_{uy}}{2}, \quad (12.12)$$

$$\text{otherwise} \quad u(x,y,0) = 0,$$

$$v(x,y,0) = v_0 \quad \text{for} \quad |x - x_v| \leq \frac{\Delta_{vx}}{2} \quad \text{and} \quad |y - y_v| \leq \frac{\Delta_{vy}}{2}, \quad (12.13)$$

$$\text{otherwise} \quad v(x,y,0) = 0,$$

where $(x_u, y_u)$ and $(x_v, y_v)$ give the patch centers; $\Delta_{ux}$, $\Delta_{vx}$, and $\Delta_{uy}$, $\Delta_{vy}$ define the patches size in the direction of $x$ and $y$, respectively; $u_0$ and $v_0$ are the initial population densities inside the patches.

Note that even the dimensionless system (12.4)–(12.5) contains a few parameters, i.e., $\alpha$, $\beta$, $\gamma$, $\delta$, and $\epsilon$, while the system in original dimensional variables contains twice as many. Therefore, a regular simulation study of its properties over a reasonably wide parameter range would imply at least several thousands simulation runs. For a two-dimensional system, where each single simulation takes hours (if done with proper accuracy, which implies large numerical grids and small time-steps), this is hardly possible. Instead, we use another strategy. We choose one controlling parameter and consider the changes in the invasion scenario subject to its variation, keeping all other parameters fixed at certain hypothetical values. In agreement with the idea of biological control (see the paragraph below Equation (12.10)), it seems that to use the (dimensionless) predator mortality $\delta$ as the controlling parameter is a convenient and biologically reasonable choice.

We begin with the case when $\delta$ is large – correspondingly, the predator is weak and predation is low. Intuitively, a weak predator would unlikely affect prey spread in any significant way. In the one-dimensional case, in the absence of a predator, the prey would normally spread as a stationary traveling wave (Fisher, 1937; Kolmogorov et al., 1937; Aronson and Weinberger, 1975; Fife, 1979); therefore, in a prey–predator system with a weak predator, the invasion of prey would likely take place through the traveling wave scenario as well. Although the plane waves do not directly apply to the two-dimensional case with the initial conditions of compact support, one can still expect the formation of a moving population front separating invaded and uninvaded areas.

These heuristic arguments agree very well with simulations. Having other parameters fixed, for the values of $\delta$ on the order of unity or larger, the initial conditions (12.12)–(12.13) lead to prey invasion through a radial expansion of the infested domain, with its shape being approximately circular except for a very early stage of the system's dynamics when the specifics of the initial distribution can be essential. The domain boundary propagates as a traveling front (although its speed is not constant now but grows monotonously to a constant value in the large-time limit), behind the front the species are distributed homogeneously. The corresponding local kinetics is of type I; see Figure 12.1.

For somewhat smaller $\delta$, system's kinetics changes to type II (stable limit cycle) and the homogeneous population distribution in the wake changes to chaotic spatiotemporal oscillations. Typical population distributions are shown in Figure 12.8. Recall that a similar pattern has been observed in a prey–predator system without Allee effect when the local dynamics is oscillatory; see Chapter 10.

A decrease in $\delta$ destroys the limit cycle and makes the system kinetics excitable; cf. Figure 12.3. In case $\delta$ is only slightly less than $\delta_{het}$, the expanding front with patterns in the wake remains to be the invasion pattern. A further decrease, however, changes it to expanding rings (see Figure 12.9), which seems to be a natural two-dimensional extension of traveling population pulses. The rings are centered around the place of original species introduction. Note that in this regime of species spread the species are absent both in front of the propagating front (i.e., outside of the ring) and behind the front (inside the ring). From the ecological standpoint, this means that the invasive species fails to establish itself in the new environment in spite of the fact that the species spread does take place on a geographical scale.

Since the growing rings scenario already means invasion failure, one might expect that smaller values of $\delta$ (stronger predator, higher predation) would likely lead to a fast eradication of invasive species, probably without any spread at all. Surprisingly, this is not the case. A further decrease in $\delta$ (local kinetics remains to be of type III) changes the rings to a completely different type of dynamics; see Figure 12.10. In this case, at a very early stage the species spread looks similar to the one with patterns in the wake of the expanding front; cf. Figures 12.10 (top, left) and 12.8. At a certain moment, however, the continuous front breaks to pieces and never reappears again.

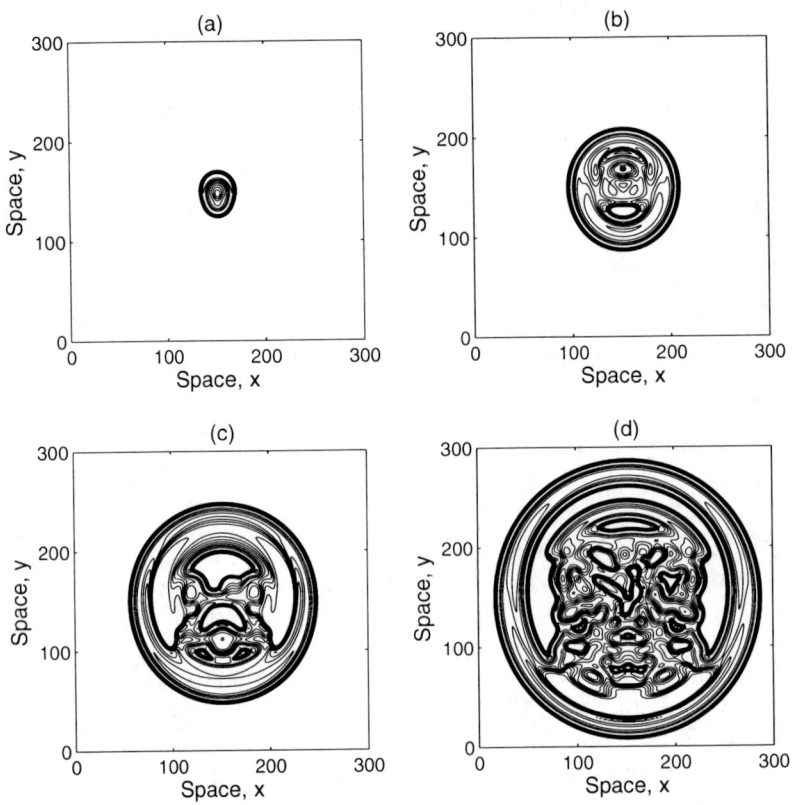

**FIGURE 12.8**: Invasion fronts with chaotic spatiotemporal oscillations in the wake: Snapshots (contour lines) of the prey density shown for $t = 25$, $t = 100$, $t = 175$, and $t = 250$, (a) to (d), respectively. Thick curves correspond to large gradients in the population density. The distribution of predator density exhibits qualitatively similar properties. Parameters are $\alpha = 0.1$, $\beta = 0.22$, $\gamma = 3$, and $\delta = 0.52$. The initial conditions are given by (12.12)–(12.13) with $x_u = 145$, $y_u = 152.5$, $x_v = 140$, $y_v = 155$, $\Delta_{ux} = 20$, $\Delta_{vx} = 10$, $\Delta_{uy} = 5$, $\Delta_{vy} = 20$, $u_0 = 1$, and $v_0 = 0.2$.

At any later time, the population spreads over space via irregular motion of separate patches. The patches move, grow, merge, split, produce new patches, etc. We emphasize that this "patchy invasion" is a self-sustained regime and it has nothing to do with the long-living transients mentioned in Section 12.1.1. Long-term simulations show that, after the population patches invade over the whole domain, the spatiotemporal dynamics of the system does not change and the spatial distribution of species at any moment is qualitatively similar (up to the position and shape of particular patches, which are changing all the time) to the one shown in Figure 12.10 (bottom, right).

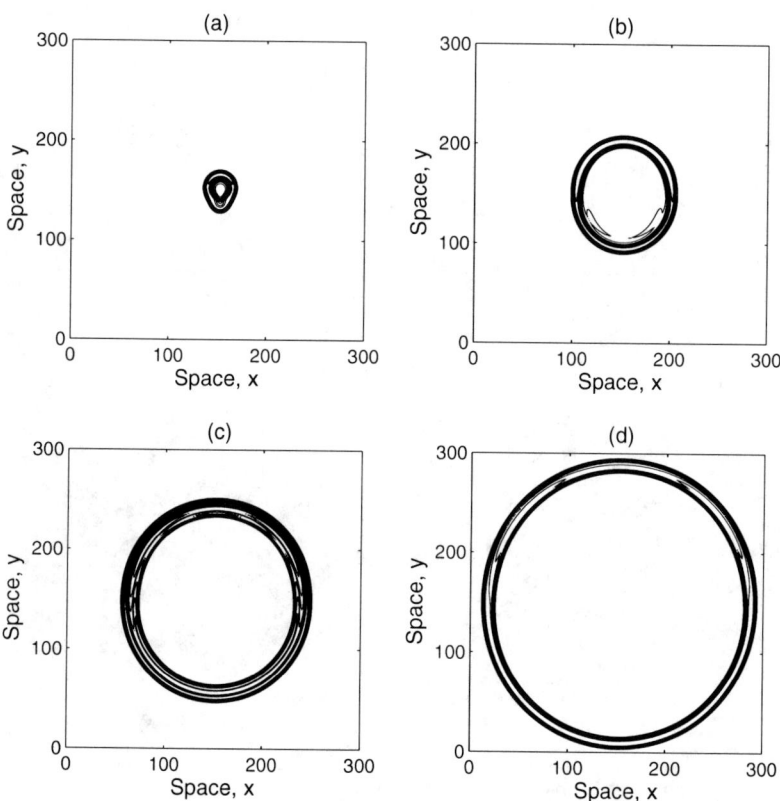

**FIGURE 12.9:** Spread of invasive species through expanding rings: Snapshots (contour lines) of the prey density shown for $t = 25$, $t = 150$, $t = 375$, and $t = 400$, (a) to (d), respectively. Thick curves correspond to large gradients in the population density. Note that the species is absent both outside and inside the rings. The distribution of predator density exhibits qualitatively similar properties. Parameters are $\alpha = 0.1$, $\beta = 0.22$, $\gamma = 3$, and $\delta = 0.44$. The initial conditions are the same as in Figure 12.8.

For still smaller values of $\delta$, predation becomes too strong and both species go extinct. Note that, since the model (12.1)–(12.2) does not allow for an alternative food source for the predator, it cannot survive after the prey disappears. Between patchy invasion and extinction, there can be a very narrow parameter range where evolution of the initial conditions may result in stationary patches; see Morozov and Li (2007) for more details and also see Section 12.3.1 for an example of similar dynamics.

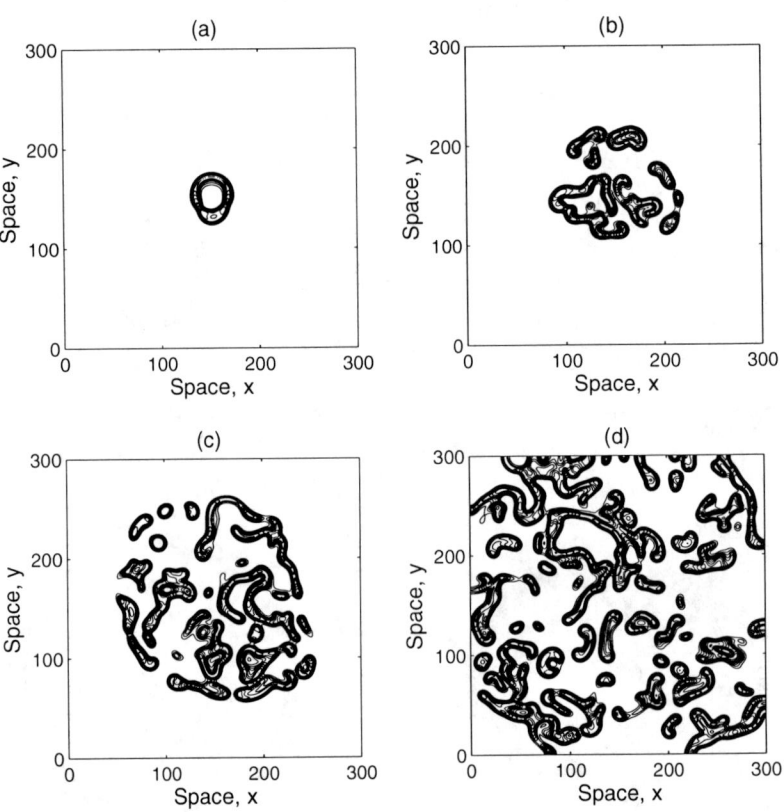

**FIGURE 12.10**: Patchy invasion: Snapshots (contour lines) of the prey density shown for $t = 50$, $t = 250$, $t = 450$, and $t = 750$, (a) to (d), respectively. Thick curves correspond to large gradients in the population density. Between the patches the species is absent, inside the patches the species is present at high density. Note that, except for some very early stage, there is no continuous boundary separating invaded and uninvaded areas. The distribution of predator density exhibits qualitatively similar properties. Parameters are $\alpha = 0.1$, $\beta = 0.22$, $\gamma = 3$, and $\delta = 0.42$. The initial conditions are the same as in Figure 12.8.

Thus, in a two-dimensional prey–predator system where prey growth is damped by the strong Allee effect, depending on the value of the predator mortality, the spatial spread of invasive species can follow a variety of scenarios. A decrease in mortality first changes the standard traveling wave scenario (with either homogeneous or patchy species distribution in the wake) to growing/expanding population rings (with no species in the wake) and then to patchy invasion. It should be mentioned that the last two scenarios are specific for the prey–predator system with the Allee effect, which makes system kinetics excitable, and cannot take place otherwise.

Let us recall that, in our approach, the predator mortality $\delta$ is a measure of the biological control effort. Therefore, the above succession of invasion scenarios arises in response to an increase in the control effort. One conclusion that can be made here is that these results make a strong argument in support of the biological control strategy: Indeed, a sufficiently strong predator will eradicate the invasive prey, although the actual scenario is much more complicated than the previously suggested blocking and reverse of the propagating population fronts (Owen and Lewis, 2001; Petrovskii and Malchow, 2005).

Now, one curious feature of this succession is that the success of biological control depends on the control effort in a nonmonotonous way. Increasing effort brings successful invasion (traveling front with patterns in the wake) to failure (expanding rings). In the latter case, the invasive species is spreading, but it will not do much harm because it is going to disappear anyway. However, a further increase in controlling effort restores successful invasion, although the scenario ("patchy invasion") will be quite different.

A relevant question is how the rate of species spread changes between the different regimes. It should be mentioned here that, while in traveling wave scenarios the rate of spread apparently coincides with the wave speed and thus can be calculated straightforwardly, in case of the patchy invasion its definition is somewhat less obvious. One way to obtain it is to introduce the domain radius $R$ as the maximum distance between the center of the invaded domain and the positions in space where the prey density exceeds a certain threshold level $u_{cut}$, i.e.,

$$R(t) = \max_{u(x,y)>u_{cut}} \left[(x - c_C)^2 + (y - y_C)^2\right]^{1/2} \qquad (12.14)$$

(where $u_{cut}$ should be chosen reasonably small), and then to calculate $dR/dt$. The position $(x_C, y_C)$ of the domain center can be defined differently; however, the results do not depend much on the exact definition (cf. Morozov et al., 2006).

The results are shown in Figure 12.11. It is readily seen that, while for the regimes associated with traveling fronts the rate of spread varies insignificantly (showing a gradual increase with an increase in $\delta$), in the regime of patchy invasion it drops to zero within a rather narrow parameter range.

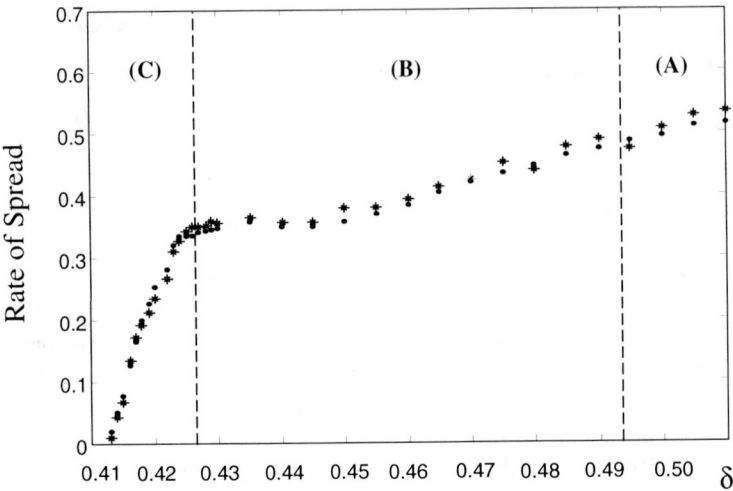

**FIGURE 12.11:** Rate of species spread for different values of the predator mortality $\delta$ and for different invasion scenarios (dots and stars correspond to different definitions of $x_C$ and $y_C$); (A) for invasion front with patterns in the wake, (B) for expanding rings, (C) for patchy invasion. Other parameters are $\alpha = 0.1$, $\beta = 0.22$, and $\gamma = 3$. With permission from Morozov et al. (2006).

### 12.2.1 Properties of the patchy invasion

From the invasion scenarios described above, probably the most curious and the most congenial to the scope of this book is the patchy invasion. Therefore, now we are going to make a somewhat deeper insight into its properties.

Perhaps the first question coming into mind when looking at the irregular patterns shown in Figure 12.10 is whether this dynamics is actually chaotic. The answer is yes; numerical simulations show that the temporal fluctuations of the population densities are remarkably irregular and the solutions of the system (12.4)–(12.5) exhibit the special type of sensitivity to the initial conditions typical for chaos when the difference between the perturbed and unperturbed solutions grows promptly (exponentially) with time. A more quantitative approach to this issue gives the value of the dominant Lyapunov exponent on the order of $10^{-2}$; in particular, for the parameters of Figure 12.10, $\lambda_D \approx 0.035$ (Morozov et al., 2006). Also, the spatial properties of the pattern, with the autocorrelation function showing fast decay at short distances, appear to be typical for spatiotemporal chaos; cf. Section 11.2 for an example of a similar behavior.

It should be mentioned here that chaos in a spatially extended prey–predator system is a typical rather than exotic phenomenon. In particular, we have mentioned earlier that the irregular patterns arising in the wake of the propagating front are chaotic as well. (Indeed, the patterns in the wake look qualitatively similar to the patterns arising as a result of patchy invasion; cf. the bottom of Figure 12.8 and the middle row of Figure 12.10.) Thus, chaos alone is not enough to distinguish between the two regimes. A question that arises is whether it is in both cases "the same" chaos (i.e., with essentially the same spatiotemporal properties) in spite of the fact that its onset follows rather different scenarios, or there is anything special or unusual about the pattern emerging as a result of patchy invasion.

In order to address this issue, we have to make a deeper insight into the properties of the corresponding dynamics. High complexity of the system behavior means that it is close to stochastic dynamics if considered on a timescale larger than the correlation time $\tau_{corr}$ of the system. This observation makes it possible to apply a statistical approach and to quantify the system dynamics in terms of the probability distribution of its different states.

To make use of the patch statistics, first, we need to estimate the correlation time of the system. That can be done in a standard way by means of calculating the autocorrelation function(s) and finding its/their first zero; for the parameters of Figure 12.10, $\tau_{corr} \approx 35$.

Second, we should construct a relevant "ensemble" of the system's states. For that purpose, from a numerical solution of the system (12.4)–(12.5) obtained at a certain fixed position $(\bar{x}, \bar{y})$, we extract a time series of the population densities where any two consecutive terms are separated with a time-lag $\tau_{corr}$. Owing to the meaning of $\tau_{corr}$, any two measurements in these series can be regarded as independent, so that we can now restore the probability distribution functions (PDF) of the system states.

Since the time series obtained for prey and predator exhibit qualitatively similar properties, below we show only the results obtained for prey. Figure 12.12 shows the PDF of the local prey density obtained for the parameters of Figure 12.8, i.e., for the chaotic pattern generated by a propagating front. The properties of the obtained histogram are heuristically clear and agree very well with intuitive expectations. A "sampling" of the system would normally bring a value of the prey density somewhere inside the limit cycle (cf. the shallow maximum in the middle); recall that the system kinetics is of type II in this case. Existence of another local maximum in vicinity of $u = 0$ reflects the fact that $(0, 0)$ is a stable steady state of the system.

Obviously, a similar analysis can also be applied to a time series of spatially averaged population density. Figure 12.13 shows the PDF of the spatially averaged prey density calculated for the parameters of Figure 12.8. Note that, in this case, the shape of the PDF is very close to that of the normal distribution (shown by the solid curve). Since in the regime of spatiotemporal chaos the whole domain appears to be dynamically split to an ensemble of mutually independent oscillators with the spatial size $\simeq L_{corr}$ (see Section 11.2), a normal distribution arises naturally as a result of the Central Limit Theorem.

However, the PDF properties appear to be rather different in case of the

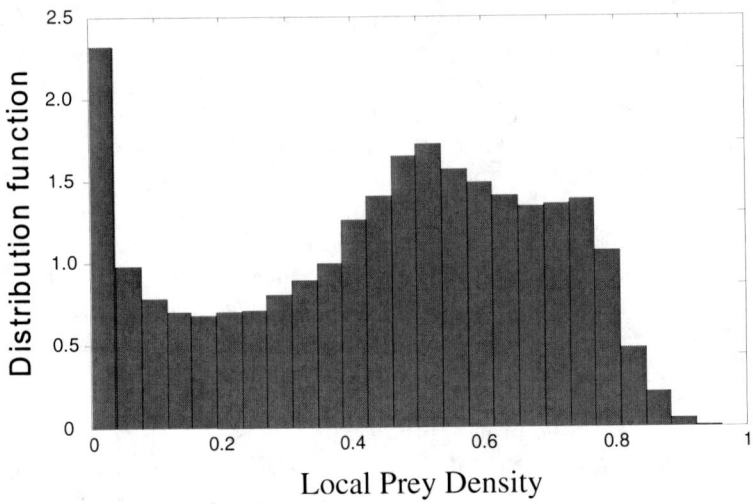

**FIGURE 12.12**: Probability distribution function of the local prey density in case of spatiotemporal chaos generated by the propagating front; parameters are the same as in Figure 12.8 (with permission from Morozov et al., 2006).

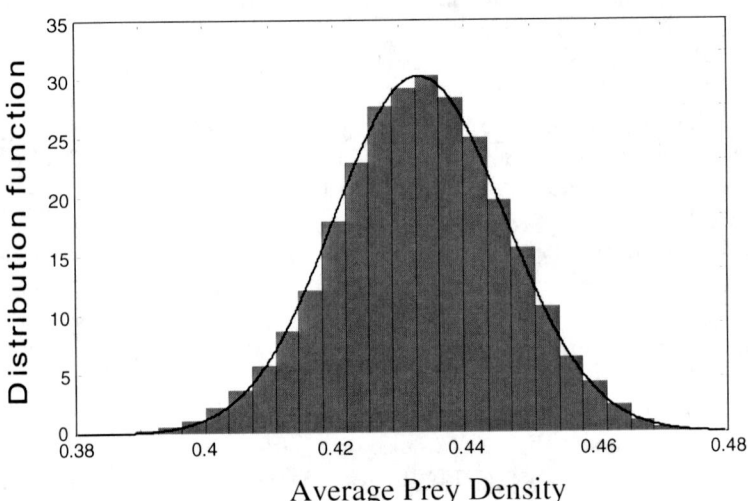

**FIGURE 12.13**: Probability distribution function of the spatially averaged prey density in case of spatiotemporal chaos generated by the propagating front; parameters are as in Figure 12.8 (with permission from Morozov et al., 2006).

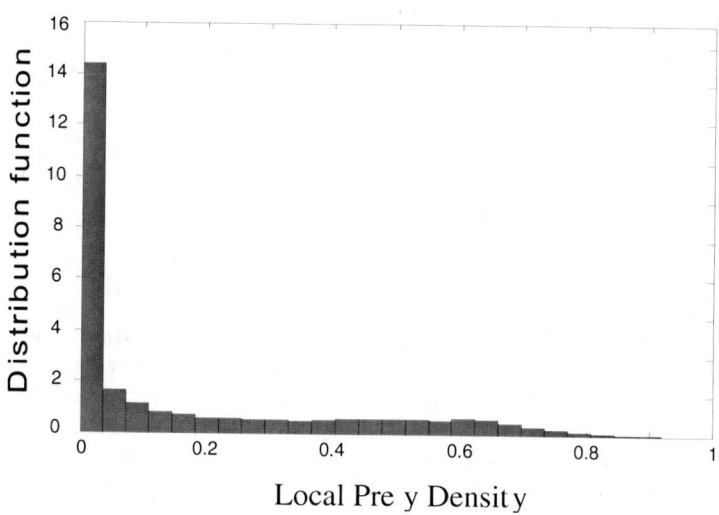

**FIGURE 12.14:** Probability distribution function of the local prey density in case of spatiotemporal chaos generated by the patchy invasion; parameters are the same as in Figure 12.10 (with permission from Morozov et al., 2006).

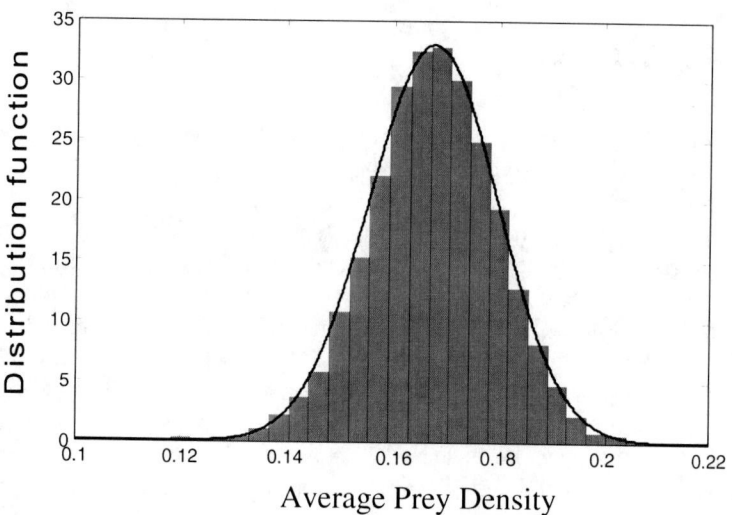

**FIGURE 12.15:** Probability distribution function of the spatially averaged prey density in case of spatiotemporal chaos generated by the patchy invasion; parameters are the same as in Figure 12.10 (with permission from Morozov et al., 2006).

patterns emerging as a result of the patchy invasion. Figure 12.14 shows the PDF of the local prey density calculated for the parameters of Figure 12.10. In this case, the PDF has the only distinct maximum at $u = 0$, which means that an expected value of the prey density is well below the survival threshold $\beta = 0.22$.

A similar result is obtained for the PDF of the spatially averaged density. Figure 12.15 shows the PDF of the spatially averaged prey density obtained for the parameters of Figure 12.10. Again, the PDF shape is very close to normal distribution. However, contrary to the case shown in Figure 12.13, now the PDF maximum is well below the survival threshold.

Therefore, in the case of pattern formation preceded by front propagation, high values of the prey density are more probable than low values. On the contrary, in the case of the patchy spread, low values of prey density are more probable than high ones. Note that, in both cases, the value of the survival threshold is the same, $\beta = 0.22$. A brief inspection of Figure 12.14 immediately leads to a rather surprising result, that, in the regime of patchy spread, the probability of detecting the prey density below the survival threshold is higher than the probability of finding it above the threshold. Moreover, as it is readily seen from comparison between Figures 12.13 and 12.15, the spatially averaged prey density appears to be well above and well below the survival threshold for the regimes of invasion with and without continuous population front, respectively.

*Impact of space dimension*

We have already shown that patchy invasion can be observed in a parameter range separating the parameter domain where the species will spread without invasion and the domain where the species will go extinct straightforwardly, i.e., without spreading at all. Therefore, patchy invasion gives a scenario of species invasion "at the edge of extinction" so that a reasonably small variation of the controlling parameter $\delta$ will turn successful invasion into failure.

Moreover, statistical analysis of the patch dynamics shows that, in the patchy regime, the species spread takes place at an amazingly low value of the average population density, i.e., well below the Allee threshold $\beta$. Obviously, this is essentially a spatiotemporal effect associated with the self-organized patchiness; a similar dynamics – i.e., species persistence below the survival threshold – could never be observed in a nonspatial system. That gives another example of crucial importance of spatial dimension(s) for population dynamics.

Note that a qualitatively similar succession of regimes has been observed in the corresponding one-dimensional case (see section 12.1.1), where the patchy invasion also takes place in a parameter range preceding species extinction. An interesting question is how the parameter ranges in one and two dimensions correspond to each other.

Simulations show that, for the parameter values when patchy invasion occurs in the one-dimensional system, the species spread in two spatial dimensions still takes place through an expanding continuous front of perfectly circular shape, i.e., two-dimensional invasion follows either the growing rings scenario or chaotic pattern formation in the wake of the front.

A decrease in $\delta$ turns the one-dimensional patchy invasion to extinction. Surprisingly, however, for the same parameter values the system dynamics in two spatial dimensions leads not to species extinction but to patchy invasion; a detailed consideration of this issue can be found in Petrovskii et al. (2005) and Morozov and Li (2007).

A heuristic description of invasion failure in the one-dimensional case is as follows. At an early stage of species spread, the traveling front of the prey density propagates into empty space, followed by a traveling front of predator density; the dynamics is similar to the so-called prey–predator pursuit scenario (Murray, 1989). The speed of prey wave appears to be lower than that of the predator wave; as a result, prey is caught up by the predator. After some oscillations, predator decreases the prey density below the survival threshold $\beta$ at every location in space, and extinction of both species takes place.

In the two-dimensional system, however, the patch border is curvilinear; thus, prey can escape through the lateral sides and create separate patches. Each new patch then starts growing through the formation and expansion of a circular population front (cf. the top left panel of Figure 12.10) until prey is caught again by the predator, and this scenario occurs again and again resulting in a prominent patchy structure.

Note that the effect of patch border curvature is positive for prey and negative for predator. Indeed, the larger curvature (i.e., the smaller the patch radius), the higher the rate of population density decrease due to its outflux through the patch border. A decrease in the population densities inside a given patch weakens the impact of predation on the prey growth and also decreases the growth rate of predator; see Equations (12.4)–(12.5), where the term describing interspecies interaction has a different sign in the equation for prey and in the equation for predator. Thus, prey has more chances to survive in small patches where the border curvature is large than in large patches where the border shape is close to a straight line. That helps us to understand how prey can possibly survive in the two-dimensional case for the parameter values when it is brought down by predation in the one-dimensional case.

In the above, we have considered how the scenario of biological invasion in a prey–predator system with a strong Allee effect changes in response to a decrease in predator mortality. We want to mention, however, that similar changes are observed if we choose another controlling parameter, e.g., the threshold density $\beta$, and keep the other parameters fixed. In this case, an increase in $\beta$ (instead of a decrease in $\delta$) eventually leads from species invasion through propagation of continuous fronts to patchy invasion and then, for a larger $\beta$, to extinction.

It should be mentioned that, apart from restrictions on parameters, patchy invasion is also somewhat sensitive to the choice of initial conditions. In particular, simulations show that the patchy invasion scenario will never take place in case of predators spreading into the area already inhabited by prey; even for the relevant parameter values, it will be the regime of traveling patches or growing rings instead. Thus, the finiteness of the species' initial distribution is an essential condition.

A question important for understanding prospective ecological implications of patchy invasion is, what could be the typical size of the patches, if any, in terms of real ecosystems? Indeed, a closer look at the spatial patterns (e.g., by means of calculating the correlation length; for details see Morozov et al., 2006) emerging as a result of patchy invasion shows that there is an intrinsic spatial scale, with its value being typically between 10 and 20 in dimensionless units. According to the definition of dimensionless variables (see the lines above Equations (12.4)–(12.5)), the relation between the dimensional $R$ and dimensionless $r$ spatial scales is given as $R = r[DH/(A\kappa K)]^{1/2}$. Note that, in our simulations, parameter $\alpha$ is always fixed at a hypothetical value of 0.1, which means that the effect of saturation on predator response is insignificant. As for $D$ and $A\kappa$ (recalling here that $A\kappa$ has the meaning of predator maximum growth rate), they can be different for different species. As an example, we consider the vole-weasels interaction (cf. Sherratt et al., 2002) with $D = 0.2$ km$^2$ year$^{-1}$ and $A\kappa$=2.7 year$^{-1}$ as typical values. We then obtain that $R$ lies between 8.5 km and 17 km, which seems to be ecologically reasonable. Note that an increase/decrease in $D$ or $A\kappa$ as much as ten times corresponds to only about three times' increase/decrease in $R$; thus, an estimate for $R$ to be between a few kilometers and a few dozens kilometers is likely to remain valid for many other terrestrial species.

## 12.3 Biological control through infectious diseases

In the previous two sections we showed that the use of predation as a factor of biological control aiming to slow down or block the spatial spread of an invasive pest may result, provided the pest's local population growth is affected by the strong Allee effect, in a new and curious scenario of spatiotemporal pattern formation. Namely, in a certain parameter range the spread can take place not via the intuitively expected traveling continuous population front but via the motion and interaction of separate patches. We then showed that the phenomenon of patchy invasion takes place "at the edge of extinction," so that a reasonably small change of controlling parameters either brings the species to extinction or restores the standard traveling front scenario.

Another recognized factor of biological control on invasive species is the impact of infectious diseases. In that case, in order to affect the spread of a harmful species (and, ideally, to lead to its eradiction) at an early stage of invasion an infectious disease is introduced into the spreading population, i.e., a number of individuals are deliberately infected by a certain dangerous or even lethal virus (Fitzgerald and Veitch, 1985; Courchamp et al., 1995). The impact of infection on invasive species spread has been studied theoretically and a possible slowdown of invasion rates was demonstrated for one-dimensional systems; e.g., see Hilker et al. (2005, 2007) and references therein. A "strong" infection (i.e., one with high transmission rate and/or high virulence) was predicted to block species invasion by means of blocking and reversing the corresponding traveling population fronts. The two-dimensional case, however, has yet remained less studied and poorly understood. A thorough consideration of this problem would carry us away far beyond the scope of this book; instead, our goal here is much more modest. Namely, we are going to check whether infection-based biological control can change the scenario of invasive species spread in a manner similar to that observed above for the prey–predator systems. We are especially interested to know whether the impact of infection may change a continuous front scenario to patchy invasion.

We begin with one of the simplest models of mathematical epidemiology, i.e., the so-called SI model:

$$\frac{\partial s(\mathbf{r},t)}{\partial t} = \left(\frac{\partial^2 s}{\partial x^2} + \frac{\partial^2 s}{\partial y^2}\right) + \gamma s(s-\beta)(1-s) - si , \qquad (12.15)$$

$$\frac{\partial i(\mathbf{r},t)}{\partial t} = \epsilon \left(\frac{\partial^2 i}{\partial x^2} + \frac{\partial^2 i}{\partial y^2}\right) + si - \delta i \qquad (12.16)$$

(e.g., see Murray, 1989), where $s$ and $i$ are the densities of the susceptible and infected individuals, respectively, at moment $t$ and position $\mathbf{r} = (x,y)$. For the sake of brevity, in Equations (12.15)–(12.16), all variables are already scaled to dimensionless values following a standard routine. Throughout this section, we assume that the diffusivity ratio $\epsilon = 1$. The term $si$ describes the disease transmission rate from infected to susceptible individuals. We assume that the disease is serious enough that infected individuals cannot produce offspring and the population can grow only due to multiplication of susceptibles.

Obviously, the initial spatial distribution of $s$ and $i$ should be described by functions of compact support; therefore, for modeling purposes, the initial conditions (12.12)–(12.13) are still appropriate, up to the corresponding change of $u$ to $s$ and $v$ to $i$.

The system (12.15)–(12.16) has been studied by means of numerical simulations. We obtain that the succession of the regimes in response to a change in the controlling parameter, e.g., in either $\delta$ or $\beta$ (cf. the remark at the end of Section 12.2), is similar to the one that has been observed for the prey–predator system. A decrease in $\delta$ eventually changes the invasion scenario via propagation of traveling fronts with pattern formation in the wake (see Figure 12.16) to patchy invasion (see Figure 12.17). For smaller $\delta$, the species

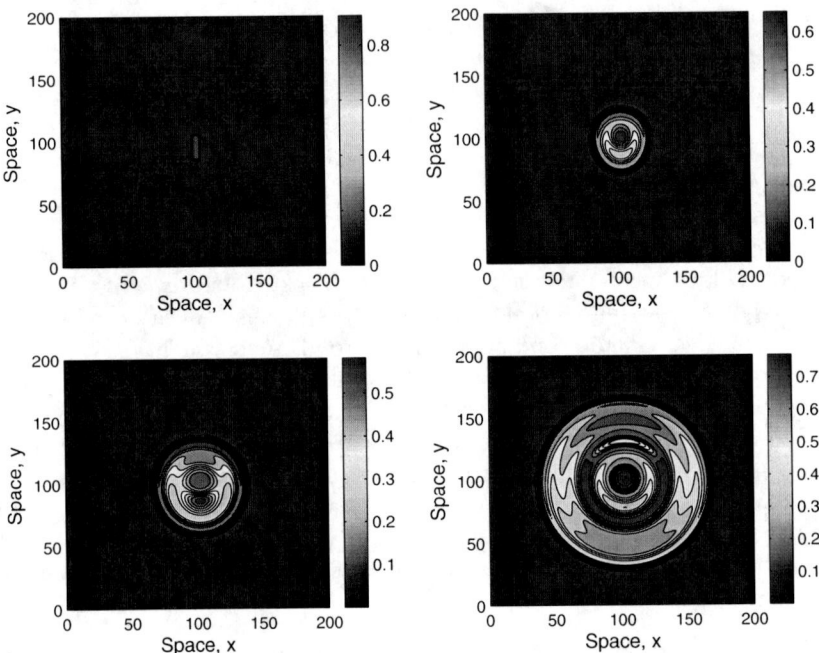

**FIGURE 12.16**: (See color insert.) Snapshots (contour lines) of the two-dimensional spatial distribution of the density of susceptibles, as given by the SI model (12.15)–(12.16), calculated at the moments $t = 0$, $t = 40$, $t = 80$, and $t = 160$ (left to right, top to bottom). The density of infected exhibits similar properties. Parameters are $\gamma = 2$, $\beta = 0.2$, and $\delta = 0.44$.

goes extinct. Thus, the patchy spread of invasive species can arise in response to controlling efforts based on the introduction of certain infectious diseases in the same way as it arises in response to predation. The patchy invasion in this case has exactly the same implication as in the prey–predator system: It describes the scenario of spatial spread at the edge of extinction so that a further small (but finite) change of the controlling parameter leads to species extinction.

Note that, from a mathematical aspect, the system (12.15)–(12.16) is not just a particular case of Equations (12.4)–(12.5) corresponding to $\alpha = 0$. The decrease from $\alpha > 0$ to $\alpha = 0$ means a certain structural change: Instead of strong nonlinearity $uv/(1+\alpha u)$, we now have a bilinear term that corresponds to the classical Lotka–Volterra model. It is well known (cf. Murray, 1989) that the dynamics of the model with a bilinear interaction term and the one with Holling type II can differ in many aspects, in particular, with regards to steady states/limit cycle(s) existence and stability. Thus, extension of the main results onto the SI model is nontrivial, although it might be intuitively expected.

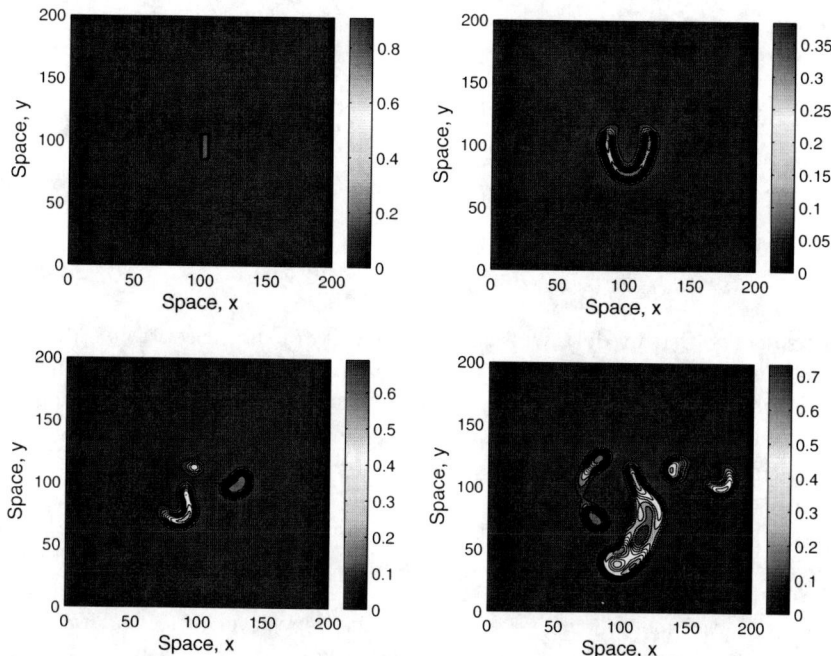

**FIGURE 12.17**: (See color insert.) The two-dimensional density of susceptibles as given by the SI model (12.15)–(12.16) calculated at the moments $t = 0$, $t = 60$, $t = 130$, and $t = 350$ (left to right, top to bottom). The density of infected exhibits similar properties. Parameters are $\gamma = 2$, $\beta = 0.2$, and $\delta = 0.4$.

It should also be mentioned that, although the succession of dynamical regimes described above when a decrease in $\delta$ eventually leads to species extinction may seem to be somewhat counter-intuitive, it is in full agreement with biological arguments. Indeed, the dimensionless parameter $\delta$ gives, up to a certain factor, a ratio of the infected mortality rate and the transmission rate. A decrease in $\delta$ thus corresponds to an increase in the transmission rate, which tends to make disease more dangerous.

### 12.3.1 Patchy spread in SIR model

The results of the previous section inspire us to look now at the dynamics of a more complicated epidemiological model. Namely, now we are going to consider the spatiotemporal dynamics of an infectious disease described by the following equations:

$$\frac{\partial s(\mathbf{r},t)}{\partial t} = \left(\frac{\partial^2 s}{\partial x^2} + \frac{\partial^2 s}{\partial y^2}\right) + \gamma s(s-\beta)(1-s) - si + \alpha i + \eta\rho, \quad (12.17)$$

$$\frac{\partial i(\mathbf{r},t)}{\partial t} = \left(\frac{\partial^2 i}{\partial x^2} + \frac{\partial^2 i}{\partial y^2}\right) + si - \delta i - \alpha i - \sigma i, \quad (12.18)$$

$$\frac{\partial \rho(\mathbf{r},t)}{\partial t} = \epsilon\left(\frac{\partial^2 \rho}{\partial x^2} + \frac{\partial^2 \rho}{\partial y^2}\right) + \sigma i - \eta\rho - \omega\rho \quad (12.19)$$

(the so-called SIR model), where $s$ is the density of the susceptible individuals of a given population, $i$ is the density of infected, and $\rho$ is the density of removed at the position $\mathbf{r} = (x,y)$ and time $t$. As above, we assume that Equations (12.17)–(12.19) are already scaled to dimensionless values. As in the case of the SI model, we assume that only the susceptibles can produce offsprings.

Apparently, the SIR model is more complicated than the SI model and contains more mechanisms and scenarios of disease development; in particular, it includes a possibility for the infected and removed individuals to become susceptible again (with the rates $\alpha$ and $\eta$, respectively). Also, it allows a somewhat different biological interpretation. For instance, $\rho$ can be treated as the density of individuals who recovered from the disease but cannot become susceptible again, e.g., because they get immunized. In this case, $i$ gives the density of sick individuals. Alternatively, however, $\rho$ can be treated as the density of sick individuals; in this case, $i$ gives the density of the individuals who have the disease in the latent stage. More details and further references can be found in Hethcote (2000) and Diekmann and Heesterbeck (2000).

Some typical simulation results are shown in Figures 12.18 to 12.20. The snapshots were obtained for the same initial conditions as for the SI model (cf. (12.12)–(12.13) with apparent change of notations), and assuming that at the beginning of the disease spread the removed subpopulation is absent, $\rho(x,y,0) = 0$.

Since the SIR model has additional feedbacks, e.g., through possible recovering of the removed individuals, the system response to variation of the mortality $\delta$ of infected is more complicated than it is in the SI model. For that reason, in our search for the regime of patchy spread, it appears more convenient to vary $\beta$, not $\delta$, and to keep all other parameters fixed. Specifically, the simulation results shown here were obtained for $\epsilon = 0.5$, $\alpha = 0$, $\gamma = 4$, $\delta = 0$, $\eta = 0.1$, $\omega = 0.8$, and $\sigma = 0.5$. Note that, in our choice of parameter values, we are more inclined to consider $\rho$ as the density of the sick subpopulation and $i$ as the latent subpopulation; that is why we neglect the mortality rate of infected ($\delta = 0$) and choose $\epsilon < 1$.

For the corresponding one-dimensional case, the species spread from the place of original introduction would occur (for sufficiently small values of $\beta$) through propagation of a traveling population front, in exactly the same

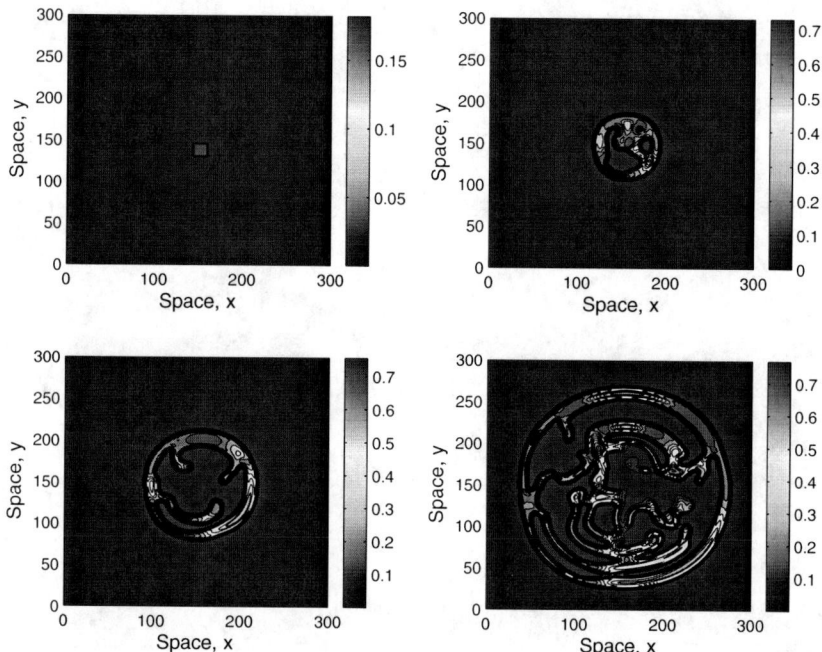

**FIGURE 12.18**: (See color insert.) The two-dimensional density of infected individuals calculated in the SIR model (12.17)–(12.19) at the moments $t = 0$, $t = 100$, $t = 200$, and $t = 400$ (left to right, top to bottom) for $\beta = 0.26$; other parameters are given in the text. The density of susceptibles and removed exhibits essentially the same properties except for very early stage of the system dynamics.

way as it takes place for the prey–predator and the SI models considered above. Depending on $\beta$, the species distribution behind the front can be either homogeneous and stationary or patchy and transient. In all those cases, the species spread in the corresponding two-dimensional system takes place through an expanding population front of circular shape.

For somewhat larger $\beta$, the regime of spread in the one-dimensional system turns to the propagation of separate patches, or groups of patches, qualitatively similar to the pattern shown in Figure 12.6; see Petrovskii et al. (2005) for further details. Remarkably, for the same value of $\beta$, the species spread in the two-dimensional system still takes place through an expanding circular front; see Figure 12.18.

A further increase in $\beta$ leads to species extinction in the one-dimensional system. In the two-dimensional system, however, for the same parameter when species goes extinct in the one-dimensional case, the evolution of the initial population distribution does not lead to species extinction but to its patchy spread; see Figure 12.19. Larger values of $\beta$ make the patchiness of

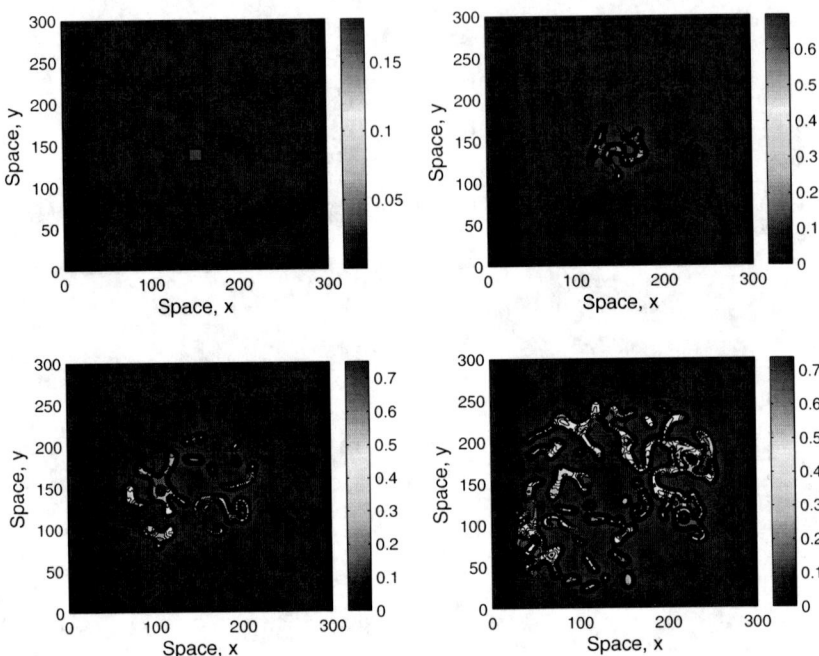

**FIGURE 12.19**: (See color insert.) The two-dimensional density of infected individuals calculated in the SIR model at the moments $t = 0$, $t = 200$, $t = 700$, and $t = 1300$ (left to right, top to bottom) for $\beta = 0.273$; other parameters are the same as in Figure 12.18. The density of susceptibles and removed exhibits essentially the same properties.

the spatial pattern even more distinct and the rate of spread notably lower. These results confirm our earlier observation that the regime of patchy spread provides a mechanism of species invasion at the edge of extinction.

A further increase in $\beta$ does not immediately lead to species extinction in two dimensions but first to formation, in the large-time asymptotic, of stationary patchy distribution of the species; see Figure 12.20. After the transient stage, which takes a considerable time, $t \simeq 1500$, a few stationary patches of the population density appear, see the bottom of the figure. Large-time simulations confirm that the patches remain stationary. The number of patches depends on the parameter values while the initial conditions may affect both the number of patches and their position. Recall that the formation of stationary patches was also observed in the two-dimensional prey–predator system in a narrow parameter range between patchy invasion and extinction.

Therefore, in this section we have shown that the "response" of the invasive species to an effort of biological control based on intentional introduction of an infection disease is similar to that observed in the case of biological control through predation. In particular, by means of numerical simulations

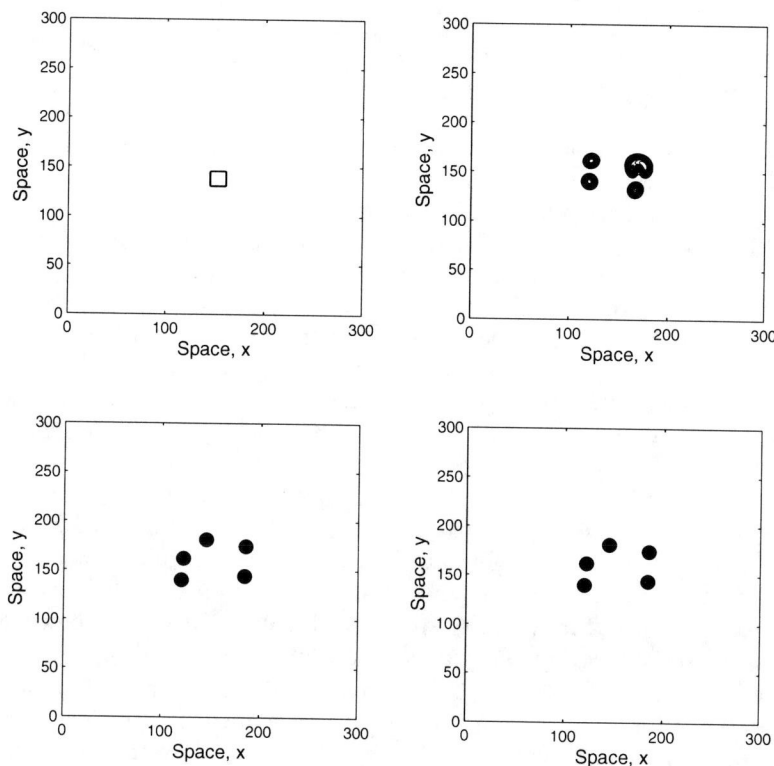

**FIGURE 12.20**: Snapshots (contour lines) of the two-dimensional spatial distribution of the density of infected individuals calculated in the SIR model at the moments $t = 0$, $t = 500$, $t = 2000$, and $t = 2500$ (left to right, top to bottom) for $\beta = 0.275$; other parameters are the same as in Figure 12.18. The density of susceptibles and removed exhibits essentially the same properties.

we showed that an increase in the disease "strength" may change the standard invasion scenario by a traveling population front to patchy invasion, provided the local population growth of the alien species is damped by the strong Allee effect.

As previously for the prey–predator model, we have shown that patchy invasion is a mechanism of species spread "at the edge of extinction" so that a small change of controlling parameters can bring a given invasive species to extinction. For those parameter values when the patchy invasion takes place in a two-dimensional system, in the corresponding one-dimensional system the species go extinct.

## 12.4 Concluding remarks

We have already shown earlier in this book (see Chapter 10) that biological invasion may trigger spatiotemporal pattern formation in the wake of the propagating population front. In this chapter, we have extended those results onto the case when the growth rate of an invasive species is affected by the strong Allee effect. We have specifically focused on the situation when the invasive species are subjected to biological control. The agents of biological control were considered to be either predation or infectious disease, which means that, respectively, either a relevant predatory species or some lethal virus are deliberately introduced in the wake of the spreading pest population. In both cases, we have shown that the interplay between the Allee effect and predation or infection then makes the system's spatiotemporal dynamics much richer, resulting in invasion scenarios that have not been observed otherwise. In particular, we have shown that invasive species spread may take the form of "patchy invasion" when formation of a distinct irregular patchy spatial population structure (where patches of high population density are separated by large areas of virtually empty space; see Figure 12.10) is not preceded by a propagation of a continuous population front.

Patchy invasion is shown to be a scenario of species invasion "at the edge of extinction," so that a reasonably small change in the value of a controlling parameter would turn the population to extinction. Moreover, it appears to possess some rather unusual and counter-intuitive properties: A closer look at the statistics of the patch dynamics shows that the corresponding species spread takes place with the typical values of population density well below the survival threshold.

We mention here that, as in the case of patterns in the wake, formation of the spatiotemporal pattern in the course of patchy invasion was observed when diffusivity was the same for both species. Therefore, the pattern formation is obviously not the result of the Turing mechanism (which requires sufficiently different diffusion coefficients; see Chapter 9) and has a completely different origin. However, we also want to emphasize that existence of patchy spread is not restricted to the case of identical diffusivity. Simulations show that all invasion scenarios considered in this chapter remain qualitatively the same when $D_2/D_1$ is not equal to unity but remains on the order of unity. In particular, patchy invasion with the properties as described above can certainly be observed in the range $0.7 < D_2/D_1 < 1.5$, although other parameters might have to be chosen slightly differently.

Also, patchy invasion appears to be robust with respect to external heterogeneity when some of the system parameters may become a function of the position in space. (Obviously, such position dependence would mimic the properties of real nature where species spread often takes place in a heterogeneous environment or fragmented landscape.) Namely, Morozov et al. (2006)

showed that, in a relevant parameter range, invasive species spread still follows the patchy invasion scenario over the space with a checkboard-type heterogeneity, where "good" patches alternate with "bad" patches on a rectangular grid, even when the environmental conditions inside the bad patches are very unfavorable so that the corresponding spatially homogeneous population would go extinct.

Another curious regime of the dynamics of the systems with a strong Allee effect is the formation of stationary patches, e.g., see Figure 12.20. Remarkably, field studies give many examples when, after introduction, invasive species remain localized inside a certain area for a long time, so that their local invasion and subsequent regional persistence is not followed by geographical spread. There exist a number of different explanations of this phenomenon such as the impact of environmental borders, the existence of time-lags related to mutations and evolutionary changes caused by adaptation in the new environment, etc. The results of this chapter provide another plausible explanation showing that invasive species can be held localized due to purely dynamical mechanisms, i.e., due to the interplay between the Allee effect and predation.

Now, an important question is whether a scenario qualitatively similar to patchy invasion can ever be observed in real nature. A traditional theoretical approach to species spread during biological invasion, which ascends to seminal papers by Kolmogorov et al. (1937) and Fisher (1937), predicts invasion through the propagation of a population front separating invaded areas (behind the front) with high population density of the alien species from uninvaded areas (in front of the front) where the species is yet absent. For a two-dimensional case, this approach has been extended to predict a continuous propagating front (Skellam, 1951; Andow et al., 1990; Okubo et al., 1989). However, there has been a growing amount of empirical evidence recently, both in published literature (Shigesada and Kawasaki, 1997; Davis et al., 1998; Kolb et al., 2004; Swope et al., 2004) and on the Web, showing that, in some cases, invasion of exotic species takes place through dynamics of separate population patches not preceded by propagation of any continuous population front at all. By way of example, Figure 12.21 shows maps of the gypsy moth invasion in the United States. It is readily seen that the infested area has a distinct patchy structure and there does not exist anything that, even with a large bit of imagination, might be regarded as a continuous boundary. Note that, if we exclude a few heavily infested areas, the typical size of the remaining patches is estimated to be on the order of a few dozen kilometers, which is in an encouraging agreement with the theoretical prediction; see the last paragraph in Section 12.2.1.

It would be rather frivolous, though, to try to relate all cases of patchy invasion to a single reason. Indeed, there have been suggested several different mechanisms for patchy invasion such as impact of environmental heterogene-

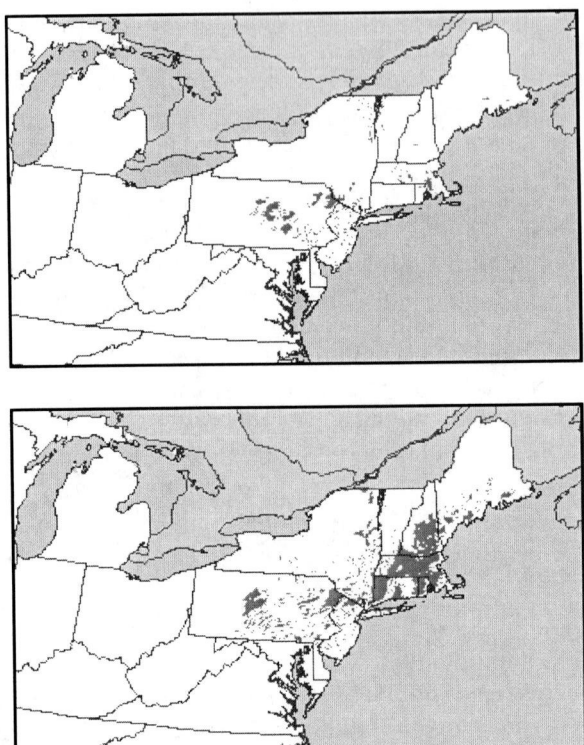

**FIGURE 12.21**: (See color insert.) Maps of gypsy moth invasion in the Northeast of the United States; top for 1977, bottom for 1981. Red color shows infested areas (By courtesy of Andrew Liebhold; from www.fs.fed.us/ne/morgantown/4557/gmoth/atlas/).

ity (Murray, 1989), "stratified diffusion"[2] (Shigesada et al., 1995), transport with either humans or vector species (Shigesada and Kawasaki, 1997), impact of stochastic factors (Lewis, 2000; Lewis and Pacala, 2000), etc. It is not always easy to distinguish between the impact of these factors in field data, even though they may sometimes act on a different spatial scale. In most cases, the information currently available does not make it possible to establish an unambiguous relation between the pattern and the underlying process or processes, and additional studies are required. Thinking about a specific example, the first idea that may come to mind when looking at the maps

---

[2] A behavioral population-level response to an overcritical increase in population density when flocks/swarms of a given species migrate out of population range in order to establish new colonies.

of gypsy moth invasion (cf. Figure 12.21) is that it is probably habitat fragmentation that results in the patchy spread. However, comparison with the corresponding vegetation maps readily reveals that the correlation is actually not that strong. On the other hand, in order to link it to the patchy invasion due to the interplay between the Allee effect and predation, one should first prove the existence of the strong Allee effect in the gypsy moth population growth and identify the relevant predatory species. Therefore, it would be premature to name the dynamical mechanism of patchy invasion considered in this chapter as an explanation of these or other similar data. Nevertheless, we think that this mechanism is important in a wider theoretical perspective because it shows that a basic prey–predator interaction in space and time is intrinsically capable of generating a patchy invasion without any additional assumptions. In that sense, it gives a "minimum model" of the phenomenon.

# Part IV

# Spatiotemporal patterns and noise

# Chapter 13

# Generic model of stochastic population dynamics

Mother nature is noisy, therefore, it is time to deal with stochasticity. The evolution of deterministic systems is fixed by the initial and boundary conditions, though a forecast is impossible for chaotic dynamics on large time scales. In stochastic systems, the noise leads to different realizations for the same initial conditions. The statistical ensemble of infinitely many realizations defines a stochastic process. For simplicity, only Markov processes will be considered here, where the present state determines the further evolution. The interplay of stochasticity and determinism can be modeled by stochastic (partial) differential equations [S(P)DE's] on the level of state variables (Langevin equations) or by dynamical equations for different probability densities (diffusion–reaction master and Fokker–Planck equations). Nice introductions to the theory of stochastic processes and its applications have been written by Gardiner (1985); Allen (2003); Anishenko et al. (2003), and with regard to spatial processes by García-Ojalvo and Sancho (1999).

On average, the system dynamics might remain unchanged. However, the noise generates a corridor in phase space for the dynamics rather than a single trajectory and only certain probabilities for steady states or oscillations and waves. But this is not the only effect of noise. It will be shown by example that it can induce transitions between steady states in systems with steady-state multiplicity. Furthermore, it can generate unexpected new structures in space and time. A side effect is that the noise blurs unrealistic distinct and symmetric spatial and spatiotemporal patterns.

Population-dynamical systems are subjected to internal demographic and external environmental noise. Demographic noise describes stochastic fluctuations caused by the random and discrete nature of individual growth, interaction, and motion. It is especially important at small population sizes, i.e., low species numbers, because any fluctuation can become dangerous and finally cause extinction. The impact of demographic noise on the system dynamics can be described by (diffusion–reaction) master equations. The latter concept originated from mathematics and physics; cf. Wax (1954); Bharucha-Reid (1960); Arnold (1974); Kuramoto (1974); and Karlin and Taylor (1975) with early and then most applications in physical chemistry, especially chemical kinetics (Bartholomay, 1958b; McQuarrie, 1967; Janssen, 1974; Haken, 1975; Matheson et al., 1975; van den Broeck et al., 1977; Feistel and Ebeling,

1978; Ebeling and Schimansky-Geier, 1979; Nicolis and Malek-Mansour, 1980; Malchow et al., 1983; Wissel, 1984a; Malchow and Schimansky-Geier, 1985; Baras and Malek-Mansour, 1996, 1997) with corresponding numerical simulation techniques (Gillespie, 1976; Feistel, 1977; Hattne et al., 2005; Isaacson and Peskin, 2006). Other applications have been found in the stochastic theory of birth and death processes (Bartholomay, 1958a; van Kampen, 1973), epidemics (Bailey, 1950; McNeil, 1972), molecular replication-selection processes (Ebeling and Feistel, 1974, 1977), or ecosystems with multiple stability (Wissel, 1984b).

For larger population sizes, the demographic noise is masked by the environmental (extrinsic, external) random variability that induces parameter fluctuations (Bonsall and Hastings, 2004). The latter can be incorporated in a straightforward manner by applying the white noise forcing directly to one or more selected parameters. The values of these parameters are then chosen randomly at each space point and each unit time step from a truncated normal distribution between a minimum and maximum fraction of their spatiotemporal mean. An example with an application to plankton dynamics can be found in Steele and Henderson (1992b). A corresponding problem will be presented in Ch. 14.

Another way to account for noise is the use of stochastic partial differential (Langevin) equations of the form

$$\frac{\partial \phi(\vec{r},t)}{\partial t} = d\nabla^2 \phi(\vec{r},t) + f(\phi(\vec{r},t),\psi) + \omega\left[\phi(\vec{r},t)\right]\xi(\vec{r},t), \qquad (13.1)$$

where $\phi(\vec{r},t)$ and $f(\phi(\vec{r},t),\psi)$ are the usual vectors or scalar fields of population densities and growth and interaction functions, respectively. $\psi$ stands for the system parameters. The last term on the right-hand side represents the stochastic force $\xi(\vec{r},t)$ that disturbs the system with intensity $\omega\left[\phi(\vec{r},t)\right]$. These equations have a long history in mathematical modeling of population dynamics (Levins, 1969; May, 1973; Capocelli and Ricciardi, 1974; Tuckwell, 1974; May et al., 1978; Braumann, 1979, 1983; Kliemann, 1983; Dennis and Patil, 1984). Throughout the chapter, it is assumed that $\xi(\vec{r},t)$ is a Gaussian white noise with zero mean and delta correlation

$$\langle \xi(\vec{r},t)\rangle = 0\,,\, \langle \xi(\vec{r}_1,t_1)\,\xi(\vec{r}_2,t_2)\rangle = \delta(\vec{r}_1-\vec{r}_2)\delta(t_1-t_2)\,. \qquad (13.1a)$$

$\omega\left[\phi(\vec{r},t)\right]$ is the density-dependent noise intensity. The postulate of parenthood (Hutchinson, 1978) in population dynamics requires this density dependence, i.e., multiplicative noise. Furthermore, it is chosen as

$$\omega\left[\phi(\vec{r},t)\right] = \omega\phi(\vec{r},t)\,;\, \omega = \text{const} \qquad (13.1b)$$

which reflects an increase of noise with growing species numbers. In particular, such noise is originated by fluctuating mortalities. But it is not intended to specify the origin of the noise. Colored noise (Kaitala et al., 1997), i.e., different shapes of Equations (13.1a) and (13.1b) would result in similar effects.

In this respect, the action of the considered fluctuations in (13.1a) and (13.1b) is structurally robust and a good approximation of environmental noise. One should keep in mind that the numerical treatment of spatially two-dimensional white noise problems introduces a spatial correlation length that is the grid spacing (Walsh, 1986; Lythe and Habib, 2001; Milstein and Tretyakov, 2004). General results state that the continuum solutions of the presented stochastic partial differential equations driven by space-time white noise in two spatial dimensions are not continuous functions but only distributions. Nevertheless, a discrete lattice can be used to resemble a spatially two-dimensional setting.

Further on, Stratonovich calculus (Stratonovich, 1967; Anishenko et al., 2003) will be applied for the interpretation of the multiplicative white noise during simulations. There has been and possibly still is a controversy on using Stratonovich or Itô (1961) calculus, however, Braumann (1999, 2007) has shown that this issue is merely semantic.

A quite general form of a model of the interplay determinism and stochasticity in population dynamics is with restriction to two populations $U, V$ at location $\vec{R}$ and time $T$:

$$\frac{\partial U(\vec{R},T)}{\partial T} = D_U \nabla^2 U + \omega U \Xi(\vec{R},T) +$$
$$+ \Phi_U(\vec{R},T) + P(U) - E(U,V) - M_{IU} I_U, \qquad (13.2)$$
$$\frac{\partial V(\vec{R},T)}{\partial T} = D_V \nabla^2 V + \omega V \Xi(\vec{R},T) +$$
$$+ \Phi_V(\vec{R},T) + Q(V) - \kappa E(U,V) - M_{IV} I_V. \qquad (13.3)$$

The function $P(U)$ describes the intrinsic growth of the prey population, usually logistic growth here. $E(U,V)$ stands for predation usually of Holling type II or III. Compared to previous chapters, the stochastic force $\Xi(\vec{R},T)$, in- and outflows $\Phi_U(\vec{R},T), \Phi_V(\vec{R},T)$, and additional mortality rates $M_{IU}, M_{IV}$ due to a possible disease as well as intrinsic growth $Q(V)$ of population $V$ have been added. Later on, an infectious disease of $U$ will be considered; then, the total population of $U$ will be split into a susceptible part $S$ and an infected $I$ with $U = S + I$. Equation (13.2) then splits into

$$\frac{\partial S(\vec{R},T)}{\partial T} = D_S \nabla^2 S + \omega S \Xi(\vec{R},T) +$$
$$+ \Phi_S(\vec{R},T) + P(S,I) - E(S,I,V) - \Sigma(S,I), \qquad (13.2a)$$
$$\frac{\partial I(\vec{R},T)}{\partial T} = D_I \nabla^2 I + \omega I \Xi(\vec{R},T) +$$
$$+ \Phi_I(\vec{R},T) + P(S,I) - \kappa E(S,I,V) + \Sigma(S,I) - M_I I; \qquad (13.2b)$$

whereas Equation (13.3) reduces to

$$\frac{\partial V(\vec{R},T)}{\partial T} = D_V \nabla^2 V + \omega V \Xi(\vec{R},T) +$$
$$+ \Phi_V(\vec{R},T) + Q(V) + \kappa E(S,I,V). \quad (13.3a)$$

The function $\Sigma(S,I)$ describes the mechanism of disease transmission (Nold, 1980; de Jong et al., 1995; Hethcote, 2000; McCallum et al., 2001). Here, only two mechanisms will be considered. One is the mass-action type

$$\Sigma(S,I) = \sigma SI, \quad (13.4)$$

the other the frequency-dependent transmission

$$\Sigma(S,I) = \sigma \frac{SI}{U}. \quad (13.5)$$

Following McCallum et al. (2001), for a directly transmitted pathogen, the infection rate is the product of three factors: first, the contact rate, second, the fraction of these contacts that take place with susceptibles, and third, the fraction of contacts that finally lead to infections. The mass-action type suggests that the contact rate of susceptibles and infected that leads to infection is directly proportional to density. On the other hand, the frequency-dependent type is independent of the host density. For randomly mixed susceptibles and infected, the transmission goes with $\sigma SI/U$. On average, each susceptible $S$ will make the same number of contacts independent of the host density, and a fraction $I/U$ of these contacts will be with infected. This frequency-dependent type is often assumed for sexually transmitted diseases (May and Anderson, 1987, 1988; Barlow, 1994). However, Beltrami and Carroll (1994) have applied this approach to the modeling of viral diseases in phytoplankton. They have assumed that the number of infected $I$ is much smaller than the number of susceptibles $S$. Therefore, the term $I$ can be neglected in the denominator and the transmission becomes proportional to the number of infected. This assumption will be applied further on to models with virally infected phytoplankton.

# Chapter 14

# Noise-induced pattern transitions

## 14.1 Transitions in a patchy environment

To begin with, the influence of parametric noise on the pattern formation in the Rosenzweig-MacArthur (1963) prey–predator model is investigated. The latter model has been used by Scheffer (1991a) for specifying the phytoplankton–zooplankton dynamics in a shallow lake under the control of nutrient density and of planktivorous fish stock. Structures in a deterministic environment have been presented by Malchow (1993, 2000a). The model reads in dimensionless quantities

$$\frac{\partial u}{\partial t} = r\,u\,(1-u) - \frac{au}{1+bu}\,v + d_u\,\Delta u\,, \qquad (14.1)$$

$$\frac{\partial v}{\partial t} = \frac{au}{1+bu}\,v - m_v\,v - \frac{g^2 v^2}{1+h^2 v^2}\,f + d_v\,\Delta v\,. \qquad (14.2)$$

The environmental heterogeneity is described through a simple approach: the considered $L \times 2L$ model area is divided into three habitats of sizes $L \times L/2$, $L \times L$ and $L \times L/2$ respectively; cf. Figure 14.1. The following model parameters have been chosen for the simulations, cf. Pascual (1993); Malchow et al. (2000, 2002):

$$\langle r \rangle = 1\,,\ a = b = 5\,,\ g = h = 10\,,\ \overline{m} = 0.6\,,\ f = 0\,, \qquad (14.3)$$

$$L = 100\,,\ x \in [0, L]\,,\ y \in [0, 2L]\,,\ d_u = d_v = 5 \times 10^{-2}\,. \qquad (14.4)$$

$\langle r \rangle$ is the spatially averaged prey growth rate $r(x,y)$ whereas $\overline{m}$ stands for the spatiotemporal mean of the noisy predator mortality rate $m_v(x,y,t)$. The value of $m_v$ is randomly chosen at each space point and each unit time step from a truncated normal distribution between $I = 0$ and 15% of $\overline{m}$, i.e. $m(x,y,t) = \overline{m}\,[\,1 + I - \mathtt{rndm}\,(2I)\,]$ with $\mathtt{rndm}\,(z)$ as a random number between 0 and $z$. The upper layer has double mean phytoplankton productivity $2\langle r \rangle$; the bottom layer only 60% of $\langle r \rangle$. Both are coupled by the middle habitat with linearly decreasing productivity. The latter gradient reflects assumptions by Pascual (1993). The chosen model parameters generate limit cycles at each space point, i.e., one has a kind of continuous chain of diffusively coupled nonlinear oscillators. The spatial setting yields a fast spatially uniform prey–predator limit cycle in the top habitat, continuously changing into quasiperiodic and chaotic oscillations and waves along the productivity gradient in the middle, coupled to slow spatially uniform limit cycle oscillations in the lower layer.

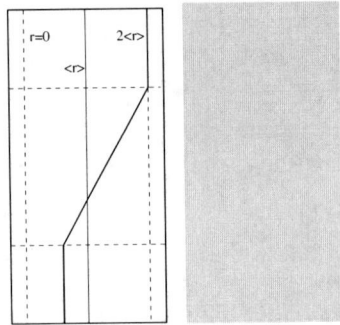

**FIGURE 14.1**: Left: Gradual change of mean prey growth rate in space. Neumann boundary conditions at bottom and top boundaries. Periodic boundary conditions at left- and right-hand sides. Right: Homogeneous initial condition.

### 14.1.1 No noise

At first, the pattern formation due to the layered spatial structure of the environment is studied. Fish and environmental noise are ignored. Five snapshots of a long-term simulation of the spatial and temporal dynamics have been taken and are presented in Figure 14.2.

The densities in the upper layer oscillate rather quickly throughout the simulation. The diffusively coupled limit cycles along the gradient in the middle layer induce a transition from periodic oscillations near the upper border to quasiperiodic in the middle part and to chaotic oscillations near the lower border (Pascual, 1993) which couple to the slowly oscillating bottom layer. The latter slow oscillator is not strong enough to fight the chaotic forcing from above. Finally, chaos prevails in the underpart of the model area.

### 14.1.2 Noise-induced pattern transition

For a weak 10% noise intensity, patterns are formed that are very similar to those in section 14.1.1 without noise. The spatial structures remain qual-

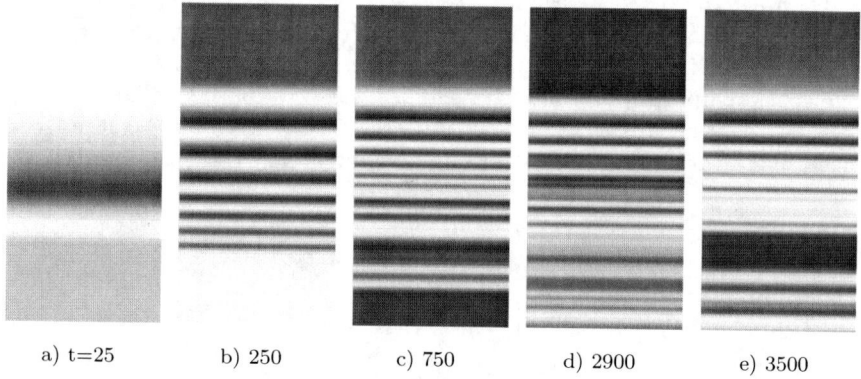

a) t=25    b) 250    c) 750    d) 2900    e) 3500

**FIGURE 14.2**: No noise. Rapid oscillations in the upper layer due to high prey growth rate. Formation of plane waves with spatiotemporal quasiperiodicity and chaos with decreasing prey growth rate. These plane waves compete with the rapid oscillations in the top and the slower in the bottom layer but can invade only the latter.

itatively the same, however, the noise enhances the spread of the wavy and chaotic part towards the upper layer. Furthermore, the boundaries of the layers become blurred. A slightly higher noise intensity of 15% shakes the results up.

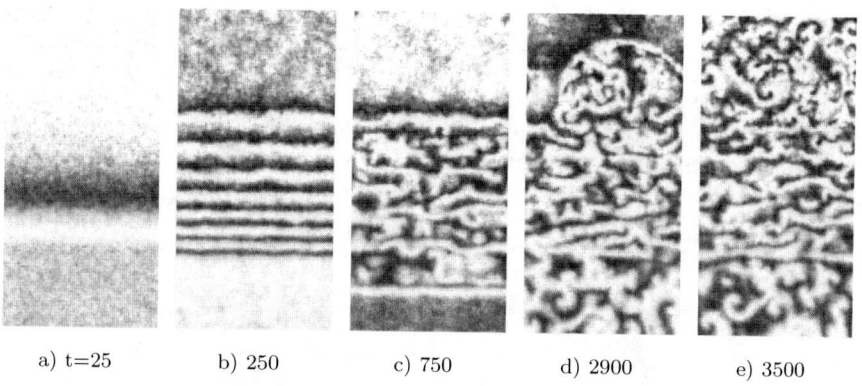

a) t=25    b) 250    c) 750    d) 2900    e) 3500

**FIGURE 14.3**: Fifteen percent noise. The noise lets the plane waves break up to spirals that first occupy the bottom layer again. However, due to the noisy forcing, the spirals are able to also invade the top layer, and the system undergoes a pattern transition.

The wavy and chaotic region on the lower side "wins the fight" against the upper regular structures and invades the whole space. This corresponds to a pronounced noise-induced transition (Horsthemke and Lefever, 1984) from one spatiotemporally structured dynamical state to another. This transition can also be seen in the local power spectra, which have been computed for the upper layer close to the upper reflecting boundary (Malchow et al., 2002).

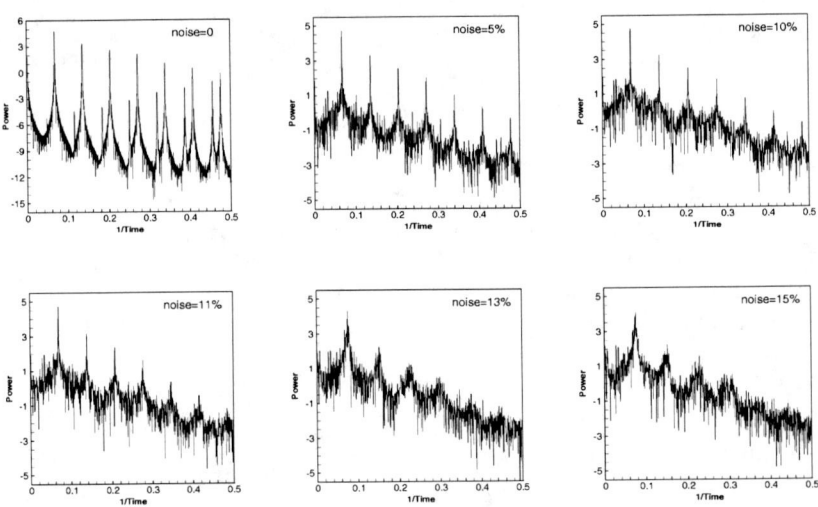

**FIGURE 14.4**: Power spectra for increasing noise intensities at $x = 50$ and $y = 5$.

The increase to 15% lets the periodicity disappear and a nonperiodic system dynamics remains. This is another proof of the noise-induced transition from periodical to aperiodical local behavior in the upper part of the model area after crossing a critical value of the external noise intensity.

## 14.2 Transitions in a uniform environment

Now, the stochastic variability is modeled through Langevin equations 13.1. At first, the influence of external noise on the formation of Turing patterns in model (14.1)–(14.2) is investigated.

## 14.2.1 Standing waves driven by noise

Parameter intervals for the occurrence of diffusive and/or advective instabilities of a spatially homogeneous stationary solution against supercritical wave perturbations have been assessed and given in detail, cf. Malchow (1995, 2000a); Satnoianu and Menzinger (2000) and Satnoianu et al. (2000).

The following set of model parameters has been chosen for the simulations described in this section, cf. Malchow (2000a):

$r = 1$, $a = 8.0$, $b = 11.905$, $g = 1.434$, $h = 0.857$, $m_v = 0.49$, $f = 0.093$,

$L = 100$, $x, y \in [0, L]$, $d_u = 2.8 \times 10^{-5}$, $d_v = 5.6 \times 10^{-3}$.

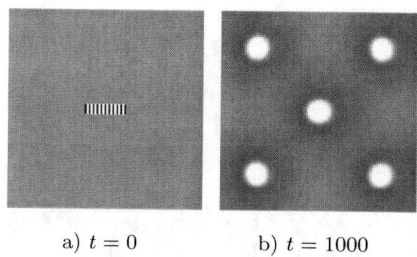

a) $t = 0$      b) $t = 1000$

**FIGURE 14.5**: Formation of a symmetric Turing structure after perturbation of the homogeneous solution with a central circular wave of supercritical wave number; cf. Malchow (2000a). Neumann boundary conditions. No noise.

The system is integrated using the explicit Euler-Maruyama scheme (Maruyama, 1955; Kloeden and Platen, 1999) for the stochastic interaction part and a Peaceman-Rachford alternating-direction implicit (ADI) scheme for diffusion (Peaceman and Rachford Jr., 1955; Thomas, 1995). For the noise term, random numbers are generated using the Mersenne Twister (Matsumoto and Nishimura, 1998), normally distributed by the Box-Muller method (Box and Muller, 1958).

The simulations start with a wave of supercritical wave number in the center that destabilizes the spatially homogeneous species distribution, resulting in that funny fully symmetric 5-eyes pattern; cf. Figure 14.5. Here and further on, the prey pattern is displayed on a greyscale from black ($u = 1$) to white ($u = 0$).

Besides the strong conditions on the difference of the diffusion coefficients, the symmetry and polarity of Turing patterns have long been a reason to question the role of those instabilities and structures in biodynamics because nature is not that symmetric in a mathematical sense. We will see now how a little noise changes the patterns completely. The same initial perturbation

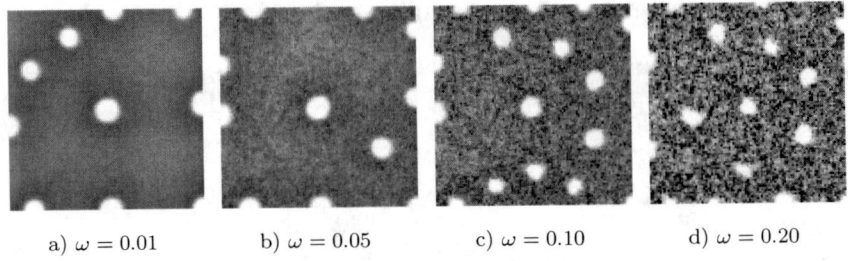

a) $\omega = 0.01$  b) $\omega = 0.05$  c) $\omega = 0.10$  d) $\omega = 0.20$

**FIGURE 14.6**: Increasing noise intensity disturbs and destroys the symmetry of the final stationary pattern. Neumann boundary conditions.

wave is used, however, the system is subject to increasing noise. The result is seen in Figure 14.6.

The pattern symmetry is immediately disturbed and the 5-eyes pattern disappears. Finally, we have an asymmetric Turing pattern even with varying sizes of the empty (white) patches. The disturbing symmetry of diffusion-induced Turing patterns has been overcome by some noise that is present everywhere in natural systems. Furthermore, noise has induced transitions between different structures that has also been shown in Section 14.1.2.

# Chapter 15

# Epidemic spread in a stochastic environment: plankton as a paradigm

There are many publications on pattern formation and chaos in minimal prey–predator models, also for phytoplankton-zooplankton dynamics with patchiness and blooming that will be in the focus of this chapter. In contrast, not so much is known on about marine viruses and their role in aquatic ecosystems and the species that they infect; for reviews cf. Fuhrman (1999); Wommack and Colwell (2000) as well as Suttle (2000, 2005). Jacquet et al. (2002) have found that viral infection might accelerate the termination of phytoplankton blooms.

There are two types of viral replication cycles. Contrary to lytic infections with destruction and without reproduction of the host cell, lysogenic infections are a strategy whereby viruses integrate their genome into the host's genome. As the host reproduces and duplicates its genome, the viral genome reproduces, too. The understanding of the importance of lysogeny is just at the beginning (Wilcox and Fuhrman, 1994; Jiang and Paul, 1998; McDaniel et al., 2002; Ortmann et al., 2002). Cochran et al. (1998) report that many environmentally important pollutants may be inducing agents for natural lysogenic viral production in the marine environment.

Mathematical models of the ecoepidemiology of virally infected phytoplankton populations are rare as well. The classical publication is by Beltrami and Carroll (1994). Recent work is of Chattopadhyay et al. (2002; 2002; 2003; 2004). The latter deals with lytic infections and mass action incidence functions. Hilker and Malchow et al. have observed oscillations and waves in a phytoplankton-zooplankton system with Holling type II (Malchow et al., 2004; Hilker and Malchow, 2006; Hilker et al., 2006) and III (Malchow et al., 2005) grazing under lysogenic viral infection and proportionate mixing incidence function (frequency-dependent transmission) (Nold, 1980; Hethcote, 2000; McCallum et al., 2001).

In this chapter, the focus is on modeling lysogenic and lytic infections and frequency-dependent transmission under multiplicative noise and its impact on the local and spatiotemporal dynamics of interacting phytoplankton and zooplankton. The Rosenzweig-MacArthur prey–predator model is used again. Starting with the local stationary behavior and for lytic infections, the local coexistence on a strange periodic attractor is shown to be possible even under nonstationary conditions. Then, lysogenic infections are considered to remarkably simplify the studies. The final section deals with noise-induced pattern formation in a slow-fast, excitable, prey–predator dynamics with structures like local coherence resonance, global synchronization, and the generation of stationary spatial patterns.

## 15.1 Model

Starting with model (14.1)–(14.2), the phytoplankton population $u$ is split into a susceptible part $s$ and an infected portion $i$. Zooplankton $v$ grazes on both susceptible and infected phytoplankton. Then, the model system reads for symmetric inter- and intra-specific competition of susceptibles and infected

$$\frac{ds}{dt} = r_s\, s\,(1-u) - \frac{a\,s}{1+b\,u}v - \sigma\frac{s\,i}{u}, \tag{15.1a}$$

$$\frac{di}{dt} = r_i\, i\,(1-u) - \frac{a\,i}{1+b\,u}v + \sigma\frac{s\,i}{u} - m_i\,i, \tag{15.1b}$$

$$\frac{dv}{dt} = \frac{a\,u}{1+b\,u}v - m_v\,v. \tag{15.1c}$$

A frequency-dependent transmission rate $\sigma$ as well as an additional disease-induced mortality of infected (virulence) with rate $m_i$ are assumed. The intrinsic growth rates of susceptibles and infected are $r_s$ and $r_i$, respectively. In the case of lysogenic infection, it holds that $0 \le r_i \le r_s$, whereas in the case of lytic infection, $r_i \le 0 \le r_s$. Then, the first term on the right-hand side of (15.1b) describes the losses due to natural mortality and competition.

Now, it is searched for stationary and oscillatory solutions of the system (15.1a)–(15.1c) which is simplified for that through a convenient transformation, then describing the dynamics of the total phytoplankton population $u = s + i$ and the prevalence $p = i/u$. The vector of population densities and prevalence is $\mathbf{X} = \{u, p, v\}$, and the model equations read

$$\frac{du}{dt} = [\,r_s(1-p) + r_i p\,](1-u)u - \frac{a\,u}{1+b\,u}v - m_i p u, \tag{15.2a}$$

$$\frac{dp}{dt} = [(r_i - r_s)(1-u) + (\sigma - m_i)](1-p)p, \tag{15.2b}$$

$$\frac{dv}{dt} = \frac{a\,u}{1+b\,u}v - m_v v. \tag{15.2c}$$

System (15.2a)–(15.2c) has the following trivial and semitrivial equilibria

$E = \{u^S, p^S, v^S\}$ with

$$\left.\frac{du}{dt}\right|_{\mathbf{X}=E} = \left.\frac{dp}{dt}\right|_{\mathbf{X}=E} = \left.\frac{dv}{dt}\right|_{\mathbf{X}=E} = 0:$$

1) $E_{00} = \{0, 0, 0\}$.

   The trivial state is always unstable.

2) $E_{01} = \{0, p_{01}^S, 0\}$ with $p_{01}^S = 1$.

   This disease-induced extinction of the total prey population is possible for $r_i < m_i < \sigma + r_i - r_s$.

3) $E_1 = \{u_1^S, 0, 0\}$ with $u_1^S = s = 1$.

   Only the susceptible prey species survive at their carrying capacity for $\sigma < m_i$ and $m_v > a/(1+b)$.

4) $E_2 = \{u_2^S, p_2^S, 0\}$ with $u_2^S > 0$, $p_2^S > 0$.

   a) $E_{21} = \{u_{21}^S, p_{21}^S, 0\}$ with $u_{21}^S = i = 1 - m_i/r_i$, $p_{21}^S = 1$.
      For $m_i < r_i$, $m_i < (r_i/r_s)\sigma$ and $au_{21}^S/(1 + bu_{21}^S) < m_v$, only the infected can survive.

   b) $E_{22} = \{u_{22}^S, p_{22}^S, 0\}$ with

   $$u_{22}^S = 1 - \frac{\sigma - m_i}{r_s - r_i}, \quad p_{22}^S = \frac{r_s}{\sigma}\frac{\sigma - m_i}{r_s - r_i}.$$

   The computer-aided stability analysis of this solution yields some the lengthy expressions for the stability intervals that do not provide further insight and are omitted here.

5) $E_3 = \{u_3^S, 0, v_3^S\}$ with

   $$u_3^S = \frac{m_v}{a - m_v b}, \quad v_3^S = \frac{r_s}{a}(1 + bu_3^S)(1 - u_3^S).$$

   The infected go extinct for a too low transmission rate or a too high virulence. The remaining $u$-$v$ prey–predator model is a well studied textbook example. The solution can be a stable node or focus. An unstable focus bound by a stable limit cycle may appear after a Hopf bifurcation.

Nontrivial equilibria, i.e., the coexistence of all species, only exist for a single combination of parameters. Then, there is a continuum of stationary states:

6) $E_{v_4^S(p)} = \{u_4^S, p, v_4^S(u_4^S, p)\}$ with $u_4^S > 0$, $0 < p < 1$, $u_4^S(u_4^S, p) > 0$.

From Equation (15.2b) and (15.2c) one finds

$$u_{41}^S = 1 - \frac{\sigma - m_i}{r_s - r_i} \quad \text{and} \quad (15.3\text{a})$$

$$u_{42}^S = \frac{m_v}{a - m_v b}, \quad (15.3\text{b})$$

which define two parallel planes independent of $p$ and $v$ in $(u\text{-}p\text{-}v)$ phase space. These planes are orthogonal to the $(u\text{-}v)$ and parallel to the $(p\text{-}v)$ plane. Both must coincide, i.e., the system parameters have to strictly obey the relationship

$$1 - \frac{\sigma - m_i}{r_s - r_i} = \frac{m_v}{a - m_v b} = u_4^S. \quad (15.4)$$

From Equation (15.2a), one obtains the plane

$$v = v(u, p) = \frac{1 + bu}{a} \left\{ r_s(1 - u) + [(r_i - r_s)(1 - u) - m_i] p \right\} \; (15.5)$$

All points $E_{v_4^S(p)}$ on the straight intersection line of planes (15.4) and (15.5),

$$v_4^S(u_4^S, p) = v(u_4^S, p) \text{ with } u_4^S \text{ as in } (15.4), \quad (15.6)$$

are stationary states independent of $p$ for $0 < p < 1$. This line is a heteroclinic connection between the semitrivial equilibria $E_2$ and $E_3$.

For $m_v < a(b-1)/(b(b+1))$, the $u\text{-}v$ subsystem exhibits an unstable focus bound by a stable limit cycle. In the stationary case (15.4), a numerical stability analysis shows that all equilibria $E_{v_4^S(p)}$ on line (15.6), including $E_2$ and $E_3$, are degenerated, i.e., their third eigenvalue is zero. The upper part of line (15.6) consists of degenerated unstable foci. A fold-Hopf (zero-pair) bifurcation point (Kuznetsov, 1995; Nicolis, 1995) separates them from the lower part closer to the $(u\text{-}p)$ plane with degenerated stable foci. The result of a corresponding numerical simulation, starting on the unstable upper branch, is shown in Figure 15.1. Such a bifurcation has also been called a *zip bifurcation* (Farkas, 1984), because a singular curve folds into periodic solutions when a parameter is varied. This parameter is the decreasing prevalence here.

The closer the initial condition to the $(u\text{-}v)$ plane, the longer is the journey along the trajectory through phase space. The final position of the stationary state on the line strongly depends on the initial conditions. Therefore, the final positions are only neutrally stable. This remains true in the case when the intersection point of (15.6) in the $u$-$v$ subsystem is a stable focus or stable node, and the line (15.6) becomes a continuum of stable solutions.

The growth rate of infected $r_i$ has been set to zero. This choice describes the cell lysis of infected phytoplankton cells and nonsymmetric competition of infected and susceptibles; the infected still have an impact on the growth of susceptibles by shading and space demand, but not vice versa. Furthermore, $m_i$ now stands for an effective mortality, i.e., the additional disease-induced mortality (virulence) plus the natural mortality of the infected.

## 15.2 Strange periodic attractors in the lytic regime

The strong parameter relationship (15.4) is surely not realistic. The probability to find such a fixed combination in nature is almost zero. Therefore, nonstationary situations will be computed now, with parameter settings when the planes (15.3a) and (15.3b) do not coincide and the intersection lines with plane (15.5) are no longer stationary.

First, the virulence is increased. A computer-assisted bifurcation and stability analysis shows that under these nonstationary conditions $E_2$ is a saddle-focus with a stable two-dimensional manifold and an unstable one-dimensional manifold. In the $(u$-$v)$ plane, $E_3$ is also a saddle-focus, but with an unstable two-dimensional manifold and a stable one-dimensional manifold. In Figure 15.2, the trajectory starts in the upper corner and approaches the lower end point of the right-hand line (15.3a) in the $(u$-$p)$ plane which is the semitrivial stationary state $E_2$. This is the mentioned saddle-focus with stable oscillation but unstable in the direction of $v$. Therefore, the trajectory is shot along the heteroclinic connection to the $(u$-$v)$ plane and gets into the sphere of influence of the end point of the left-hand line (15.3b). This is also a semitrivial saddle-focus, namely $E_3$ with unstable oscillation and stable in $v$ direction. Thus, the trajectory bounces back, spirals down the lines and "tube-rides" up again and again. In other terms, on the way up, it is "reinjected" (Nicolis, 1995) and tunnels through the two formed funnels. This is illustrated in Figure 15.2a. It resembles the movement on a torus, where the centre hole of the torus is shrinked to a thin tube. However, the precessing trajectory gets "phase-locked" and finally, for long times, approaches a periodic attractor (Langford, 1983, pp. 233). This is shown in Figure 15.2b. The oscillation takes place in a plane that is orthogonal to the $(p$-$v)$ plane. The attractor surrounds the two intersection points of the two lines of nonstationary points and the plane of oscillation.

The attractor is called a *strange periodic attractor* (Hilker and Malchow,

2006) because of the long-lasting very peculiar approach towards the asymptotic oscillations. Obviously, the attractor is not chaotic which is sometimes called "strange." But in order to underline the impressive long-term transient dynamics, it is named "strangely periodic."

As for a limit cycle, the position of the asymptotic periodic attractor is independent of the initial conditions. For further increasing values of $m_i$, the behavior of the system becomes simpler. The distinct funnel formation disappears and the periodic attractor is stabilizing faster and faster. For too high virulence, the infected go extinct and the system oscillates in the $u$-$v$ subsystem. This is illustrated in Figure 15.3.

For virulences below the stationary value given in Figure 15.1, $E_2$ becomes a stable and $E_3$ an unstable focus, respectively. Numerical simulations yield that zooplankton dies out and the dynamics relaxes to $E_2$ in the $u$-$p$ subsystem; cf. Figure 15.4.

Summarizing, $E_2$ has undergone a bifurcation from an unstable saddle-focus to a stable focus. At this bifurcation point, the continuum of degenerated nontrivial equilibria $E_{v_4^S(p)}$ appeared simultaneously. A zero-pair Hopf bifurcation took place along this continuum line.

The strange periodic attractor allows for the coexistence of all three species

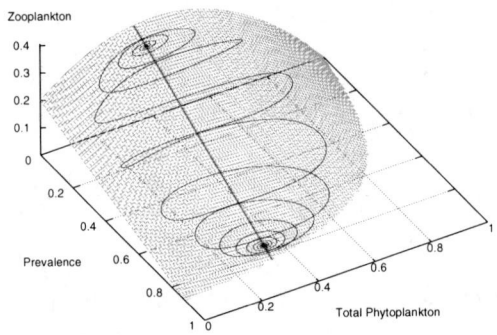

**FIGURE 15.1:** (See color insert.) Stationary dynamics of system (15.2a)–(15.2c) with coexistence of all three populations. The trajectory starts at the upper unstable part of the line of stationary points, passes the fold-Hopf bifurcation point, and finally relaxes on the neutrally stable lower part. Parameters: $r_s = 1$, $r_i = 0$, $a = b = 5$, $\sigma = 4/5$ from Equation (15.4), $m_i = 2/15$, $m_v = 5/8$. Initial conditions: $u_0 = 1/3$, $p_0 = 3/40$, $v_0$ from Equation (15.6). The straight line is the continuum of equilibria $E_{v_4^S(p)}$, lying on the shaded plane (15.5). With permission from Hilker and Malchow (2006), http://www.informaworld.com.

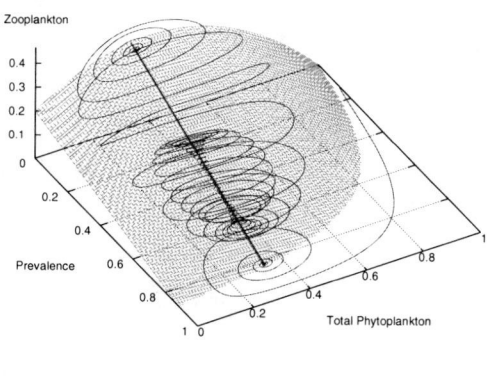

a) $t = 0 - 3000$

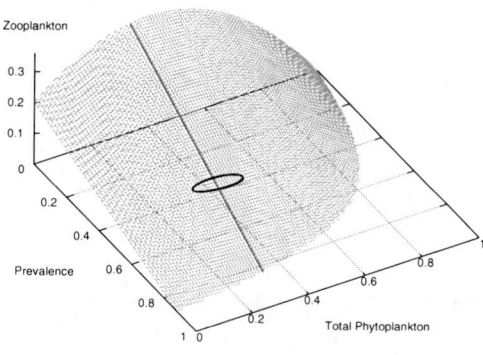

b) $t = 6500 - 8000$

**FIGURE 15.2**: (See color insert.) Nonstationary dynamics of system (15.2a)–(15.2c) with coexistence of all three populations; see details in the text. Parameters: $m_i = 7/50$, all others as in Figure 15.1. Initial condition: $u_0 = 1.0$, $p_0 = 0.125$, $v_0 = 0.0015$. With permission from Hilker and Malchow (2006), http://www.informaworld.com.

positive in a substantial range of parameters. This makes the model much more realistic and biologically interesting, because the special condition (15.4) unlikely holds exactly in reality. The nontrivial oscillations in the interior of the first octant are remarkable, because there does not exist a nontrivial stationary state.

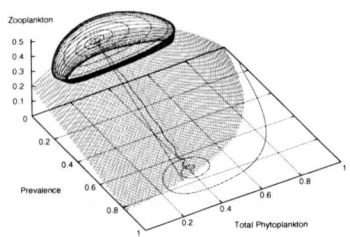 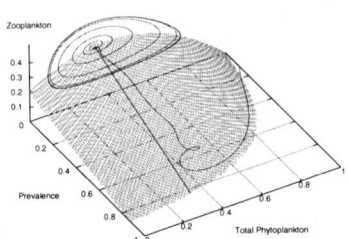

a) $m_i = 1/5$, coexistence of all three populations

b) $m_i = 3/10$, extinction of infected

**FIGURE 15.3**: (See color insert.) Dynamics of system (15.2a)–(15.2c), see details in the text. Parameters like in Figure 15.1 except for $m_i$. Inital condition: $u_0 = 1.0$, $p_0 = 0.125$, $v_0 = 0.0015$. With permission from Hilker and Malchow (2006), http://www.informaworld.com.

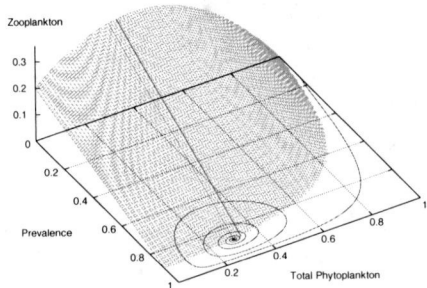

**FIGURE 15.4**: (See color insert.) Dynamics of system (15.2a)–(15.2c), see details in the text. Parameters: $m_i = 0.1$, all others like in Figure 15.1. Initial condition: $u_0 = 1.01$, $p_0 = 0.1$, $v_0 = 0.001$. With permission from Hilker and Malchow (2006), http://www.informaworld.com.

## 15.3 Local dynamics in the lysogenic regime

The richness of the system dynamics is reduced when considering lysogenic infections with $r_s = r_i = r$. For $m_i > \sigma$, the infected go extinct, and for $m_i < \sigma$, the susceptibles do. In the case of $m_i = \sigma$, susceptibles and infected coexist. Because of the symmetry of the growth terms of susceptibles and infected, the initial conditions determine their final dominance in the endemic state, i.e., if $s(t=0) > i(t=0)$ then $s(t) > i(t) \; \forall \; t$. A corresponding example is presented in Figure 15.5 for $r = 1$ and $a = b = 5$. These three parameter values will be kept for all simulations.

It is readily seen that the transformation of model (15.1a)–(15.1c) to system (15.2a)–(15.2c) with $r_s = r_i = r$ reduces the considerations of stationarity and stability to a pseudo-two-dimensional problem because the prevalence can take only three values, i.e., zero for $\sigma < m_i$, unity for $\sigma > m_i$, or its initial value for $\sigma = m_i$. The computations are remarkably simplified because the $u$-$v$ dynamics only proceeds in the plane $p = const$.

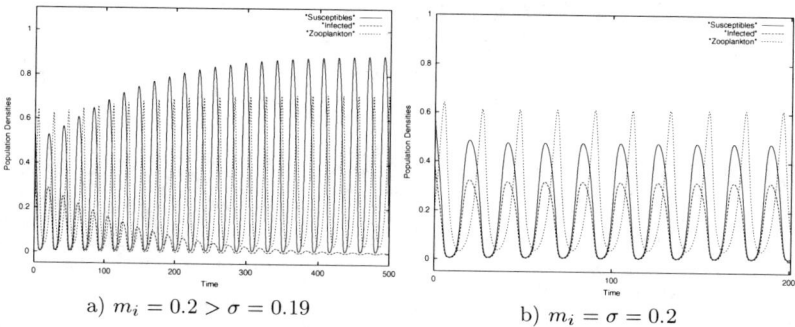

a) $m_i = 0.2 > \sigma = 0.19$  b) $m_i = \sigma = 0.2$

**FIGURE 15.5:** Local dynamics with (a) extinction of infected for $m_i > \sigma$, (b) coexistence of susceptibles $s$, infected $i$ and zooplankton $v$ for $m_i = \sigma$, $m_v = 0.5$. For $m_i < \sigma$, the susceptibles would go extinct. With permission from Malchow et al. (2004).

For $\sigma = m_i$, the initial value of the prevalence is an additional control parameter that might drive the system to different dynamic behavior. An example for bistability is shown in Figure 15.6.

After the deterministic local behavior, the spatial dynamics is studied now.

## 15.4 Deterministic and stochastic spatial dynamics

There is a vast number of publications on the spatiotemporal self-organization in prey–predator communities, modeled by (advection–)diffusion–reaction equations; cf. the references in the introduction. Much less is known about equation-based modeling of the spatial spread of epidemics, a small collection of pa-

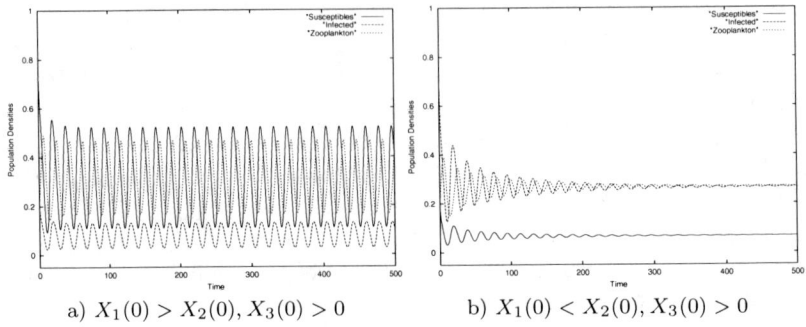

**FIGURE 15.6**: Endemic state with initial conditions in different basins of attraction: (a) stable and (b) damped oscillation. $m_i = \sigma = 0.2, m_v = 0.625$. With permission from Malchow et al. (2004).

pers includes Grenfell et al. (2001); Abrams et al. (2003); Lin et al. (2003); Zhdanov (2003); Malchow et al. (2004); Hilker (2005); Hilker et al. (2005); Malchow et al. (2005); Petrovskii et al. (2005); Hilker et al. (2006) and Hilker et al. (2007).

In this section, the space-time dynamics of the plankton model (15.1a)–(15.1c) is considered, i.e., zooplankton, grazing on susceptible and virally infected phytoplankton, both diffusing in horizontally two-dimensional space, under the influence of environmental noise. The diffusion terms have been integrated using the Peaceman-Rachford alternating direction implicit (ADI) scheme, cf. Peaceman and Rachford Jr. (1955); Thomas (1995). For the source and noise terms, the explicit Euler-Maruyama scheme has been applied (Maruyama, 1955; Kloeden and Platen, 1999; Higham, 2001). The spatial grid is a $99 \times 99$ point square with spacing $\Delta x = \Delta y = 1$.

The following series of Figures 15.7–15.10 shows the results of computations of the spatiotemporal dynamics of growth and interaction with parameters from section 15.3, but now including diffusion and noise.

In order to avoid boundary effects, periodic boundary conditions have been chosen for all simulations. Localized patches in empty space have been taken as initial conditions. They are the same for deterministic and stochastic simulations. The first two rows of figures show the dynamics of the susceptibles for deterministic and stochastic conditions, the two middle rows show the infected and the two lower rows the zooplankton.

For Figures 15.7, 15.8, and 15.9, there are two initial patches, one with zooplankton surrounded by susceptible phytoplankton in the upper part of the model area, and one with zooplankton surrounded by infected on the right-hand side of the model area. For Figure 15.10, there are initial central patches of all three species. Susceptibles are ahead of infected which are ahead of zooplankton. This special initial configuration leads at first to the propagation of concentric waves for the deterministic case in rows 1, 3 and 5. These naturally unrealistic waves are immediately blurred and only a leading diffusive front remains for the stochastic case in rows 2, 4 and 6. A stochastic theory of diffusive waves has been developed by Schimansky-Geier et al. (1983); Schimansky-Geier and Zülicke (1991); and van Saarloos (2003).

In Figure 15.7, the final spatial coexistence of all three species for $m_i = \sigma$ is presented. The localized initial patches generate concentric waves that break up after collision and form spiral waves in a deterministic environment. The noise only blurs these unrealistic patterns. The greyscale changes from high population densities in black color to vanishing densities in white.

This changes for $m_i > \sigma$ and $m_i < \sigma$ in Figures 15.8 and 15.9, respectively. While infected or susceptibles die out in the deterministic case, the noise supports their survival and spread under unfavourable conditions.

In Figure 15.10, one can readily see the so-called dynamic stabilization of the locally unstable focus in space for a deterministic environment. A long plateau is formed with a leading diffusive front ahead; cf. Petrovskii and Malchow (2000); Malchow and Petrovskii (2002). Furthermore, the infected seem to be trapped in the center and become almost extinct. The noise fosters the escape, spread and survival of the infected.

The equal growth rates of susceptibles and infected have led to the situation where, in a constant environment, the ratio of the mortality of the infected and the transmission rate of the infection determines coexistence, survival, or extinction of susceptibles and infected. A fluctuating environment supports the survival and the spatial spread of the "endangered" species. Furthermore, noise has not only enhanced the spatiotemporal coexistence of susceptibles and infected but it has been necessary to blur distinct artificial population structures like concentric or spiral waves and to generate more realistic fuzzy patterns.

## 15.5 Local dynamics with deterministic switch from lysogeny to lysis

Lysogenic viral replication is very sensitive to environmental variability. First, the switch from lysogeny to lysis in the prey is studied for a constant environment. Only one local example for such a switch is drawn in Figure 15.11. After the transition, there is no further replication of infected. $r_i$ is simply set to zero when the switch occurs; cf. Section 15.1. A more technical assumption for the simulation is that the remaining natural mortality of the infected is added to the virulence, leading to a higher effective mortality of the infected, i.e., the parameter $m_i$ increases. Furthermore, the lytic cycle

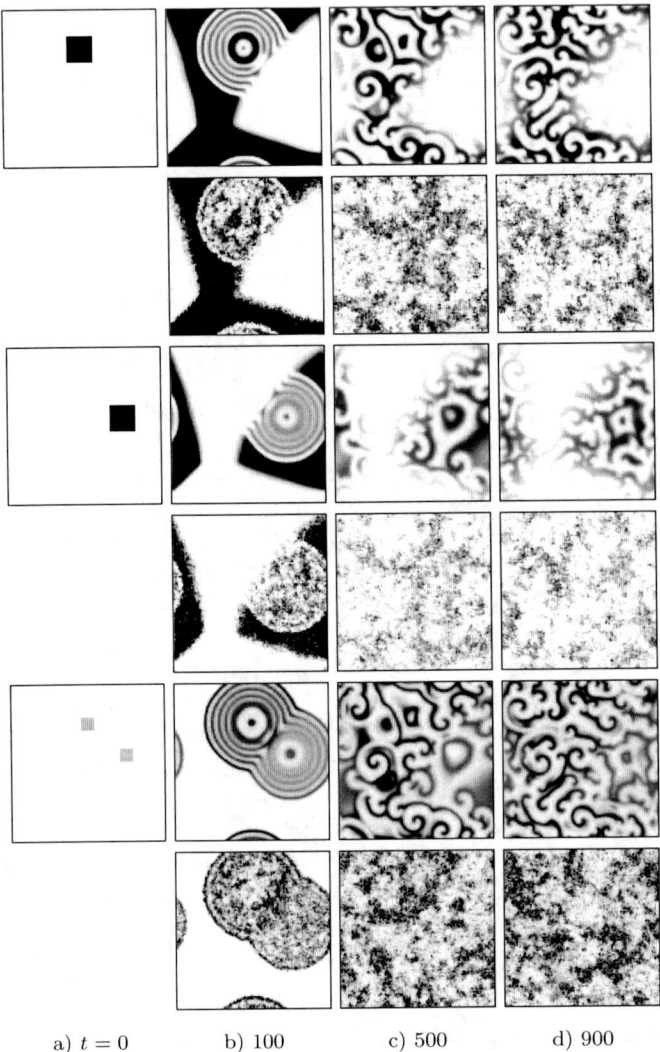

a) $t = 0$    b) 100    c) 500    d) 900

**FIGURE 15.7**: Spatial coexistence of susceptibles (two upper rows), infected (two middle rows), and zooplankton (two lower rows) for $m_i = \sigma = 0.2$, $m_v = 0.5$, $d = 0.05$. No noise $\omega = 0$ and 0.25 noise intensity, respectively. Periodic boundary conditions.

generates many more viruses, i.e., the transmission rate $\sigma$ increases as well. And, finally, the intra-specific competition of the dying infected phytoplankton cells vanishes whereas the interspecific competition of susceptibles and infected becomes nonsymmetric, i.e., the dead and dying infected still influ-

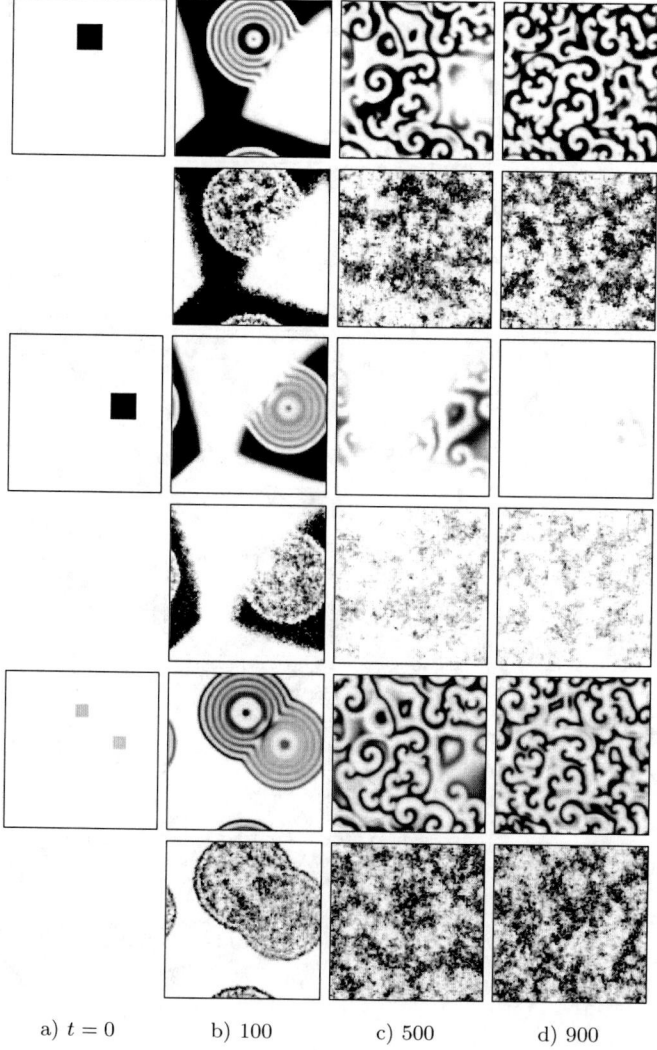

**FIGURE 15.8**: Spatial coexistence of susceptibles (two upper rows) and zooplankton (two lower rows). Extinction of infected (third row) for $m_i = 0.2 > \sigma = 0.19$, $m_v = 0.5$, $d = 0.05$, and no noise. Survival of infected for $\omega = 0.25$ noise intensity (fourth row).

ence the growth of the susceptibles and contribute to the carrying capacity, but not vice versa.

The switch from lysogenic to lytic virus replication results in a much lower

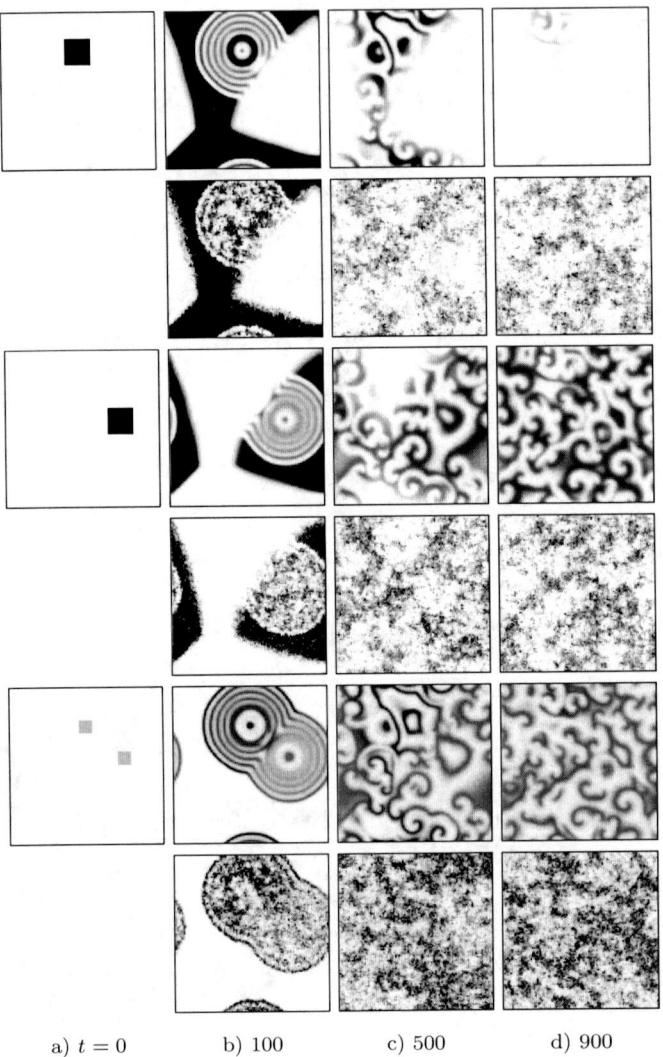

a) $t = 0$  b) 100  c) 500  d) 900

**FIGURE 15.9**: Spatial coexistence of infected (two middle rows) and zooplankton (two lower rows). Extinction of susceptibles (first row) for $m_i = 0.2 < \sigma = 0.21$, $m_v = 0.5$, $d = 0.05$, and no noise. Survival of susceptibles for $\omega = 0.25$ noise intensity (second row).

mean abundance of infected, though endemicity is still stable. However, the system responds rather sensitively to parameter changes, especially to variations of virulence and transmission rate, and the infected can easily become

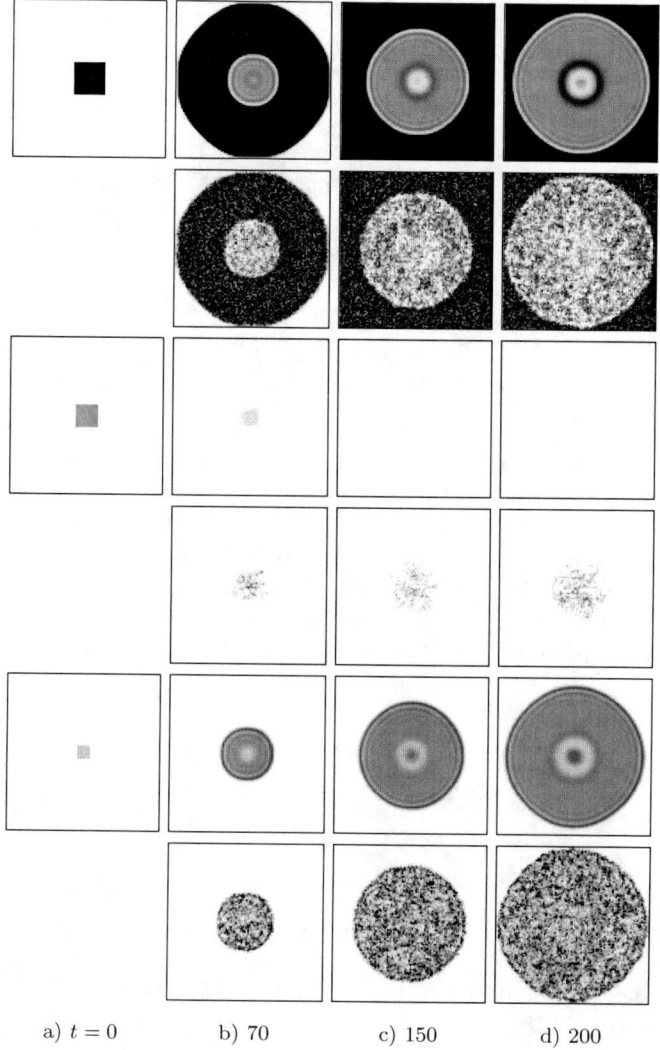

a) $t = 0$  b) 70  c) 150  d) 200

**FIGURE 15.10**: Spatial coexistence of susceptibles (two upper rows), infected (two middle rows), and zooplankton (two lower rows) for $m_i = \sigma = 0.2$, $m_v = 0.625$, $d = 0.05$. Without noise, trapping and almost extinction of infected in the center (third row). With $\omega = 0.25$ noise intensity, noise-enhanced survival, and escape of infected (fourth row). Phenomenon of dynamic stabilization of a locally unstable equilibrium (first and fifth rows).

extinct. As in the preceding Section, multiplicative noise supports the survival

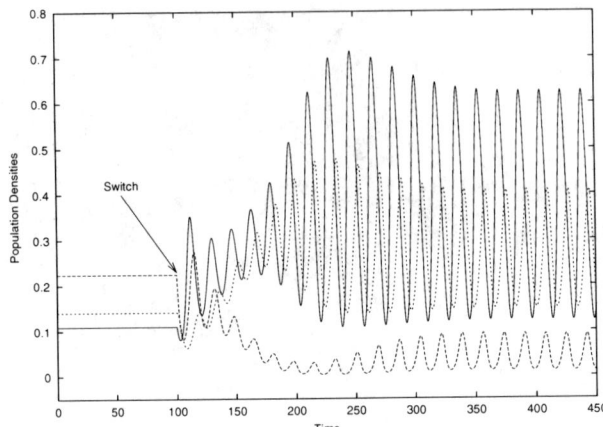

**FIGURE 15.11**: The nonoscillating endemic state with lysogenic infection, switching at t=100 to an oscillating endemic state with lytic infection. The growth rate of infected $r_i$ is set from $r_i^{max} = 0.4$ to zero, the virulence $m_i$ from $m_i^{min} = 0.2$ to $m_i^{max} = 0.3$, and the transmission rate $\sigma$ from $\sigma^{min} = 0.6$ to $\sigma^{max} = 0.9$; see details in the text. Other parameter values: $r_s = 1$, $a = b = 5$, $m_v = 0.625$. Susceptibles are plotted with a solid line, infected with a dashed line, and zooplankton with dots. With permission from Hilker et al. (2006).

of the endangered species, i.e., there is always some probability to survive in a noisy environment while the deterministic setting inevitably leads to extinction.

## 15.6 Spatiotemporal dynamics with switches from lysogeny to lysis

Now, the switch dynamics of the plankton model (15.1a)–(15.1c) with horizontal diffusion in two-dimensional space in a constant and a fluctuating environment is considered.

Periodic boundary conditions have been chosen again for all simulations in order to avoid boundary effects. The initial conditions are as follows: The space is filled with the non-oscillating endemic state $(u^S, i^S, v^S) = (0.109, 0.224, 0.141)$; cf. Figure 15.11. Furthermore, there are two localized patches in space. They can be seen in Figure 15.12. The greyscale changes again from high population densities in black color to vanishing densities in white.

As in Section 15.4, one patch is located in the upper middle of the model area with susceptibles $u = 0.550$, four grid points further away from zooplank-

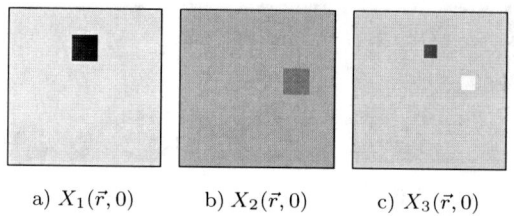

a) $X_1(\vec{r}, 0)$  b) $X_2(\vec{r}, 0)$  c) $X_3(\vec{r}, 0)$

**FIGURE 15.12**: Initial conditions for further spatiotemporal computations. With permission from Hilker et al. (2006).

ton $v = 0.450$ in each direction. The infected are at $i^S$. In the other patch at the right, the infected $i = 0.333$ are further away from zooplankton $v = 0.036$, whereas the susceptibles are at $u^S$. These initial conditions are the same for deterministic and stochastic computations.

The chosen system parameters generate oscillations in the center of the patches. The latter act as leading centers for concentric waves that collide and break up to spirals. Increasing noise again blurs this naturally unrealistic patterning; see also Malchow et al. (2004).

### 15.6.1  Deterministic switching from lysogeny to lysis

At first, switching begins in the area with the highest initial density of infected, i.e., in the right-hand patch. The growth rate of infected $r_i$ vanishes whereas virulence and natural mortality of infected increase to a higher effective virulence $m_i$. The transmission rate $\sigma$ also rises as described in Section 15.3. It is assumed that these parameter changes diffuse through space like a Fisher wave (1937). If a subsidiary quantity $r_a$ with Fisher dynamics is introduced,

$$\frac{\partial r_a(\vec{r}, t)}{\partial t} = r_a(1 - r_a) + d\Delta r_a ,  \qquad (15.7)$$

$$\text{then} \quad r_i(\vec{r}, t) = r_i^{max}(1 - r_a) , \qquad (15.7a)$$

$$m_i(\vec{r}, t) = m_i^{min} + (m_i^{max} - m_i^{min})r_a , \qquad (15.7b)$$

$$\sigma(\vec{r}, t) = \sigma^{min} + (\sigma^{max} - \sigma^{min})r_a . \qquad (15.7c)$$

The initial conditions are $r_a = 1$ in the right patch and zero elsewhere. For simplicity, the diffusivity $d$ is assumed to be the same as for all the populations. Its value of $d = 0.05$ has been chosen from Okubo's (1971) diffusion diagrams in order to model processes on a kilometer scale. The spatial propagation of $r_i = 0$ is displayed in white in Figure 15.13.

The arising dynamics of susceptibles is presented in Figure 15.14. The latter graphics has been chosen because of richer contrast, however, the patterns of infected are similar.

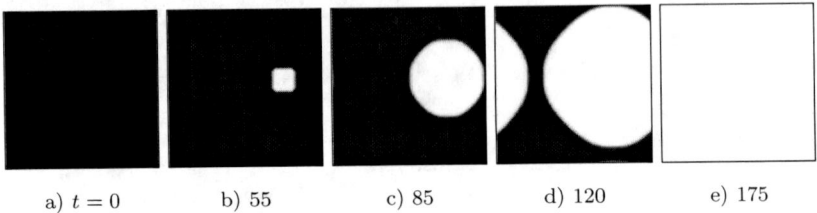

a) $t = 0$   b) 55   c) 85   d) 120   e) 175

**FIGURE 15.13**: Spatial propagation of zero replication rate of the infected. With permission from Hilker et al. (2006).

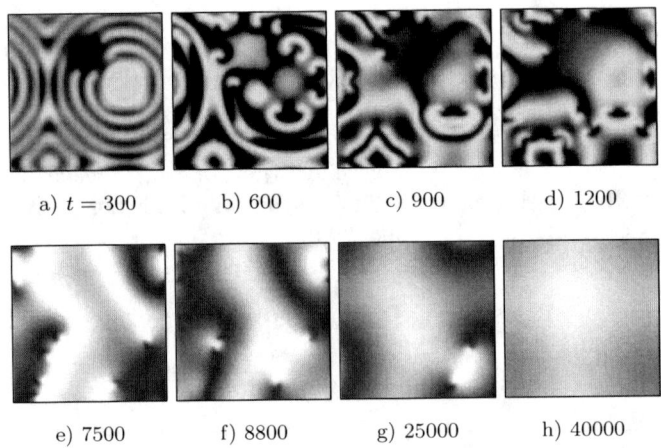

a) $t = 300$   b) 600   c) 900   d) 1200

e) 7500   f) 8800   g) 25000   h) 40000

**FIGURE 15.14**: Dynamics of susceptibles. No noise. With permission from Hilker et al. (2006).

The breakup of concentric waves to a rather complex structure with spirals is nicely seen. In the long run, a pinning-like behavior of pairs of spirals is found. This effect is well-known from excitation waves in cardiac muscles; cf. the classical publications by Davidenko et al. (1992) and Pertsov et al. (1993). Here, the biological meaning remains unclear. The almost fixed pair-forming and rapidly rotating spirals approach each other extremely slowly, collide, and burst. A weak multiplicative noise accelerates this process, a fact shown in Figure 15.15. Stronger noise, i.e., higher environmental variability, suppresses the generation of pins. For the long run, the homogeneously oscillating endemic state prevails.

The system parameters have been selected in order to guarantee the survival of all three populations under deterministic conditions. The simulation for Figure 15.15 with 5% noise also yields this final endemic coexistence. However,

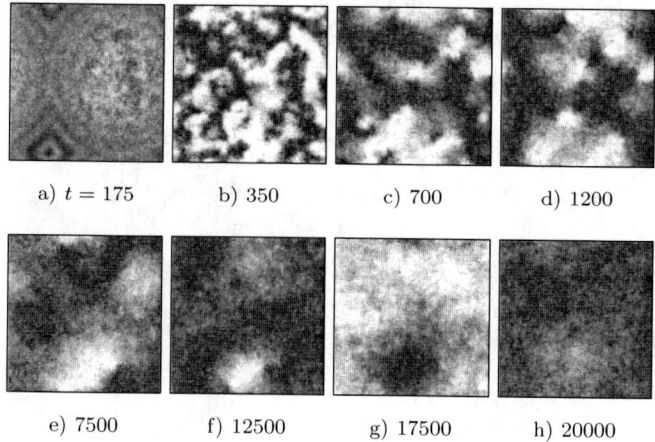

**FIGURE 15.15**: Dynamics of susceptibles. Noise intensities $\omega_1 = \omega_2 = \omega_3 = 0.05$. With permission from Hilker et al. (2006).

after the switch to cell lysis, it should be noted that there is only a certain survival probability for all three populations, the lowest for the infected.

### 15.6.2 Stochastic switching

The deterministic once-for-ever switching mechanism is very unrealistic. Furthermore, lytic infections might become lysogenic again (Herskowitz and Hagen, 1980; Moebus, 1996; Wilson et al., 1996; Oppenheim et al., 2005). Therefore, it is considered that only a certain fraction of viruses locally begins with the lysogenic replication and then switches. In order to model this, the subsidiary quantity $r_a$ is redefined to obey bistable kinetics and multiplicative noise, i.e.,

$$\frac{\partial r_a(\vec{r},t)}{\partial t} = (r_a - r_a^{min})(r_a - r_a^{crit})(r_a^{max} - r_a) + \omega_a r_a \cdot \xi(\vec{r},t) \ . \quad (15.8)$$

The noise stimulates system (15.8) to jump between its stable stationary states $r_a^{min}$ and $r_a^{max}$ (Nitzan et al., 1974; Ebeling and Schimansky-Geier, 1980; Malchow and Schimansky-Geier, 1985). It is assumed that the replication rate of the infected changes accordingly, i.e.,

$$\text{if } r_a > r_a^{crit} \quad \text{then } m_i = m_i^{min}, \sigma = \sigma^{min}, r_i = r_i^{max} \text{ (lysogeny)}, \quad (15.9)$$
$$\text{if } r_a \leq r_a^{crit} \quad \text{then } m_i = m_i^{max}, \sigma = \sigma^{max}, r_i = 0 \text{ (lysis)}. \quad (15.10)$$

This noise-induced dynamics of $r_a$ and $r_2$ as well as the temporal development of the spatial mean and the spatiotemporal pattern of $r_2$ are drawn

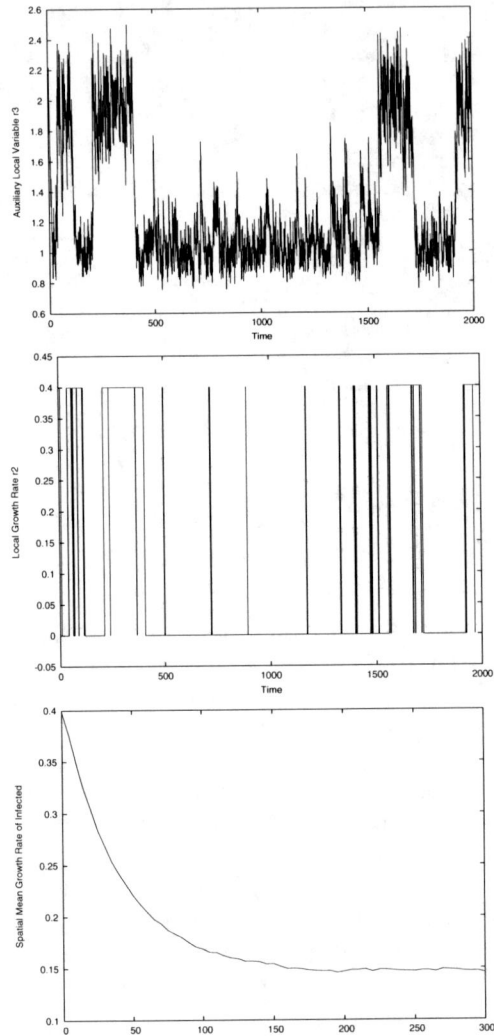

**FIGURE 15.16**: Noisy bistable dynamics of $r_a$ and resulting local switches of $r_i$ for $r_a^{max} = 2$, $r_a^{crit} = 1.5$, $r_a^{min} = 1$, and $\omega_a = 0.1$. The spatial mean of $r_i$ decreases from the maximum as a homogeneous initial condition to a value of approximately 0.15. The growth rate of infected $r_i$ switches from $r_i^{max} = 0.4$ to zero, the virulence $m_i$ from 0.2 to 0.3, and the transmission rate $\sigma$ from 0.6 to 0.9; see details in the text. Other parameter values: $r_1 = 1$, $a = b = 5$, $m_3 = 0.625$. With permission from Hilker et al. (2006).

in Figures 15.16 and 15.17, respectively. Initially, the whole system is in the lysogenic state (15.9).

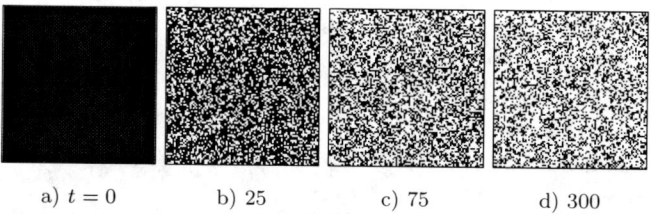

**FIGURE 15.17**: Spatiotemporal pattern of $r_i$. With permission from Hilker et al. (2006).

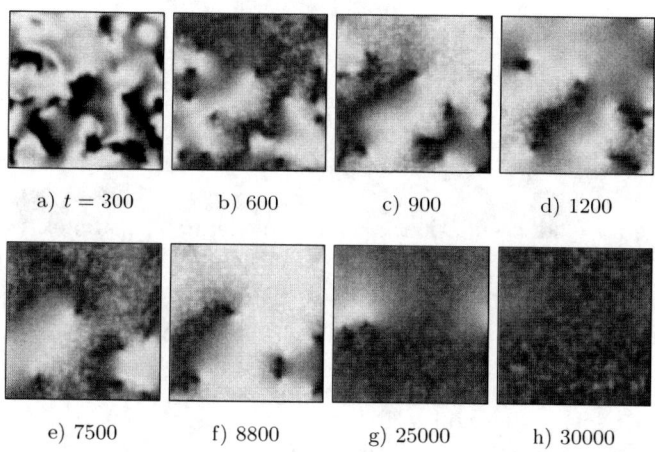

**FIGURE 15.18**: Deterministic dynamics of susceptibles with noisy switch. $\omega_u=\omega_i=\omega_v=0$, $\omega_a=0.1$. The almost stationary spiral pairs do still exist. With permission from Hilker et al. (2006).

At first, the simulation is run with noisy switches of $r_i$, $\sigma$ and $m_i$ but deterministic population dynamics. In this unrealistic setting, the pins can still be seen; cf. Figure 15.18.

If also the population dynamics is subject to noise, the result becomes more realistic. The pins are suppressed and the plankton forms a rather complex noise-induced patchy structure, cf. Figure 15.19.

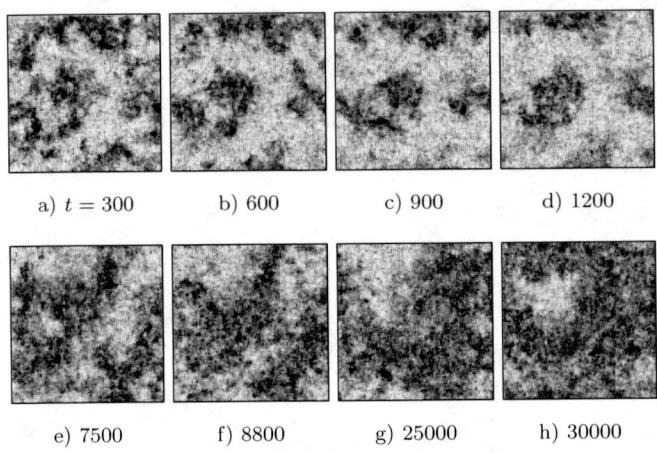

a) $t = 300$    b) 600    c) 900    d) 1200

e) 7500    f) 8800    g) 25000    h) 30000

**FIGURE 15.19**: Noisy dynamics of susceptibles and noisy switch. $\omega_j = 0.1$, $j = u, i, v, a$. The "pins" can no longer be found. With permission from Hilker et al. (2006).

Noisy jumps and population dynamics generate a complex patchy spatiotemporal structure that is typical for natural plankton populations. The presented sample results have led to a final endemic state with the coexistence of susceptibles, infected and zooplankton like in the deterministic case. However, one should be aware that the survival probability of the system with all nonzero populations is smaller than 1. Thus, there are good chances that the three-component system will switch to one of its subsystems. Again, noise does not only influence the spatiotemporal coexistence of the populations but also blurs distinct artificial population structures.

# Chapter 16

# Slow-fast dynamics and noise-induced pattern formation in an excitable prey–predator plankton system with infected prey

In this chapter, we continue considering the lysogenic case but focus on modeling the pattern-generating impact of multiplicative noise (Spagnolo et al., 2002, 2004; Allen, 2003; Anishenko et al., 2003; Valenti et al., 2004, 2006; Sieber et al., 2007) on excitable dynamics, i.e., noise-induced effects on interacting phytoplankton and zooplankton with Holling type III grazing, in the excitable parameter interval, close to a Hopf bifurcation, in time and space. A review on processes and models of spatiotemporal ordering out of noise has been provided by Sagués et al. (2007).

The local excitable prey–predator model with virally infected prey and lysogeny reads

$$\epsilon \frac{ds}{dt} = r_s s (1 - u) - \frac{a^2 s u}{1 + b^2 u^2} v - \sigma \frac{si}{u}, \qquad (16.1)$$

$$\epsilon \frac{di}{dt} = r_i i (1 - u) - \frac{a^2 i u}{1 + b^2 u^2} v + \sigma \frac{si}{u} - m_i i, \qquad (16.2)$$

$$\frac{dv}{dt} = \frac{a^2 u^2}{1 + b^2 u^2} v - m_v v. \qquad (16.3)$$

Slow-fast prey–predator cycles in this model have been studied by Fernández et al. (2002). Such processes with longer and shorter turnover times are wellknown in ecosystem dynamics, e.g., forest-pest interactions with periodic massive outbreaks of insect pests (Ludwig et al., 1978; Rinaldi and Muratori, 1992a,b) or cyclic grazing systems with periodic collapses and recoveries of the vegetation (Noy-Meir, 1975; Rietkerk, 1998). Other sudden catastrophic regime shifts in ecosystems with long return times have been reviewed by Scheffer et al. (2001), Scheffer and Carpenter (2003), and Rietkerk et al. (2004); see also Carpenter and Turner (2000) and the whole *Ecosystems* issue including the work by Rinaldi and Scheffer (2000) on prey–predator food chain models.

The small parameter $\epsilon \ll 1$ even pushes the fast prey dynamics modeling the high sensitivity and much faster response of the phytoplankton population to environmental changes like fluctuations of nutrient supply, radiation, or temperature.

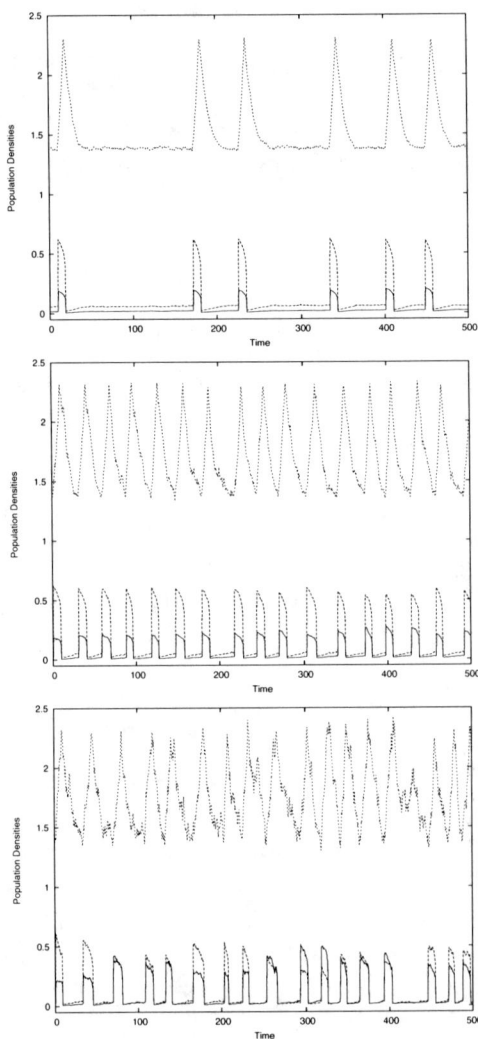

**FIGURE 16.1**: Noise-driven regimes for $\omega = 0.006$ and $0.0475$ (upper and lower figure, resp.), as well as coherence resonance for $\omega = 0.02$ (middle figure). Parameters: $r_s = r_i = 1.0$, $a = 4.0$, $b = 12.0$, $\sigma = m_i = 0.01$, $m_v = 0.0525$, $\epsilon = 10^{-3}$.

One of the remarkable effects discovered in stochastic excitable dynamics is the generation of oscillatory behavior by noisy perturbations (Treutlein and Schulten, 1986; Ebeling et al., 1986; Sakaguchi et al., 1988; Sigeti and Horsthemke, 1989; Hillenbrand, 2002; Zaks et al., 2003) and without an external periodic force. Excitable systems driven by noise have a noise-induced eigenfrequency, and, hence, they are able to show a stochastic limit cycle in phase space. These noise-sustained oscillations are often accompanied by coherence resonance (Gang et al., 1993; Longtin, 1997; Neiman et al., 1997; Pikovsky and Kurths, 1997; Pradines et al., 1999), which corresponds to the existence of an optimal noise intensity at which noise-induced oscillations are most coherent. Examples for coherence resonance have been found in neural models, laser models, models of excitable biomembranes, and even in climate models; cf. Lindner et al. (2004) for a comprehensive review and the references therein. Kuske et al. (2007) report on sustained oscillations through coherence resonance in a non-excitable SIR model. In system (16.1)–(16.3) one finds the wanted coherence resonance, i.e., noise-induced prey–predator oscillations, for medium noise intensities (Sieber et al., 2007); see Figure 16.1.

In space, one finds correspondingly interesting scenarios: For low noise intensities, the system is biologically controlled; for high noise intensities it is noise-driven. However, again for medium noise intensities, there is global synchronization of all local oscillators. This is demonstrated in Figures 16.2 and 16.3.

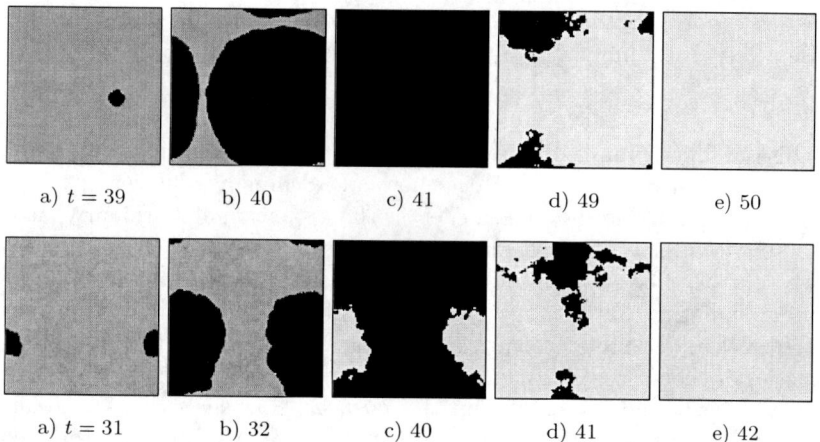

**FIGURE 16.2**: Formation of globally synchronized oscillations. Only the density distribution of susceptibles is drawn. Parameters: $r_s = r_i = 1.0$, $a = 4.0$, $b = 12.0$, $\sigma = m_i = 0.01$, $m_v = 0.0525$, $\epsilon = 10^{-3}$, $\omega = 0.015$ (upper row) and $0.04$ (lower row). Spatially uniform initial condition. Periodic boundary conditions.

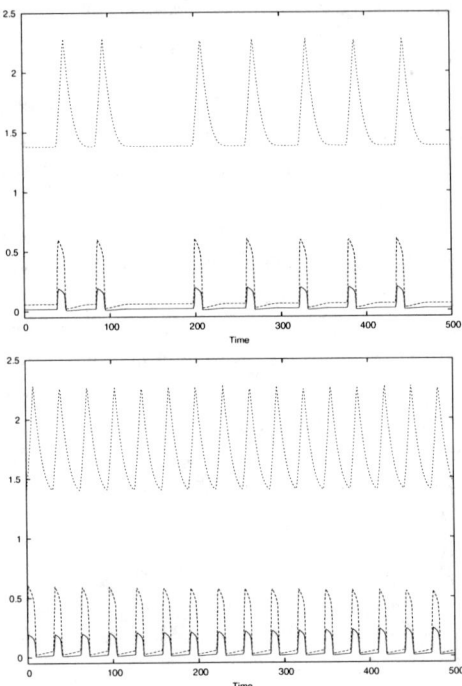

**FIGURE 16.3**: Corresponding dynamics of the spatial mean population densities for $\omega = 0.015$ (upper figure) and 0.04 (lower figure). Dashed line for susceptibles, solid line for infected, dotted line for zooplankton.

This is a first example of noise-induced prey–predator oscillations in an eco-epidemiological model system (Malchow and Schimansky-Geier, 2006). An even more striking noise-induced effect is the formation of stationary patterns in a small window of noise intensities between global synchronization and noise control. The process of pattern formation starts with local patches of higher population densities surviving after a global excitation due to combined effects of slow-fast prey–predator interaction, strong noise, and diffusion. The densities of the susceptibles and infected surrounding the patch do not relax back to their actual rest states, but are pushed below these due to the higher zooplankton density in the vicinity of those patches. Since the phytoplankton densities in this area are now too low to become excited again by noisy perturbations, a neighbourhood of an initial patch does not take part in the next global excitation. This leads to an interspace with low population densities. The phytoplankton densities are highest at the edges of the excited parts of the domain and once these parts drop back to the rest state, the edges also remain locked in the excited state. This phenomenon of population densities being highest at the edge of a sharply bounded excited patch has already been investigated by Hastings et al. (1997) in order to explain field observations of outbreaks in an insect host-parasitoid system (Maron and Harrison, 1997; Brodmann et al., 1997). Two examples of this pattern formation are shown in Figure 16.4 an 16.5.

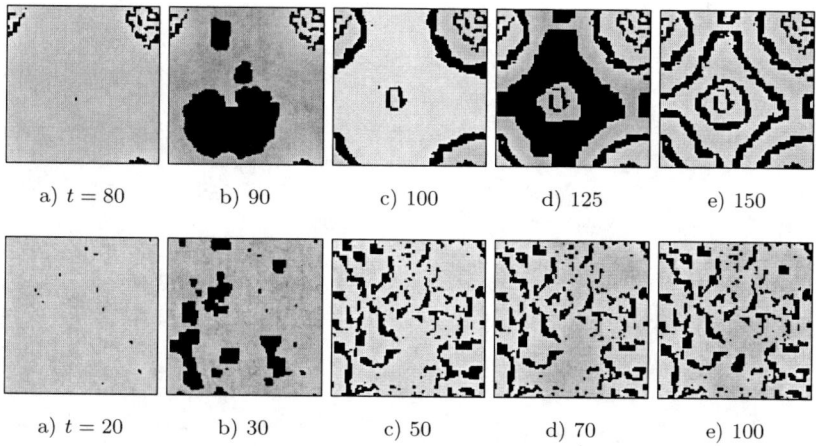

a) $t = 80$   b) 90   c) 100   d) 125   e) 150

a) $t = 20$   b) 30   c) 50   d) 70   e) 100

**FIGURE 16.4**: Formation of stationary spatial structures. Parameters: $\omega = 0.06$ (upper row) and 0.1 (lower row); all others parameters and conditions as in Figure 16.2.

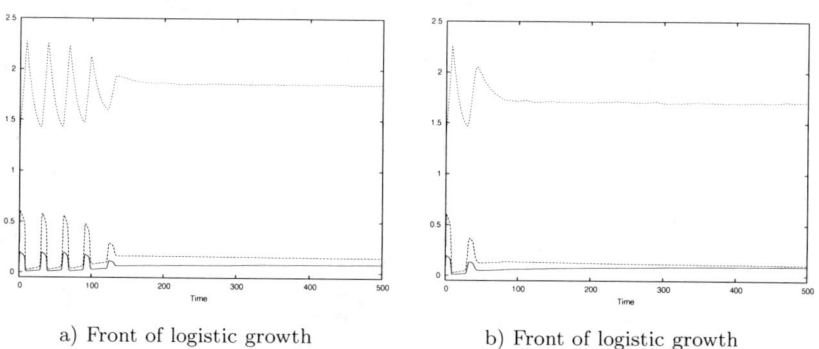

a) Front of logistic growth

b) Front of logistic growth

**FIGURE 16.5**: Corresponding dynamics of the spatial mean population densities for $\omega = 0.06$ (upper figure) and 0.1 (lower figure). Lines as in Figure 16.3.

Environmental stress is obviously capable of stimulating catastrophic shifts in the dynamics of ecosystems, here, the recurrent generation of phytoplankton blooms. This is also seen in space. The spatiotemporal lattice dynamics of the presented model exhibits several interesting features at different ranges of noise intensity. This includes spatially synchronized oscillations, enhanced coherence resonance, and the formation of stationary spatial patterns.

# References

Abarbanel, H. D. I. (1996). *Analysis of observed chaotic data*. Institute for Nonlinear Science Series. New York: Springer.

Abbott, M. R. and P. M. Zion (1985). Satellite observations of phytoplankton variability during an upwelling event. *Continental Shelf Research 4*, 661–680.

Abraham, E. R. (1998). The generation of plankton patchiness by turbulent stirring. *Nature 391*, 577–580.

Abrams, P. A., C. E. Brassil, and R. D. Holt (2003). Dynamics and responses to mortality rates of competing predators undergoing predator–prey cycles. *Theor. Popul. Biol. 64*, 163–176.

Abrams, P. A. and C. J. Waters (1996). Invulnerable prey and the paradox of enrichment. *Ecology 77*, 1125–1133.

Agger, P. and J. Brandt (1988). Dynamics of small biotopes in danish agricultural landscapes. *Landscape Ecology 1*, 227–240.

Agladze, K., L. Budriene, G. Ivanitsky, V. Krinsky, S. V, and M. Tsyganov (1993). Wave mechanisms of pattern formation in microbial populations. *Proc. R. Soc. Lond. B 253*, 131–135.

Allee, W. C. (1938). *The social life of animals*. New York: Norton and Co.

Allen, J. C., W. M. Schaffer, and D. Rosko (1993). Chaos reduces species extinction by amplifying local population noise. *Nature 364*, 229–232.

Allen, L. J. S. (2003). *An introduction to stochastic processes with applications to biology*. Upper Saddle River NJ: Pearson Education.

Allen, R. B. and W. G. Lee (Eds.) (2006). *Biological invasions in New Zealand*, Volume 186 of *Ecological Studies*. Berlin: Springer.

Alonso, D., F. Bartumeus, and J. Catalan (2002). Mutual interference between predators can give rise to Turing spatial patterns. *Ecology 83*, 28–34.

Alt, W. and G. Hoffmann (Eds.) (1990). *Biological motion*, Volume 89 of *Lecture Notes in Biomathematics*. Berlin: Springer.

Amarasekare, P. (1998). Interactions between local dynamics and dispersal: insights from single species models. *Theor. Popul. Biol. 53*, 44–59.

Anderson, J. (1974). Responses to starvation in the spiders Lycosa lenta Hentz and Filistata hibernalis (Hentz). *Ecology 55*, 576–585.

Anderson, R. M. and R. M. May (1986). The invasion, persistence and spread of infectious diseases within animal and plant communities. *Phil. Trans. R. Soc. London B 314*, 533–570.

Anderson, R. M. and R. M. May (1991). *Infectious diseases of humans. Dynamics and control.* Oxford: Oxford University Press.

Andow, D. A., P. Kareiva, S. A. Levin, and A. Okubo (1990). Spread of invading organisms. *Landscape Ecology 4*(2/3), 177–188.

Andresen, P., M. Bache, E. Mosekilde, G. Dewel, and P. Borckmans (1999). Stationary space-periodic structures with equal diffusion coefficients. *Phys. Rev. E 60*, 297–301.

Anishenko, V. S., V. V. Astakov, A. B. Neiman, T. Vadivasova, and L. Schimansky-Geier (2003). *Nonlinear dynamics of chaotic and stochastic systems. Tutorial and modern developments.* Springer Series in Synergetics. Berlin: Springer.

Arditi, R. and L. Ginzburg (1989). Coupling in predator–prey dynamics: ratio-dependence. *J. Theor. Biol. 139*, 311–326.

Arino, O., A. El abdllaoui, J. Mikram, and J. Chattopadhyay (2004). Infection in prey population may act as biological control in ratio-dependent predator–prey models. *Nonlinearity 17*, 1101–1116.

Arneodo, A., P. Coullet, and C. Tresser (1980). Occurence of strange attractors in three-dimensional Volterra equations. *Phys. Lett. A 79*(4), 259–263.

Arnold, L. (1974). *Stochastic differential equations: Theory and applications.* New York: Wiley.

Aronson, D. G. and H. Weinberger (1975). Nonlinear diffusion in population genetics, combustion, and nerve propagation. In J. A. Goldstein (Ed.), *Partial differential equations and related topics*, Volume 446 of *Lecture Notes in Mathematics*, pp. 5–49. Berlin: Springer.

Aronson, D. G. and H. Weinberger (1978). Multidimensional nonlinear diffusion arising in population genetics. *Advances in Mathematics 30*(1), 33–76.

Ascioti, F. A., E. Beltrami, T. O. Carroll, and C. Wirick (1993). Is there chaos in plankton dynamics? *J. Plankt. Res. 15*, 603–617.

Bailey, N. T. J. (1950). A simple stochastic epidemic. *Biometrika* 37, 193–202.

Baras, F. and M. Malek-Mansour (1996). Reaction-diffusion master equation: A comparison with microscopic simulations. *Phys. Rev. E 54*, 6139–6148.

Baras, F. and M. Malek-Mansour (1997). Microscopic simulations of chemical instabilities. *Advances in Chemical Physics 100*, 393–475.

Barenblatt, G. I. (1996). *Scaling, self-similarity, and intermediate asymptotics.* Cambridge UK: Cambridge University Press.

Barenblatt, G. I. and Y. B. Zeldovich (1971). Intermediate asymptotics in mathematical physics. *Russian Mathematics Surveys 26*, 45–61.

Barlow, N. D. (1994). Predicting the effect of a novel vertebrate biocontrol agent: a model for viral-vectored immunocontraception of New Zealand possums. *J. Appl. Ecol. 31*, 454–462.

Bartholomay, A. F. (1958a). On the linear birth and death processes of Biology as Markoff chains. *Bull. Math. Biophys. 20*, 97–118.

Bartholomay, A. F. (1958b). Stochastic models for chemical reactions, I. Theory of the unimolecular reaction process. *Bull. Math. Biophys. 20*, 175–190.

Beard, J. S. (1967). A study of patterns in some West Australian heath and Mallee communities. *Australian Journal of Botany 15*, 131–139.

Becks, L., F. M. Hilker, H. Malchow, K. Jürgens, and H. Arndt (2005). Experimental demonstration of chaos in a microbial food web. *Nature 435*, 1226–1229.

Begon, M. and R. Bowers (1995a). Beyond host pathogen dynamics. In B. Grenfell and A. Dobson (Eds.), *Ecology of Infectious Diseases in Natural Populations*, pp. 478–509. Cambridge Univ. Press.

Begon, M. and R. Bowers (1995b). Host-host-pathogen models and microbial pest control: the effect of host self-regulation. *J. Theor. Biol. 169*, 275–287.

Beltrami, E. and T. O. Carroll (1994). Modelling the role of viral disease in recurrent phytoplankton blooms. *J. Math. Biol. 32*, 857–863.

Ben-Jacob, E., I. Cohen, and H. Levine (2000). Cooperative self-organization of microorganisms. *Advances in Physics 49*(4), 395–554.

Beretta, E. and Y. Kuang (1998). Modeling and analysis of a marine bacteriophage infection. *Math. Biosci. 149*, 57–76.

Bernd, J. (1978). The problem of vegetation stripes in semi-arid Africa. *Plant Research and Development 8*, 37–50.

Bernoulli, D. (1760). Essai d'une nouvelle analyse de la mortalité causée par la petite vérole et des avantages de l'inoculation pour la prévenir. *Mémoires de Mathématique et de Physique de l'Académie Royale des Sciences*, 1–45.

Berryman, A. A. (1981). *Population systems: A general introduction.* New York: Plenum Press.

Berryman, A. A. and J. A. Millstein (1989). Are ecological systems chaotic - and if not, why not? *Trends in Ecology & Evolution 4*, 26–28.

Bharucha-Reid, A. T. (1960). *Elements of the theory of Markov processes and their applications.* McGraw-Hill Series in Probability and Statistics. New York: McGraw-Hill.

Biktashev, V. N., I. V. Biktasheva, A. V. Holden, M. A. Tsyganov, J. Brindley, and N. A. Hill (1999). Spatiotemporal irregularity in an excitable medium with shear flow. *Phys. Rev. E 60*(2), 1897–1900.

Birnie, L., K. Shaw, B. Pye, and I. Denholm (1998). Considerations with the use of multiple dose bioassays for assessing pesticide effects on non-target arthropods. In *Proceedings of Brighton Crop Protection Conference – Pests and Diseases*, pp. 291–296. Brighton, UK.

Boerlijst, M. C., M. Lamers, and P. Hogeweg (1993). Evolutionary consequences of spiral waves in a host-parasitoid system. *Proc. R. Soc. Lond. B 253*, 15–18.

Bohannan, B. J. M. and R. E. Lenski (1997). Effect of resource enrichment on a chemostat community of bacteria and bacteriophage. *Ecology 78*, 2303–2315.

Bohr, T., M. H. Jensen, G. Paladin, and A. Vulpiani (1998). *Dynamical systems approach to turbulence.* Cambridge: Cambridge University Press.

Bonsall, M. B. and A. Hastings (2004). Demographic and environmental stochasticity in predator–prey metapopulation dynamics. *J. Anim. Ecol. 73*, 1043–1055.

Boucher, D. H. (1985). *The biology of mutualism: ecology and evolution.* London: Croom Helm.

Box, G. E. P. and M. E. Muller (1958). A note on the generation of random normal deviates. *Annals of Mathematical Statistics 29*(2), 610–611.

Brauer, F. and C. Castillo-Chavez (2001). *Mathematical models in population biology and epidemiology*, Volume 40 of *Texts in Applied Mathematics*. New York: Springer.

Brauer, F. and A. Soudack (1978). Response of predator–prey nutrient enrichment and proportional harvesting. *Int. J. Control 27*, 65–86.

Braumann, C. A. (1979). *Population growth in random environments*. Ph. D. thesis, State University of New York, Stony Brook.

Braumann, C. A. (1983). Population growth in random environments. *Bull. Math. Biol. 45*, 635–641.

Braumann, C. A. (1999). Applications of stochastic differential equations to population growth. In D. Bainov (Ed.), *Proceedings of the Ninth International Colloquium on Differential Equations. Plovdiv, Bulgaria, 18-23 August, 1998*, pp. 47–52. Utrecht: VSP.

Braumann, C. A. (2007). Itô versus Stratonovich calculus in random population growth. *Math. Biosci. 206*, 81–107.

Braza, P. (2003). The bifurcations structure for the Holling–Tanner model for predator–prey interections using two-timing. *SIAM. J. Appl. Math. 63*, 889–904.

Britton, N. F. (2003). *Essential mathematical biology*. Berlin: Springer.

Brockmann, D., L. Hufnagel, and T. Geisel (2006). The scaling laws of human travel. *Nature 439*, 462–465.

Brodmann, P. A., C. V. Wilcox, and S. Harrison (1997). Mobile parasitoids may restrict the spatial spread of an insect outbreak. *J. Anim. Ecol. 66*, 65–72.

Bronshtein, I. and K. Semendyayev (1964). *Guide book to mathematics for technologists and engineers*. Pergamon Press.

Brown, K., J. Lawton, and S. Shires (1994). Effects of insecticides on invertebrate predators and their cereal aphid (Hemiptera: Aphidae) prey: laboratory experiments. *Environ. Entomol. 12*, 1747–1750.

Brown de Colstoun, E., C. L. Walthall, A. T. Cialella, E. R. Vermote, R. N. Halthore, and J. R. Irons (1996). Variability of BRDF with land cover type for the west central HAPEX-Sahel super site. In *IGARSS'96 – International Geoscience and Remote Sensing Symposium. Remote Sensing for a Sustainable Future.*, Volume 2, Lincoln, NE, pp. 1904–1907.

Busenberg, S. N. and K. Cooke (1981). *Differential equations and applications in ecology, epidemics, and population problems*. New York: Academic Press.

Capasso, V. (1993). *Mathematical structures of epidemic systems*, Volume 97 of *Lectures Notes in Biomathematics*. New York: Springer.

Capasso, V. and G. Serio (1978). A generalization of the Kermack-McKendrick deterministic epidemic model. *Math. Biosci. 42*, 41–61.

Capocelli, R. M. and L. M. Ricciardi (1974). A diffusion model for population growth in random environment. *Theor. Popul. Biol. 5*, 28–41.

Carpenter, S. R. and M. G. Turner (2000). Hares and tortoises: Interactions of fast and slow Variables in ecosystems. *Ecosystems 3*, 495–497.

Caughley, G. (1970). Eruption of ungulate populations with emphasis on Himalayan thar in New Zealand. *Ecology 51*, 53–72.

Chattopadhyay, J. and O. Arino (1999). A predator–prey model with disease in the prey. *Nonlinear Analysis 36*, 747–766.

Chattopadhyay, J. and S. Pal (2002). Viral infection on phytoplankton-zooplankton system - a mathematical model. *Ecological Modelling 151*, 15–28.

Chattopadhyay, J., R. R. Sarkar, and G. Ghosal (2002). Removal of infected prey prevent limit cycle oscillations in an infected prey–predator system - a mathematical study. *Ecological Modelling 156*, 113–121.

Chattopadhyay, J., R. R. Sarkar, and S. Mandal (2002). Toxin-producing plankton may act as a biological control for planktonic blooms - field study and mathematical modelling. *J. Theor. Biol. 215*(3), 333–344.

Chattopadhyay, J., R. R. Sarkar, and S. Pal (2003). Dynamics of nutrient-phytoplankton interaction in the presence of viral infection. *BioSystems 68*, 5–17.

Chattopadhyay, J., R. R. Sarkar, and S. Pal (2004). Mathematical modelling of harmful algal blooms supported by experimental findings. *Ecological Complexity 1*(3), 225–235.

Cochran, P. K., C. A. Kellogg, and J. H. Paul (1998). Prophage induction of indigenous marine lysogenic bacteria by environmental pollutants. *Marine Ecology Progress Series 164*, 125–133.

Collie, J. S., K. Richardson, and J. H. Steele (2004). Regime shifts: can ecological theory illuminate the mechanisms? *Progress in Oceanography 60*, 281–302.

Comins, H. N., M. P. Hassell, and R. M. May (1992). The spatial dynamics of host-parasitoid systems. *J. Anim. Ecol. 61*, 735–748.

Costantino, R. F., R. A. Desharnais, J. M. Cushing, and B. Dennis (1997). Chaotic dynamics in an insect population. *Science 275*, 389–391.

Courchamp, F., L. Berec, and J. Gascoigne (2008). *Allee effects in ecology and conservation*. Oxford: Oxford University Press.

Courchamp, F., T. Clutton-Brock, and B. Grenfell (1999). Inverse density dependence and the Allee effect. *Trends in Ecology & Evolution 14*, 405–410.

Courchamp, F., T. Clutton-Brock, and B. Grenfell (2000). Multipack dynamics and the Allee effect in the african wild dog, *Lycaon pictus*. *Animal Conservation 3*, 277–285.

Courchamp, F., P. Pontier, M. Langlais, and M. Artois (1995). Population dynamics of Feline Immunodeficiency Virus within cat populations. *J. Theor. Biol. 175*(4), 553–560.

Couteron, P. and O. Lejeune (2001). Periodic spotted patterns in semi-arid vegetation explained by a propagation-inhibition model. *J. Ecol. 89*, 616–628.

Cushing, J. (1998). *An introduction to structured population dynamics*. Philadelphia: SIAM.

Cushing, J. M., R. Costantino, B. Dennis, R. A. Desharnais, and S. Henson (2003). *Chaos in ecology. Experimental nonlinear dynamics*. Theor. Ecol. Series. Amsterdam: Academic Press.

Cushing, J. M., S. M. Henson, R. A. Desharnais, B. Dennis, R. F. Costantino, and A. King (2001). A chaotic attractor in ecology: theory and experimental data. *Chaos, Solitons & Fractals 12*, 219–234.

Czárán, T. (1998). *Spatiotemporal models of population and community dynamics*, Volume 21 of *Population and Community Biology Series*. London: Chapman & Hall.

D'Ancona, U. (1954). *The struggle for existence*. Leiden: Brill.

Davidenko, J. M., A. M. Pertsov, R. Salomonsz, W. Baxter, and J. Jalífe (1992). Stationary and drifting spiral waves of excitation in isolated cardiac muscle. *Nature 355*, 349–351.

Davidson, F. A. (1998). Chaotic wakes and other wave-induced behavior in a system of reaction-diffusion equations. *International J. Bifurcation and Chaos 8*, 1303–1313.

Davis, M. B., R. R. Calcote, S. Sugita, and H. Takahara (1998). Patchy invasion and the origin of a Hemlock-Hardwoods forest mosaic. *Ecology 79*, 2641–2659.

de Jong, M., O. Diekmann, and H. Heesterbeek (1995). How does transmission of infection depend on population size? In D. Mollison (Ed.), *Epidemic models. Their structure and relation to data.*, Publications of the Newton Institute, pp. 84–94. Cambridge: Cambridge University Press.

DeAngelis, D. and L. J. Gross (1992). *Individual-based models and approaches in ecology. Populations, communities and ecosystems.* New York: Chapman & Hall.

Denman, K. L. (1976). Covariability of chlorophyll and temperature in the sea. *Deep-Sea Res. 23*, 539–550.

Denman, K. L. and T. Platt (1976). The variance spectrum of phytoplankton in a turbulent ocean. *J. Mar. Res. 34*, 593–601.

Dennis, B. (1989). Allee effects: population growth, critical density, and the chance of extinction. *Natural Resource Modeling 3*, 481–538.

Dennis, B., R. A. Desharnais, J. M. Cushing, and R. F. Costantino (1997). Transitions in population dynamics: equilibria to periodic cycles to aperiodic cycles. *J. Anim. Ecol. 66*, 704–729.

Dennis, B., R. A. Desharnais, J. M. Cushing, S. M. Henson, and R. F. (2001). Estimating chaos and complex dynamics in an insect population. *Ecological Monographs 71*(2), 277–303.

Dennis, B. and G. P. Patil (1984). The gamma distribution and weighted multimodal gamma distributions as models of population abundance. *Math. Biosci. 68*, 187–212.

Dent, C. L., G. S. Cumming, and S. R. Carpenter (2002). Multiple states in rivers and lake ecosystems. *Phil. Trans. R. Soc. Lond. B 357*, 635–645.

Dieckmann, U., J. Metz, M. Sabelis, and K. Sigmund (Eds.) (2002). *Adaptive dynamics of infectious diseases: in pursuit of virulence management.* Cambridge Studies in Adaptive Dynamics. Cambridge: Cambridge University Press.

Diekmann, O. and J. Heesterbeck (Eds.) (2000). *Mathematical epidemiology of infectious diseases. Model building, analysis and interpretation.* Wiley Series in Mathematical and Computational Biology. Chichester: Wiley.

Dillon, R., P. K. Maini, and H. G. Othmer (1994). Pattern formation in generalized Turing systems I. steady-state patterns in systems with mixed boundary conditions. *J. Math. Biol. 32*, 345–393.

Dobson, A. (1985). The population dynamics of competition between parasites. *Parasitology 91*, 317–347.

d'Onofrio, A. (2002). Stability properties of pulse vaccination strategy in seir epidemic model. *Math. Biosc. 179*, 57–72.

Drake, J. A. and H. A. Mooney (Eds.) (1989). *Biological invasions: a global perspective*, Volume 27 of *SCOPE*. Chichester: Wiley.

Dunbar, S. R. (1983). Travelling wave solutions of diffusive Lotka–Volterra equations. *J. Math. Biol. 17*, 11–32.

Dunbar, S. R. (1984). Travelling waves solutions of diffusive Lotka–Volterra equations: A heteroclinic connection in $\mathbf{R}^4$. *Transactions of the American Mathematical Society 286*, 557–594.

Dunbar, S. R. (1986). Travelling waves in diffusive predator–prey equations: periodic orbits and point-to-periodic heteroclinic orbits. *SIAM J. Appl. Math. 46*, 1057–1078.

Durrett, R. and S. Levin (1994). The importance of being discrete (and spatial). *Theor. Popul. Biol. 46*, 363–394.

Durrett, R. and S. Levin (2000). Lessons on pattern formation from planet WATOR. *J. Theor. Biol. 205*, 201–214.

Ebeling, W. and R. Feistel (1974). Zur Kinetik molekularer Replikationsprozesse mit Selektionscharakter. *studia biophysica 46*, 183–195.

Ebeling, W. and R. Feistel (1977). Stochastic theory of molecular replication processes with selection character. *Annalen der Physik (Leipzig) 34*, 81–90.

Ebeling, W., H. Herzel, W. Richert, and L. Schimansky-Geier (1986). Influence of noise on Duffing–van der Pol oscillators. *Zeitschrift für Angewandte Mathematik und Mechanik 66*, 141–146.

Ebeling, W. and L. Schimansky-Geier (1979). Stochastic dynamics of a bistable reaction system. *Physica A 98*, 587–600.

Ebeling, W. and L. Schimansky-Geier (1980). Nonequilibrium phase transitions and nucleation in reaction systems. In *Proceedings of the 6th International Conference on Thermodynamics*, Merseburg, pp. 95–100.

Edelstein-Keshet, L. (1988). *Mathematical models in biology*. Birkhäuser Mathematics Series. New York: McGraw-Hill.

Edgar, W. (1969). Prey and predators of the wolf spider, Lycosa lugubris. *J Zool 159*, 405–411.

Edgar, W. (1971). The life-cycle, abundance and seasonal movement of the wolf spider, Lycosa (Pardosa) lugubris, in central Scotland. *J. Anim. Ecol. 40*, 303–322.

Edwards, A. M. and A. Yool (2000). The role of higher predation in plankton population models. *J. Plankt. Res. 22*(6), 1085–1112.

Eftimie, R., G. de Vries, and M. A. Lewis (2007). Complex spatial group patterns result from different animal communication mechanisms.

*Proceedings of the National Academy of Sciences of the United States of America 104*, 6974–6979.

Ellner, S. and P. Turchin (1995). Chaos in a noisy world: New methods and evidence from time-series analysis. *Am. Nat. 145*(3), 343–375.

Elton, C. S. (1924). Periodic fluctuations in the numbers of animals: Their causes and effects. *British Journal of Experimental Biology 2*, 119–163.

Elton, C. S. (1942). *Moles, mice and lemmings. Problems in population dynamics.* Oxford: The Clarendon Press.

Engbert, R. and F. R. Drepper (1994). Chance and chaos in population biology - models of recurrent epidemics and food chain dynamics. *Chaos, Solitons & Fractals 4*, 1147–1169.

Evans, G. T. and J. S. Parslow (1985). A model of annual plankton cycles. *Biological Oceanography 3*(3), 327–347.

Fagan, W. F. and J. G. Bishop (2000). Trophic interactions during primary succession: herbivores slow a plant reinvasion at Mount St. Helens. *Am. Nat. 155*, 238–251.

Fagan, W. F., M. A. Lewis, M. G. Neubert, and P. van den Driessche (2002). Invasion theory and biological control. *Ecol. Lett. 5*, 148–158.

Farkas, M. (1984). Zip bifurcation in a competition model. *Nonlinear Analysis. Theory, Methods & Applications 8*(11), 1296–1309.

Feistel, R. (1977). Betrachtung der Realisierung stochastischer Prozesse aus automatentheoretischer Sicht. *Wissenschaftliche Zeitschrift der Wilhelm-Pieck-Universität Rostock XXVI*, 663–670.

Feistel, R. and W. Ebeling (1978). Deterministic and stochastic theory of sustained oscillations in autocatalytic reaction systems. *Physica A 93*, 114–137.

Fernández, I., J. M. Lopéz, J. M. Pacheco, and C. Rodríguez (2002). Hopf bifurcations and slow-fast cycles in a model of plankton dynamics. *Revista de la Academia Canaria de Ciencias 14*(1-2), 121–137.

Field, R. J. and M. Burger (Eds.) (1985). *Oscillations and traveling waves in chemical systems.* New York: Wiley.

Fife, P. C. (1979). *Mathematical aspects of reacting and diffusing systems*, Volume 28 of *Lecture Notes in Biomathematics.* Berlin: Springer.

Fisher, R. A. (1937). The wave of advance of advantageous genes. *Annals of Eugenics 7*, 355–369.

Fitzgerald, B. M. and C. R. Veitch (1985). The cats of Herecopare Island, New Zealand: Their hystory, ecology and effects on birdlife. *N.Z. J. of Zoology 12*, 319–330.

Foley, J. A., M. T. Coe, M. Scheffer, and G. Wang (2003). Regime shifts in the Sahara and Sahel: Interactions between ecological and climatic conditions in northern Africa. *Ecosystems 6*, 524–539.

Frauenthal, J. C. (1980). *Mathematical modeling in epidemiology*. Universitext. Berlin: Springer.

Freedman, H. and G. Wolkowicz (1986). Predator-prey systems with group defence: the paradox of enrichment revisited. *Bull. Math. Biol. 48*, 493–508.

French, N., D. Clancy, H. Davison, and A. Trees (1999). Mathematical models of Neospora caninum infection in dairy cattle: transmission and options for control. *Int. J. for Parasitology 29*, 1691–1704.

Freund, J. A., S. Mieruch, B. Scholze, K. Wiltshire, and U. Feudel (2006). Bloom dynamics in a seasonally forced phytoplankton-zooplankton model: Trigger machanisms and timing effects. *Ecological Complexity 3*, 129–139.

Freund, J. A., L. Schimansky-Geier, B. Beisner, A. Neiman, D. F. Russell, T. Yakusheva, and F. Moss (2002). Behavioral stochastic resonance: How the noise from a Daphnia swarm enhances individual prey capture by juvenile paddlefish. *J. Theor. Biol. 214*, 71–83.

Fromont, E., D. Pontier, and M. Langlais (1998). Dynamics of a feline retrovirus (FeLV) in host populations with variable spatial structure. *Proc. R. Soc. Lond. B 265*, 1097–1104.

Fuhrman, J. A. (1999). Marine viruses and their biogeochemical and ecological effects. *Nature 399*, 541–548.

Furuta, K. (1977). Evaluation of spiders, Oxyopes sertatus and O. badius (Oxyopidae) as a mortality factor of gipsy moth, Lymantria dispar (Lepidoptera, Lymantriidae) and pine moth, Dendrolimus spectabilis (Lepidoptera: Lasiocampidae). *Applied Entomology and Zoology 12*, 313–324.

Gafiychuk, V. V. and B. Y. Datsko (2006). Pattern formation in a fractional reaction-diffusion system. *Physica A 365*, 300–306.

Gang, H., T. Ditzinger, C. Z. Ning, and H. Haken (1993, Aug). Stochastic resonance without external periodic force. *Phys. Rev. Lett. 71*(6), 807–810.

Gao, Q. and H. Hethcote (1992). Disease transmission models with density dependent demographics. *J. Math. Biol. 30*, 717–731.

García-Ojalvo, J. and J. M. Sancho (1999). *Noise in spatially extended systems*. Institute for Nonlinear Science. New York: Springer.

Gardiner, C. W. (1985). *Handbook of stochastic methods*, Volume 13 of *Springer Series in Synergetics*. Berlin: Springer.

Garvie, M. R. (2007). Finite-difference schemes for reaction-diffusion equations modeling predator–prey interactions in MATLAB. *Bull. Math. Biol. 69*, 931–956.

Gause, G. F. and A. A. Vitt (1934). O periodicheskikh kolebaniyakh chislennosti populyatsii: matematicheskaya teoriya relaksatsionogo vzaimodeistviya mezhdu khishchnikami i zhertvami i ee primenenie k populyatsiyam dvukh prosteishikh. *Izvestiya Akademii Nauk SSSR, Otdelenie Matematicheskikh i Estestvennykh Nauk 10*, 1551–1559.

Genieys, S., V. Volpert, and P. Auger (2006). Patterns and waves for a model in population dynamics with nonlocal consumption of resources. *Mathematical Modelling of Natural Phenomena 1*, 65–82.

Genkai-Kato, M. and N. Yamamura (1999). Unpalatable prey resolves the paradox of enrichment. *Proc. R. Soc. Lond. B 266*, 1215–1219.

Gerisch, G. (1968). Cell aggregation and differentiation in *Dictyostelium*. In A. A. Moscona and A. Monroy (Eds.), *Current Topics in Developmental Biology*, Volume 3, pp. 157–197. New York: Academic Press.

Gerisch, G. (1971). Periodische Signale steuern die Musterbildung in Zellverbänden. *Naturwissenschaften 58*, 430–438.

Gilg, O., I. Hanski, and B. Sittler (2003). Cyclic dynamics in a simple vertebrate predator–prey community. *Science 302*, 866–868.

Gilg, O., B. Sittler, B. Sabard, A. Hurstel, R. Sané, P. Delattre, and I. Hanski (2006). Functional and numerical responses of four lemming predators in high arctic Greenland. *Oikos 113*, 193–216.

Gillespie, D. T. (1976). A general method for numerically simulating the stochastic time evolution of coupled chemical reactions. *J. Computational Phys. 22*(4), 403–434.

Gilpin, M. E. (1972). Enriched predator–prey systems: theoretical stability. *Science 177*, 902–904.

Gilpin, M. E. (1979). Spiral chaos in a predator–prey model. *Am. Nat. 113*, 306–308.

Gilpin, M. E. and I. Hanski (Eds.) (1991). *Metapopulation dynamics: Empirical and theoretical investigations*. London: Academic Press.

Glansdorff, P. and I. Prigogine (1971). *Thermodynamics of structure, stability, and fluctuations*. New York: Wiley-Interscience.

Goel, N. and N. Richter-Dyn (1974). *Stochastic models in biology.* New York: Academic Press.

Gray, P. and S. K. Scott (1990). *Chemical oscillations and instabilities.* Oxford: Oxford University Press.

Greene, C. H. and A. J. Pershing (2007). Climate drives sea change. *Science 315*, 1084–1085.

Greenhalgh, D. (1992). Vaccination in density-dependent epidemic models. *Bull. Math. Biol. 54*, 733–758.

Greenstone, M. (1979). Spider feeding behaviour optimises dietary essential amino acid composition. *Nature 181*, 501–503.

Greenstone, M., C. Morgan, A.-L. Hultsch, R. Farrow, and J. Dowse (1987). Ballooning spiders in Missouri, USA, and New South Wales, Australia: family and mass distribution. *J. Arachnol. 15*, 163–170.

Greig-Smith, P. (1979). Patterns in vegetation. *J. Ecol. 67*, 755–779.

Grenfell, B. T., O. Bjørnstad, and J. Kappey (2001). Travelling waves and spatial hierarchies in measles epidemics. *Nature 414*, 716–723.

Grimm, V. and S. F. Railsback (2005). *Individual-based Modeling and Ecology.* Princeton Series in Theoretical and Computational Biology. Princeton and Oxford: Princeton University Press.

Gurtin, M. and R. McCamy (1974). Nonlinearly age-dependent population dynamics. *Archs. Ration. Mech. Analysis 54*, 281–300.

Gyllenberg, M., J. Hemminki, and T. Tammaru (1999). Allee effects can both conserve and create spatial heterogeneity in population densities. *Theor. Popul. Biol. 56*, 231–242.

Hadeler, K. P. and H. I. Freedman (1989). Predator-prey populations with parasitic infection. *J. Math. Biol. 27*, 609–631.

Hagstrum, D. (1970). Ecological energetics of the spider Tarentula kochi (Araneae: Lycosidae). *Ann. Entomol. Soc. 63*, 1297–1304.

Haken, H. (1975). Statistical physics of bifurcation, spatial structures, and fluctuations of chemical reactions. *Zeitschrift für Physik B 20*, 413–420.

Halley, J., C. Thomas, and P. Jepson (1996). A model for the spatial dynamics of linyphiid spiders in farmland. *J. App. Ecol. 33*, 471–492.

Hanski, I., P. Turchin, E. Korpimäki, and H. Henttonen (1993). Population oscillations of boreal rodents: regulation by mustelid predators leads to chaos. *Nature 364*, 232–235.

Hanski, I. A. (1999). *Metapopulation ecology.* Oxford Series in Ecology and Evolution. New York: Oxford University Press.

Haque, M. and E. Venturino (2006a). Increase of the prey may decrease the healthy predator population in presence of a disease in the predator. *HERMIS 7*, 39–60.

Haque, M. and E. Venturino (2006b). The role of transmissible diseases in holling–tanner predator–prey model. *Theor. Popul. Biol. 70*, 273–288.

Haque, M. and E. Venturino (2007). An ecoepidemiological model with disease in the predator; the ratio-dependent case. *Math. Meth. Appl. Sci. 30*, 1791–1809.

Harrison, S. (1997). Persistent localized outbreaks in the western tussock moth (*Orgyia vetusta*): the roles of resource quality, predation and poor dispersal. *Ecological Entomology 22*, 158–166.

Hassell, M., H. Comins, and R. M. May (1991). Spatial structure and chaos in insect population dynamics. *Nature 353*, 255–258.

Hastings, A. (2001). Transient dynamics and persistence of ecological systems. *Ecol. Lett. 4*, 215–220.

Hastings, A. (2004). Transients: the key to long-term ecological understanding? *Trends in Ecology & Evolution 19*, 39–45.

Hastings, A., K. Cuddington, K. F. Davies, C. J. Dugaw, S. Elmendorf, A. Freestone, S. Harrison, M. Holland, J. Lambrinos, U. Malvadkar, B. A. Melbourne, K. Moore, C. Taylor, and D. Thomson (2005). The spatial spread of invasions: new developments in theory and evidence. *Ecol. Lett. 8*, 91–101.

Hastings, A., S. Harrison, and K. McCann (1997). Unexpected spatial patterns in an insect outbreak match a predator diffusion model. *Proc. R. Soc. Lond. B 264*, 1837–1840.

Hastings, A., C. L. Hom, S. Ellner, P. Truchin, and H. C. J. Godfray (1993). Chaos in ecology: Is mother nature a strange attractor? *Annual Review of Ecology and Systematics 24*, 1–33.

Hastings, A. and T. Powell (1991). Chaos in a three-species food chain. *Ecology 72*, 896–903.

Hattne, J., D. Fange, and J. Elf (2005). Stochastic reaction-diffusion simulation with MesoRD. *Bioinformatics 21*, 2923–2924.

Hempel, H., L. Schimansky-Geier, and J. Garcia-Ojalvo (1999). Noise-sustained pulsating patterns and global oscillations in subexcitable media. *Phys. Rev. Lett. 82*(18), 3713–3716.

Herskowitz, I. and D. Hagen (1980). The lysis-lysogeny decision of phage lambda: explicit programming and responsiveness. *Annual Review of Genetics 14*, 399–445.

Hethcote, H. and S. Levin (1989). Periodicity in epidemiological models. In S. Levin, T. Hallam, and L. Gross (Eds.), *Applied Mathematical Ecology*, Volume 18 of *Biomathematics*, pp. 193–211. Heidelberg: Springer.

Hethcote, H. W. (2000). The mathematics of infectious diseases. *SIAM Rev. 42*(4), 599–653.

Higgins, P. A. T., M. D. Mastrandrea, and S. H. Schneider (2002). Dynamics of climate and ecosystem coupling: abrupt changes and multiple equilibria. *Phil. Trans. R. Soc. Lond. B 357*, 647–655.

Higham, D. J. (2001). An algorithmic introduction to numerical simulation of stochastic differential equations. *SIAM Rev. 43*(3), 525–546.

Hilker, F. M. (2005). *Spatiotemporal patterns in models of biological invasion and epidemic spread.* Berlin: Logos Verlag.

Hilker, F. M., M. Langlais, S. V. Petrovskii, and H. Malchow (2007). A diffusive SI model with Allee effect and application to FIV. *Math. Biosci. 206*, 61–80.

Hilker, F. M., M. A. Lewis, H. Seno, M. Langlais, and H. Malchow (2005). Pathogens can slow down or reverse invasion fronts of their hosts. *Biological Invasions 7*, 817–832.

Hilker, F. M. and H. Malchow (2006). Strange periodic attractors in a prey–predator system with infected prey. *Mathematical Population Studies 13*(3), 119–134.

Hilker, F. M., H. Malchow, M. Langlais, and S. V. Petrovskii (2006). Oscillations and waves in a virally infected plankton system. Part II: Transition from lysogeny to lysis. *Ecological Complexity 3*, 200–208.

Hillenbrand, U. (2002). Subthreshold dynamics of the neural membrane potential driven by stochastic synaptic input. *Phys. Rev. E 66*(2), 021909.

Hirsch, M. and S. Smale (1974). *Differential equations, dynamical systems and linear algebra.* New York: Academic Press.

Hochberg, M., M. Hassel, and R. May (1990). The dynamics of host-parasitoid-pathogen interactions. *Am. Nat. 135*, 74–94.

Hochberg, M. and R. Holt (1990). The coexistence of competing parasites. 1. The role of cross-species infection. *Am. Nat. 136*, 517–541.

Hofbauer, J. and K. Sigmund (1988). *The theory of evolution and dynamical systems.* Cambridge: Cambridge University Press.

Höfer, T., J. A. Sherratt, and P. K. Maini (1995). Cellular pattern formation during Dictyostelium aggregation. *Physica D 85*, 425–444.

Holt, R. and J. Pickering (1986). Infectious disease and species coexistence: A model of Lotka–Volterra form. *American Naturalist 126*, 196–211.

Holt, R. D. (1977). Predation, apparent competition and the structure of prey community. *Theor. Popul. Biol. 12*, 197–229.

Holt, R. D., T. H. Keitt, M. A. Lewis, B. A. Maurer, and M. L. Taper (2005). Theoretical models of species' borders: single species approaches. *Oikos 108*, 18–27.

Holyoak, M. (2000). Effects of nutrient enrichment on predator–prey metapopulation dynamics. *J. Anim. Ecol. 69*, 985–997.

Horiuchi, J., B. Prithiviraj, H. Bais, B. Kimball, and J. Vivanco (2005). Soil nematodes mediate positive interactions between legume plants and rhizobium bacteria. *Planta 222*, 848–857.

Horsthemke, W. and R. Lefever (1984). *Noise-induced transitions. Theory and applications in physics, chemistry, and biology*, Volume 15 of *Springer Series in Synergetics*. Berlin: Springer.

Hosono, Y. (1998). The minimal speed of travelling fronts for a diffusive lotka–volterra competition model. *Bull. Math. Biol. 60*, 435–448.

Hsu, S.-B. and T.-W. Huang (1995). Global stability for a class of predator–prey systems. *SIAM J. Appl. Math. 55*(3), 763–783.

Hudson, P. J. and O. N. Bjørnstad (2003). Vole stranglers and lemming cycles. *Science 302*, 797–798.

Huisman, J. and F. J. Weissing (1999). Biodiversity of plankton by oscillations and chaos. *Nature 402*, 407–410.

Hukusima, S. and K. Kondo (1962). Further evaluation in the feeding potential of the predaceous insects and spiders in association with aphids harmful to apple and pear growing and the effect of pesticides on predators. *Jpn. J. Appl. Entomol. Zool. 6*, 274–280.

Hutchinson, G. E. (1978). *Introduction to population ecology.* New Haven: Yale University Press.

Huth, A. and C. Wissel (1994). The simulation of fish schools in comparison with experimental data. *Ecological Modelling 75/76*, 135–145.

Huusela-Veistola, E. (1998). Effects of perennial grass strips on spiders (Araneae) in cereal fields and impact on pesticide side-effects. *J. Appl. Ent. 122*, 575–583.

Iannelli, M. (1995). *Mathematical theory of age-structured population dynamics*. Pisa: Giardini.

Isaacson, S. A. and C. S. Peskin (2006). Incorporating diffusion in complex geometries into stochastic chemical kinetics simulation. *SIAM J. Sci. Comput. 28*(1), 47–74.

Isaia, M., F. Bona, and G. Badino (2006a). Comparison of polyethylene bubble wrap and corrugated cardboard traps for sampling tree-inhabiting spiders. *Environ. Entomol. 35*, 1654 – 1660.

Isaia, M., F. Bona, and G. Badino (2006b). Influence of landscape diversity and agricultural practices on spider assemblages in Italian vineyards of Langa Astigiana (NW-Italy). *Environ. Entomol. 35*, 297–307.

Itô, K. (1961). *Lectures on stochastic processes*. Bombay: Tata Institute of Fundamental Research.

Ivanitskii, G., A. B. Medvinskii, and M. A. Tsyganov (1994). From the dynamics of population autowaves generated by living cells to neuroinformatics. *Physics–Uspekhi 37*, 961–989.

Jacquet, S., M. Heldal, D. Iglesias-Rodriguez, A. Larsen, W. Wilson, and G. Bratbak (2002). Flow cytometric analysis of an *Emiliana huxleyi* bloom terminated by viral infection. *Aquatic Microbial Ecology 27*, 111–124.

Jansen, V. and A. Lloyd (2000). Local stability analysis of spatially homogeneous solutions of multi-patch systems. *J. Math. Biol. 41*, 232–252.

Jansen, V. A. A. (1995). Regulation of predator–prey systems through spatial interactions: a possible solution to the paradox of enrichment. *Oikos 74*, 384–390.

Jansen, V. A. A. (2001). The dynamics of two diffusively coupled predator–prey populations. *Theor. Popul. Biol. 59*, 119–131.

Janssen, H. K. (1974). Stochastisches Reaktionsmodell für einen Nichtgleichgewichts-Phasenübergang. *Zeitschrift für Physik 270*, 67–73.

Jantzen, D., P. D. Vries, D. Gladstone, M. Higgins, and T. Levinsohn (1980). Self- and cross-pollination of Encyclia cordigera (Orchidaceae) in Santa Rosa National Park, Costa Rica. *Biotropica 12*, 1398–1406.

Jeltsch, F., C. Wissel, S. Eber, and R. Brandl (1992). Oscillating dispersal patterns of tephritid fly populations. *Ecological Modelling 60*, 63–75.

Jensen, C. X. J. and L. R. Ginzburg (2005). Paradoxes or theoretical failures? The jury is still out. *Ecological Modelling 188*, 3–14.

Jiang, S. C. and J. H. Paul (1998). Significance of lysogeny in the marine environment: studies with isolates and a model of lysogenic phage production. *Microbial Ecology 35*, 235–243.

Kaitala, V., J. Ylikarjula, E. Ranta, and P. Lundberg (1997). Population dynamics and the colour of environmental noise. *Proc. R. Soc. Lond. B 264*, 943–948.

Kapral, R. and K. Showalter (1995). *Chemical waves and patterns*, Volume 10 of *Understanding chemical reactivity*. Dordrecht: Kluwer.

Karlin, S. and H. M. Taylor (1975). *A first course in stochastic processes*. New York: Academic Press.

Kawasaki, K., A. Mochizuki, M. Matsushita, T. Umeda, and N. Shigesada (1997). Modeling spatio-temporal patterns generated by *Bacillus subtilis*. *J. Theor. Biol. 188*, 177–185.

Kawasaki, K., A. Mochizuki, and N. Shigesada (1995). A mathematical model of pattern formation in a bacterial colony (in Japanese). *Control & Measurement 34*, 811–816.

Keeler, K. (1985). Cost: benefit models of mutualism. In D. Boucher (Ed.), *The Biology of Mutualism: Ecology and Evolution*, pp. 100–127. London: Croom Helm.

Keller, E. F. and L. A. Segel (1970). Initiation of slime mold aggregation viewed as an instability. *J. Theor. Biol. 26*, 399–415.

Keller, E. F. and L. A. Segel (1971a). Model for chemotaxis. *J. Theor. Biol. 30*, 225–234.

Keller, E. F. and L. A. Segel (1971b). Traveling bands of chemotactic bacteria: A theoretical analysis. *J. Theor. Biol. 30*, 235–248.

Keller, P. and E. Venturino (2007). The use of the aggregation method for an ecoepidemic model. *Far East Journal of Applied Mathematics 26*, 13–23.

Khan, Q., E. Balakrishnan, and G. Wake (2004). Analysis of a predator–prey system with predator switching. *Bull. Math. Biol. 66*, 109–123.

Khan, Q., B. Bhatt, and R. Jaju (1998). Switching model with two habitats and a predator involving group defence. *J. Nonlin. Math. Phys. 5*, 212–223.

Khan, Q. J. A., E. V. Krishnan, and M. A. Al-Lawatia (2002). A stage structure model for the growth of a population involving switching and cooperation. *ZAMM 82*, 125–135.

Kierstead, H. and L. B. Slobodkin (1953). The size of water masses containing plankton blooms. *J. Mar. Res. XII(1)*, 141–147.

Klafter, J., M. Shlesinger, and G. Zumofen (1996). Beyond Brownian motion. *Physics Today 49*, 33–39.

Klausmeier, C. A. (1999). Regular and irregular patterns in semiarid vegetation. *Science 284*, 1826–1828.

Kliemann, W. (1983). Qualitative theory of stochastic dynamical systems – applications to life sciences. *Bull. Math. Biol. 45*, 483–506.

Kloeden, P. E. and E. Platen (1999). *Numerical solution of stochastic differential equations*, Volume 23 of *Applications of Mathematics*. Berlin: Springer.

Kolb, A., P. Alpert, D. Enters, and C. Holzapfel (2004). Environmental stress and plant community invasibility in a coastal grassland in California. ESA Annual Conference (Portland, Oregon, July 30-August 6, 2004).

Kolmogorov, A., I. Petrovskii, and N. Piskunov (1937). Étude de l'equation de la diffusion avec croissance de la quantité de matière et son application à un problème biologique. *Bulletin de l'Université de Moscou, Série Internationale, Section A 1*, 1–25.

Kopell, N. and L. N. Howard (1973). Plane wave solutions to reaction-diffusion equations. *Studies in Appl. Math. 52*, 291–328.

Kronk, A. and S. Riechert (1979). Parameters affecting the habitat choice of a desert wolf spider, Lycosa santrita Chamberlin and Ivie. *J. Arachnol. 7*, 155–166.

Kuramoto, Y. (1974). Effects of diffusion on fluctuations in open chemical systems. *Progr. Theor. Phys. 52*, 711–713.

Kuramoto, Y. (1984). *Chemical oscillations, waves, and turbulence*, Volume 19 of *Springer Series in Synergetics*. Berlin: Springer.

Kuske, R., L. F. Gordillo, and P. Greenwood (2007). Sustained oscillations via coherence resonance in SIR. *J. Theor. Biol. 245*, 459–469.

Kuznetsov, Y. A. (1995). *Elements of applied bifurcation theory*, Volume 112 of *Applied Mathematical Sciences*. Berlin: Springer.

Lande, R. (1993). Risks of population extinction from demographic and environmental stochasticity and random catastrophes. *American Naturalist 142*, 911–922.

Langford, W. F. (1983). A review of interactions of Hopf and steady-state bifurcations. In G. I. Barenblatt, G. Iooss, and D. D. Joseph (Eds.), *Nonlinear Dynamics and Turbulence*, Interaction of Mechanics and Mathematics Series. Boston: Pitman.

Lefever, R. and O. Lejeune (1997). On the origin of tiger bush. *Bull. Math. Biol.* 59(2), 263–294.

Lefever, R. and O. Lejeune (2000). Generic modelling of vegetation patterns. A case study of *Tiger Bush* sub-saharian sahel. In P. K. Maini and H. G. Othmer (Eds.), *Mathematical models for biological pattern formation*, Volume 121 of *The IMA Volumes in Mathematics and its Applications*, pp. 83–112. New York: Springer.

Lejeune, O., P. Couteron, and R. Lefever (1999). Short range co-operativity competing with long range inhibition explains vegetation patterns. *Acta Oecologica* 20, 171–183.

Leslie, P. (1945). On the use of matrices in certain population mathematics. *Biometrika* 33, 183–212.

Leslie, P. H. (1948). Some further notes on the use of matrices in population mathematics. *Biometrika* 35(3&4), 213–245.

Levin, S. A. (1990). Physical and biological scales and the modelling of predator–prey interactions in large marine ecosystems. In K. Sherman, L. M. Alexander, and B. Gold (Eds.), *Large marine ecosystems: patterns, processes and yields*, pp. 179–187. Washington: American Association for the Advancement of Science.

Levin, S. A. (1992). The problem of pattern and scale in ecology. *Ecology* 73, 1943–1967.

Levin, S. A. and L. A. Segel (1976). Hypothesis for origin of planktonic patchiness. *Nature* 259, 659.

Levins, R. (1969). The effect of random variations of different types on population growth. *Proceedings of the National Academy of Sciences of the United States of America* 62, 1061–1065.

Lewis, M. A. (2000). Spread rate for a nonlinear stochastic invasion. *J. Math. Biol.* 41, 430–454.

Lewis, M. A. and P. Kareiva (1993). Allee dynamics and the spread of invading organisms. *Theor. Popul. Biol.* 43, 141–158.

Lewis, M. A. and S. Pacala (2000). Modeling and analysis of stochastic invasion processes. *J. Math. Biol.* 41, 387–429.

Li, T. and A. Yorke (1975). Period three implies chaos. *Amer. Math. Monthly* 82, 985–992.

Lin, J., V. Andreasen, R. Casagrandi, and S. Levin (2003). Traveling waves in a model of influenza A drift. *J. Theor. Biol.* 222, 437–445.

Lindner, B., J. García-Ojalvo, A. Neiman, and L. Schimansky-Geier (2004). Effects of noise in excitable systems. *Physics Reports* 392, 321–424.

Liu, W. (1994). Criterion of Hopf bifurcations without using eigenvalues. *J. Math. Anal. Appl. 182*, 250–256.

Liu, W., H. Hethcote, and S. Levin (1987). Dynamical behavior of epidemiological models with nonlinear incidence rates. *J. Math. Biol. 25*, 359–380.

Liu, W., S. Levin, and Y. Iwasa (1986). Influence of nonlinear incidence rates upon the behavior of SIRS epidemiological models. *J. Math. Biol. 23*, 187–204.

Longtin, A. (1997, Jan). Autonomous stochastic resonance in bursting neurons. *Phys. Rev. E 55*(1), 868–876.

Lorenz, E. N. (1963). Deterministic nonperiodic flow. *Journal of Atmospheric Sciences 20*, 130–141.

Lotka, A. (1956). *Elements of mathematical biology.* New York: Dover.

Lotka, A. J. (1925). *Elements of physical biology.* Baltimore: Williams and Wilkins.

Lubina, J. A. and S. A. Levin (1988). The spread of reinvading species: range expansion in the California sea otter. *Am. Nat. 131*(4), 526–543.

Luckinbill, L. L. (1974). The effects of space and enrichment on a predator–prey system. *Ecology 55*, 1142–1147.

Ludwig, D., D. D. Jones, and C. S. Holling (1978). Qualitative analysis of insect outbreak systems: the spruce budworm and forest. *J. Anim. Ecol. 47*, 315–332.

Luther, R. (1906). Räumliche Ausbreitung chemischer Reaktionen. *Zeitschrift für Elektrochemie 12*, 596–600.

Lythe, G. and S. Habib (2001). Stochastic PDEs: convergence to the continuum? *Computer Physics Communications 142*, 29–35.

MacFadyen, W. A. (1950a). Soil and vegetation in British Somaliland. *Nature 165*, 121.

MacFadyen, W. A. (1950b). Vegetation patterns in the semi-desert plains of British Somaliland. *The Geographical Journal 116*, 199–211.

Malchow, H. (1993). Spatio-temporal pattern formation in nonlinear nonequilibrium plankton dynamics. *Proc. R. Soc. Lond. B 251*, 103–109.

Malchow, H. (1995). Flow- and locomotion-induced pattern formation in nonlinear population dynamics. *Ecological Modelling 82*, 257–264.

Malchow, H. (1996). Nonlinear plankton dynamics and pattern formation in an ecohydrodynamic model system. *J. Mar. Sys. 7*(2-4), 193–202.

Malchow, H. (1998). Flux-induced instabilities in ionic and population-dynamical interaction systems. *Zeitschrift für Physikalische Chemie 204*, 95–107.

Malchow, H. (2000a). Motional instabilities in predator–prey systems. *J. Theor. Biol. 204*, 639–647.

Malchow, H. (2000b). Nonequilibrium spatio-temporal patterns in models of nonlinear plankton dynamics. *Freshwater Biology 45*, 239–251.

Malchow, H., W. Ebeling, R. Feistel, and L. Schimansky-Geier (1983). Stochastic bifurcations in a bistable reaction-diffusion system with Neumann boundary conditions. *Annalen der Physik (Leipzig) 40*(2/3), 151–160.

Malchow, H., F. M. Hilker, and S. V. Petrovskii (2004). Noise and productivity dependence of spatiotemporal pattern formation in a prey–predator system. *Discrete and Continuous Dynamical Systems B 4*(3), 707–713.

Malchow, H., F. M. Hilker, S. V. Petrovskii, and K. Brauer (2004). Oscillations and waves in a virally infected plankton system. Part I: The lysogenic stage. *Ecological Complexity 1*(3), 211–223.

Malchow, H., F. M. Hilker, R. R. Sarkar, and K. Brauer (2005). Spatiotemporal patterns in an excitable plankton system with lysogenic viral infection. *Mathematical and Computer Modelling 42*(9-10), 1035–1048.

Malchow, H., S. V. Petrovskii, and A. B. Medvinsky (2002). Numerical study of plankton-fish dynamics in a spatially structured and noisy environment. *Ecological Modelling 149*, 247–255.

Malchow, H. and S. V. Petrovskii (2002). Dynamical stabilization of an unstable equilibrium in chemical and biological systems. *Mathematical and Computer Modelling 36*, 307–319.

Malchow, H., B. Radtke, M. Kallache, A. B. Medvinsky, D. A. Tikhonov, and S. V. Petrovskii (2000). Spatio-temporal pattern formation in coupled models of plankton dynamics and fish school motion. *Nonlinear Analysis: Real World Applications 1*, 53–67.

Malchow, H. and L. Schimansky-Geier (1985). *Noise and diffusion in bistable nonequilibrium systems*, Volume 5 of *Teubner-Texte zur Physik*. Leipzig: Teubner-Verlag.

Malchow, H. and L. Schimansky-Geier (2006). Coherence resonance in an excitable prey–predator plankton system with infected prey. In T. Pöschel, H. Malchow, and L. Schimansky-Geier (Eds.), *Irreversible Prozesse und Selbstorganisation*, pp. 293–301. Berlin: Logos Verlag.

Malchow, H. and N. Shigesada (1994). Nonequilibrium plankton community structures in an ecohydrodynamic model system. *Nonlinear Processes in Geophysics* $1$(1), 3–11.

Maloney, D., F. Drummond, and R. Alford (2003). Spider predation in agroecosystems: can spiders effectively control pest population? *MAFES Technical Bulletin 190*, www.umaine.edu/mafes/elec_pubs/techbulletins/tb190.pdf.

Mansour, F., D. Rosen, and A. Shulov (1980). A survey of spider populations (araneae) in sprayed and unsprayed apple orchards in Israel and their ability to feed on larvae of Spodoptera littoralis (Boisd.). *Acta Oecol., Oecol. Appl. 1*, 189–197.

Marc, P. and A. Canard (1997). Maintaining spider biodiversity in agroecosystems as a tool in pest control. *Agric. Ecosyst. Environ. 62*, 229–235.

Marc, P., A. Canard, and F. Ysnel (1999). Spiders (Aranee) useful for pest limitation and bioindication. *Agric. Ecosyst. Environ. 74*, 229–273.

Maron, J. L. and S. Harrison (1997). Spatial pattern formation in an insect host-parasitoid system. *Science 278*, 1619–1621.

Martin, A. P. (2003). Phytoplankton patchiness: the role of lateral stirring and mixing. *Progress in Oceanography 57*, 125–174.

Maruyama, G. (1955). Continuous Markov processes and stochastic equations. *Rendiconti del Circolo Matematico di Palermo 4*, 48–90.

Matheson, I., D. F. Walls, and C. W. Gardiner (1975). Stochastic models of first-order nonequilibrium phase transitions in chemical reactions. *J. Stat. Phys. 12*, 21–34.

Matsumoto, M. and T. Nishimura (1998). Mersenne Twister: a 623-dimensionally equidistributed uniform pseudorandom number generator. *ACM Transactions on Modeling and Computer Simulation 8*(1), 3–30.

Matsushita, M. and H. Fujikawa (1990). Diffusion-limited growth in bacterial colony formation. *Physica A 168*, 498–506.

Mauchamp, A., S. Rambal, and J. Lepart (1994). Simulating the dynamics of a vegetation mosaic: a spatialized functional model. *Ecological Modelling 71*, 107–130.

May, R. M. (1972a). Limit cycles in predator–prey communities. *Science 177*, 900–902.

May, R. M. (1972b). Will a large complex system be stable? *Nature 238*, 413–414.

May, R. M. (1973). *Stability and complexity in model ecosystems*, Volume 6 of *Monographs in Population Biology*. Princeton: Princeton University Press.

May, R. M. (1974). Biological populations with nonoverlapping generations: stable points, stable cycles, and chaos. *Science 186*, 645–647.

May, R. M. and R. M. Anderson (1987). Transmission dynamics of HIV infection. *Nature 326*, 137–142.

May, R. M. and R. M. Anderson (1988). The transmission dynamics of human immunodeficiency virus (HIV). *Proc. R. Soc. Lond. B 321*, 565–607.

May, R. M., J. R. Beddington, and J. W. Horwood (1978). Exploiting natural populations in an uncertain world. *Math. Biosci. 42*, 219–252.

May, R. M. and W. Leonard (1975). Nonlinear aspects of competition between three species. *SIAM J. Appl. Math. 29*, 243–252.

McCallum, H., N. Barlow, and J. Hone (2001). How should pathogen transmission be modelled? *Trends in Ecology & Evolution 16*(6), 295–300.

McCauley, E. and W. Murdoch (1990). Predator-prey dynamics in rich and poor environments. *Nature 343*, 455–457.

McDaniel, L., L. A. Houchin, S. J. Williamson, and J. H. Paul (2002). Lysogeny in *Synechococcus*. *Nature 415*, 496.

McKendrick, A. (1926). Applications of mathematics to medical problems. *Proc. Edinburgh Math. Soc. 44*, 98–130.

McNair, J. (1987). A reconciliation of simple and complex models of age-dependent predation. *Theor. Popul. Biol. 32*, 383–392.

McNeil, D. R. (1972). On the simple stochastic epidemic. *Biometrika 59*, 494–497.

McQuarrie, D. A. (1967). *Stochastic approach to chemical kinetics*, Volume 8 of *Supplementary Review Series in Applied Probability. Methuen's Monographs on Applied Probability and Statistics*. London: Methuen.

Medvinsky, A. B., S. V. Petrovskii, I. A. Tikhonova, H. Malchow, and B.-L. Li (2002). Spatiotemporal complexity of plankton and fish dynamics. *SIAM Rev. 44*(3), 311–370.

Meinhardt, H. (1982). *Models of biological pattern formation*. London: Academic Press.

Mena-Lorca, J. and H. Hethcote (1992). Dynamic models of infectious diseases as regulator of population sizes. *J. Math. Biol. 30*, 693–716.

Merkin, J. H., V. Petrov, S. K. Scott, and K. Showalter (1996). Wave-induced chemical chaos. *Phys. Rev. Lett. 76*, 546–549.

Merkin, J. H. and M. A. Sadiq (1996). The propagation of travelling waves in an open cubic autocatalytic chemical system. *IMA J. Appl. Math. 57*, 273–309.

Merriam, G. (1988). Landscape dynamics in farmland. *TREE 3*, 16–20.

Milstein, G. N. and M. V. Tretyakov (2004). *Stochastic numerics for mathematical physics*. Scientific Computation. New York: Springer.

Miyashita, K. (1968). Quantitative feeding biology of Lycosa T-insignita Boes et Str (Araneae: Lycosidae) under different feeding conditions. *Applied Entomology and Zoology 3*, 81–88.

Moebus, K. (1996). Marine bacteriophage reproduction under nutrient-limited growth of host bacteria. II. Investigations with phage-host system [H3:H3/1]. *Marine Ecology Progess Series 144*, 13–22.

Morozov, A. and B.-L. Li (2007). On the importance of dimensionality of space in models of space-mediated population persistence. *Theor. Popul. Biol. 71*, 278–289.

Morozov, A., S. Petrovskii, and B.-L. Li (2004). Bifurcations and chaos in a predator–prey system with the allee effect. *Proceedings of Royal Society of London B 271*, 1407–1414.

Morozov, A., S. Petrovskii, and B.-L. Li (2006). Spatiotemporal complexity of patchy invasion in a predator–prey system with the Allee effect. *J. Theor. Biol. 238*, 18–35.

Morozov, A., S. Petrovskii, and N. Nezlin (2007). Towards resolving the paradox of enrichment: the impact of zooplankton vertical migrations on plankton systems stability. *J. Theor. Biol.*, in press.

Morse, D. and R. Fritz (1982). Experimental and observational studies of patch choice at different scales by the crab spider Misumena vatia. *Ecology 63*, 172–182.

Murdoch, W. W. and A. Oaten (1975). Predation and population stability. *Advances in Ecological Research 9*, 1–131.

Murray, J. D. (1989). *Mathematical biology*, Volume 19 of *Biomathematics Texts*. Berlin: Springer.

Namba, T. and M. Mimura (1980). Spatial distribution of competing populations. *J. Theor. Biol. 87*, 795–814.

Nayfeh, A. H. and B. Balachandran (1995). *Applied Nonlinear Dynamics*. Wiley Series in Nonlinear Science. New York: Wiley.

Neiman, A., P. I. Saparin, and L. Stone (1997, Jul). Coherence resonance at noisy precursors of bifurcations in nonlinear dynamical systems. *Phys. Rev. E 56*(1), 270–273.

Neiman, A., L. Schimansky-Geier, A. Cornell-Bell, and F. Moss (1999). Noise-enhanced phase synchronization in excitable media. *Phys. Rev. Lett.* 83(23), 4896–4899.

Neubert, M. G., H. Caswell, and J. D. Murray (2002). Transient dynamics and pattern formation: reactivity is necessary for Turing instabilities. *Math. Biosci.* 175, 1–11.

Neufeld, Z. (2001). Excitable media in a chaotic flow. *Phys. Rev. Lett.* 87, 108301(4).

Newel, P. C. (1983). Attraction and adhesion in the slime mold Dictyostelium. In J. E. Smith (Ed.), *Fungal differentiation. A contemporary synthesis*, Volume 43 of *Mycology Series*, pp. 43–71. New York: Marcel Dekker.

Nicolis, G. (1995). *Introduction to nonlinear science*. Cambridge: Cambridge University Press.

Nicolis, G. and M. Malek-Mansour (1980). Systematic analysis of the multivariate master equation for a reaction-diffusion system. *J. Stat. Phys.* 22, 495–512.

Nicolis, G. and I. Prigogine (1977). *Self-organization in nonequilibrium systems*. New York: Wiley-Interscience.

Nisbet, R. M., S. Diehl, W. G. Wilson, S. Cooper, D. Donaldson, and K. Kratz (1997). Primary-productivity gradients and short-term population dynamics in open systems. *Ecological Monographs* 67, 535–553.

Nisbet, R. M., E. McCauley, A. M. de Roos, W. Murdoch, and W. C. Gurney (1991). Population dynamics and element recycling in an aquatic plant-herbivore system. *Theor. Popul. Biol.* 40, 125–147.

Nitzan, A., P. Ortoleva, and J. Ross (1974). Nucleation in systems with multiple stationary states. *Faraday Symposia of the Chemical Society* 9, 241–253.

Niwa, H.-S. (1996). Newtonian dynamical approach to fish schooling. *J. Theor. Biol.* 181, 47–63.

Nold, A. (1980). Heterogeneity in disease-transmission modeling. *Math. Biosci.* 52, 227–240.

Noy-Meir, I. (1975). Stability of grazing systems: an application of predator–prey graphs. *J. Ecol.* 63, 459–481.

Nyffeler, M. and G. Benz (1987). Spiders in natural pest control: a review. *J. App. Entomol.* 103, 321–339.

Okubo, A. (1971). Oceanic diffusion diagrams. *Deep-Sea Res.* 18, 789–802.

Okubo, A. (1980). *Diffusion and ecological problems: Mathematical models*, Volume 10 of *Biomathematics Texts*. Berlin: Springer.

Okubo, A. (1986). Dynamical aspects of animal grouping: Swarms, schools, flocks, and herds. *Adv. Biophys. 22*, 1–94.

Okubo, A. and S. Levin (2001). *Diffusion and ecological problems: Modern perspectives*, Volume 14 of *Interdisciplinary Applied Mathematics*. New York: Springer.

Okubo, A., P. K. Maini, M. H. Williamson, and J. D. Murray (1989). On the spatial spread of the grey squirrel in britain. *Proc. R. Soc. Lond. B 238*, 113–125.

Olsen, L. F. and W. M. Schaffer (1990). Chaos versus noisy periodicity: Alternative hypotheses for childhood epidemics. *Science 249*, 499–504.

Olsen, L. F., G. L. Truty, and W. M. Schaffer (1988). Oscillations and chaos in epidemics: a nonlinear dynamic study of six childhood diseases in copenhagen. *Theor. Popul. Biol. 33*, 344–370.

Oppenheim, A. B., O. Kobiler, J. Stavans, D. L. Court, and S. Adhya (2005). Switches in bacteriophage lambda development. *Annual Review of Genetics 39*, 409–429.

Ortmann, A. C., J. E. Lawrence, and C. A. Suttle (2002). Lysogeny and lytic viral production during a bloom of the cyanobacterium *Synechococcus* spp. *Microbial Ecology 43*, 225–231.

Owen, M. R. and M. A. Lewis (2001). How predation can slow, stop or reverse a prey invasion. *Bull. Math. Biol. 63*, 655–684.

Parrish, J. K. and L. Edelstein-Keshet (1999). Complexity, pattern, and evolutionary trade-offs in animal aggregation. *Science 284*, 99–101.

Parrish, J. K., W. M. Hamner, and C. T. Prewitt (Eds.) (1997). *Animal aggregations: Three-dimensional measurement and modeling*. Cambridge: Cambridge University Press.

Pascual, M. (1993). Diffusion-induced chaos in a spatial predator–prey system. *Proc. R. Soc. Lond. B 251*, 1–7.

Pascual, M. and H. Caswell (1997). Environmental heterogeneity and biological pattern in a chaotic predator–prey system. *J. Theor. Biol. 185*, 1–13.

Pascual, M. and S. A. Levin (1999). From individuals to population densities: searching for the intermediate scale of nontrivial determinism. *Ecology 80*, 2225–2236.

Peaceman, D. W. and H. H. Rachford Jr. (1955). The numerical solution of parabolic and elliptic differential equations. *Journal of the Society for Industrial and Applied Mathematics 3*, 28–41.

Pearson, J. E. (1993). Complex patterns in simple systems. *Science 261*, 189–192.

Pertsov, A. M., J. M. Davidenko, R. Salomonsz, W. Baxter, and J. Jalífe (1993). Spiral waves of excitation underlie reentrant activity in isolated cardiac muscle. *Circulation Research 72*, 631–650.

Petrovskii, S. V. and H. Malchow (1999). A minimal model of pattern formation in a prey–predator system. *Mathematical and Computer Modelling 29*, 49–63.

Petrovskii, S. V., K. Kawasaki, F. Takasu, and N. Shigesada (2001). Diffusive waves, dynamical stabilization and spatio-temporal chaos in a community of three competitive species. *Japan J. Industr. Appl. Math. 18*(2), 459–481.

Petrovskii, S. V. and B.-L. Li (2001). Increased coupling between subpopulations in a spatially structured environment can lead to population outbreaks. *J. Theor. Biol. 212*, 549–562.

Petrovskii, S. V. and B.-L. Li (2006). *Exactly solvable models of biological invasion*. CRC Mathematical Biology and Medicine Series. Boca Raton: CRC Press.

Petrovskii, S. V., B.-L. Li, and H. Malchow (2003). Quantification of the spatial aspect of chaotic dynamics in biological and chemical systems. *Bull. Math. Biol. 65*(3), 425–446.

Petrovskii, S. V., B.-L. Li, and H. Malchow (2004). Transition to spatiotemporal chaos can resolve the paradox of enrichment. *Ecological Complexity 1*(1), 37–47.

Petrovskii, S. V. and H. Malchow (2000). Critical phenomena in plankton communities: KISS model revisited. *Nonlinear Analysis: Real World Applications 1*, 37–51.

Petrovskii, S. V. and H. Malchow (2001a). Spatio-temporal chaos in an ecological community as a response to unfavourable environmental changes. *Advances in Complex Systems 4*(2 & 3), 227–249.

Petrovskii, S. V. and H. Malchow (2001b). Wave of chaos: new mechanism of pattern formation in spatio-temporal population dynamics. *Theor. Popul. Biol. 59*(2), 157–174.

Petrovskii, S. V. and H. Malchow (2005). An exact solution of a system of diffusive Lotka–Volterra equations. *Proc. R. Soc. Lond. A 461*, 1029–1053.

Petrovskii, S. V., H. Malchow, F. M. Hilker, and E. Venturino (2005). Patterns of patchy spread in deterministic and stochastic models of biological invasion and biological control. *Biological Invasions 7*, 771–793.

Petrovskii, S. V., A. Morozov, and B.-L. Li (2005). Regimes of biological invasion in a predator–prey system with the Allee effect. *Bull. Math. Biol. 67*, 637–661.

Petrovskii, S. V., A. Y. Morozov, and E. Venturino (2002). Allee effect makes possible patchy invasion in a predator–prey system. *Ecol. Lett. 5*, 345–352.

Petrovskii, S. V., M. E. Vinogradov, and A. Y. Morozov (1998). Spatial-temporal dynamics of a localized populational "burst" in a distributed prey–predator system. *Oceanology 38*, 881–890.

Pikovsky, A. S. and J. Kurths (1997, Feb). Coherence resonance in a noise-driven excitable system. *Phys. Rev. Lett. 78*(5), 775–778.

Pimentel, D. (Ed.) (2002). *Biological invasions. Economic and environmental costs of alien plant, animal, and microbe species.* Boca Raton: CRC Press.

Platt, T. (1972). Local phytoplankton abundance and turbulence. *Deep-Sea Res. 19*, 183–187.

Polezhaev, A. A., C. Hilgardt, T. Mair, and S. C. Mller (2005). Transition from an excitable to an oscillatory state in *Dictyostelium discoideum*. *IEE Proceedings Systems Biology 152*, 75–79.

Pool, R. (1989). Ecologists flirt with chaos. *Science 243*, 310–313.

Popova, E. E., M. J. R. Fasham, A. V. Osipov, and V. A. Ryabchenko (1997). Chaotic behaviour of an ocean ecosystem model under seasonal external forcing. *J. Plankt. Res. 19*, 1495–1515.

Porter, K. (1976). Enhancement of algal growth and productivity by grazing zooplankton. *Science 192*, 1332–1134.

Porter, K. (1977). The plant-animal interface in fresh water ecosystems. *American Scientist 65*, 159–170.

Powell, T. M. (1995). Physical and biological scales of variability in lakes, estuaries and the coastal ocean. In T. M. Powell and J. H. Steele (Eds.), *Ecological Time Series*, pp. 119–138. New York: Chapman & Hall.

Powell, T. M. and A. Okubo (1994). Turbulence, diffusion and patchiness in the sea. *Proc. R. Soc. Lond. B 343*, 11–18.

Powell, T. M., P. J. Richerson, T. M. Dillon, B. A. Agee, B. J. Dozier, D. A. Godden, and L. O. Myrup (1975). Spatial scales of current speed and phytoplankton biomass fluctuations in Lake Tahoe. *Science 189*, 1088–1090.

Pradines, J. R., G. V. Osipov, and J. J. Collins (1999, Dec). Coherence resonance in excitable and oscillatory systems: The essential role of slow and fast dynamics. *Phys. Rev. E 60*(6), 6407–6410.

Pugliese, A. (1990). Population models for disease with no recovery. *J. Math. Biol. 28*, 65–82.

Putman, W. and D. Herne (1966). The role of predators and other biotic agents in regulating the population density of phytophagous mite in ontario peach orchards. *Can. Entomol. 98*, 808–820.

Qian, H. and J. D. Murray (2003). A simple method of parameter space determination for diffusion-driven instability with three species. *Appl. Math. Lett. 14*, 405–411.

Radakov, D. V. (1973). *Schooling in the ecology of fish*. New York: Wiley.

Rai, V. and W. M. Schaffer (2001). Chaos in ecology. *Chaos, Solitons & Fractals 12*, 197–203.

Rand, D. A. and H. B. Wilson (1995). Using spatio-temporal chaos and intermediate-scale determinism to quantify spatially extended ecosystems. *Proc. R. Soc. Lond. B* (259), 111–117.

Ranta, E., V. Kaitala, and P. Lundberg (1997). The spatial dimension in population fluctuations. *Science 278*, 1621–1623.

Rasmussen, K. E., W. Mazin, E. Mosekilde, G. Dewel, and P. Borckmans (1996). Wave-splitting in the bistable Gray-Scott model. *International J. Bifurcations and Chaos 6*, 1077–1092.

Reigada, R., R. M. Hillary, M. A. Bees, J. M. Sancho, and F. Sagues (2003). Plankton blooms induced by turbulent flows. *Proc. Roy. Soc. Lond. B 270*, 875–880.

Renshaw, E. (1991). *Modelling biological populations in space and time*. Cambridge, UK: Cambridge University Press.

Reuter, H. and B. Breckling (1994). Selforganization of fish schools: an object-oriented model. *Ecological Modelling 75/76*, 147–159.

Riechert, S. and L. Bishop (1990). Prey control by an assemblage of generalist predators: spiders in garden test systems. *Ecology 71*, 1441–1450.

Rietkerk, M. (1998). *Catastrophic vegetation dynamics and soil degradation in semi-arid grazing systems*, Volume 20 of *Tropical Resource Management Papers*. Wageningen: Wageningen Agricultural University.

Rietkerk, M., M. C. Boerlijst, F. van Langevelde, R. HilleRisLambers, J. van de Koppel, L. Kumar, H. H. T. Prins, and A. M. de Roos (2002).

Self-organization of vegetation in arid ecosystems. *Am. Nat.* *160*(4), 524–530.

Rietkerk, M., S. C. Dekker, P. C. de Ruiter, and J. van de Koppel (2004). Self-organized patchiness and catastrophic shifts in ecosystems. *Science* *305*, 1926–1929.

Rietkerk, M. and J. van de Koppel (2002). Alternate stable states and threshold effects in semi-arid grazing systems. *Oikos* *79*(1), 69–76.

Rinaldi, S. and O. D. Feo (1999). Top-predator abundance and chaos in tritrophic food chains. *Ecol. Lett.* *2*, 6–10.

Rinaldi, S. and S. Muratori (1992a). Limit cycles in slow-fast forest-pest models. *Theor. Popul. Biol.* *41*, 26–43.

Rinaldi, S. and S. Muratori (1992b). Slow-fast limit cycles in predator–prey models. *Ecological Modelling* *61*, 287–308.

Rinaldi, S. and M. Scheffer (2000). Geometric analysis of ecological models with slow and fast processes. *Ecosystems* *3*, 507–521.

Rovinsky, A. B. and M. Menzinger (1992). Chemical instability induced by a differential flow. *Phys. Rev. Lett.* *69*, 1193–1196.

Root, R. and K. P.M. (1984). The search for resources bycabbage butterflies (Pieris rapae): ecological consequences and adaptive significance of Markovian movements in a patchy environment. *Ecology* *65*, 147–165.

Rosenzweig, M. L. (1971). Paradox of enrichment: Destabilization of exploitation ecosystems in ecological time. *Science* *171*, 385–387.

Rosenzweig, M. L. and R. H. MacArthur (1963). Graphical representation and stability conditions of predator–prey interactions. *Am. Nat.* *97*, 209–223.

Rovinsky, A. B., H. Adiwidjaja, V. Z. Yakhnin, and M. Menzinger (1997). Patchiness and enhancement of productivity in plankton ecosystems due to differential advection of predator and prey. *Oikos* *78*, 101–106.

Ryabchenko, V. A., M. J. R. Fasham, B. Kagan, and E. Popova (1997). What causes short-term oscillations in ecosystem models of the ocean mixed layer? *J. Mar. Sys,* *13*, 33–50.

Rypstra, A. (1995). Spider predators reduce herbivory; both by direct consumption and by altering the foraging behavior of insect pests. *Bull. Ecol. Soc. Am.* *76*, 383.

Saffre, F. and J. L. Deneubourg (2002). Swarming strategies for cooperative species. *J. Theor. Biol.* *214*, 441–451.

Sagués, F., J. M. Sancho, and J. García-Ojalvo (2007). Spatiotemporal order out of noise. *Rev. Mod. Phys. 79*, 829–882.

Sakaguchi, H., S. Shinomoto, and Y. Kuramoto (1988). Phase transitions and their bifurcation analysis in a large population of active rotators with mean-field coupling. *Progr. Theor. Phys. 79*(3), 600–607.

Satnoianu, R. A., P. K. Maini, and M. Menzinger (2001). Parameter space analysis, pattern sensitivity and model comparison for Turing and stationary flow-distributed waves (FDS). *Physica D 160*, 79–102.

Satnoianu, R. A. and M. Menzinger (2000). Non-Turing stationary patterns in flow-distributed oscillators with general diffusion and flow rates. *Phys. Rev. E 62*(1), 113–119.

Satnoianu, R. A., M. Menzinger, and P. K. Maini (2000). Turing instabilities in general systems. *J. Math. Biol. 41*, 493–512.

Sax, D. F., J. J. Stachowicz, and S. D. Gaines (Eds.) (2005). *Species invasions. Insights into ecology, evolution, and biogeography.* Sunderland: Sinauer.

Schaffer, W. M. and M. Kot (1986). Chaos in ecological systems: The coals that newcastle forgot. *Trends in Ecology & Evolution 3*, 58–63.

Scheffer, M. (1991a). Fish and nutrients interplay determines algal biomass: a minimal model. *Oikos 62*, 271–282.

Scheffer, M. (1991b). Should we expect strange attractors behind plankton dynamics – and if so, should we bother? *J. Plankt. Res. 13*, 1291–1305.

Scheffer, M. (1998). *Ecology of shallow lakes*, Volume 22 of *Popul. and Community Biology Series*. London: Chapman & Hall.

Scheffer, M., S. Carpenter, J. A. Foley, C. Folke, and B. Walker (2001). Catastrophic shifts in ecosystems. *Nature 413*, 591–596.

Scheffer, M. and S. R. Carpenter (2003). Catastrophic regime shifts in ecosystems: linking theory to observation. *Trends in Ecology & Evolution 18*(12), 648–656.

Schimansky-Geier, L., A. S. Mikhailov, and W. Ebeling (1983). Effect of fluctuation on plane front propagation in bistable nonequilibrium systems. *Annalen der Physik (Leipzig) 40*, 277–286.

Schimansky-Geier, L. and C. Zülicke (1991). Kink propagation induced by multiplicative noise. *Zeitschrift für Physik B 82*, 157–162.

Schlögl, F. (1972). Chemical reaction models for nonequilibrium phase transitions. *Zeitschrift für Physik 253*, 147–161.

Segel, L. A. (1977). A theoretical study of receptor mechanisms in bacterial chemotaxis. *SIAM J. Appl. Math. 32*, 653–665.

Segel, L. A. and J. L. Jackson (1972). Dissipative structure: an explanation and an ecological example. *J. Theor. Biol. 37*, 545–559.

Segel, L. A. and B. Stoeckly (1972). Instability of a layer of chemotactic cells, attractant and degrading enzyme. *J. Theor. Biol. 37*, 561–585.

Shapiro, J. A. and C. Hsu (1989). *Escherichia coli* k-12 cell-cell interactions seen by time-lapse video. *Journal of Bacteriology 171*, 5963–5974.

Shapiro, J. A. and D. Trubatch (1991). Sequential events in bacterial colony morphogenesis. *Physica D 49*, 214–223.

Sherratt, J. A. (1994a). Irregular wakes in reaction-diffusion waves. *Physica D 70*, 370–382.

Sherratt, J. A. (1994b). On the evolution of periodic plane waves in reaction-diffusion equations of lambda-omega type. *SIAM J. Appl. Math. 54*, 1374–1385.

Sherratt, J. A. (1998). Invading wave fronts and their oscillatory wakes are linked by a modulated travelling phase resetting wave. *Physica D 117*, 145–166.

Sherratt, J. A. (2001). Periodic travelling waves in cyclic predator–prey systems. *Ecol. Lett. 4*(1), 30–37.

Sherratt, J. A., B. T. Eagan, and M. A. Lewis (1997). Oscillations and chaos behind predator–prey invasion: mathematical artifact or ecological reality? *Phil. Trans. R. Soc. Lond. B 352*, 21–38.

Sherratt, J. A., X. Lambin, C. J. Thomas, and T. N. Sherratt (2002). Generation of periodic waves by landscape features in cyclic predator–prey systems. *Proceedings of Royal Society of London B 269*, 327–334.

Sherratt, J. A., M. A. Lewis, and A. Fowler (1995). Ecological chaos in the wake of invasion. *Proceedings of the National Academy of Sciences of the United States of America 92*, 2524–2528.

Shigesada, N. and K. Kawasaki (1997). *Biological invasions: Theory and practice*. Oxford: Oxford University Press.

Shigesada, N., K. Kawasaki, and Y. Takeda (1995). Modeling stratified diffusion in biological invasions. *Am. Nat. 146*(2), 229–251.

Shoji, H. and Y. Iwasa (2003). Pattern selection and the direction of stripes in two-dimensional Turing systems for skin pattern formation of fishes. *Forma 18*, 3–18.

Sieber, M., H. Malchow, and L. Schimansky-Geier (2007). Constructive effects of environmental noise in an excitable prey–predator plankton system with infected prey. *Ecological Complexity 4*, 223–233.

Siegert, F. and C. J. Weijer (1991). Analysis of optical density wave propagation and cell movement in the cellular slime mould *Dictyostelium discoideum*. *Physica D 49*, 224–232.

Sigeti, D. and W. Horsthemke (1989). Pseudo-regular oscillations induced by external noise. *J. Stat. Phys. 54*(5/6), 1217–1222.

Singh, B. K., J. Chattopadhyay, and S. Sinha (2004). The role of virus infection in a simple phytoplankton-zooplankton system. *J. Theor. Biol. 231*, 153–166.

Skellam, J. G. (1951). Random dispersal in theoretical populations. *Biometrika 38*, 196–218.

Smith, G. (1979). *Numerical solution of partial differential equations: finite difference methods, 2nd Ed.* Oxford: Clarendon Press.

Solé, R. V. and J. Bascompte (2006). *Self-organization in complex ecosystems*. Princeton: Princeton University Press.

Spagnolo, B., M. Cirone, A. La Barbera, and F. de Pasquale (2002). Noise-induced effects in population dynamics. *J. Phys.: Cond. Matt. 14*, 2247–2255.

Spagnolo, B., D. Valenti, and A. Fiasconaro (2004). Noise in ecosystems: a short review. *Math. Biosci. Engn. 1*(1), 185–211.

Steele, J. H. (Ed.) (1977). *Fisheries mathematics*. London: Academic Press.

Steele, J. H. (Ed.) (1978). *Spatial patterns in plankton communities*, Volume 3 of *NATO Conf. Series IV (Marine Sciences)*. New York: Plenum Press.

Steele, J. H. (2004). Regime shifts in the ocean: reconciling observations and theory. *Progress in Oceanography 60*, 135–141.

Steele, J. H. and E. W. Henderson (1992a). The role of predation in plankton models. *J. Plankt. Res. 14*, 157–172.

Steele, J. H. and E. W. Henderson (1992b). A simple model for plankton patchiness. *J. Plankt. Res. 14*, 1397–1403.

Steinbock, O., H. Hashimoto, and S. C. Müller (1991). Quantitative analysis of periodic chemotaxis in aggregation patterns of *Dictyostelium discoideum*. *Physica D 49*, 233–239.

Stenseth, N. C., K.-S. Chan, H. Tong, R. Boonstra, S. Boutin, C. J. Krebs, E. Post, M. O'Donoghue, N. G. Yoccoz, M. C. Forchhammer, and J. W. Hurrell (1999). Common dynamic structure of canada lynx populations within three climatic regions. *Science 285*, 1071–1073.

Stenseth, N. C., A. Mysterud, G. Ottersen, J. W. Hurrell, K.-S. Chan, and M. Lima (2002). Ecological effects of climate fluctuations. *Science 297*, 1292–1296.

Stephens, P. A. and W. J. Sutherland (1999). Consequences of the Allee effect for behaviour, ecology and conservation. *Trends in Ecology & Evolution 14*(10), 401–405.

Stephens, P. A., W. J. Sutherland, and R. P. Freckleton (1999). What is the Allee effect? *Oikos 87*, 185–190.

Stöcker, S. (1999). Models for tuna school formation. *Math. Biosci. 156*, 167–190.

Stratonovich, R. L. (1967). *Topics in the theory of random noise*, Volume 3(1-2) of *Mathematics and Its Applications*. New York: Gordon and Breach.

Strogatz, S. H. (1994). *Nonlinear dynamics and chaos with applications to physics, biology, chemistry, and engineering*. Studies in Nonlinearity. Reading MA: Addison-Wesley.

Suttle, C. A. (2000). Ecological, evolutionary, and geochemical consequences of viral infection of cyanobacteria and eukaryotic algae. In C. J. Hurst (Ed.), *Viral ecology*, pp. 247–296. San Diego: Academic Press.

Suttle, C. A. (2005). Viruses in the sea. *Nature 437*, 356–361.

Swope, S., R. Carruthers, G. Anderson, and D. Bubenheim (2004). USDA and NASA collaborate to use hyperspectral imagery to detect invasive species through time and space. ESA Annual Conference (Portland, Oregon, July 30-August 6, 2004).

Tanner, J. (1975). The stability and intrinsic growth rates of prey and predator populations. *Ecology 56*, 855–867.

Tansky, M. (1978). Switching effect in prey–predator system. *J. Theor. Biol. 70*, 263–271.

Taylor, C. M. and A. Hastings (2005). Allee effects in biological invasions. *Ecol. Lett. 8*, 895–908.

Thomas, C., P. Brain, and P. Jepson (2003). Aerial activity of linyphiid spiders: modelling dispersal distances from meteorology and behaviour. *J. Appl. Ecol. 40*, 912–927.

Thomas, C., E. Hol, and J. Everts (1990). Modelling the diffusion component of dispersal during recovery of a population of linyphiid spiders from exposure to an insecticide. *Funct. Ecol. 4*, 357–368.

Thomas, J. W. (1995). *Numerical partial differential equations: Finite difference methods*, Volume 22 of *Texts in Applied Mathematics*. New York: Springer.

Thorbek, P. and C. Topping (2005). The influence of landscape diversity and heterogeneity on spatial dynamics of agrobiont linyphiid spiders: An individual-based model. *BioControl 50*, 1–33.

Topping, C. (1999). An individual-based model for dispersive spiders in agroecosystems: simulations of the effects of landscape structure. *J. Arachnol. 27*, 378–386.

Topping, C., T. Hansen, T. Jensen, J. Jepsen, F. Nikolajsen, and P. Odderskaer (2003). ALMaSS an agent-based model for animals in temperate European landscapes. *Ecol. Modelling 167*, 65–82.

Topping, C. and K. Sunderland (1994). A spatial population dynamics model for Lepthyphantes tenuis (Araneae: Linyphiidae) with some simulations of the spatial and temporal effects of farming operations and land-use. *Agr. Ecosyst. Environ. 48*, 203–217.

Treutlein, H. and K. Schulten (1986). Noise-induced neural impulses. *European Biophysics Journal 13*(6), 355–365.

Truscott, J. E. (1995). Environmental forcing of simple plankton models. *J. Plankt. Res. 17*, 2207–2232.

Truscott, J. E. and J. Brindley (1994). Ocean plankton populations as excitable media. *Bull. Math. Biol. 56*, 981–998.

Tuckwell, H. C. (1974). A study of some diffusion models of population growth. *Theor. Popul. Biol. 5*, 345–357.

Turchin, P. (2003). *Complex population dynamics*, Volume 35 of *Monographs in Population Biology*. Princeton and Oxford: Princeton University Press.

Turchin, P., A. D. Taylor, and J. D. Reeve (1999). Dynamical role of predators in population cycles of a forest insect: an experimental test. *Science 285*, 1068–1071.

Turing, A. M. (1952). On the chemical basis of morphogenesis. *Phil. Trans. R. Soc. Lond. B 237*, 37–72.

Upadhyay, R. K., N. Kumari, and V. Rai (2007). Wave of chaos in a diffusive system: Generating realistic patterns of patchiness in plankton–fish dynamics. *Chaos, Solitons and Fractals*, in press.

Valenti, D., A. Fiasconaro, and B. Spagnolo (2004). Stochastic resonance and noise delayed extinction in a model of two competing species. *Physica A 331*, 477–486.

Valenti, D., L. Schimansky-Geier, X. Sailer, and B. Spagnolo (2006). Moment equations for a spatially extended system of two competing species. *The European Physical Journal B 50*, 199–203.

Valiela, I. (1995). *Marine ecological processes*. New York: Springer.

van den Bosch, F., A. M. de Roos, and W. Gabriel (1988). Cannibalism as a life boat mechanism. *J. Math. Biol. 26*, 619–633.

van den Broeck, C., W. Horsthemke, and M. Malek-Mansour (1977). On the diffusion operator of the multivariate master equation. *Physica A 89*, 339–352.

van Kampen, N. G. (1973). Birth and death processes in large populations. *Biometrika 60*, 419–420.

van Leeuwen, J. L., A. A. Jansen, and W. Bright (2007). How population dynamics shape the functional response in a one-predator–two-prey system. *Ecology 88*, 1571–1581.

van Saarloos, W. (2003). Front propagation into unstable states. *Physics Reports 386*, 29–222.

Vandermeer, J. (2006). Oscillating populations and biodiversity maintenance. *BioScience 56*(12), 967–975.

Vasiev, B. N., P. Hogeweg, and A. V. Panfilov (1994). Simulation of *Dictyostelium discoideum* aggregation via reaction-diffusion model. *Phys. Rev. Lett. 73*, 3173–3176.

Venturino, E. (1992). The influence of diseases on lotka–volterra systems. Volume 913 of *IMA preprint series*. Minneapolis, MN, USA.

Venturino, E. (1994). The influence of diseases on Lotka–Volterra systems. *Rocky Mountain Journal of Mathematics 24*, 381–402.

Venturino, E. (2001). The effect of diseases on competing species. *Math. Biosci. 174*, 111–131.

Venturino, E. (2002). Epidemics in predator–prey models: disease in the predators. *IMA J. Math. Appl. Med. Biol. 19*, 185–205.

Venturino, E. (2004). A stage-dependent ecoepidemic model. *WSEAS Transactions on Biology and Biomedicine 1*, 449–454.

Venturino, E. (2006). On epidemics crossing the species barrier in interacting population models. *Varahmihir J. Math. Sci. 6*, 247–263.

Venturino, E. (2007). How diseases affect symbiotic communities. *Math. Biosci. 206*, 11–30.

Verhulst, P. (1845). Recherches mathématiques sur la loi d'accroissement de la population. *Mém. Acad. Roy. Bruxelles 18*.

Verhulst, P. (1847). Recherches mathématiques sur la loi d'accroissement de la population. *Mém. Acad. Roy. Bruxelles 20*.

Vilar, J. M. G., R. V. Solé, and J. M. Rubi (2003). On the origin of plankton patchiness. *Physica A 317*, 239–246.

Volpert, A. I., V. A. Volpert, and V. A. Volpert (1994). *Traveling wave solutions of parabolic systems*, Volume 140 of *Translations of Mathematical Monographs*. Providence, Rhode Island: American Mathematical Society.

Volterra, V. (1926a). Fluctuations in the abundance of a species considered mathematically. *Nature 118*, 558–560.

Volterra, V. (1926b). Variazioni e fluttuazioni del numero d'individui in specie animali conviventi. *Atti della Reale Accademia Nazionale dei Lincei, Memorie della Classe di Scienze Fisiche, Matematiche e Naturali, Serie 6, Volume II*(3), 31–113.

Volterra, V. (1931). *Théorie mathématique de la lutte pour la vie*. Paris: Gauthier - Villars.

von Foerster, H. (1959). Some remarks on changing populations. In F. Stohlman (Ed.), *The Kinetics of Cellular Proliferation*, pp. 382–407. Grune and Stratton, New York.

Walker, R., C. Ferguson, N. Booth, and E. Allan (2002). The symbiosis of Bacillus subtilis L-forms with Chinese cabbage seedlings inhibits conidial germination of Botrytis cinerea. *Lett. Appl. Microbiol. 34*, 42–45.

Walsh, J. B. (1986). An introduction to stochastic partial differential equations. In R. Carmona, H. Kesten, and J. B. Walsh (Eds.), *École d'été de probabilités de Saint-Flour XIV - 1984*, Volume 1180 of *Lecture Notes in Mathematics*, pp. 265–437. Berlin: Springer.

Wang, M.-H. and M. Kot (2001). Speeds of invasion in a model with strong or weak Allee effects. *Math. Biosci. 171*(1), 83–97.

Wax, N. (Ed.) (1954). *Selected papers on noise and stochastic processes*. New York: Dover Publications.

Webb, G. (1985). *Theory of nonlinear age-dependent population dynamics*, Volume 89 of *Monographs and Textbooks in Pure & Applied Mathematics Series*. New York: Dekker.

Weber, L. H., S. Z. El-Sayed, and I. Hampton (1986). The variance spectra of phytoplankton, krill and water temperature in the Antarctic ocean south of Africa. *Deep-Sea Res. 33*, 1327–1343.

White, L. P. (1969). Vegetation arcs in Jordan. *J. Ecol. 57*, 461–464.

White, L. P. (1970). Brousses Tigrees patterns in Southern Niger. *J. Ecol. 58*, 549–553.

White, T. C. R. (2001). Opposing paradigms: regulation or limitation of populations? *Oikos 93*, 148–152.

White, T. C. R. (2004). Limitation of populations by weather-driven changes in food: a challenge to density-dependent regulation. *Oikos 105*, 664–666.

Wickens, G. E. and F. W. Collier (1971). Some vegetation patterns in the Republic of the Sudan. *Geoderma 6*, 43–59.

Wiens, J., N. Stenseth, B. V. Horne, and R. Ims (1993). Ecological mechanisms and landscape ecology. *Oikos 66*, 369–380.

Wilcox, R. M. and J. A. Fuhrman (1994). Bacterial viruses in coastal seawater: lytic rather than lysogenic production. *Marine Ecology Progress Series 114*, 35–45.

Wilson, W. H., N. G. Carr, and N. H. Mann (1996). The effect of phosphate status on the kinetics of cyanophage infection in the oceanic cyanobacterium *Synechococcus* sp. WH7803. *J. Phycol. 32*(4), 506–516.

Wise, D. (1993). *Spiders in ecological webs*. Cambridge, UK: Cambridge Studies in Ecology.

Wissel, C. (1981). Lassen sich ökologische Instabilitäten vorhersagen? *Verhandlungen der Gesellschaft für Ökologie IX*, 143–152.

Wissel, C. (1984a). Solution of the master equation of a bistable reaction system. *Physica A 128*, 150–163.

Wissel, C. (1984b). Stochastische Einflüsse auf Ökosysteme mit multipler Stabilität; die vollständige Lösung der Master-Gleichung. *Verhandlungen der Gesellschaft für Ökologie XII*, 447–458.

Wissel, C. (1985). Zur Wirkung zufälliger Umwelteinflüsse auf die periodischen Massenvermehrungen eines Tannentriebwicklers. *Verhandlungen der Gesellschaft für Ökologie XIII*, 305–312.

Wissel, C. (1989). *Theoretische Ökologie. Eine Einführung*. Berlin: Springer.

Witten, T. A. and L. M. Sander (1981). Diffusion-limited aggregation, a kinetic critical phenomenon. *Phys. Rev. Lett. 47*, 1400–1403.

Wommack, K. E. and R. R. Colwell (2000). Virioplankton: Viruses in aquatic ecosystems. *Microbiology and Molecular Biology Reviews 64*(1), 69–114.

Wyatt, T. (1973). The biology of *Oikopleura dioica* and *Fritillaria borealis* in the Southern Bight. *Marine Biology 22*, 137–158.

Yachi, S., K. Kawasaki, N. Shigesada, and E. Teramoto (1989). Spatial patterns of propagating waves of fox rabies. *Forma 4*, 3–12.

Zachmanoglou, E. and D. Thoe (1976). *Introduction to partial differential equations*. Baltimore, MD: Williams & Wilkins.

Zaks, M. A., A. B. Neiman, S. Feistel, and L. Schimansky-Geier (2003). Noise-controlled oscillations and their bifurcations in coupled phase oscillators. *Phys. Rev. E 68*, 066206.

Zeldovich, Y. B., G. I. Barenblatt, V. B. Librovich, and G. M. Makhviladze (1985). *Mathematical theory of combustion and explosions*. New York: Consultants Bureau, Plenum Press.

Zhdanov, V. P. (2003). Propagation of infection and the prey–predator interplay. *J. Theor. Biol. 225*, 489–492.

# Index

Allee effect, 3, 6, 7, 24, 29, 78, 212, 213, 276, 278, 279, 312, 325–327, 329, 331, 332, 334, 336–339, 343, 349, 350, 357–359, 361

biological control, 123, 325, 326, 330, 335, 338, 339, 343, 350, 351, 356, 358

correlation length, 294, 296, 298–300, 303–305, 307, 310, 323, 367

disease control, 195, 196

epidemic spread, 10, 278, 375

heteroclinic connection, 261–263, 265, 329, 378, 379

instability, 7, 8, 32, 84, 114, 119, 130, 174, 175, 177, 185, 186, 192, 205, 206, 210–216, 218–224, 226–228, 230, 231, 233–236, 242, 244–248, 260, 261, 282–284, 325

invasion, 6, 10, 12, 203, 247, 248, 251, 252, 256, 258, 277, 278, 282, 325, 326, 331, 332, 335–339, 341–345, 347–352, 356–361

Lyapunov exponent, 255, 275, 344
Lyapunov function, 50, 70, 72, 77, 144, 146

nonlocal, 236, 237

patchiness, 8, 236, 245, 252, 260, 281–283, 294, 300, 308–312, 318, 325, 348, 355, 375

periodic wake, 292

persistence, 283, 290, 292, 314, 315, 320, 326, 332, 348, 359

plankton, 3, 4, 11, 12, 245, 246, 260, 308–311, 366, 375, 384, 390, 396, 397

random walk, 199, 200

spirals, 9, 56, 296, 371, 379, 391, 392

switching, 66–68, 252, 262, 390, 391, 393

wave of chaos, 283, 287, 290, 293, 325

**RENEWALS 458-4574**
DATE DUE